AF279176

Joseph Moretz is an independent researcher and author whose primary focus is the evolution of naval thought and development of joint doctrine focusing further on the British from 1914 onwards. Following a career in computing and telecommunications with the United States Department of Defense culminating as the Director of Information Assurance with the Defense Contract Management Agency, the author retired in 2010. Holding both an MA and PhD in War Studies from King's College, London, he graduated with distinction from the United States Naval War College in 1989. Dr Moretz is a member of the British Commission for Military History, the Army Records Society, the Navy Records Society and the Society of Military History. *Thinking Wisely, Planning Boldly* is his second book.

Thinking Wisely, Planning Boldly

The Higher Education and Training of Royal Navy Officers, 1919–1939

Joseph Moretz

Helion & Company Limited

Helion & Company Limited
26 Willow Road
Solihull
West Midlands
B91 1UE
England
Tel. 0121 705 3393
Fax 0121 711 4075
Email: info@helion.co.uk
Website: www.helion.co.uk
Twitter: @helionbooks
Visit our blog http://blog.helion.co.uk/

Published by Helion & Company 2014

Designed and typeset by Bookcraft Ltd, Stroud, Gloucestershire
Cover designed by Paul Hewitt, Battlefield Design (www.battlefield-design.co.uk)
Printed by Lightning Source, Milton Keynes, Buckinghamshire

Text © Joseph Moretz
Cover: Naval Officers of World War I by Sir Arthur Stockdale Cope, oil on
canvas, 1921 © National Portrait Gallery, London

ISBN 978 1 909982 90 1

British Library Cataloguing-in-Publication Data.
A catalogue record for this book is available from the British Library.

For details of other military history titles published by Helion & Company
Limited contact the above address, or visit our website: http://www.helion.co.uk.

We always welcome receiving book proposals from prospective authors.

Joseph Moretz is an independent researcher and author whose primary focus is the evolution of naval thought and development of joint doctrine focusing further on the British from 1914 onwards. Following a career in computing and telecommunications with the United States Department of Defense culminating as the Director of Information Assurance with the Defense Contract Management Agency, the author retired in 2010. Holding both an MA and PhD in War Studies from King's College, London, he graduated with distinction from the United States Naval War College in 1989. Dr Moretz is a member of the British Commission for Military History, the Army Records Society, the Navy Records Society and the Society of Military History. *Thinking Wisely, Planning Boldly* is his second book.

Thinking, Wisely, Planning Boldly examines the style, content and manner of Royal Navy executive officer higher education and training between the World Wars. Based on official and private archival records, oral histories and the secondary literature extant, this book traces the changes the Navy made in how it prepared its mid-level officers following the First World War, contrasts this approach with that of the British Army and Royal Air Force and addresses the use the Royal Navy made of the officers so trained. In the process, the work offers a fundamental reappraisal of the interwar Royal Navy challenging many of the accepted conclusions rendered by earlier authors who failed to actually examine the style and content of officer education and did not weigh the many competing factors the service had to balance in any professional development program. Along the way, it offers insight into the relative centrality of the Battle of Jutland in interwar training and concludes that contrary to received wisdom its role was a secondary one at best and that the experience of most relevance in the Navy's educational efforts was the Dardanelles campaign.

This work is original in scope and original in interpretation with no other book-length volume in print now or previously covering the subject. Beyond saying something valuable about the 1919-39 Royal Navy, it discusses issues that resound with contemporary military officers faced with the eternal question of what to teach, how to teach it, and the pitfalls faced in preparing officers in an uncertain world. It sheds fresh light on such noted figures as Admiral Sir Herbert Richmond and Major General J. F. C. Fuller and offers insight into such events as the Washington Naval Treaty and the Invergordon Mutiny not previously considered.

Though many writers have had much to say about interwar training, none actually took the time to examine what was taught, how instruction was imparted, and the aims that the Navy sought to achieve. Thinking Wisely, Planning Boldly fills the void and in the process speaks to the continuing issues facing professional military education.

To my parents

Contents

Acknowledgements

Writing any work is a singular task, but it is often an endeavour where the support, suggestions, criticism, and encouragement of others play no small part in its eventual completion. Such has been the case in the efforts of this work and the author would be remiss not to note those who have been of especial assistance during its research, writing, editing, and publication. To Chuck Steele my appreciation is noted for encouraging me to present a subset of the conclusions reached at a meeting of the North American Society for Oceanic History. To Robin Broadhurst, a fellow member of the British Commission for Military History, warm thanks are extended for reading selected chapters relating to the service of Admiral Dudley Pound. Likewise, a similar dose of thanks is extended to Gary Greco and Commander Youssef Aboul-Enein, USN, for ensuring that what follows is of interest to those not so closely concerned with the workings of the interwar Royal Navy nor of educational problems in general. Meanwhile, Peter Robinson, another fellow associate of the BCMH, has been of the greatest support in so many ways, but his timely intervention and sage advice ensured that the project found its way to a willing publisher. Meanwhile, to Professor Geoffrey Till a singular debt remains. Your teaching, encouragement, and guidance to this writer, as to so many others, over the years has been the foundation upon which my tepid efforts have progressed.

The assistance of those toiling to the public and private records of the period has been truly wonderful. To the staffs in general of The National Archives, Kew, the Liddell Hart Centre for Military Archives, the National Maritime Museum, Greenwich, the Churchill Archives Centre, Cambridge, the Imperial War Museum, London, and the National Army Museum, Chelsea, your kindness and assistance is greatly appreciated. The work you perform are the essential ingredients in any historian's achievements and little progress would follow without your support and efforts. Difficult though it is to cite all who have been of help with their time and patience, the author would be remiss not to call attention to Diane Manipud and Lianne Smith of the Liddell Hart Centre, Francesca Alves at the Churchill Archives Centre and Dr Simon Robbins and Anthony Richards of the Imperial War Museum.

Though this work is not a history of *The Naval Review*, the private, professional journal of the Royal Navy, it could never have been written without recourse to that excellent forum. To the editor, Vice Admiral Sir Jeremy Blackham, my deepest appreciation is extended for allowing material to be quoted, the absence of which would have made the story appear but a shadow of the truth. Similarly, the author is indebted for allowing a form of the tag that adorned the cover of *The Naval Review* for so many years to be adopted in the work's title.

For permission to quote from private copyrighted material, the following acknowledgments are gratefully extended. To the Master, Fellows and Scholars of Churchill College, Cambridge for kind permission for use of the papers of Lord Hankey of the Chart. To Lieutenant Commander Henry Drax my thankfulness is extended for allowing portions of the papers of Admiral Sir Reginald Plunkett-Ernle-Erle-Drax to appear. To the Trustees of the Imperial War Museum, my heartfelt thanks are extended for allowing extracts to be quoted from the papers of Admiral Sir Gerald Dickens. To Captain Stephen Harwood, RN, and Henry Harwood the sincere gratitude of the author is offered for allowing their recollections of their father, Admiral Sir Henry Harwood, to be cited. Additionally, to the copyright holder of the papers of Admiral Sir Vernon Haggard my sincere appreciation is offered. The Imperial War Museum possess a most remarkable collection of oral histories including those interviews with Captain Henry Lockhart St John Fancourt, RN, and Field Marshal Lord Harding of Petherton. This work would be the poorer without reference to and usage of these records and the gratitude of the author is offered for permission to cite extracts from these holdings.

To the Trustees of the Liddell Hart Centre for Military Archives sincere thanks are extended for allowing use of the papers of Field Marshal Viscount Alanbrooke of Brookeborough, Air Chief Marshal Sir Robert Brooke-Popham, Sir Julian Corbett, Admiral Sir Gerald Dickens, Field Marshal Sir John Dill, Sir Basil Liddell Hart and Lieutenant General Sir Henry Pownall. Likewise, my gratitude to use previously published material is extended to David Higham Associates for J. F. C. Fuller's *Memoirs of an Unconventional Solider* originally published by Nicolson and Watson, to Grub Street Publishing for extracts originally found in Kenneth Mackesy's *Armoured Crusader: The Biography of Major-General Sir Percy 'Hobo' Hobart* and to the Oxford University Press for citing material which first appeared in *From the Dardanelles to Oran: Studies of the Royal Navy in War and Peace 1915-1940* and *The Evolution of the Operational Art: From Napoleon to the Present*.

As a member of both the Navy Records Society and the Army Records Society, the author fully acknowledges the important work that both have played in expanding his knowledge and understanding of the Royal Navy and the British Army. Both perform a valuable and needed service in collating material around a central theme and providing scholarly commentary to the private and official papers published. Meanwhile, information reproduced from Crown copyright material is used by permission of The Controller, Her Majesty's Stationery Office. While every effort has been made to trace the owners of copyright, the vagaries of mergers, acquisitions and the death of immediate kin take their toll and I trust any who have escaped my searches will accept my sincere apologies. Finally, it remains that what follows is the responsibility of the author and the author alone. If the goal posts of understanding of the interwar Royal Navy have been advanced and prompts others to take a second-look at the received wisdom of the period and to think anew, then the effort will have been as worthwhile as it has been pleasurable.

List of Abbreviations

ACNS	Assistant Chief of the Naval Staff
ACQ	Call Sign for Commander Battle Cruiser Squadron
ASW	Anti-Submarine Warfare
BR	Books of Reference
CB	Confidential Book
CCC	Churchill Archives Centre Cambridge
CID	Committee of Imperial Defence
CIGS	Chief of the Imperial General Staff
COS	Chiefs of Staff or Chiefs of Staff Sub-Committee
DCNS	Deputy Chief of the Naval Staff
DIO	District Intelligence Officer
DNI	Director of Naval Intelligence
DSC	Distinguished Service Cross
DSO	Distinguished Service Order
DTSD	Director of Training and Staff Duties
FAA	Fleet Air Arm
FSR	Field Service Regulations
HMAS	His Majesty's Australian Ship
HMCS	His Majesty's Canadian Ship
HMS	His Majesty's Ship
IDC	Imperial Defence College
idc	Passed Imperial Defence College
ISTDC	Inter-Service Training and Development Centre
IWM	Imperial War Museum
JIC	Joint Intelligence Sub-Committee
JOWC	Junior Officers' War Course
JPC	Joint Planning Sub-Committee
LHCMA	Liddell Hart Centre for Military Archives
NCS	Naval Control of Shipping
NAM	National Army Museum
NMM	National Maritime Museum
OU	Official Use (Book)
psa	Passed Staff College Air Force

psc	Passed Staff College
qs	Qualified for War Staff without examination
RAF	Royal Air Force
RCN	Royal Canadian Navy
RCNR	Royal Canadian Naval Reserve
RM	Royal Marines
RNAS	Royal Naval Air Service
RNR	Royal Naval Reserve
RNVR	Royal Naval Volunteer Reserve
RUSI	Royal United Services Institution
SMS	Supreme Majesty's Ship
SOI	Staff Officer Intelligence
SOO	Staff Officer Operations
SOO&I	Staff Officer Operations and Intelligence
SOTC	Senior Officers' Technical Course
SOWC	Senior Officers' War Course
TEWT	Tactical Exercise Without Troops
TNA	The National Archives
TSD	Training and Staff Duties Division
ws	War Staff eligible through examination

A Note on Sources

The Navy Lists, the monthly compendia specifying the assignments of Royal Navy and Royal Marine officers for the period 1919-39, are a major source of the conclusions drawn by the author. Though official in publication, they are not definitive as errors are present and a caveat to that effect was frequently inserted into the publication reminding all that the Admiralty remained the sole authority for an officer's relative standing. This is entirely understandable. In an age when the collection, collation and printing of information occurred manually, errors were ever a problem. Some errors were of a genuine type where an honest mistake was made by the editor. Yet, portions were of a deliberate kind being obfuscations inserted to mislead the greater public of the true posting of an officer. This practice contemporary readers will understand as affording a measure of operational security as the Admiralty sought to protect the essential elements of friendly information. Reference to officers attending the Tactical School is one such example. In other cases, the information presented was simply wrong. Here, the example of listing officers, at times, as studying at the Royal Naval Staff College when, in truth, they were attending the co-located Intelligence Course can be cited. Reference to officer personnel records found in the Admiralty 196 series of the National Archives can help resolve such matters, but here, too, a caveat is required. For the most part, officers' records were maintained in longhand and subject to the same human error that affected the accuracy of *The Navy List*. More significantly, the subsidiary problem of legibility arises. The author has attempted to reduce such transcription errors to the absolute minimum, but cannot vouchsafe that his own interpretation is flawless for all instances. Likewise, information for the *Navy Estimates* shares a similar caveat, as the publication remained a projection of anticipated spending for the financial year indicated.

Where author citation is provided for articles appearing in *The Naval Review*, the source in most instances is based on the summary specified in James Goldrick and John B. Hattendorf, eds., *Mahan is Not Enough: The Proceedings of a Conference on the Works of Sir Julian Corbett and Admiral Sir Herbert Richmond*. The presumed authorship of a limited number of articles is based on textual analysis of the article compared to the style and content of similar pieces made by a known writer. In these cases, a footnote reflects this conclusion.

For officers where an amplifying footnote is provided, birth details are as specified in the relevant official service file held by the National Archives of the United

Kingdom. The date of death is as recorded in the same source, or if not present, then as found in the online repository www.ancestry.co.uk. Finally, when providing a quotation where a degree of confusion exists over the meaning of a word or phrase, the use '???' appears within brackets to indicate uncertainty of rendering.

Introduction

Education ought to receive prodigious attention after the war, for our failures have been largely due directly to our lack of education.[1]

Captain Herbert Richmond

This study, like many historical works before it, began with the author asking a question. During a period of initial enquiries into the 1919-39 Royal Navy, a recurring theme raised by several contemporary naval officers in letters, diaries, and memoirs was that the Service deprecated learning—at least critical thinking about the higher aspects of war and naval policy—so that when it came time for one to assume a position of greater responsibility, such as the command of a fleet or an independent squadron, too much was 'left to chance' to borrow a phrase from Admiral Lord Nelson. Winston Churchill,[2] himself a man of not inconsiderable talents and with direct experience of the Navy during the recent war, too, commented that 'there was no moment in the career training of a naval officer when he was obliged to read a single book about naval war, or even pass a rudimentary examination in naval history.'[3] The implication of this argument, if not always explicitly voiced by all, was that the Royal Navy was negligent in its constitutional duties by failing to prepare for war prior to its initiation, and once engaged in any conflict, performed poorly as it did not understand the linkages between policy, strategy, and the operational art.

As the author continued to read the literature governing the Royal Navy for the early twentieth century, the omission of any study addressing how the Service trained its officers in the higher aspects of war was palpable.[4] True, Arthur Marder, the sage of

---

1   Richmond diary, 19 March 1917 cited in Arthur J. Marder, ed., *Portrait of an Admiral: The Life and Papers of Sir Herbert Richmond* (London: Jonathan Cape, 1952), p. 241.

2   Sir Winston Leonard Spencer Churchill (1874-1965). First Lord of the Admiralty, 1911-15 and 1939-40.

3   Winston Churchill, cited in Anthony R. Wells, 'Staff Training and the Royal Navy, 1918-1939,' in Brian Bond and Ian Roy, eds., *War and Society, Vol. II* (London: Croom Helm, 1977), p. 86.

4   This omission for the pre-World War era has been rectified by Robert L. Davison, *The Challenges of Command: The Royal Navy's Executive Branch Officers, 1880-1919* (Farnham: Ashgate, 2011).

the Edwardian and Georgian Navy discussed the topic in passing, but for a scholar who could write to such lengths the lack of rigorous examination on staff and war training and education remains striking.[5] A school of later historians a generation or more removed from the events in question has revised Marder's conclusions if not defeated them outright. Still, even amongst this group—of which Professor Jon Sumida, Drs Andrew Gordon and Nicholas Lambert are at the fore in writing exacting, new, and even provocative scholarship—the place of higher level education and training in the preparation of a naval officer's career remains largely void. This is not a criticism, but only an observation for it remains that historians write towards that which interests them and what they can document. Where the story has been surveyed previously, brevity has circumscribed the picture with the effect that the Staff College and War Course are divorced from the broader lines of naval education and the several avenues available for preparing officers for greater responsibilities. Such essays excellent in their way have but scratched the surface of a topic deep in complexity and rich in nuance failing to explore the backgrounds of the officers trained, the uses made of their talents, and weighing the import of the several curricula.

Most naval historians, including this one, accept the general thrust of the argument that British naval performance, especially during the earliest period of the World War was deficient. Certainly, ample evidence exists to not refute the charges. Whether considering the lamentable execution of the Dardanelles campaign of 1915-16, failing to prevent the German ships SMS *Goeben* and SMS *Breslau* escape to Turkish waters in 1914, the forlorn Coronel engagement of Rear Admiral Sir Christopher Cradock's squadron, the poor coordination of vital information between the Operations and Intelligence Divisions of the Naval Staff, or the failure to convoy its merchant shipping earlier in the face of the commerce war waged by German submarines, the record of Royal Navy performance during the Great War was not in the best traditions of the Service and the staff work and command performance demonstrated were not all that could be desired.

Such missteps, if regrettable, are part of the ebb and flow of war. Particularly, when one is engaged in a test against another first class military power. The expectation of another Trafalgar, or *der Tag* to employ the German parlance frequently used by naval officers, was a common belief before the war making what followed all the more lamentable. Perhaps if observers and participants had remembered that victory against

5   Marder's primary works addressing the Royal Navy include *The Anatomy of British Seapower: A History of British Naval Policy in the Pre-Dreadnought Era, 1880-1905* (New York: Alfred A. Knopf, 1940); *Portrait of an Admiral: The Life and Papers of Sir Herbert Richmond* (London: Jonathan Cape, 1952); *Fear God and Dread Nought: The Correspondence of Admiral of the Fleet Lord Fisher of Kelverstone* 3 Vols. (London: Jonathan Caper, 1952-59); *From the Dreadnought to Scapa Flow.* 5 Vols. (London: Oxford University Press, 1961-70); *From the Dardanelles to Oran: Studies of the Royal Navy in War and Peace* (London: Oxford University Press, 1974); and *Old Friends, New Enemies: The Royal Navy and the Imperial Japanese Navy*, 2 Vols. (Oxford: Clarendon Press, 1981-90).

Napoleonic France was secured only after another ten years hard fighting following the success of Trafalgar, the expectations of contemporaries would have been more circumscribed. Nevertheless, British naval performance in the Great War was deficient, not critically deficient, but deficient enough and those deficiencies in organization, education, and training demonstrated in the contest perhaps loomed largest as these were elements most subject to purely naval control. Compounding such deficiencies was an ever changing and varying leadership. Civilians such as Churchill and Edward Carson serving as First Lords, whether as an active or a passive minister of the Service, must share some of the blame for the Navy's fortunes, but poor administration did not stop with the First Lord's remit.[6] As the late Brian Schofield noted, the office of First Sea Lord had five occupants between the years 1910-18.[7] Surely, a circumstance not conducive to formulating an effective naval policy and seeing it implemented to a successful conclusion.

Aggravating the situation is that the Service placed a premium on practical and not theoretical knowledge during the period preceding the Great War and lacked an effective staff system. True, tentative steps were in hand to put this right, but the outbreak of war in August 1914 came before sufficient officers had been trained and appointed to the nascent Admiralty War Staff. Still, it was not merely a case of numbers. The culture of the Navy deprecated the staff system with its emphasis on division of labour and specialization so much at variance to its traditions of executive command where a Commanding Officer of a ship no less than a fleet was master of all that he surveyed. This aversion to the staff system consequently resulted in the policy and doctrine of the Royal Navy being understood by few, overseen by fewer, and tending to be whatever the senior leadership of the day held it to be.[8]

Maurice Hankey, a former officer of the Royal Marine Artillery, acolyte of Charles Ottley and serving as Secretary of the Committee of Imperial Defence (CID) and to the War Cabinet in its various guises, recognised that the administration of the Royal Navy was troublesome. A man of much discretion but not adverse to influencing affairs if events required it, Hankey relayed to the prime minister, Herbert Asquith, his fears later recording in his diary:

> I saw the Prime Minister and talked to him at great length about the defects of the Admiralty. I point out – (1) that there was not a single naval officer on the Board of the Admiralty who had actual war experience at sea; (2) that the Admiralty had assumed the function of Commander-in-Chief of all forces in home waters outside the Grand Fleet, but nevertheless that none of the Board ever went near the fleet or

6   Geoffrey Lewis, *Carson: The Man Who Divided Ireland* (London: Hambledon and London, 2005), p. 211.

7   Brian Schofield, *British Sea Power: Naval Policy in the Twentieth Century* (London: B. T. Batsford, 1967), p. 71.

8   Reginald Plunkett-Ernle-Erle-Drax, 'The Staff College,' *The Naval Review* (London: The Naval Society, 1915), pp. 97-98.

inspected anything; (3) that the Chief of the War Staff not only acted practically as C.inC. and Chief of Staff simultaneously of the forces in home waters, writing every single telegram personally, but was also charged with the distribution of the fleet as a whole, arrangements with allies, and the main policy of the war tasks far too great for any one man, however industrious & capable he might be, to perform. I urged drastic changes.[9]

Apart from the rich irony that Hankey, himself an arch-centralizer, could render such an observation, his evidence must weigh heavy in the modern historian's assessment of the efficacy of Royal Navy administration during the war given the degree of access he possessed to principals and events both. Changes in personnel and organization followed and 1917 proved to be a watershed year in the development of a proper staff system. However, changing reporting lines on an organizational chart and removing less effective personages from positions of oversight were easier to effect than changing the culture of a navy and the willingness of officers to accept the place of the staff and to make use of the benefits it could bring.

When examining the interwar period the issue of sound administration is present too. Direct comparisons are invidious as the challenges arising and the political context supporting are so different. Perhaps the soundest approach is to compare the Navy to its sister Services during the period in question to gauge how well it rose to the tasks it faced. On that score, the Royal Navy's military leadership as represented in the First Sea Lords of the era, with the exception of Admiral Sir Frederick Field,[10] proved highly competent and carried much weight in their counsels. Even to Field, this author is hesitant to criticize too strongly. Remembered for being the First Sea Lord at the time of the Invergordon Mutiny (a disaffection occurring in the Atlantic Fleet in September 1931 over the planned reductions in pay to bring balance to Britain's finances) much of the responsibility for that episode rested squarely with the Government. To be sure, commanding officers of the ships in turmoil also had a degree of culpability. Field's culpability was that he failed to anticipate events, did not probe his flag officers for the sentiment prevailing amongst the men of the fleet in light of the findings of the *Report* issued by Sir George May's committee issued in July 1931 and once events occurred failed to take a lead.

Invergordon was not the only disaffection to be displayed in the interwar Navy. Episodes of previous mutinies included a Royal Fleet Reserve Battalion activated

---

9   Diary entry of 28 February 1916, Churchill Archives Centre, Lord Hankey of the Chart (Maurice Hankey), HNKY/1/1.

10   Later Admiral of the Fleet Sir Frederick Laurence Field (1871-1945). Captain, 1907; War Course, 1909-10; Director of Torpedo and Mining, 1918-20; Rear Admiral, 1919; Third Sea Lord and Controller, 1920-23; Commander, Battle Cruiser Squadron, 1923; Commander, Special Service Squadron, 1923-24; Vice Admiral, 1924; Deputy Chief of the Naval Staff, 1925-28; Admiral, 1928; Commander-in-Chief, Mediterranean, 1928-30; First Sea Lord, 1930-33 and Admiral of the Fleet, 1933.

during a strike of coalminers in April 1921 refusing to assemble leaving little stain on Admiral of the Fleet Earl Beatty's[11] reputation. Nor did the outbreak in ill-discipline arising in HMS *Guardian* and HMS *Warspite*[12] affect the standing of Admiral of the Fleet Lord Chatfield.[13] Field's lot was that the Invergordon Mutiny had a political import missing in these other outbreaks. Moreover, the Navy had long anticipated that problems were apt to arise if the pay of the men serving in the lower deck was reduced.[14] Roskill's assessment that the Boards of Admiralty sitting from 1927-33 were amongst the weakest ever formed may be a correct finding on its own merits, but when weighed against the administrations governing the British Army and the Royal Air Force (RAF) for the period was not noticeably poorer.[15] The Chiefs of the Air Staff, Hugh Trenchard[16] and John Salmond,[17] made the best of the bad hands dealt the RAF and Trenchard benefited upon Beatty's retirements, when contrary to custom, he became the next chairman of the Chiefs of Staff (COS) Sub-Committee and not Admiral Sir Charles Madden who was the senior most officer of the three.[18] Yet, Trenchard was increasingly exhausted and Salmond lacked the stature of the former.[19] Field Marshal Sir George Milne is notable for his lack of mark on British

11   Admiral of the Fleet Sir David Beatty, later Earl Beatty (1871-1936). Captain, 1900; Rear Admiral, 1910; War Course, 1911; Naval Secretary to First Lord, 1912-13; Commander, Battle Cruiser Fleet, 1913-16; Vice Admiral, 1915; Commander-in-Chief, Grand Fleet, 1916-19; Admiral, 1919; Admiral of the Fleet, 1919, First Sea Lord, 1919-27, and retired, 1927.

12   *Guardian* in October 1936 and *Warspite* in June 1937 suffered incidents of incitement to mutiny and insubordination, respectively. The events surrounding are discussed in *C.B. 3027(A), Mutiny in the Royal Navy, v. II, 1921-1937*, Admiralty, Tactical and Staff Duties Division, T.S.D. 76/55. Privately held.

13   Admiral of the Fleet Sir Alfred Ernle Montacute Chatfield, later Baron Chatfield of Ditchfield (1873-1967). Captain, 1909; Gunnery, Torpedo, Navigation and Signal Courses, 1910-11; War Course, 1911; Flag Captain in HMS *Medina*, 1911-12 and HMS *Lion*, 1913-16; Flag Captain and Fleet Gunnery Officer in Grand Fleet, 1917-19; Fourth Sea Lord, 1919-20; Rear Admiral, 1920; Assistant Chief of Naval Staff, 1920-22; Third Sea Lord and Controller, 1925-28; Vice Admiral, 1926; Commander-in-Chief, Atlantic Fleet, 1929-30; Admiral, 1930; Commander-in-Chief, Mediterranean, 1930-32; First Sea Lord, 1933-38; Admiral of the Fleet, 1935; Minister for the Co-ordination of Defence, 1939-40.

14   ADM 203/69, Captain Edward Astley-Rushton, 'The Staff College', Senior Officers' War Course lecture, no date, but *c.* 1924.

15   Stephen Roskill, *Hankey: Man of Secrets, Volume II* (London: Collins, 1972), p. 508.

16   Marshal of the Royal Air Force Hugh Montague Trenchard, later Viscount (1873-1956). Chief of the Air Staff, 1918 and 1919-30.

17   Marshal of the Royal Air Force Sir John Maitland Salmond (1881-1968). Chief of the Air Staff, 1930-33.

18   Admiral of the Fleet Admiral Sir Charles Edward Madden, Bart. (1862-1935), Captain, 1901; Fourth Sea Lord, 1910-1911; Rear Admiral, 1911; Chief of Staff to Commander-in-Chief, Grand Fleet, 1914-16; Vice Admiral, 1916; Commander, First Battle Squadron, 1916-19; Admiral, 1919; Commander-in-Chief, Atlantic Fleet, 1919-22; Admiral of the Fleet, 1924; and First Sea Lord, 1927-30.

19   Andrew Boyle, *Trenchard* (London: Collins, 1962), p. 573.

military policy during the period, but given the financial stringency ever present it was beyond his powers to do more.[20] He, too, was a tired officer and one cannot but feel that collectively the senior officers of the period were being asked to make bricks without straw and were retained in office when younger officers offering greater energy were required.

Turning to the civil side of affairs borne by the First Lords of the 1919-39 period, the story is less satisfactory. Too frequently the office was occupied by personages of the second rank or statesmen in the twilight of their careers. Sir Austen Chamberlain, the former Foreign Secretary, is a telling example. Stanley Baldwin, the Conservative prime minister thought him the dimmest of persons (Chamberlain held an equal view of Baldwin's abilities), and the tag whether first aired by Churchill or F. E. Smith that he 'always played the game and always lost it' coming from his peers and party rivals, if an unfair description of the man, contains an essential truth.[21]

Recent scholarship has shed a welcoming light on the many adjustments that the Royal Navy adopted during the course of the First World War, and, indeed, has challenged the very tenets of British naval policy and its supporting strategy in the years immediately preceding the conflict.[22] This is welcomed. Still, notwithstanding any errors of commission and omission in planning, execution, and the matériel employed, it remains that the Service achieved its broad strategic aims of ensuring allied use of the seas, denying the same to the Central Powers, and played a major role in the victory reached in 1918. The latter most clearly highlighted through its sustaining the British Expeditionary Force on the continent and in supporting the many subsidiary deployments in other theatres, in enforcing an increasingly more determined and effective blockade, and—with echoes of support provided in the next war—its provision of sustenance to allies.

Others echoed Churchill's lament about the poor state of pre-1914 naval officer training and did not need to wait for the onset of a global war before raising such doubts. Rear Admiral Charles Ottley, a former Director of Naval Intelligence, Hankey's predecessor as Secretary to the Committee of Imperial Defence, and an officer ironically highly regarded by Admiral Sir John Fisher,[23] spoke in a similar

20   Brian Bond, *British Military Policy between the Two World Wars* (Oxford: Clarendon Press, 1980), pp. 158-60.
21   Robert C. Self, ed., *The Austen Chamberlain Diary Letters: the Correspondence of Sir Austen Chamberlain with his Sisters Hilda and Ida, 1916-1937* (Cambridge: Royal Historical Society, 1995), pp. 13-14, 18 and 29.
22   See Jon T. Sumida, *In Defence of Naval Supremacy: Finance, Technology and British Naval Policy, 1889-1914* (Boston: Unwin Hyman, 1989), Nicholas A. Lambert, *Sir John Fisher's Naval Revolution* (Columbia: University of South Carolina Press, 1999) and Nicholas Black, *The British Naval Staff During the First World War* (Woodbridge: Seaforth Publishing, 2009).
23   Admiral of the Fleet Sir John Fisher, later Baron Fisher of Kilverstone (1841–1920). Captain, 1874; Rear Admiral, 1890; Vice Admiral, 1896; Admiral, 1901; First Sea Lord, 1904-10 and 1914-15; Admiral of the Fleet, 1905 and Baron, 1909.

vein. Whilst serving at the CID he asserted on more than one occasion that naval affairs were not in order in two crucial areas: training and war planning.[24] Meanwhile, Captain Herbert Richmond,[25] from whom much more remains to be said, argued in the opening essay of a new journal aspiring to provoke naval thought:

> For several years our young officers are taught no inconsiderable number of facts, but no stage in the process can be found at which they are taught to think and reason: and while much time is devoted to instruction in the technique of their work—to learning the mechanical details of engines, guns and torpedoes for example—none is allotted to instruction in the art and conduct of war.[26]

Indeed, the very movement of Churchill from the Home Office portfolio to the Admiralty was as a palliative by Asquith to bring a firmer hand on naval administration and defuse what was becoming increasingly a partisan political issue.[27]

It is this writer's contention that if officers such as Richmond and Captain Kenneth Dewar[28] were too critical in their views on the performance of the Royal Navy during the late war, it was mainly due to their being advocates for reform and not dispassionate historians of the recent past. Such officers, fired with a desire to improve their Service—a Service it might be noted which they had dedicated their lives to—even prior to the outbreak of war in 1914, must have felt deep angst in seeing their worst pronouncements proved right by the test of war. The upshot was that it moved many of the zealous officers of reform to over-emphasize the failings they observed without applying due consideration of the many facets that the Navy actually managed to get correct. To them, sufficient was, indeed, the enemy of the better, and if victory was, at last won, it had taken longer and cost more in treasure, blood, and strategic balance than need be the case.

24   Lambert, *Fisher's Naval Revolution*, pp. 167 and 181.
25   Later Admiral Sir Herbert Richmond (1871–1946). Naval Ordnance Department, 1903-04; Naval Assistant to First Sea Lord, 1906; Captain, 1908, Commanding Officer, *Dreadnought*, 1909-11 and HMS *Furious*, 1911-12; Assistant Director, Operations Division, 1913-15; Commanding Officer, HMS *Commonwealth*, 1915-17 and HMS *Conqueror*, 1917-18; Director, Training and Staff Duties Division, 1918; Commanding Officer, *Erin*, 1919; Rear Admiral, 1920; Royal Naval War College, 1920-22; Admiral-President, Royal Naval College Greenwich, 1922-23; Commander-in-Chief, East Indies, 1923–25; Vice Admiral, 1925; Commandant, Imperial Defence College, 1926-29; Admiral, 1929 and retired, 1931.
26   Herbert Richmond, 'Introductory,' *The Naval Review*, Vol. 1, No.1, p. 1.
27   Richard Burdon Haldane, *An Autobiography* (London: Hodder and Stoughton, 1929), pp. 227-31 and Randolph S. Churchill, *Winston S. Churchill: Young Statesman, 1901–1914* (Boston: Houghton Mifflin Company, 1967), pp. 518-26.
28   Vice Admiral Kenneth Gilbert Balmain Dewar (1879-1964). Royal Naval War College, Portsmouth, 1911 and 1912-13; Admiralty War Staff, 1912; Assistant Director, Plans Division, 1917; Commanding Officer, HMS *Capetown*, 1923-24; Senior Officers' War Course, 1924; Assistant Director, Naval Intelligence Division; 1925-27; Commanding Officer, HMS *Royal Oak*, 1928-29; Rear Admiral and retired, 1929 and Vice Admiral, 1934.

If familiarity does breed contempt, then Richmond's posting to the Operations Division of the Admiralty in 1913 is candidate for the best naval example. Possessing a brilliant mind with a contrarian disposition, he looked askance at the level of understanding about war displayed by members of the Naval Staff. One outcome to the Service's and historians lasting benefit to the angst felt by officers such as Richmond and Dewar was the creation of a private forum, the Naval Society and its related journal *The Naval Review*, where members could discuss, indeed, argue maritime matters in the broadest meaning of the term without fear of professional censure.[29] Anonymity of authors writing for the journal was the norm, though, of course, the style of discourse and the positions taken meant that anonymity was more in the guise of plausible deniability than actual mystery for subscribers to a quarterly publication that did not speak to a broader audience. If the Navy as an institution initially looked askance at the society's efforts, it is to its credit that it was not suppressed outright given the Service's recent history. The public airing of the strategic differences between Fisher, the First Sea Lord and Lord Charles Beresford,[30] the Commander-in-Chief, Channel Fleet had not shown the Royal Navy in its most favourable light, had necessitated a Cabinet-level review of the issues outstanding, and demonstrated what may well happen if the Pandora's box of subordinate criticism was taken too far.[31] Though promulgation of *The Naval Review* was suspended for most of the First World War due to questions of operational security voiced by the Admiralty and the Commander-in-Chief, Grand Fleet, Admiral John Jellicoe,[32] it flourished during the interwar period and remains an excellent source of the intellectual milieu surrounding the Royal Navy.

The absence of analysis on the role and place of higher education in the 1919-39 Navy is surprising if for no other reason than the financial stringency commonly cited as lying at the heart of the Service's problems during the period leading to the Second World War. Not always able to conduct sustained tactical evolutions with the fleet, the gaming of new concepts ashore and instructing officers in the many doctrinal changes caused by the increasing capabilities of gunnery, torpedo fire, and airpower became a necessity. This study examines the measures the Royal Navy adopted to

29   The first issue was released in 1913 and the founding members of the Naval Society were Admiral W. H. Henderson (Honorary Editor); Lieutenant R. M. Bellairs; Commander K.G.B. Dewar; Lieutenant T. Fisher; Captain E.W. Harding, RM; Commander the Honourable R.A.R. Plunkett, Captain H.W. Richmond and Lieutenant H. G. Thursfield. See http://www.naval-review.org/about.asp.

30   Admiral Lord Charles William Delapoer Beresford (1846-1919). Captain, 1882; Rear Admiral, 1897; Vice Admiral, 1902 and Admiral, 1906.

31   Barry M. Gough, 'Admiral Sir (later Baron) John Arbuthnot Fisher (1904-1910; 1914-1915),' Malcolm H. Murfett, ed., *The First Sea Lords: From Fisher to Mountbatten* (Westport: Praeger, 1995), pp. 24-25.

32   Admiral of the Fleet Sir John Rushworth Jellicoe, later Earl Jellicoe (1859–1935). Captain, 1897; Rear Admiral, 1907; Vice Admiral, 1911, Commander-in-Chief, Grand Fleet, 1914-16; Admiral, 1915; First Sea Lord, 1916-17; Viscount, 1918; Admiral of the Fleet, 1919; Governor General New Zealand, 1920-24 and Earl, 1925.

improve the higher-level training and education of its officers and associated Imperial naval members during the interwar period. It focuses on the Royal Naval College Greenwich, home of the War and Staff Courses, and the institutions and courses residing in Portsmouth. Amongst the latter were the Tactical School borne in HMS *Victory*, and the Senior Officers' Technical Course where selected officers received instruction in the broader aspects of their profession, the evolving nature of war and the lessons of past operational experience. Additionally, it investigates the increasing dependence of naval operations on the support provided by the other forces of the Crown—in today's parlance what is termed 'jointness'—and how the Royal Navy adopted to this phenomenon.

Constrained by finance and perhaps more importantly by Cabinet fiat, the Service could no longer look merely to its own devices to secure maritime dominance. Thus, the instruction of key officers at the Staff Colleges of the British and Indian Armies— Camberley and Quetta—and at Andover, the venue of Royal Air Force staff education and the pinnacle for war studies within the British context, the Imperial Defence College, became increasingly essential. The reader may question the inclusion of the Tactical School, a body devoted to the perfection of the art of the naval engagement, and, thus, the very antithesis of the stated aim of what is to follow. Yet, tactical acts at times have had an import beyond the immediacy of the present skirmish; the sinking of the *Lusitania* being perhaps the most noted example from the recent past in the context of this study. Whether officers of the Royal Navy became acquainted with the growing body of legal restrictions placed on armed conflict and the linkages between tactical actions and the policy implications arising are further points worth considering.

This work examines the types of executive officers (e.g., Gunnery, Torpedo, Navigating, Signals, etc.) trained in staff duties and reviews their subsequent career assignments to ascertain whether they suffered from want of promotion and whether the Service made proper use of their newly acquired skills. To wit, how frequently did an officer trained at the Air Force Staff College, for instance, see a future posting to an aircraft carrier or a squadron of the Fleet Air Arm (FAA)? Conversely, when assigning a naval officer to the Admiralty's Naval Air Division or Naval Air Material branch or to RAF Coastal Command or the Air Ministry, itself, was a period at Andover considered a requisite for the posting? In a like manner, in sending an officer on exchange or liaison duties to the War Office, is there evidence that a period of instruction at Camberley or the Senior Officers' School was the norm for a candidate before assuming the assignment.

An additional consideration weighed is whether pilots with their need to maintain proficiency in aviation skills had the opportunity to attend the Staff Course or the more abbreviated War Course. Professor Geoffrey Till has ably treated the story of British naval aviation during the interwar period.[33] The divided responsibility exercised over

---

33   Geoffrey Till, *Air Power and the Royal Navy 1914-1945: A Historical Survey* (London: Jane's Publishing Company, 1979).

British naval aviation by the Air Force and the Navy gave vent to a most contentious bureaucratic battle, and the solution finally reached was even then one not fully in accord with the Admiralty's desires. In the absence of real operational control, whether the Navy developed a cadre of trained staff officers conversant with tactical aviation from its establishment of pilots and observers is examined. Conversely, the work will examine whether Army and Air Force officers, after a period of instruction at the Royal Naval Staff College, spent time afloat with the fleet or at an ashore establishment of the Royal Marines. Lastly, it was seemingly an article of faith that the higher ranks of the Service was dominated by technical officers—particularly those specializing in gunnery or torpedo work. Is this view sustained based on the numbers actually trained in the higher aspects of war? If not, what does this say about the status of higher education in the Service and its ethos amongst a core constituency of naval officers?[34]

The curricula developed between the wars is investigated to discern the major themes of Staff and War College education and training and note is made of any significant omissions based on the experience of previous operations, and, most significantly, the First World War. A criticism often heard is that the Royal Navy, recognizing that its one test of fleet engagement during World War One, Jutland, was inconclusive at best and poorly handled at worst, became the crux of naval learning between the wars. The writer will measure the correctness of this assertion becoming as it has conventional wisdom—accepted with nary a question by historians—and identified most prominently with the late Captain Stephen Roskill.[35]

A concern commonly voiced by naval officers of the era, as opposed to their Army brethren, was that Staff College attendance was a hindrance to career advancement and not an aid as it took one away from the true path to professional advancement, the sea.[36] Thus, did they eschew obtaining the *psc*, the post-nominal designation for Passed Staff College, awarded for successful completion of the Staff Course or by serving as a member of the course's directing staff for at least a year. This view will be confirmed or refuted by systematically analysing the career prospects of graduating officers whilst a survey of the general ethos surrounding naval education and training during the period in question is made. Likewise, this work examines the officers assigned to the Navy's higher educational establishments to gauge whether the Service short-changed their career prospects or used the institutions as weigh-stations of the mediocre. Conversely, those officers who failed to benefit from an advanced period of study and yet achieved positions of great responsibility with commensurate rank is considered with an eye to assessing the circumstances that allowed them to do so. Admiral of the Fleet Lord Tovey[37] is a type best exemplifying the former instance. Attending at times

34   Anon., 'Specialization', *The Naval Review*, Vol. XXII, No. 3, August 1934, p. 447.
35   Stephen Roskill, *Naval Policy between the Wars: Volume I, The Period of Anglo-American Antagonism, 1919-1929* (New York: Walker and Company, 1968), p. 533.
36   Wells, 'Staff Training', in Bond and Roy, eds., *War and Society*, p. 97.
37   Admiral of the Fleet John Cronyn Tovey, later Baron Tovey of Langton Maltravers (1885-1971). Commanding Officer, HMS *Onslow*, 1916-17; Staff Course, 1919-20; Operations

the Staff Course, the British Army Senior Officers' School, the Imperial Defence College, the Senior Officers' Technical Course, the Tactical Course, and, finally, the Senior Officers' War Course, Tovey comes close to approaching the status of a professional student and this when thought an of officer of only average intellectual ability and rather a plodding type.[38]

Meanwhile, Admiral of the Fleet Charles Forbes[39] exemplifies an officer who eschewed the rigors of formal training failing to attend the Staff College, the Tactical Course,[40] the Senior Officers' Technical Course, or the Imperial Defence College. His one concession to higher education was to attend the Senior Officers' War Course in 1921 and though holding the naval *psc*, his was the award of one who had served on the directing staff of the Staff College. Recognized as an able administrator, an excellent gunnery specialist, but held to lack imagination,[41] Forbes was thought by one knowledgeable critic to be something of a bully who had advanced his career by following 'the well worn orthodox track with great competence.'[42] As both officers ultimately served as Commander-in-Chief, Home Fleet and were near contemporaries, staff education and training must be adjudged only one factor in determining an officer's prospects for higher command and the degree to which either represented the norm will be assessed.

However, a cautionary note is registered. Unlike naval ratings subject to orders, officers in the main were responsible for the progression of their careers through the requests that they made to the Admiralty for future assignments including training.

Division, 1920-22; Commanding Officer, HMS *Seawolf*, 1922-24; Captain, 1923; Senior Officer's School Sheerness, 1924; Captain (D), Eight Flotilla, 1924-25 and Sixth Flotilla, 1926-27; Imperial Defence Course, 1927; Senior Officers' Technical Course, 1928 and 1934; Tactical Course, 1928 and 1937; Tactical School, 1928-30; Naval Assistant to Second Sea Lord, 1930-32; Rear Admiral, 1935; Senior Officers' War Course, 1937; Vice Admiral, 1939; Admiral, 1942; Admiral of the Fleet, 1943 and Baron, 1946.

38   ADM 196/143 (Tovey). The verdict was applied by Captain Frederic Dreyer in 1913.
39   Admiral of the Fleet Sir Charles Morton Forbes (1880-1960). Five firsts in examinations for Lieutenant. Flag Commander and War Staff duties in HMS *Queen Elizabeth*, 1915-16; Appointed to War Staff, 1916; Flag Commander and War Staff duties in HMS *Marlborough*, 1916-17 and in HMS *Revenge*, 1917; Captain, 1917; Ordnance Committee, 1920; Senior Officers' War Course, 1921; Deputy Director, Royal Naval Staff College, 1921-23; Qualified War Staff, 1922; Flag Captain in *Queen Elizabeth*, 1923-25; Director, Naval Ordnance Department, 1925-28; Rear Admiral, 1928; Rear Admiral (D) in Mediterranean Fleet, 1930-31; Third Sea Lord, 1932-34; Vice Admiral, 1933; Vice Admiral Commanding First Battle Squadron and Second-in-Command, Mediterranean, 1934-36; Admiral, 1936; Commander-in-Chief, Home Fleet, 1938-40; Admiral of the Fleet, 1940; Commander-in-Chief, Plymouth, 1941-43.
40   Forbes had been scheduled to attend the May 1930 sitting of the Tactical Course but the appointment was cancelled when he became Rear Admiral (D) in the Mediterranean Fleet.
41   ADM 196/90 (Forbes).
42   Russell Grenfell to Liddell Hart letter of 2 June 1937, Sir Basil Liddell Hart Papers, Liddell Hart Centre for Military Archives, King's College, London, LH 1/330/9.

At least, this was case for the initial period under review. Many are the factors influencing the choices made—or not made—by an officer seeking any assignment that the mere reading of a personnel file and confidential reports may not distill. Underlying health issues of the officer or a family member, conflicts with a superior whose positive endorsement was required for any application to the Admiralty for the desired course, or the perceived financial burden of an assignment in the London area remain but three of the reasons for forestalling an officer's prospects to a desired posting.[43] The last point perhaps of more concern to a flag officer seeking an educational or training opportunity, as initially during the period under investigation a revision to half-pay for Admirals attending courses was required.[44] Moreover, financial considerations were not without influence over the staff assigned to the Royal Naval College Greenwich, as Kenneth Dewar discovered to his dismay. When investigating the chance of securing the directorship of the Staff College in 1924, a caution that private means were necessary to meet the entertaining responsibilities associated with the office put paid any chances of netting the appointment.[45] In Dewar's case and as will become clearer, other factors may also have been at play indicating that aptitude minus money was far from the only disqualification.

The role and place of higher naval education was set within the context of the corollary changes manifested in British academic and adult education circles during the period. Officers such as Admiral Richmond, though every bit a part of the Service, had broader horizons and contacts to draw upon in fashioning their opinions. Thus, in a moment of introspection in the waning days of the First World War, he recorded an encounter with a leading Liberal light, Viscount Haldane:

> Dined last night with Lord Haldane. He returned to London last week & asked me to come & discuss education. A most interesting evening – he & I alone...Of education we talked for three hours. H. strongly supported my views, which I explained at length. The use of the public schools, a naval college, the subsequent course at the university, & finally the Staff course to be held at a college in the vicinity of Camberley were all concurred in. Also my views that it was wrong to try & educate Admirals who had no proper groundwork. He did in fact use the same terms as I had written down myself about the absurdity of teaching elderly men the first elements of their work by lecturing to them, & that we wanted our Admirals to be our teachers & not our pupils...He recognises fully how greatly the High Command of the Navy has failed in this war, & that our victory at sea is due solely to crushing superiority & not at all to skill.[46]

43   Articles 329 and 332, *The King's Regulations and Admiralty Instructions, Volume I* (London: His Majesty's Stationary Officer, 1938), pp. 113-14.

44   'Courses for Senior Officers,' *The Navy List*, July 1920, p. 2402.

45   Leslie Gardiner, *The Royal Oak Courts Martial* (Edinburgh: William Blackwood & Sons, Ltd., 1965), p. 83.

46   Richmond cited in Marder, ed., *Portrait of an Admiral*, p. 325.

Given the age of entry for officer-candidates into the Royal Navy—thirteen or fourteen years old typically the norm and absent the direct experience of British secondary and university education—this ability to reach out to statesmen, members of the academy, or professional counterparts in the British Army and Royal Air Force was of no little importance to a reformer such as a Richmond.[47] Meanwhile, in seeking out Haldane, a former Secretary of State for War with a zeal for reform, whether of a military or educational flavour or both, Richmond was reaching wide indeed.

Still, beyond being an incubator in the craft of warfare and its associated planning, the Naval Staff looked to the War and Staff Colleges to investigate current strategic and tactical issues and sought their support in framing problems to be tested in the exercise programmes of the fleet. Too often over-looked or even discounted by historians, a systematic analysis of the place of higher-level instruction in the Royal Navy for the 1919-39 period and the benefits, and, yes, problems encountered along the way follows. All history is revision save for the first draft. Thus, what follows is revisionist to the extent that it challenges the conclusions of previous chroniclers who accepted the arguments of contemporary naval officers without going deeper into the subject.[48] And, yet, what follows is but the first draft as it remains that so little has been written about Greenwich, the Tactical School, and the Imperial Defence College previously.

The military profession is unique among disciplines in that the ends of training and education it pursues are directed to endeavors which may never arise during one's active career, or it may even be the case that the concepts and methods one has spent a career mastering are found wanting at the time of reckoning. Yet, the education of officers remains vital because the consequences of failure in war are indeed so great. How militaries raise, train, and equip their forces are timeless issues and the past, whilst never a predicator of the future, is worthy of investigation because it sheds valuable light on problems that continue to perplex us. This work examines the interwar Royal Navy and in doing so raises issues that a contemporary reader may feel are all too familiar. Specialist or general service officers, skills for war or the skills for the most likely tasks to be executed, the proper size of staffs, and when to introduce joint education into a member's career remain as relevant in the first part of the twenty-first century as they were in the early part of the last. The record that follows shows that if there is much the Royal Navy got wrong, there was even more that it got right. Having officers who truly desired to learn from the errors of their predecessors was the first step in the process of learning.

Finally, the story told is but a part of a larger tale of the many changes that were occurring within the Royal Navy, Britain and the Empire set between the wars. The

47   Brian Lavery, *In Which They Served: The Royal Navy Officer Experience in the Second World War* (London: Conway, 2008), pp. 73-74.

48   See Barry Gough, *Historical Dreadnoughts: Arthur Marder, Stephen Roskill, and Battles for Naval History* (Barnsley: Seaforth Publishing, 2010), p. 42 for a recent affirmation that naval officers and the Royal Navy, in general, deprecated intellectual development.

Royal Naval College Greenwich predated the establishment of the War and Staff Colleges and the education and training of naval and marine officers had undergone much revision in the period immediately preceding the Great War. Much of this change was in response to the new matériel being adopted within the fleet such as steam and turbine propulsion, the adoption of electricity and non-visual signaling, the replacing of muzzle-loading naval rifles with breech-loading guns, and the introduction of a reliable and much improved self-propelled torpedo. These implements fostered degrees of specialization for officers unheard of previously and gave rise to the founding of new training establishments to complement the technological revolution underway, and, in time, required staff officers familiar with the tools of naval warfare to be attached to the fleets and to the Admiralty.[49]

Specialization required to master the new naval technologies came with a price. It was thought to narrow an officer's outlook and the adage 'not being able to see the forest for the trees' is an apt one capturing, as it does, the risk that an officer in mastering one element of his trade might lose sight of its ultimate purpose of use. Captain Gerald Dickens,[50] grandson of the famous author and later to be an editor of *The Naval Review*, writing soon after the First World War argued that:

> The science of war is high above mere technical knowledge of either gunnery, torpedo, navigation, seamanship, engineering or of all these put together. Efficiency in these things alone, as has been proved times without number, is of little use when leaders and their staff possess but a *dilettante* knowledge of the art of waging war, and the vastest amount of gun, torpedo horsepower, however beautifully and efficiently set in motion, will fail to win a war if set in the wrong direction.
>
> An officer's training must therefore first of all be such to produce a man capable of undertaking the functions of a leader in war, whose main task is to deal in broad policies, "think in oceans," and be thoroughly imbued with correct strategical principles.[51]

Dickens was not arguing against specialization, but, rather, a new type of specialist: the Staff Officer.[52]

49   Liberal Unionist, 'Conservatives and Reformers', *The Naval Review*, Vol. XXII, No. 2, May 1939, p. 228. The author is probably Commander the Honourable Walter Carson.

50   Later Admiral Sir Gerald Charles Dickens (1879–1962). Qualified for War Staff Duties. Flag Commander for War Staff Duties in HMS *Egmont*, 1917-19; Naval Intelligence Division, 1919; Captain, 1919; Deputy Director, Plans Division, 1920-22; Commanding Officer, HMS *Carlisle*, 1922-24; Tactical Course, 1925, 1932 and 1938; Commanding Officer HMS *Harebell*, 1925-26; Imperial Defence College, 1926-29; Commanding Officer, HMS *Repulse*, 1929-31; Senior Officers' War Course, 1931-32; Rear Admiral, 1932; Director, Naval Intelligence Division, 1932-35; Commander Reserve Fleet, 1935-36; Vice Admiral, 1936; Senior Officers' Technical Course, 1938 and retired, 1940. Recalled for war service.

51   Gerald C. Dickens, 'The Training of Naval Officers', *The Naval Review*, Vol. VIII, No. 1, p. 61.

52   *Ibid*, p. 64.

Agreeing with the thrust of Dickens' argument, but tackling it from its polar opposite, in 1923, Lieutenant Commander Russell Grenfell[53] believed the problem for the Royal Navy was that specialization came at the expense of an officer's understanding of tactical theory. Specialization allowed one to be a master of weapon efficiency, but it did little to develop the knowledge how such a weapon should be used. Indeed, that was the very problem. In battle, a fleet was required to develop its maximum fire potential and not rest upon the efficiency of a single type of weapon. Grenfell argued for the creation of a tactical school and in doing so anticipated others more senior and influential.

At the time Grenfell was writing, the Navy operated schools providing instruction for executive officers in gunnery, torpedo, navigation, mining, signals, submarine operations, anti-submarine warfare, physical and recreational training, gas, air observation and intelligence.[54] Experts in kind were everywhere. One need only review the *Navy List* to see how many officers counted the 'G' for Gunnery, 'T' for Torpedo or 'N' for Navigation against their name, but where in the *Navy List* did one find the tacticians? The Navy's response, in time, was to create the Tactical School as a venue where senior officers could learn current tactical theory and practice. The relationship of this new school of naval development to the already established Senior Officers' War Course and the Staff Course are topics to be considered. Moreover, if the Service was indeed consumed about learning the lessons of Jutland, then it is at the Tactical School where one would expect to see this focus of study played out given that the Navy's tactical performance was at the heart of the controversy surrounding that engagement.

Grenfell might also have noted that the Service allowed officers time to pursue languages at university or in a relevant country and wrote before the Meteorology course's founding. Compounding the issue even further was the very nature of war was changing. Flight had moved its consideration into a third dimension and a new military arm, the Royal Air Force, formed in 1918 from the merger of the Royal Naval Air Service and the Royal Flying Corps was now a factor in British defence planning. Moreover, issues of war and peace for Britain and her Empire were evolving because the Empire itself was evolving. This, too, had implications for the Navy. From these considerations and others flowed the decision to create an Imperial Defence College, a school for studying war in its broadest sense for Britain and the Empire.

Still, other changes experienced owed something to societal influences. Here, one may count the age at which an officer candidate should be enrolled for instruction. Thus, the focus and purpose of the initial training provided to cadets and midshipmen

---

53   Later Captain Russell Grenfell (1892-1954). War Staff Course, 1921-22; Assistant War Staff Officer, First Battle Squadron, 1922; War Staff duties in HMS *Julius*, 1922-23; Tactical Course, 1932; Royal Naval Staff College, 1932-34; Captain and retired, 1937. Recalled for war service.

54   Russell Grenfell, 'Training in Tactics', *The Naval Review*, Vol. XI, No. 4, p. 686.

were issues very much at the centre in the reforms pursued by Admiral Fisher. A matter still unresolved and generating much controversy following the Great War when practical experience was available and continuing well into the next war.[55] Though many believed the Navy was correct in its approach in officer selection and development by accepting adolescents for training before other ideas had taken hold, others questioned such an approach for the twentieth century. It was not simply fashioning an officer for the Navy and war, but leaders had to be adept at dealing with statesmen, civil servants and their peers from the other armed forces of the Crown who had been schooled to different standards and in different methods. Thus, some began to question whether it was time to engage officers from the universities where skills honed in the Union might be of more relevance when dealing with the Treasury than those acquired in the gunroom.[56]

Economic and budgetary considerations, too, played a major factor in the scale and scope of training and education afforded to officers. The Service had to cut its coat to the cloth available and for much of the time between the years 1919-39 the *Navy Estimates* were severely constrained. Accordingly, the *Estimates* are reviewed with an eye to weighing the relative importance attached to the higher-level education of officers by the Board of Admiralty. In a time of strict financial stringency, the funding afforded in successive years to naval education will be analysed to measure where compromises in duration, numbers and types of instruction were implemented. Higher-level instruction desirable though it may have been considered in the abstract was conducted for a purpose and was not an end in and of itself. Billets, whether afloat or ashore, where the skills recently instilled at the Staff and War Courses had to exist. Little purpose was served if an officer so recently possessing a *psc* found himself redundant due to the effects of the 'Geddes Axe'—of which, more latter—or spent extended periods on half-pay because ships were laid up in reserve or now manned with reduced ships' companies.

Notwithstanding the ebb of flow of these many influences, what emerges in the story that follows is that the officer corps of the Royal Navy became a more professional body between the two world wars. More professional, in the sense that the education and training of its executive officers, as they advanced along their career paths, became more regularised and orderly, as the Admiralty directed not only what was to be taught, but also when it was to be taught. Perquisites in staff and war training

---

55 The contemporary literature on the issue is vast occupying many pages of *The Naval Review* and *The Times* newspaper both. Representative is Gerald C. Dickens, 'Naval Education—Admirals or Lieutenants?', *The Naval Review*, Vol. XIV, No. 1, February 1926, pp. 138-43 and Duke of Montrose *et al*, 'Naval Education', *The Naval Review*, Vol. XV, No. 1, February 1927, pp. 146-54..

56 J.M.G.G, 'The University in Education', *The Naval Review*, Vol. XXXII, No. 2, May 1944, pp. 149. The gunroom was the common room and social milieu of Sub Lieutenants and Midshipmen.

increasingly became the norm before one could aspire to assume command of a ship or a station or serve in a critical staff position.

In light of the Service's rapid and unparalleled expansion during the second conflict and the resulting dilution of officer training introduced to meet the needs of a second global conflict in less than 20 years, Admiral Richmond's argument that one could not defend a two-hemisphere empire with a one-hemisphere navy is telling.[57] This work will examine the scope of higher-level training afforded to officers of the Royal Naval Reserve—it remains that the Navy largely failed to train officers holding a commission in the Royal Naval Volunteer Reserve[58] at its Staff and War Colleges—and conclude whether this shortfall was an opportunity missed and failed to provide value for money to the Service.

In an influential monograph, nearly fifty years ago Donald Schurman examined the influence of six major personages on the Royal Navy and the debt owed to them in developing the strategic rationale of navies. Collectively, a by-product of their work was that naval history became more than just the recounting of epic battles. True, there had been writers on naval warfare in the past, but such exposition in the hands of Alfred Mahan, John Laughton, Julian Corbett, Herbert Richmond, and the brothers Colomb was aimed towards a higher purpose than mere storytelling and sought to codify the principles that underscored naval warfare. Whilst Schurman's treatise can be faulted for its brevity and omission of key contemporaries such as Major-General Sir George Aston, it is still a very useful piece of historical instruction for the survey that it provides of the crystallization of naval thinking that was happening during the period from the mid-Victorian age to the time at which this study begins. The intellectual foundations that the six writers laid were the bedrock upon which the higher education of naval officers in the Royal Navy occurred. It is worth recalling that Laughton was instrumental in the founding of the Navy Records Society, Richmond was a doyen of *The Naval Review*, and Mahan, the American officer who influenced naval thinking miles beyond his own shores, served Rear Admiral Stephen Luce admirably during the first days of the U.S. Naval War College with his public writings arising from the lectures provided to the officers undergoing instruction at Newport. If the writers surveyed in *The Education of a Navy* were as important, indeed, as Professor Schurman concludes one would expect to see their influences reflected in the curricula of the period and this facet too will be explored.

57   Brian Bond and Williamson Murray, 'The British Armed Forces, 1918-39', in Allan R. Millett and Williamson Murray, eds., *Military Effectiveness, Volume II, The Interwar Period* (London: Unwin Hyman, 1990), p. 108.

58   Royal Naval Volunteer Reserve. There was no Admiralty prohibition barring an officer of the RNVR from attending the Staff College Course. However, meeting the course perquisites and attending a course lasting ten months was not a practical proposition for members of the force. The single officer known to have attended the course was Lieutenant Charles B. Sanders, RNVR, who began the 1934 session but did not complete the course.

Chapter 1 surveys the Great War, the writing of the Official Histories, and the reformation of naval war and staff education. Chapter 2 reviews the broad theme of finance and naval policy from 1919 to the outbreak of the Second World War as it affected officer education. Chapters 3 and 4 review the role of both Greenwich and Portsmouth in the style of officer education that was fashioned between the wars examining the purpose and themes that the Staff Course, the Senior Officers' War Course, the Senior Officers' Technical Course, and the Tactical Course pursued. Chapters 5 and 6 survey the joint educational environment focusing particularly on the Staff Colleges of the other Services, the Army's higher educational courses frequented by naval and marine officers, and the apex of joint education for the period under review: the Imperial Defence College. Chapter 7 examines how the Navy employed its officers now highly trained and addresses the centrality of education in officer selection and advancement while a final chapter offers some concluding remarks including possible lessons for the present.

Those officers who gave birth to *The Naval Review* came to the fore in the interwar period. Though none achieved the highest gifts of naval prominence—Admiral of the Fleet, First Sea Lord and Chief of the Naval Staff, or even command of one of the major fleets—collectively they left a mark on the Service through their influence in staff training, the higher education of the Navy's officers, and in their many writings. Those who did achieve such honours might well envy the intellectual legacy they bestowed and what follows is the record of the strides that the Royal Navy made in better preparing its executive officers to fight a better war and remains their strongest legacy.

1

Regeneration: The Great War, the Official Histories it Begot and the Higher-Level Education and Training of Officers

It is by no means too early to think about things that should be done after the war. The first requirement is that we should have a full and accurate history of all the naval operations; a history revealing every step and showing the cause and effect of every failure and every success. On this we can base a system of training, and the construction of official manuals, which should ensure the lessons learnt at such heavy cost shall never be lost to us, and the errors made shall never be repeated.[1]

Commander the Hon. Reginald Plunkett-Ernle-Erle-Drax, 1915

Of naval staff work in the Great War what can one say?

Whatever it was, good, bad or indifferent, it was almost wholly improvised. There were no adequately trained staff officers at all.[2]

Captain Walter Koe, 1924

History is as essential to a War Staff as a ledger is to a business and in its ulterior aspect constitutes the crucible of doctrine.[3]

Captain Alfred Dewar, 1919

Though it is conveniently claimed that the 'War to End All Wars' ended on the eleventh hour, of the eleventh day, of the eleventh month, a temporary cessation of hostilities between the Allied and Associated Powers and a disintegrating German Empire and her partners is closer to the truth. The Armistice reached confirmed the military ascendancy of the allies, but it awaited the imprint of statesmen and diplomatists to

---

1   Commander the Hon. Reginald Plunkett-Ernle-Erle-Drax, 'With the Grand Fleet (4th October, 1915.),' *The Naval Review* (London: The Naval Society, 1915), p. 544.
2   Walter P. Koe, 'Some Observations Upon the Naval Staff System and Organisation', *The Naval Review*, Vol. XII, No. 4, November 1924, p. 667.
3   A. C. Dewar, 'Lecture on Historical Method,' p. 7, June 1919, Royal Naval Staff College, Dewar Papers, NMM/DEW/4.

put flesh to bones of the truce reached in a railway siding in Compiègne. Another seven months lapsed before Germany received the final peace terms confirming perhaps for Britain that making a peace acceptable to her allies was only slightly less daunting than waging war with her partners.

For Britain and the Royal Navy, the victory secured in time gave rise to questions whether the price paid had been too rich in treasure and body for the results obtained. To the historian, an acceptable answer is almost impossible without soon falling into the trap of the counter-factual. Yet, for the Navy distilling the lessons of the conflict were important if only to understand the effectiveness of its operations, to master shortfalls in technique, means, and doctrine, and to avoid a repetition in any perceived mistakes made and to fight a better war if called upon to do so again. Though naval and marine officers were not short of opinions in these matters as evidenced in the *Journal of the Royal United Services Institution* or in *The Naval Review*, institutional analysis awaited formulation. To the Historical Section of the Committee of Imperial Defence resided the tasks of preparing the Official Histories of the war receiving in the process the financial assistance of the Admiralty.

The decision to write an officially sanctioned series covering land and maritime operations in their fullest meaning was reached at an early date though whether they would be limited to internal Government use and not made available to the public was still an open question.[4] In 1916, Herbert Asquith, the prime minister, announced to parliament that the Official Histories would be released to the public though as discussed below this was to happen in at least one case only following a period of 24 years.[5] In time, the series expanded to cover the operations of the recently created Royal Air Force, other new agencies of government such as the Ministry of Munitions, and to recount the contribution of railways. The series covering the Ministry of Munitions reached twelve volumes, of which the final three were classified editions, and, thus, not suitable for public release. These, then, were meant for internal use only and they were made available to the Service Ministries and Staff Colleges as ready references for considering similar problems at a future date.[6]

Doubtlessly, those who decided to compile a history of the war were not to know the cost and the scale of the effort as the war was still ongoing. Time would demonstrate that cost was ever a concern, would influence the appointment of those selected to write the volumes, and would increasingly receive the attention of the Treasury. By 1933, a sum of £200,000 was thought the expense to date in writing the histories and many years remained before Brigadier Edmonds, the officer overseeing these efforts, could call it a day.[7]

4   *The Hansard*, House of Commons Debate, 27 April 1917, v. 71, c. 560.
5   *Ibid.*, House of Commons Debate, 28 June 1916, v. 83, cc. 838-9.
6   CAB 24/158, C.P. 66(23) minute of 1 February 1923.
7   'With the Colours', *Naval and Military Record*, 3 August 1932, Brigadier Frederick James Moberly Papers, National Army Museum, London, 2000-06-139.

Writing a sanctioned history was not new for Britain, as previous efforts covering the Second Afghan War, the South African War of 1899-1902, and the Russo-Japanese War of 1904-05 demonstrate. Thus, the decision to prepare studies covering Britain's greatest military effort was in keeping with earlier practices to record significant military endeavours and would have been the more remarkable if not attempted at all. These endeavours had not always been successful or timely. G.F.R. Henderson's volume on South Africa was reduced to pulp upon completion,[8] the Second Afghan War history was suppressed due to criticisms of certain officers,[9] and the second and third volumes of the Russo-Japanese War history, though completed in 1914, were denied immediate publication, as the resources of the Ordnance Office and the Stationery Office were required elsewhere.[10] Time would again show that writing such histories for the First World War was easier to contemplate than actually accomplish. Partially, this was because those commissioning such works failed to define adequately their purpose and audience, but mainly it was because Official History in the early twentieth century was never independent history. Whether the histories contemplated were to be unvarnished expositions of the British role in the recent war remained unsaid and just as significantly the ability to write a worthwhile operational treatise without considering the broader strategic and political dimensions they were meant to support remained problematic. One putative author of the Gallipoli campaign, Lieutenant General Sir Gerald Ellison, fought unsuccessfully against this restriction leading him to warn of the risks of repeating the very same mistakes present in the series recounting the South African War.[11]

Ellison's warning was not wrong if true, but was it, indeed, correct? From the start, the Official History series contemplated addressing questions of policy and strategy though these areas were largely reserved for Julian Corbett and the naval editions. Further, the Foreign Office was preparing its own study of the war, particularly of the blockade and the many questions of policy surrounding it, under Carless Davis of Balliol, Oxford and more recently of the War Trade Intelligence Department, an agency originally created by Admiral Hall to quantify enemy commercial activities in light of allied contraband control.[12] Of the Official Histories, the military series

8   'Fourth Course. The Work at the R.A.F. Staff College,' address by Air Vice Marshal H. R. M. Brooke-Popham, 5 May 1925, Air Chief Marshal Sir Henry Robert Moore Brooke-Popham Papers, Liddell Hart Centre for Military Archives, King's College London, United Kingdom, 1/5/10.
9   Field Marshal Viscount Alanbrooke Papers, Liddell Hart Centre for Military Archives, King's College, London, United Kingdom, 3/4.
10   CAB 24/4, 'Historical Section of the Committee of Imperial Defence, note by Major E.Y. Daniel, 8 June 1917.
11   Gerald Ellison, *The Perils of Amateur Strategy: As Exemplified by the Attack on the Dardanelles Fortress in 1915* (London: Longmans, Green and Co. Ltd., 1926), p. 137.
12   CAB 24/42, Official History of the War Sub-Committee, Committee of Imperial Defence minutes of 7 February 1918 and Erik Goldstein, *Winning the Peace: British Diplomatic Strategy, Peace Planning, and the Paris Peace Conference 1916-1920* (Oxford: Clarendon Press, 1991), p. 47.

was fundamentally to be an operational assessment. This decision made sense in early 1916 when it was taken, as the major questions of strategy to that date had primarily focused on where to send an expeditionary army in the first instance and the ongoing Dardanelles campaign. This was also before the events of 1 July 1916, Third Ypres, and even the fall of Kut in Mesopotamia.[13] The wisdom of this view became less sustainable as the British Expeditionary Force assumed growing proportions. Questions governing the allocation of resources and the weight of effort were no longer simply debates over the priorities of the Navy and the Army concerning as they did issues of industrial mobilization, manpower, and relations between allies and associates. The wiser course would have been to modify the terms of reference governing the respective series to account for the issues of strategy and policy germane to their areas. This evidently was never done and as the original authors writing the series were replaced by newer blood, the relationship of the constituent parts of the Official History series and the terms of reference governing the overall effort now receded from view. This was for the CID to resolve, but its make-up and priorities were ever changing. Hankey, as its Secretary and the power behind the throne, was in the best position to correct matters but this he failed to do.

One reason may have been that while the war was ongoing, the problem was not so apparent. After all, the Army was desirous to have the histories available as soon as possible—indeed, to have some editions issued even as the war was still waging to inform the instruction provided to its units in France. This, the Navy would not concede and here the degree to which intelligence, foremost, special intelligence governed its operations was the primary reason for the divergence of views. The Admiralty could not afford to compromise its sources of intelligence whilst they remained effective. Accordingly, it was decided that the histories would be prepared and be released as soon as practical once hostilities ceased.[14] This explains why the first volumes covering the opening phases of the war appeared so quickly following the conflict for they had been written previously and only awaited final governmental clearance for publication.[15] Securing this posed its own set of difficulties as the events surrounding the first volume of the naval history attest. This was largely ready for release by March 1919 when copies were provided for review and comment. Amongst those reviewing was Churchill and he registered objections to that portion of the work covering the Coronel action of 1914. As First Lord at the time of the action, Churchill was a central figure to these events. Now he was Secretary of State for War and Air and his reservations were sufficient to delay publication until his complaints were addressed. This Corbett was able to meet by adding an appendix, but it does highlight that the

---

13   CAB 24/4, 'Historical Section of the Committee of Imperial Defence' note by Major E.Y. Daniel, 8 June 1917.
14   CAB 24/60, Official History of the War Sub-Committee, Committee of Imperial Defence minutes of 24 July 1918.
15   *Ibid.*

Official History series operated under constraints (and also advantages) not facing any independent survey written.[16]

Once the war was over it was natural that its authors would seek to understand the policies and strategies controlling the various campaigns, particularly the more controversial and less successful ones. Simply telling the operational side of the story, if appropriate for the Staff Colleges who could be relied upon to provide any required context, was insufficient for the larger public. They were bound to read these volumes and ask, How? Why? Ellison fell afoul of these restrictions when writing the military side of the Gallipoli campaign eventually being replaced by Cecil Aspinall-Oglander. Corbett, in preparing Volume II of the naval series, which recounted the Dardanelles campaign, addressed the strategic context of those events basing his conclusions, in part, on the minutes of the War Council. Though not at liberty to cite the nexus of the decisions reached, nor the individual views of the principals concerned—restricting himself to generic formulations—Corbett was able to place the operations in their proper context.[17]

Hankey was the official responsible for ensuring that attribution of decisions could not be tied to a specific individual or to a particular body from Cabinet records though Corbett was free to refer to such if a conclusion could be traced to a public record. Some might view the restriction as an attempt to protect Ministers and to shy away from controversy, but the rule was in keeping with Cabinet government where decisions were taken collectively and responsibility resided with the whole and not with the individual. Ministers needed to be free to vent their views in Cabinet and specific attribution might circumscribe future deliberations if an official felt that criticism might rebound at a later date by those not appreciating the true context in which decisions were reached. This issue rose anew from another angle in 1927 when the Australian Government sought verbatim copies of evidence given to the Dardanelles Commission, in part, to revise their own Official History covering the campaign which had been prepared without it. As Hankey's evidence had been prepared based on his access to the Cabinet records, release of this evidence had the potential of unmasking Cabinet decision making. Unlike, Corbett, though, who had actually seen Cabinet minutes in preparing his histories, release was not made to Australia in this instance.[18]

Yet, the Cabinet was not afraid to face controversy—witness its willingness to face Labour's ire when portions of the history recounting the Ministry of Munitions were published. In this instance, the measures taken to dilute skilled labour during

16   CAB 24/87, Secretary to the War Cabinet note of 19 August 1919 and CAB 24/94, Secretary of the Admiralty to Secretary to the Cabinet letter M.04380 of 25 November 1919.
17   CAB 24/121, 'The Use of Proceedings of the War Council, War Committee and War Cabinet', note by the Secretary, 1 April 1921.
18   CAB 24/185, Chairman, Control of Official Histories Sub-Committee note C.P. 51 (27) of 15 February 1927 and CAB 23/54, Cabinet minutes 13 (27) of 23 February 1927.

the war were related, but it was desirous of not needlessly offending prominent officials.[19] Moreover, whilst the cost of the Official Histories was ever a concern of the Treasury, their reception by the public was always the more salient matter for the greater Government. Thus, Cabinet sanction was initially required before a volume appeared and publication of one edition did not guarantee approval of subsequent ones in a series, though in the case of the naval history and following the very favourable reception of the first volume, a blanket approval was agreed.[20]

A fair conclusion is that a certain lack of overall grip governed the preparation of the Official Histories. The sacking of John Fortescue, of which more anon, the death of Corbett, and the demise of the Coalition government in the autumn of 1922 may have been contributing factors, but effective direction from the centre should have ameliorated these effects. By late 1922, a decision was required on publishing the history of the Ministry of Munitions. A new body was established under Edward Wood, the President of the Board of Education, to weigh the matter but it did not report on the question until the summer of 1923. The delay arose as Wood broadened his deliberations to consider the value of the CID's Historical Section itself and whether the Official Histories should continue as planned. Wood had to report that the original body formed to direct the Historical Section no longer performed its function and that it had simply lapsed is a difficult conclusion not to reach. The creation of a new Sub-Committee overseeing the section and the Official Histories featured as the primary recommendations from Wood's review along with retaining additional historians to tackle the daunting tasks that remained.[21] Culling the military histories back from private publishers was a further recommendation; a consideration that need not apply for the outstanding naval, maritime, and air volumes as they were not being written with the primary purpose of supporting officer education.[22]

Thus, the Official Histories were now of two-types: a popular and literate recounting of the war for the naval, maritime, and air volumes, and, secondly, a set of basic textbooks to support officer education for the military series. The selection of retired officers to compose the military series in the wake of Fortescue's dismissal accentuated this separation, but Corbett's death probably added to the equation as considerations of strategy and policy, including foreign policy, waned in the naval series. This was unfortunate, but not decisively so. The histories were of immense value in distilling the threads of complex operations and based, as they were, on a wealth of primary source material available, employing historians and officers of repute, and running to over fifty volumes, they cannot be considered mere popular histories. Nor is it fair to castigate them as only 'trade union history' as the critic and thinker Basil

19   CAB 23/37, Cabinet minute of 31 March 1920.
20   CAB 23/21, Cabinet minutes 20 (20) of 15 April 1920, CAB 23/24, Cabinet minute of March 1921 and CAB 23/25, Cabinet minute of 5 May 1921.
21   CAB 24/161, Committee on the Historical Section of the Committee of Imperial Defence, *Report*, 9 August 1923.
22   *Ibid.*

Liddell Hart was wont to do.[23] The French Navy, for one, marveled at the efforts being pursued and sought British assistance in crafting their own account.[24] Additionally, though compiled by government commission the ones deemed of widest interest were published by private houses—they were, indeed, an early example of a public-private finance initiative. Thus, at an early date, a broad audience for the works was a consideration else recourse to employing the traditional arm of Crown publishing, His Majesty's Stationery Office, would have been sufficient. Finally, the publishers desire to realize the greatest sales possible by ensuring that their product brought new perspectives to the war came to conflict with the Admiralty's desire to relay their version of events in one very important instance.

To Julian Corbett, a member of the bar by training but more recently a writer of naval history of the first rank, the task promised to be the crowning achievement to a life dedicated to understanding maritime warfare. He was an acknowledged expert on naval tactics in the age of sail, had lectured frequently on the Battle of Trafalgar, the Russo-Japanese War, and the First Anglo-Dutch War at the War College prior to 1914 and had several volumes in print covering some of the most important periods in British naval history.[25] His experience in writing a contemporary, operational naval history most recently demonstrated in the classified record compiled of the Russo-Japanese War in cooperation with Rear Admiral Edmund Slade[26] stood Corbett in good stead. Yet, for that, Corbett was not neutral nor could any history by him be neutral as he, too, had an interest in how the maritime war was related. Lord Sydenham, for one, held Corbett partly responsible for the Navy's lack of success during the war mainly because of the influence of his teachings and said as much in 1916.[27]

Entrusted to C. Ernest Fayle, of who much more is to be heard, was the story of the commerce war, whilst Archibald Hurd, a popular writer on naval affairs and correspondent for *The Daily Telegraph* newspaper, was charged with recounting the efforts of the Merchant Navy during the World War. Finally, to Lieutenant Commander Archibald Bell, previously of the Historical Section of the Training and Staff Duties Division, resided the task of preparing a new monograph analyzing the offensive side of the British commerce war as accomplished through contraband control against the

23   Alex Danchev, *Alchemist of War: The Life of Basil Liddell Hart* (London: Weidenfeld & Nicolson, 1998), p. 59.

24   CAB 24/60, Historical Section, Committee of Imperial Defence, memorandum of 22 July 1918.

25   See Eric Grove, ed., 'Introduction,' to Corbett's *Some Principles of Maritime History* (Annapolis: Naval Institute Press, 1988) for an excellent overview of his career. Copies of Corbett's lectures are held at the Liddell Hart Centre for Military Archives, King's College, London, United Kingdom.

26   Later Vice Admiral Sir Edmund John Warre Slade (1859-1928). Captain, 1899; Commander, Royal Naval War College, 1904; Rear Admiral, 1908 and Vice Admiral, 1914.

27   Schurman, Donald, *The Education of a Navy: The Development of British Naval Strategic Thought, 1867-1914* (London: Cassell, 1965), p. 150.

Central Powers. This was a follow-on assessment from the work begun by Carless Davis whose own efforts were never publicly promulgated once completed though they were lodged with the British Museum's Copyright Office.[28] Bell's efforts dated from 1931 when the Foreign Office looked to complete its survey of the blockade story turning to an officer whose previous experience had been working as an assistant to Corbett during his period of writing the naval series.[29]

Though the full gamut of Bell's assistance to Corbett is unknown, it is significant that he was a qualified Hydrographer indicating that matters of navigation and chart interpretation were questions needing resolution before any history could be prepared. Still, one area where Bell did provide unique service was in examining the records of Room 40 that were used, but not acknowledged, in the compilation of the Official History. In this, he was strongly suspected of writing an article appearing in the *Journal of the Royal United Services Institution* in February 1924, under the pseudonym of John Leighton. This essay provided dangerous hints of the role of special intelligence during the Battle of Jutland; a matter still shrouded in secrecy.[30] If there was some doubt concerning Bell's actual authorship of the offending article, then there remained no question that he had accepted editorship of the *Journal* without seeking Admiralty approval. This the Admiralty would not countenance whilst still employed on the active list forcing Bell to sever his ties. This hindrance would not last much longer, as Bell received Their Lordships severe displeasure for approaching a member of parliament in 1925 with a view to having an issue raised in the House. On surer ground, the Admiralty informed Bell that upon conclusion of his present appointment no further naval employment would follow in the Service.[31]

Fayle's efforts culminated in three volumes (with the first volume having a separate appendix for maps) published between 1920-24 as the *Seaborne Trade*; Hurd's series titled *The Merchant Navy* appeared between 1921-29. More interestingly, Bell's *A History of The Blockade of Germany and of the Countries Associated with Her in the Great War, Austria-Hungary, Bulgaria and Turkey, 1914-1918* though completed in 1937 was only released to the general public in 1961. This though the Germans had secured a copy during the Second World War and at least one postwar researcher had reviewed it before its broader publication.[32] Meanwhile, Corbett issued the first volume of his naval treatise in 1920 with a second monograph quickly following in 1921. The third volume of the naval history, though largely completed by Corbett, appeared only in 1923, the year following his death. Henry Newbolt, a writer more widely known

28   Marion C. Siney, 'British Official Histories of the Blockade of the Central Powers during the First World War', *The American Historical Review*, Vol. 68, No. 2, January 1963, p. 392.
29   'Lieut-Com. Archibald Colquhoun Bell', *Who's Who 1945* (London: Adam and Charles Black, 1945), p. 195.
30   Director of Naval Intelligence minute of 22 June 1926, Dewar Papers, NMM/DEW/4.
31   ADM 196/143 (Bell).
32   Siney, 'Blockade of the Central Powers', p. 393.

for his poetical works, continued the naval series through to its conclusion issuing a fourth volume in 1928. A final installment released in 1931 completed the series originally entrusted to Corbett of relating the official story of the Royal Navy in the World War. Thus, of the primary authors engaged in writing the Official History of the naval and maritime war, only Bell was a serving or retired officer which may be contrasted in manner to how the military histories of the war were written.[33] This difference in authorship is key to understanding how the Services viewed and intended to use the Official History series.

Corbett's commission to write the series' naval operational history coming as it did prior to the Battle of Jutland and the crisis of the German submarine campaign of 1917 even at that early date still offered reason for caution. The Dardanelles campaign, the Coronel engagement, the sinking of the His Majesty's Ships *Cressy*, *Hogue* and *Aboukir*, and the defence of Antwerp were white-hot issues in their own right. More problematic for any historian was providing commentary on the loss of HMS *Audacious*. She was a casualty to a naval mine sown near the Irish coast in 1914, had yet to be confirmed by the Admiralty as sunk, and was carried still in *The Navy List* as an active unit of the fleet.[34]

There remains a curious irony in the historiography of the First World War. In its day, the naval story was always the more controversial treatment rather than the military series. Later generations schooled in the war's legacy by classics of literature such as *Good-bye to All That, Testament of Youth, All Quiet on the Western Front, Journey's End*, and, of course, a whole genre of war poetry composed by very learned and *temporary* British Army officers whose facileness of pen owed more to *Melpomene* than *Clio* explain much the reason why this is true as a cultural response.[35] That it is also true for many of the serious discourses following may only indicate that historians write from a context and readers exist not in a void. In 1914, the Royal Navy was viewed as England's and the Empire's sure shield. Respect existed for the British Army and it was even lauded in popular works by writers such as Kipling and Henty. Yet, the South African and Crimean Wars had shown it to be an all too fallible instrument. That the victory secured in 1918 only came after the actions fought on the Somme, at Third Ypres, or Kut-al-Amara was not without precedent. Success in South Africa only followed the earlier setbacks at Spion Kop, Colenso, and the 'Black Week' at the hands of a lesser foe. Memory of the Royal Navy failing in battle—at least living

33   Bell's appointment to the Historical Section of the Training and Staff Duties Division dated from 10 February 1919. He retired from naval service in his rank in 1927 and through the intervention of Newbolt was retained by the Historical Section of the Committee of Imperial Defence in a civilian capacity. He joined the Air Ministry in 1939 becoming Staff Officer Intelligence at RAF Mount Batten.

34   Hugh Cleland Hoy, *40 O. B.: How the War Was Won* (London: Hutchinson & Co., 1932), pp. 72-73.

35   For the British legacy of the World War see Brian Bond, *The Unquiet Western Front: Britain's Role in Literature and History* (Cambridge: Cambridge University Press, 2002).

memory—did not exist and goes far to explain why the naval history being drafted was to generate the controversy that it did and why its publication was so eagerly awaited.

Meanwhile, the Admiralty decided to prepare a series of internal histories covering specific technical developments during the war. This decision was both sensible and practical and the CID recognized that the Services would still likely need to prepare their own staff histories of the war notwithstanding the place of the Official History efforts.[36] The Official Histories covering naval operations would not touch upon all of these subjects, and the Navy, in any case, could little afford to wait even if they did. Covering as they did many of the material improvements fielded during the war, the preparation of the Admiralty Technical History series was viewed as key to the Service's operational improvement. These improvements gave the postwar Service its tactical edge, and, as such, the series indicates what topics were deemed important to the Royal Navy and worth attempting to codify for future lessons or application. At least 51 separate Technical Histories were prepared during 1919-21 under the guidance of a civilian engineer covering the administrative, logistical, technical and tactical issues arising during the war.

Convoy and mercantile protection featured most prominently appearing in eleven of these histories and the rise and development of the Admiralty Naval Staff covered a further three separate papers. These addressed the creation of the Anti-Submarine Division, the role of the Hydrographic Department, and the control and administration of Royal Navy submarine forces. Other topics featured in the series covered chemical warfare and its means of defence, oil fuel to the fleet, signaling and communications, and the alteration of merchant ships and warships to adapt the newer implements of war.

Yet, beyond these Technical Histories the Admiralty sought to provide a record of the Battle of Jutland for the public at large. Given the intended audience of this work and the fact that Corbett was handling this matter already under the auspices of the CID, little good could come from this effort. It was bound to upset the CID and most certainly would upset the intended publisher of the Official History, Longmans, engaged to publish the naval series. While it promised to be a quicker read than Corbett's work, and, so, likely to reach a broader audience, any success it enjoyed would necessarily operate at the expense of Corbett and Longmans. This record, foregoing commentary and criticism, was to provide a bare summary of the battle along with diagrams depicting the movements of both fleets at key moments. Captain John Harper,[37] a Navigating Officer, at the request Admiral Sir Rosslyn Wemyss,[38] the First

---

36   CAB 24/4, 'Historical Section of the Committee of Imperial Defence, note by Major E.Y. Daniel, 8 June 1917.

37   Later Vice Admiral John Ernest Troyte Harper (1874-1949). Captain, 1913; Director of Navigation, 1919-21; Senior Officers' Technical Course, 1922 and 1924; Senior Officers' War Course, 1922; Commanding Officer, *Resolution*, 1922-24; Rear Admiral, 1924; Senior Officers' War Course, 1925; Tactical Course, 1926; retired, 1927 and Vice Admiral, 1929.

38   Vice Admiral Sir Rosslyn Erskine Wemyss, later Baron Wester Wemyss (1864-1933).

Sea Lord, attempted to square the circle of making sense from the often conflicting evidence available. His charter, alas, proved a thankless task. Assembling his team in February 1919 and completing a draft in October of the same year, Harper was not able to secure approval of his draft before Wemyss' departure from the Admiralty.[39] This was to prove critical, for the work initiated under Wemyss, perforce had to satisfy Admiral of the Fleet Earl Beatty, the incoming First Sea Lord and hardly a disinterested party to any treatment of the battle.[40]

If Wemyss could not predict the actual path of the controversy, he must have retained an inkling of what was apt to follow given his own experiences of dealing with Beatty in the time leading to his own departure from office. Matters were not helped by the anomaly of rank. For though Wemyss was First Sea Lord and the professional head of the Service, he only held the rank of Admiral whilst Beatty was an Admiral of the Fleet having been promoted in recognition of his war service.[41] There is an unwholesome aspect about the exodus of Wemyss and the accession of Beatty. Coinciding with the period of the Harper's efforts, the vanity of two senior officers had repercussions for others perhaps not recognised by the principals. Leaving the Admiralty, Wemyss confided to the Walter Long, the First Lord, that the atmosphere between Beatty and himself had not always been healthy, wondered how much of the turmoil was due to his failings (none), and thought Beatty over-wrought. Still, Wemyss believed the affairs of the Service were largely in order as there were only two issues of pressing importance: the provision of personnel to support the air requirements of the Navy and Jellicoe's report on the naval requirements of the Empire.[42]

Wemyss' selection of Harper, an officer serving at the Admiralty in its Navigation Branch, but residing outside of the Naval Staff proper might be thought unusual. After all, both the Operations and Training and Staff Duties Divisions possessed officers who in the normal course of events might have been tasked to draft such a record. Yet, the choice was perfectly rational for two reasons. First, and foremost, the Admiralty needed to present the operational details surrounding the battle to the Staff College so that it could be covered intelligently in its curriculum without waiting for the fuller, critical treatment that the Staff Appreciation and subsequent Staff Histories promised to provide. Until this was accomplished, the Staff Course, the War Course, and even the internal Staff Appreciation to be written would have

Captain, 1901; War Course, 1908-09 and 1911; Rear Admiral, 1911; Vice Admiral, 1916; First Sea Lord, 1917-19; Admiral, 1919 and Admiral of the Fleet, 1919.

39   Lady Wester Wemyss, *The Life and Letters of Lord Wester Wemyss, G.C.B, C.M.G., M.V.O. Admiral of the Fleet* (London: Eyre and Spottiswoode, 1935), p. 447.

40   J. E. T. Harper, *The Truth About Jutland* (London: John Murray, 1927), pp. 5-15.

41   Though Wemyss entered the Navy in 1876 and Beatty only in 1884, the latter's star had been ascendant since his promotion to Commander in 1898.

42   Wemyss to Long letter of 31 October 1919 cited in Bryan Ranft, ed., *The Beatty Papers: Selections from the Private and Official Correspondence and Papers of Admiral of the Fleet Earl Beatty, Volume II, 1916-1927* (Aldershot: Scolar Press, 1993), p. 65.

serious gaps in their understanding of the war's most important naval engagement and contain possible conflicts in even the basic facts surrounding the battle. Indirect confirmation of this deficiency exists based on the actual topics covered at Greenwich during the initial postwar period and discussed more fully below. Whilst much source data existed for any appreciation or history, resolving the multitude of discrepancies in time and place found in signals, charts, and plots was essential. Thus, the care demonstrated by the Admiralty in finding the wreckage of the *Invincible* in 1919 as its true location formed a constant to compare against the frequently conflicting contemporary records. Secondly, the Admiralty was desirous of correcting the most egregious deficiencies of the public's understanding of the battle and did not wish to wait the years that Corbett's efforts promised.

Interest outside the Service on the Battle of Jutland was ever present and early questions raised in parliament centred on when the first volume in the Official History would be forthcoming and whether it would cover the battle. Thomas Macnamara, the Parliamentary and Financial Secretary of the Board of Admiralty, dealt with both points advising members that a first volume would be available in late 1919 but, alas, would not address the Jutland encounter.[43] News that the Royal Navy was preparing a separate treatment of the battle to that which Corbett was endeavouring to complete under the auspices of the CID soon became known in parliament with questions arising on that score. The history of the Battle of Jutland's history is of relevance to this study only insofar as it provides the context to the pitfalls facing those attempting to discern its lessons and the reader desiring to understand the complete story could do little better than reading Andrew Gordon's *The Rules of the Game*.[44] A greater tale of woe, pathos, and vanity in naval historiography is harder to imagine ending as it did the lives of two men prematurely, ruining the career of a third, and sullying the reputations of several.

Harper's task of working against the Charybdis of Beatty and the Scylla of Jellicoe was not assisted by parliamentarians who questioned the purpose of the record he was preparing. Commander Carlyon Bellairs, a former lecturer at the Naval War College and a long-time *bête noir* of Admiral Sir John Fisher in his day, questioned the value of preparing such a record compared to writing a proper staff appreciation.[45] Notwithstanding Harper's efforts had yet to see the light of day, the Service took measures to prepare such an appreciation. This decision was entirely reasonable for the reasons previously noted. Moreover, with requests for such a study arising from the fleet and the Staff College, such a study appeared pressing. It would allow operational matters to be discussed in a confidential manner with appropriate lessons to be derived and could avail itself of the growing body of German literature now

43   *Hansard*, House of Commons Debate, 14 March 1919, v. 113, c. 1616W.
44   Andrew Gordon, *The Rule's of the Game: Jutland and British Naval Command* (London: John Murray, 1996).
45   *Hansard*, House of Commons Debate, 18 February 1920, v. 125, cc870-1.

appearing.[46] Yet, even here the Navy faced constraints by higher authority as the role that signals intelligence played in the battle remained a topic that the CID and other departments did not wish to expose even in an internal study with limited distribution. Lieutenant John Pollen serving in the Historical Section of the CID was specifically charged with reviewing the Staff Appreciation to ensure that any references to cryptology, Room 40, Special Intelligence reports or to Frank Birch and William Clarke were omitted.[47] This pair had catalogued the relevant source material pertaining to Room 40's role during the war with Clarke now working for the nascent Government Code and Cypher School—an organization the Government had no desire to publicize nor the capabilities that it represented.[48] Moreover, Pollen's task was made that much easier as the Historical Section of the Admiralty's Training and Staff Duties Division was physically located with the CID's Historical Section in Whitehall Gardens.[49]

Meanwhile, Harper's work remained and Beatty, calling attention to the tone of his subordinate's treatise sought Corbett's assistance in qualifying the many blemishes that that account promised to expose. Thus, the First Sea Lord told the historian:

> The Official Record of the Battle of Jutland, prepared by the Harper Committee, although accurate in its facts so far as they can be established, reads somewhat as a record of disasters to the British Fleet. Expressing only, as it does, the experiences of the British Fleet, this cannot be helped, but a forward is desirable in order to correct the erroneous impression which may thus be conveyed.
>
> From the large amount of evidence received from enemy sources since the conclusion of peace, it is clear that the gunnery of the British Fleet was of a very high standard. One German Officer in particular states that in the first phase of the action it was excellent, and marvels that the "Lutzow" did not sink ...
>
> I, therefore, ask you to be good enough to draft a suitable forward.[50]

That Corbett should receive such a request from a First Sea Lord was not without precedent, as Admiral Sir John Fisher had sung the historian's praises incessantly before the war, most particularly, when an article of his featured in the journal *Nineteenth*

<hr>

46   W. S. Chalmers, *The Life and Letters of David, Earl Beatty* (London: Hodder and Stoughton, 1951), pp. 357-59.

47   John Pollen to Kenneth Dewar letter of 30 November 1921, Dewar Papers, NMM/DEW/4. Room 40 was the Admiralty office where signals intelligence work was primarily conducted, special intelligence was the euphemism for information acquired through electronic interception and cryptanalysis whilst Lieutenant Commander Frank Birch, RNVR, and Paymaster Lieutenant Commander William Clarke, RNVR, were two of the principal officers employed in such work.

48   Patrick Beesly, *Room 40: British Naval Intelligence 1914-18* (New York: Harcourt Brace Janovich Publishers, 1982), p. 321.

49   '*Historical Section of the Naval Staff*', no date, NMM/DEW4.

50   Beatty letter to Corbett of 12 August 1920, Corbett Papers, Box 1, LHCMA.

Century endorsing Fisher's then ongoing and controversially received programme of reform being implemented within the Service.[51] Yet, the circumstances now were much different, as Corbett, writing the Official History under the care of the CID, was no longer an actor operating solely within naval circles and had a responsibility to protect the independence of the Official History. As the Navy funded a substantial portion of the Official History series through a grant-in-aid to the CID, Beatty may have believed that his calling on Corbett was only natural. Yet, Corbett was not an Admiralty employee and any association with the Harper survey threatened the independent status of the history.[52]

Here, the case of Corbett's military counterpart preparing a portion of the military Official History series is instructive. John Fortescue was then perhaps the preeminent scholar of the British Army and in stature was the military equivalent to Corbett. Engaged to write the military history of the war, Fortescue unwisely, as it transpired, wrote a critical article of Field Marshal Viscount French appearing in the October 1919 edition of *The Quarterly Review*. The article reviewed French's recent study of the war's early campaigns and concluded by claiming:

> Lord French is, it is true, still the recipient of honours and rewards; but no accumulation of titles, bâtons, grants, orders or decorations can ever fit him to stand in the company of such men as Ralph Abercromby, John Moore, Rowland Hill and Thomas Graham. Let these, and not Lord French, stand before the youth of Britain as the models upon which to train themselves to be officers and gentlemen.[53]

Regardless of the actual merits presented in his essay, Fortescue's personal attack on French—an officer very much still alive, a member of the Cabinet and presently Lord Lieutenant of Ireland—was not a line of argument the Government could sanction, as it could only lead to controversy and controversy was something the Official Histories eschewed. The response was immediate and severe. Fortescue's association with the CID and the Official History ended at once though his salary was honoured for the remainder of the year.[54]

How much of this was known to Corbett at the time is unknown. Hankey and the Cabinet handled the matter privately with most of the business occurring orally. Others, who presumably should have known, such as Major Edward Daniel[55] and

---

51   Arthur J. Marder, ed., *Fear God and Dread Nought: The Correspondence of Admiral of the Fleet Lord Fisher of Kilverstone, Volume II, Years of Power, 1904-1914* (London: Jonathan Cape, 1956), p. 113.

52   'Historical Section of the Naval Staff', undated minute but *c.* 1939-45, Dewar Papers, NMM/DEW/4.

53   The Honourable J. W. Fortescue, 'Field-Marshal French's "1914"', *The Quarterly Review*, No. 461, October 1919, p. 363.

54   CAB 24/92, Secretary of the Cabinet to the Hon. John Fortescue letter of 28 Oct 1919.

55   Later Lieutenant Colonel Edward Yorke Daniel (1865-1941). Naval Intelligence

Brigadier James Edmonds, were bypassed in the affair, as Edmonds recorded in his diary that Fortescue's dismissal owed everything to his lack of progress and his critical views of certain senior officers taken in his chapters of the Official History written to date.[56] If Edmonds and Daniel did not know of the actual reasons for Fortescue's removal, then can Corbett have known? Whilst the conclusion is speculative, this writer believes so. Hankey and Corbett had a relationship of long-standing and Hankey was not afraid to seek the latter's historical assistance when faced with a pressing problem. The Dardanelles offers one such example with Corbett providing Hankey in early February 1915 a summary of the British failure in 1807 to force the Straits and capture Constantinople. This, too, saw a supporting military force divorced from the assaulting fleet with naval operations alone proving insufficient to the task.[57] Likewise, Hankey shared privileged information with Corbett that he did not trust to others—witness the access to War Council minutes provided by Hankey whilst preparing his second volume of the naval series covering the events leading up to the Dardanelles campaign. Moreover, denying a friendly request from the First Sea Lord, in view of his close association with the Navy, was not a step to be taken without strong reasons.

Not able to secure Corbett's assistance, Long, the First Lord and the Minister ultimately responsible for the good of the Service, decided to forego publishing Harper's account and place his files at Corbett's disposal to be used as he saw fit in preparing his own work. A statement by Sir James Craig, the Parliamentary and Financial Secretary, to the House of Commons confirming the above was made on 27 October 1920.[58] Soon Joseph Kenworthy, somewhat of a thorn in the flesh to his former Service, sought clarification on whether Harper's efforts and other relevant war files were being shared with the Staff College as aids in their instruction. Craig, again speaking for the Government, while confirming that the directing staff had access to Admiralty information would not be drawn out on specifics.[59] To this question, Kenworthy continued to press the Government during the coming weeks and the degree to which officers at Greenwich prompted these interventions, while likely, remains speculative.

Certainly, some in the Press shared reservations that Harper's effort would not be published, and amongst these journals, the *Daily Mail* arguing for its release held that an understanding of the continued operational utility of the battlefleet was vital if

Department, 1901-05; Seconded to CID for writing naval history of the Russo-Japanese War, 1907-09; retired, 1909; Recalled for war service and Admiralty War Staff, 1914-19; Lieutenant Colonel and retired, 1919.

56   Andrew Green, *Writing the Great War: Sir James Edmonds and the Official Histories, 1915–1948* (London: Frank Cass, 2003), pp. 9-10.

57   CAB 24/1, Corbett to Hankey letter of 4 February 1915.

58   *Hansard*, House of Commons Debate 27 October 1920, v. 133, cc1716-17.

59   *Ibid*, House of Commons Debate of 1 December 1920, v. 135, cc1209-10.

British naval supremacy was to be assured.[60] Nor was it the case that only the Press believed publication of Harper's account was the best course of action. Churchill, the Secretary of State for War, argued within the Cabinet that having promised parliament that a report was forthcoming it was not possible to withhold it now or alter its form. Eventually, all would be known and it was in the interests of the Government to let those facts come out now.[61]

As for the Staff Appreciation, Captain Walter Ellerton,[62] the Director of Training and Staff Duties, looked to Alfred Dewar, a retired officer employed in his directorate's Historical Section, and his brother, Kenneth, to write it with assistance provided by Pollen of the CID. Both Dewars were fluid and original in their writing with recognition of their abilities having come from the Royal United Services Institution.[63] Each was a Prize Essay Gold Medallist with Alfred winning the honour in 1908 and Kenneth netting the award in 1912 for his survey of commerce in war and published in two parts in *The Naval Review*.[64] The appreciation when seen by Jellicoe was criticised vehemently for reasons discussed below and was never released by the Admiralty in its original form. Others including Rear Admiral Sir Ernle Chatfield, the Assistant Chief of the Naval Staff (ACNS), and Vice Admiral Sir Roger Keyes,[65] the Deputy Chief of the Naval Staff (DCNS), though agreeing with its overall conclusions, found fault in its tone and advised withholding it in even in redacted form as dissension amongst officers was apt to follow as matters now stood.[66] Admiral Madden, Beatty's successor as First Sea Lord, moved to have all copies in extant destroyed though presumably overlooking the editions retained by Beatty's personal secretary, Frank

60   *Beatty Papers, Vol. II,* pp. 452-53.
61   CAB 24/114, Secretary of State for War to Cabinet memorandum of 2 November 1920.
62   Later Admiral Walter Maurice Ellerton (1870–1948). Captain, 1910; War Course, 1911; Director, Training and Staff Duties Division, 1919-22; Rear Admiral, 1921; Admiral Superintendent Gibraltar, 1923-25; Commander-in-Chief, East Indies, 1925-27; Vice Admiral, 1926; retired, 1929 and Admiral, 1930.
63   *The Navy List,* January 1919, p. 2417.
64   Kenneth Dewar, 'The Influence of Commerce in War,' *The Naval Review, Volume II* (London: The Naval Society, 1914), pp. 245-53.
65   Later Admiral of the Fleet Sir Roger John Brownlow Keyes, Baron Keyes of Zeebrugge and Dover (1872–1945). Naval Intelligence Division, 1903-05; Captain, 1905; War Course, 1908; Commanding Officer, HMS *Centurion,* 1916-17; Director, Plans Division, 1917; Rear Admiral, 1917; Acting Vice Admiral and Commanding Dover Patrol, 1918-19; Commander, Battle Cruiser Squadron, 1919-21; Vice Admiral, 1921; DCNS, 1921-25; Commander-in-Chief, Mediterranean, 1925-28; Admiral, 1926; Commander-in-Chief, Portsmouth, 1929-31; Admiral of the Fleet, 1930 and retired, 1935.
66   DCNS and ACNS memorandum of 14 August 1922 cited in Paul G. Halpern, ed., *The Keyes Papers: Selections from the Private and Official Correspondence of Admiral of the Fleet Baron Keyes of Zeebrugge, Volume II, 1919–1938* (London: George Allen & Unwin, 1980), pp. 75-76.

Spickernell, and the copy possessed by his brother-in-law, Jellicoe, and now forming a part of his papers at the British Library.[67]

Unwilling to publish the Staff Appreciation in its current form, the Admiralty issued a revised work in 1924 titled the *Narrative of the Battle of Jutland*. This edition, drafted by Pollen while meeting many of the complaints that Jellicoe had registered against the original appreciation prepared by the Dewars still was unsatisfactory from the Admiral's viewpoint.[68] As a sop to Jellicoe, an appendix to the *Narrative* was included providing Jellicoe's reservations to the work, but not without the Admiralty providing its own rebuttal every bit as detailed.[69]

Corbett's history, too, was to prove problematic. Not able to prepare a treatment of the Battle of Jutland to its own satisfaction, the Admiralty gave to Corbett all the files that Harper and the Dewars had amassed in their efforts to write a record and an appreciation. Circumstances now intervened and Corbett succumbed before the fruits of his efforts were realized. Henry Newbolt now completed the manuscript and the balance of the naval series, but not before the publisher added a caveat adjoining the inside cover reading:

NOTE BY THE LORDS COMMISIONERS OF THE ADMIRALTY
The Lords Commissioners of the Admiralty have given the author access to the official documents in the preparation of this work, but they are in no way responsible for its production or for the accuracy of its statements.[70]

The question then arises whether the several attempts to fashion an Official History of the recent war and of its most famous battle were worth the efforts expended. The question is not measurable in terms of simple cost-benefit analysis. For against the angst felt by the principals and the fits and starts of the works attempted (with not all leading to publication) remains the vital argument that understanding the past is fundamental to the naval professional where direct experience of fleet action is rare and the price of failure is measured in greater terms than reputations lost. Bad history has it pitfalls and historical example is frequently misused, but the criticism would rightly be the greater if the CID and the Royal Navy had never made a start to understanding the most important of naval events in living memory. Where criticism can be leveled is that the Navy owed it to the country, to the CID, and to itself

---

67   Stephen Roskill, *Churchill and the Admirals* (New York: William Morrow and Company, 1978), p. 59-60. The draft appreciation, *C.B. 0938. Naval Staff Appreciation of Jutland*, with Jellicoe's comments to it can be found in the Jellicoe Papers, Additional MSS 49,042, British Library, London, United Kingdom.

68   'A Narrative of Service of Captain V. H. Haggard, R.N.' pp. 8-9, 85/21/2, Admiral Sir Vernon Haggard Papers, Imperial War Museum, London, United Kingdom.

69   *Narrative of the Battle of Jutland* (London: His Majesty's Stationery Officer, 1924), pp. 106-13.

70   Julian Corbett, *Naval Operations, Vol. III* (London: Longmans & Co., 1923).

that Corbett, Newbolt, and those others engaged in preparing the maritime series received their full support. This was not always forthcoming or came only grudgingly with the Permanent Secretary and his office being a particular offender.[71] This was important because though the Navy intended to produce its own operational histories where fuller and more technical questions were to be addressed, this did not happen largely as a result of the Jutland imbroglio and recourse was yet made in its schools to the Official History volumes written with a public and not a Service audience in mind.

A point often overlooked by those writing and reading the histories attempted is that the war operated at different levels and the lessons sought very much depended on whether one was addressing its strategic, operational, or tactical dimensions. A sense of this dichotomy is apparent in a 1923 letter Kenneth Dewar sent to the DCNS, Vice Admiral Sir Roger Keyes, following the publication of the Official History's *Volume III*. Criticizing the account, Dewar wrote:

> It always appeared to me that Corbett and his staff at Whitehall Gardens had a strong pro-Jellicoe and anti-Beatty bias. This was very noticeable during the time *Indomitable* and *Inflexible* joined them, although the question of whether the *Lion* turned 32 points or not seemed to me of no importance.
>
> Personally I attach no importance to Corbett's opinion on the battle, but unfortunately the public will. He did very valuable research work during his lifetime in a field which had been much neglected, but he was very ignorant of the technical aspects of naval warfare and perhaps for this reason he seemed incapable of drawing correct analogies between the past and present.[72]

Dewar's criticism was self-serving, to say the least, for though Corbett's name adorned the book's spine, he was assisted by serving officers in the preparation of the work.

Much of Jellicoe's perceived faults resided in the nature of the tactics adopted (i.e, turning away from the torpedo threat or deploying in a single column) and yet some of the very critics leveling these charges were of the school admonishing that battle was merely a means to an end of achieving command of the sea. Churchill, guilty of a hyperbole when cautioning that Jellicoe was the only man who could have lost the war in an afternoon, hit upon the central truth that for the Commander-in-Chief of the Grand Fleet strategy was more important than tactics.[73] Nelson, Britannia's God War, could fight as he did at Trafalgar because behind him stood another British fleet.[74] Standing behind Jellicoe sailed not another fleet but the prospects of ruin.

---

71 Stephen Roskill to Liddell Hart letter of 12 July 1957, Liddell Hart Papers, LH/610/2, LHCMA.
72 Kenneth Dewar to Keyes letter of 24 May 1923 cited in Halpern, ed., *Keyes Papers, Vol. II*, p. 88.
73 Winston S. Churchill, *The Great War, Volume II* (London: George Newnes, 1934), p. 837.
74 The appellation comes from Andrew Lambert's *Nelson: Britannia's God of War* (London: Faber and Faber, 2004). On the Trafalgar and Jutland analogy, see Commander Charles S.

And, yet, for Jellicoe and Beatty both it must be the case that a reason any criticism leveled stung as it did was the recognition that a portion of the blame was theirs and the crushing victory desired escaped their grasp. Fine fleet commanders and capable naval administrators they may have been, amongst the ranks of a Nelson they were decidedly not. Whilst they lived, a history sanctioned by the Admiralty that sought to be equally true and fair was a history not possible to write. Thus, given the multiple controversies generated by the entire episode, the Navy never completed its internal Staff History series covering Jutland and the greater war. Though volumes were prepared covering specialist operations, such as minesweeping, minelaying, and convoy and the war's opening phases through the conclusion of 1916, the effort terminated in 1927 and neglect greeted the latter phases of the war.[75] It is to Beatty's credit that he strove to provide a record of the Navy's place in the recent war by attempting to capture its many lessons. If only that had proved sufficient. Exactly why Madden cancelled the series with the story but half told remains conjecture. That it now promised to focus on the most critical phase of the naval war where even greater controversies loomed was probably one. In a later war, other disasters would befall the Royal Navy: the naval operations off Dakar in 1940, the loss of HMS *Hood* in 1941, and the sinking of HMS *Prince of Wales* and *Repulse* at the close of 1941. Their stories could be told in Staff Histories without the rancor of Jutland for the very reason that the greater context they portended was not so perilous for Britain as the loss of a fleet action in 1916 in the primary theatre of war.[76]

Still, for the Service writing the story of the naval war was but a part of the question. Other editions in the Official History series, such as Aspinall-Oglander's two-volume *Military Operations Gallipoli*, were of importance, too, as the type of operation it surveyed represented a major feature of the Staff College curriculum between the wars. In general, the reception the Official History series received was praiseworthy. Behind the scenes, though, there had been much gnashing of teeth as drafts were circulated amongst interested parties, who being major players in the events described, often fought to shape the story in a light more positive to their reputations. Moreover, publication of those volumes guaranteed to generate the most controversy typically came many years after the war's ending. The story of the first day of the Somme offensive appeared in 1932, the recounting of the March 1918 German attack came only in

Daniel lecture 'Jutland II,' Air Force Staff College, Andover, 1932, Admiral William George Tennant Papers, National Maritime Museum, Greenwich, NMM/TEN/41/5.

75   'Post War Naval Histories', no date, but *c.* 1939-45, Vice Admiral Kenneth Gilbert Balmain Dewar Papers, National Maritime Museum, Greenwich, DEW/4.

76   *B.R. 1736(1), Naval Operations Off Dakar, July–September, 1940*, Admiralty, Historical Section, 1959; *B.R. 1736(8), Loss of HM Ships Prince of Wales and Repulse, 10th December, 1941*, Admiralty, Historical Section, 1955, and *B.R. 1736(5), Chase and Sinking of German Battleship Bismarck, 23–27 May 1941*, Admiralty, Tactical and Staff Duties Division, Historical Section, 1950.

1935, whilst the telling of Third Ypres waited until 1948 before publication.[77] A more orderly and sedate process it would be harder to imagine. Thus, the white-hot glare awaiting Corbett was avoided by Edmonds and the CID for those volumes recounting the British Army's role during the war. The slow pace with which the volumes appeared was a by-product of the limited team available to write the series and the mountain of materials requiring cataloguing and assessment before writing could begin. Only the Navy's anti-submarine operations and contraband control duties approached in length, the scale of the land campaigns of the British Army. Even this does not do justice to the task facing Edmonds and his cohorts as the battles of the Army were of an intensity and orders of magnitude greater than that facing the Royal Navy.

The value of the Official History series as history remains a matter of controversy and lies outside the remit of this survey. Supporters and detractors have sufficient evidence to make their respective cases. This observer will only note that having instructors at the Staff Colleges who frequently participated in the operations, campaigns, and events covered by the series lessened the burden of the war's chroniclers from having to specify every lesson on offer. Internally, what was required was a factual telling of the events leaving it to the respective directing staffs to elucidate the lessons in the many schemes and exercises conducted. Here, the military volumes benefited from having former officers writing of the campaigns and may be contrasted to the civilian authors who largely penned the maritime volumes. The trade-off was that employing a known writer was likely to result in a highly readable account whilst former officers, perhaps not so gifted, might write in a more tendentious style. A number of false starts occurred necessitating the replacement of authors. Given the number of works commissioned and the increasing amount of time required to complete, this was perhaps inevitable. Whether a presumed author was fit to complete the task ahead was a question the CID never appears to have considered. Both Corbett, the primary naval historian, and Sir Walter Raleigh, the first air historian, succumbed before completing their works. If such events remain unpredictable, planning for such contingencies are not and a stronger hand at the centre could have lessened the disruptions.

Concurrent with the drafting of an Official History, returning the Service to a peacetime establishment became a pressing order of business, as Treasury sanction to spend without recourse to financial constraint was set to terminate on 31 March 1919 with the beginning of the new fiscal year.[78] Of course, this presented its own set of issues, as the Navy remained engaged in a series of lesser actions commonly known today as 'operations other than war', which could not easily be terminated or transferred to other parties. From the Baltic to the Black Seas, British naval forces found themselves soon in a war in all but name only with Soviet Russia whilst the remnants of the Ottoman Empire were proving equally troublesome in the Eastern Mediterranean environ. Friends to assist with policing the war's unsettled legacies

---

77   Green, *Writing the Great War*, pp. 61, 188, and 198.
78   ADM 1/8564/210, Operations Committee minutes of 20 November 1918.

were noted by their absence—Admiral Hugh Rodman's contribution to the strength of the British Grand Fleet did not last to 1919 and French and British interests in the Near East quickly parted.

Germane to this history is that the war was no less disruptive to the training and career progression of the Service's officers. The Royal Naval War College, Portsmouth, where selected officers of the rank of Commander and above on the Active List attended lectures in naval history, international law, and national economics, had closed for the duration of the First World War and the officers forming its directing staff reassigned to the Naval Staff or to the ships within the fleet. The Portsmouth curriculum lasting 15 weeks allowed for two sessions a year. Addressing the additional subjects of wireless telegraphy and coast defence, it tested fleet concepts through war games played at both the strategic and tactical levels and provided officers the opportunity for preparing assessments of contemporary naval issues. A quota of two Royal Marine officers was reserved for the senior course and Army officers attended on a reciprocal basis 'with the object of promoting the co-operation of the two Services.'[79]

Surviving documentary evidence whether a competitive examination to ascertain if the material covered had been mastered is unclear. Still, officers completing the War Course were assessed as either securing a first, second class or third class pass and with the curriculum covering raids, invasions, and wireless communications—issues at the forefront of prewar naval concerns—demonstrates that the college was founded to tackle issues of the first order.[80] To wit, the frequently acrimonious exchanges between the War Office and the Admiralty regarding the feasibility of a 'bolt from the blue' assault on Britain proper had featured in more than one review by the Committee of Imperial Defence. Benefiting the Army's case was the ability of the General Staff to provide Haldane, the Secretary State for War, with cogent arguments at a time when the Navy's view, even if largely correct, was the preserve of the principal naval member.[81] As for wireless telegraphy, its importance to naval warfare and those directing its use was confirmed by the establishment of a unified Signals Branch, merging both visual and wireless communications into a single discipline, in 1914.[82]

A more abbreviated offering open to junior naval and marine officers of the Active List as well as members of the Royal Naval Reserve and Royal Naval Volunteer Reserve and lasting but three weeks was conducted at Devonport touching upon the principles of strategy, tactics, *The Prize Manual*, aspects of merchant shipping, and international law. Completion of the Gunnery or Torpedo Short Course or another like course of instruction were perquisites for attending and Sub Lieutenants and Acting Sub Lieutenants attended by exception.[83] Finally, Chatham and Sheerness,

79   *Navy List*, January 1919, p. 2402.
80   ADM 196/44 (Hope) and 196/141 (Denison).
81   Richard Holmes, *The Little Field Marshal: A Life of Sir John French* (London: Weidenfeld & Nicolson, 2004), pp. 122-23.
82   Gerald B. Villiers, 'The New Scheme Lieutenant (S),' *Naval Review*, Vol. II, p. 357.
83   *Navy List*, January 1919, p. 2402.

making up the balance of the naval homeports, provided *ad hoc* lectures of a similar nature to its officers.

Many potential students who in the normal course of circumstances might have enrolled in one of the courses remained with the fleet for the war's duration. With operational effectiveness the watchword, an officer serving the entire conflict in a single warship was not without precedent. Commander Douglas Faviell[84] remains a case in point. Assuming command of HMS *Oak* on 12 November 1912, he served in his torpedo boat destroyer throughout the war and his application to attend the Intelligence Course, now sitting at Greenwich, only came to fruition in May 1920.[85] Likewise, Lieutenant Commander Francis Goolden,[86] serving in the battleship HMS *Iron Duke*, was an officer of similar circumstances. Appointed to his ship in March 1914, he served throughout the war in his ship until January 1919. He then entered the War Staff Course in September 1920 following promotion to Commander and time spent in the cruiser HMS *Cumberland*.[87] Nor was the practice unique to the Navy. Captain Charles Lucas,[88] a marine, served in the *Marlborough* from 1 August 1914 until 21 March 1919.[89]

Spending the duration of the war, though, in a single ship for officers was atypical. With the centre of gravity in naval operations occurring in Home Waters where the opportunities for short periods of leave remained possible, its occasion owed much to an officer's own wishes married with a judicious degree of man-management by the Second Sea Lord's office. When it is remembered that serving for an extended period in the same ship was frequently required for vessels commissioned for foreign service before 1914, its practice for a naval war focused nearer to home becomes more understandable.

84   Later Captain Douglas Faviell (1884-1947). Commanding Officer, *Oak*, 1912-19; Intelligence Course, 1920-21; Commanding Officer, HMS *Saumarez*, 1924-26 and HMS *Winchester*, 1927-29; Tactical Course, 1929; Captain, and retired, 1931.

85   ADM 196/126 (Faviell) and *The Navy List*, July 1920, p. 1865c. Though the *Navy List* specifies Faviell as attending the War Staff Course, he service record reflects his true appointment to the Intelligence Course.

86   Later Rear Admiral Francis Hugh Walter Goolden (1885-1950). War Staff Course, 1920-21; Second Sea Lord's Office, 1921-23; Tactical Course, 1925; Senior Officers' Technical Course, 1925-26 and 1937; Captain, 1926; to Royal Australian Navy, 1926-29; Royal Naval War College, 1929-31; Director, Operations Division, 1931-32; Tactical Course, 1934; Flag Captain in HMS *London* and Chief Staff Officer, First Cruiser Squadron, 1934-37; Rear Admiral and retired, 1937. Recalled for war service.

87   *Navy List*, January 1919, p. 821 and *The Navy List*, (London: His Majesty's Stationery Office, December 1920), p. 1865c.

88   Later Colonel Second Commandant Charles Anthony Cecil Lucas (1881-1962). Combined Operations Course, 1926; Attended Senior Officers' Course School of Land Artillery, 1929; Battery Commanders' Course, 1929; Lieutenant Colonel, 1935; Senior Officers' School Sheerness, 1935; Senior Officers' Course Netheravon, 1936; Senior Officers' War Course, 1938; Colonel Second Commandant, 1938 and retired, 1942. Recalled for war service.

89   ADM 196/64 (Lucas).

A by-product of officers spending such extensive time at sea and in the service of a single ship was when at last time permitted attending a course of advanced training they possessed a wealth of practical experience directly relevant to their studies. Even officers of comparatively junior rank exercised significant responsibility early in their careers. As the Navy commissioned many new vessels and small craft to enforce the maritime blockade of the Central Powers or screen the heavy ships of the Grand Fleet, officers in the rank of Lieutenant often found themselves commanding destroyers and other minor craft. With the need to shield allied shipping from the predations of German submarine attacks, execute her own submarine campaign, or meet the host of ancillary duties required of Britain's first global maritime war in a century, opportunities to exercise responsibility and judgment were often at hand. Lieutenants Loben Maund,[90] serving in HMS *Scorpion* from March 1918, and Ralph Kerr,[91] are just two of the many officers who faced the responsibilities of command at such an early stage in their careers. Nor was this exercise limited merely to officers holding the King's Commission. Judgment whether to allow a vessel to proceed or not featured in the duties of even midshipmen doing duty on the blockade line. Alexander Scrimgeour, a midshipman serving for a time in the Tenth Cruiser Squadron, the naval force responsible for investigating the merchant shipping passing through the northern waters of the United Kingdom, recorded how a fellow midshipman was tasked with leading the prize crew taking the SS *Overdia* to Kirkwall in the Orkneys for further investigation.[92]

90   Later Rear Admiral Loben Edward Harold Maund (1892–1957). Training and Staff Duties Division, 1923 and 1934; War Staff Officer in HMS *Southampton*, 1923-24; Staff Officer Operations, East Indies Fleet 1924-1925; Assistant Secretary, Committee of Imperial Defence, 1928-31; Tactical Course, 1931; Senior Officers' War Course, 1933-34 and 1938; Captain, 1934; Plans Division, 1934-35; Assistant Director, Plans Division, 1935-36; Commanding Officer, HMS *Danae*, 1936-37; Commandant, Inter-Service Training and Development Centre, 1938-39; Higher Commanders' Course School of Army Co-operation, 1939; Chief Staff Officer to Flag Officer, Narvik; 1940; Assistant Director, Combined Operations, 1940; Imperial Defence College, 1941; Deputy Director, Training and Staff Duties Division, 1941; Commanding Officer, HMS *Ark Royal*, 1941; Commanding Officer, Combined Operations Middle East, 1942-43; retired, 1943 and Rear Admiral, 1946.
91   Later Captain Ralph Kerr (1891–1941). Lieutenant, in Command HMS *Sable*, 1920-21; Commanding Officer, HMS *Vanity*, 1923-24, HMS *Witch*, 1925-26, HMS *Worcester*, 1926-27 *Seawolf*, 1927-28 and HMS *Windsor*, 1929; Tactical Course, 1929-30 and 1935; Commanding Officer, HMS *Thruster*, 1930-31; Anti-Submarine School in HMS *Osprey*, 1931-32; Senior Officers' Technical Course, 1935; Captain, 1935; Captain (D), Twenty-first Destroyer Flotilla, 1935-36; Senior Officers' War Course, 1937-38. Captain (D), Second Destroyer Flotilla, 1938-39; Flag Captain in *Hood* and lost with his ship on 24 May 1941 in action against the *Bismarck*.
92   Richard Hallam and Mark Beynon, eds., *Scrimgeour's Small Scribbling Diary, 1914-1916, The Truly Astonishing Wartime Diaries and Letters of an Edwardian Gentleman, Naval Officer, Boy and Son* (London: Conway, 2008), p. 152. Scrimgeour left blockading duties and transferred to the battle cruiser HMS *Invincible* where he came to grief when his ship was lost at the Battle of Jutland on 31 May 1916.

The dilution of experience necessary before assuming command of the many small craft used in the Great War is exemplified best, though, by the extensive use of those officers belonging to the Royal Naval Reserve and the Royal Naval Volunteer Reserve engaged for hostilities. Simply put, the prewar Navy was not manned to meet the demands of the war that arose in 1914 and the expansion required was to be mirrored to an even greater degree in the World War to follow. Though the Service soon contracted after the close of hostilities, the officers remaining possessed a type of knowledge not available from any course. Amid so many officers having both command and direct experience of the naval war, the seminars conducted between the wars at Greenwich and elsewhere benefited from a degree of realism not possible before 1914. Thus, fourteen students attending the Staff College in 1932 had been present at Jutland whilst sixteen officers of the 1931 course had been.[93]

In the presence of such experience, staff members and other lecturers having a pet hobbyhorse theory to explain British operational failings needed to tread carefully before expounding on any topic. Such caveats weighed, too, on writing the war's history and no more object demonstration was available than the Naval Staff's efforts to release an appreciation of the war's most controversial battle. Here, Jellicoe took strong exception to the line adopted in the appreciation when forwarded to him for review with his most strident criticism being the advocacy of divided tactics expressed in the study. This was something of a bugbear for certain officers and the question of divided tactics came to be a kind of shorthand for gauging an officer's general receptiveness to a whole range of modernizing issues. For Jellicoe, now removed from active naval employment, this was not the case. He simply believed that given the visibility conditions prevailing, such tactics were not possible at the battle.[94]

The issue of divided tactics was not a new one. Criticism of the fleet's preference for the single line-ahead formation was a signature debate of *The Naval Review* even during its very limited period of prewar publication.[95] Centering on the tendency of any fleet handled as a single tactical unit to ultimately find itself soon on a parallel course to its opposite number, the critics of the single-line ahead evolution advocated the use of divided tactics instead. They argued that history had shown that victory in battle was more easily achieved when one's own fleet concentrated on a portion of the enemy fleet and allowed a superior, decisive fire to be registered. Whether this fire concentration was delivered on the enemy fleet's van or rear portions was not a point upon which all the advocates of divided tactics agreed. Nor were they clear how the British fleet commander could maintain tactical control of the encounter and prevent the loss of his own forces in detail as the enemy fleet commander effected his

93   'Jutland I,' Tennant Papers, NMM/TEN/41/1.
94   Admiral of the Fleet Earl Jellicoe Papers, British Library, Additional Manuscript 49042.
95   See Edward Harding's three essays in *The Naval Review* of 1913 penned under the penname 'C.Q.I.' but especially 'Studies in the Theory of Naval Tactics. III,' *Naval Review*, 1913, pp 208-23.

own concentration of fire on a portion of the divided British fleet. They were agreed, however, that encounters fought employing the single-line-ahead formation devolved into British ships failing to achieve decisive results as they were reduced to firing on their opposite number and the concentration of fire so desired was lost.

These issues, investigated at some length at the Staff College and the Tactical School in Portsmouth during 1919-39, are examined fully in a later chapter. Noted, here, is that when an appreciation came to be written for the Battle of the River Plate fought in 1939, the question still featured as a topic of some prominence.[96] Moreover, if the Battle Cruiser Fleet under Vice Admiral Sir David Beatty and the temporarily attached Fifth Battle Squadron at Jutland is taken as a case study, then the advocates of divided tactics had much hard work to hoe to prove its efficacy in 1916 given the limits of the day's command and control system.

The need to immediately reconstitute a War College following the conflict's end was not a view universally held. Richmond, whilst Director of Training and Staff Duties and the staff element most directly responsible, advised the First Sea Lord that the first priority was to better prepare officers of the rank of Commander and Lieutenant Commander in staff duties. He stressed the creation of a Staff College; a view that the Board of Admiralty at first eschewed.[97] Likewise, his view that such a Staff College should be housed near Camberley, home of the equivalent British Army course, received short-shrift by Admiral Wemyss due primarily to the antici-pated sums involved: £30,000.[98] Richmond believed the funds required were trifling, however, when set against other costs the Navy was happy to incur. These included the rebuilding of the officer-cadet preparatory school housed at Osborne and the recon-struction of the battle cruisers *Repulse* and HMS *Renown*.[99]

Wemyss' concerns of economy were far from irrational, though, given the urgent need to get a grip on naval spending. Indeed, his successor as First Sea Lord, Beatty, resolved to close Osborne outright shifting its personnel to Dartmouth to be trained with their older brethren for such was the need to find savings in the period directly after the war.[100] Moreover, it was not simply a case of providing a building suitable to house the classrooms and library of a reconstituted War Staff Course. Quarters neces-sary to house the students were a factor, as single officers were expected to live and

96 'Tactical Lessons of the River Plate and BISMARCK Action,' Dewar Papers, NMM/ DEW/19.

97 Marder, ed., *Portrait*, p. 359.

98 Comparison of Sterling to its modern equivalence is deprecated by this writer, as it is very much a case of mixing apples and oranges when applied to modern military systems and organizations. It is given that Sterling has decreased in value, but any attempt to depict present value would not do justice to the true cost of building a comparable warship today as the threat and means to be defeated are so more complex. For the reader desirous to see such computations, they are referred to: www.nationalarchives.gov.uk/currency/.

99 Marder, ed., *Portrait*, p. 327.

100 Michael Partridge, *The Royal Naval College Osborne: A History, 1903-21* (Stroud: Sutton Publishing, 1999), pp. 141-50.

work at the college during weekdays—certainly, for the initial part of the course.[101] Additionally, messing facilities and staff to support the school were a consideration including typists, as officers largely prepared their work in longhand relying on the typists to polish the final product. Finally, offices for the directing staff, recreational facilities and grounds, a chapel, and a dispensary needed to be nearby. Acquiring a new freehold or leasehold whilst the Navy retained much surplus capacity was not a practical proposition in the near term. Whether Richmond thought of such constraints is doubtful, but a cost conscious Financial Secretary certainly had to and forsaking Greenwich for a Camberley location was not the simple matter assumed.

Never mind that Richmond was stymied in his attempt to see staff training conducted near Camberley, his plea for the early creation of a Staff College over the reinstitution of the War College proved more successful. The former course began fully nine months before the latter and was expanded in content to now run for a full year where its previous counterpart had lasted from three to ten months.[102] Accordingly, in early February 1919 the Admiralty wrote to Admiral Sir Henry Jackson,[103] the Admiral-President, Royal Naval College Greenwich and the parent naval establishment for the reformed War Staff Course, advising of the their intentions:

> The Staff College is to be entirely for the instruction of younger Officers qualifying for War Staff duties and will be a distinct organization from the War course which it is proposed to institute later for senior Officers.
>
> Normally, the Officers who will undergo the course will be selected from those who volunteer, and the selection will be confined to those between the ages of 25 and 30. In order, however, to make up for the shortage caused through the War, in the number of Trained War Staff Officers, the first course will include Commanders and Lieutenant Commanders, who have been employed in important Staff work during the War or who have been otherwise employed but are strongly recommended for the course.[104]

101 At Greenwich, rooms were reserved for qualifiers within the buildings of Queen Anne Block overlooking the Thames. As the Thames, then, was very much a working river noise was ever a concern, but Charles Drage during his time at the Staff College was quite content with his accommodation.

102 ADM 196/142 (Thursfield) and ADM 196/143 (Hall).

103 Later Admiral of the Fleet Sir Henry Bradwardine Jackson (1855-1929). Captain, 1896; Third Sea Lord and Controller, 1905-08; Rear Admiral, 1906; Vice Admiral, 1911; Director, Royal Naval War College, 1911-13; Chief of the War Staff, 1912-14; Admiral, 1914; First Sea Lord, 1915-16; Admiral President, Royal Naval College, Greenwich, 1916-19; Admiral of the Fleet, 1919 and retired, 1924.

104 ADM 1/8549/20, Admiralty to Admiral President Royal Naval College, Greenwich letter of 2 February 1919.

Jackson's presence at Greenwich at this moment was itself fortuitous, as he previously directed the War College and then led the nascent War Staff at the Admiralty before the onset of the World War.

Richmond, though, was hardly a lone voice in seeking to establish a naval college near the Army Staff College. In December 1919, Kenworthy, a former naval officer with staff experience during the Great War, also argued for a Camberley location during a House of Commons debate on the *Navy Estimates*.[105] An argument pursued with equal vigor by him, again, the following March when he complained:

> Might I repeat the criticism I made last year about the Staff College, the brain of the future Navy, being at Greenwich, a most unsuitable place? It ought to be at Camberley, where naval officers could meet and discuss and collaborate with the military and air Staff Officers and thus get that clash of ideas out of which progress and development comes. I hope that as soon as possible Greenwich will be used for other purposes. In any case, it is too near London. As my right hon. Friend knows, there is a pretty strong body of opinion in the Navy that thinks with me on this matter.[106]

Having the Staff College situated at Greenwich became a particular bugbear for Kenworthy and proved to be a recurring theme in his parliamentary interventions. His desire for the naval course to move closer to its Army and Air Force counterparts was not limited merely to the criticism that the intended site was too near London; he cited a Camberley venue as being even more preferable to one nearer the fleet at Portsmouth.

The issue, then, was how best to prepare naval and marine officers given the rise of a third military arm which in time was likely to be Britain's primary means of defence supplanting the Royal Navy. Three solutions suggested themselves. The first option was to delve into the implications of the air threat in a traditional naval school augmenting the naval and marine students attending with a couple of representatives from the British Army and the nascent Royal Air Force using Greenwich or a similar naval establishment as its home. A second choice was to create a naval school in the Camberley area where constant touch could be maintained with the faculty and students of the Army and, in time, Air Staff Colleges. This solution offered the promise of sharing academic resources (but not physical plant) between the Services and allowing operational precepts to be argued at length by mid-level officers and fostering early professional links between officers of comparable grades. The final course available was to establish a new joint institution under the control of no single Service to replace the existing colleges. Such a school would reside beneath a newly

105 *Hansard*, House of Commons Debate, 10 December 1919, v. 122, cc1481-2.
106 *Ibid*, House of Commons Debate, 18 March 1920, v. 126, cc. 2441-551.

minted Ministry of Defence or under the existing Committee of Imperial Defence to instruct future staff officers of all three Services.

To parliamentarians such as Kenworthy and Viscount Curzon, a former naval officer later achieving fame as a motor racing enthusiast, any option but the first would have been the preferable course to follow and time would yet vindicate them. In defence of the Admiralty's present position, finance was the limiting factor. Still, finance was not the only issue at play. Creating a new school to foster the joint training and education of all officers was a matter of policy beyond the remit of the Navy to implement. In the absence of an effective Chiefs of Staff Committee or a Ministry of Defence, creating such a school was problematic as the unified doctrine upon which the necessary curriculum must follow would be difficult to achieve. Absent such over-arching structures, instructors of any directing staff would teach to their own beliefs at some peril to the new institution's viability. Parliamentary critics of the Service's approach overlooked these issues and had a tendency to place the cart before the horse in their approach to staff training. Moreover, arguing that change of necessity must come from below was not a tenet that leaders in a hierarchical environment such as the Royal Navy were wont to follow.[107]

The Admiralty appreciated that the Greenwich location was not an ideal one and use of the facilities residing there was the result of the stark economic realities facing the Royal Navy. Lord Lee, the First Lord of the Admiralty, addressing the broader issue of Naval Staff organization and the training of its personnel raised by Haldane in a House of Lords' debate conceded that the existing lines of staff training was too abbreviated remarking:

> The Staff College is an institution which happens also to be at Greenwich at present—largely for financial reasons...I hope it may be possible to lengthen the course, and also to move the Staff College to Camberley in close proximity to the Army Staff College, so that there may be an even closer liaison between the two trainings than there is at the present time. But this is an age of economy, and for the moment the very heavy expenditure that would be incurred in building and equipping a new Staff College at Camberley has compelled the scheme to be deferred. But I can assure your Lordships it is only deferred, and it will be proceeded with at the earliest possible opportunity.[108]

107 *Ibid*, House of Commons Debate, 22 March 1923, v.161, cc. 2809-930. Not surprisingly, Kenworthy was an early advocate for a Ministry of Defence and believed that the Services had a vested interest in not reducing their stature through its establishment likening their endorsement to its formation to the jockeys, bookmakers and trainers of Britain advocating the abolition of the turf. See Charles Burney, *The World, the Air and the Future* (London: Alfred Knopf, 1929), p. 332.
108 *Hansard*, House of Lords Debate, 4 May 1921, v. 45, cc.151-80.

Though defeated in their desire to see the Staff College formed near Camberley where the training would be to the mutual benefit of all Services, the views of Richmond, Kenworthy and Curzon yet prevailed. In 1947, the year following Richmond's death, a Joint Services Staff College formed at Latimer, Buckinghamshire moving to the Royal Naval College, Greenwich in 1983 being then constituted as the Joint Services Defence College.[109]

Still, in 1919 the Royal Naval Staff College began its first class of instruction under the appellation then in vogue, the 'War Staff Course'. Captain the Honourable Reginald Plunkett-Ernle-Erle-Drax,[110] late commanding officer of the light cruiser HMS *Dublin*, was named the college's Director. Captain Guy Bigg-Wither,[111] specially commended for service in the flagship HMS *Hercules* at Jutland, had been serving as deputy to Richmond and Ellerton in the Training and Staff Duties Division, acted as the Deputy Director. His time at the Admiralty made him *au courant* with both Richmond's and the Service's views concerning staff training and his prior time as an instructor at the War College meant that he was no stranger to the classroom. He was a good choice to augment Drax, an officer whose war service had been spent entirely afloat with the Grand Fleet and mainly performing general staff duties in HMS *Lion*, flagship of the Battle Cruiser Fleet. Bigg-Wither's forte was tactical analysis (something anticipated when he had designed a Torpedo Tactical Board) and staff organization. Beatty recommended him especially for promotion in 1917—an event that soon followed.[112]

Drax, though, was a perfectly sound choice to oversee the reforming of the War Staff Course having attended the pre-1914 course. This was after he had already completed the War Course and then spent an additional six months in residence at the Army Staff College.[113] Drax, though, was not content merely to offer a rehash of the prewar course or spout platitudes from Camberley. He possessed original ideas

109 William Jackson and Lord Bramall, *The Chiefs: The Story of the United Kingdom Chiefs of Staff* (London: Brassey's, 1992), p. 461.
110 Later Admiral the Honourable Sir Reginald Aylmer Ranfurly Plunkett-Ernle-Erle-Drax (1880–1967). Five firsts in examinations for Lieutenant; War Course, 1908-09; Staff Course Camberley, 1909; War Staff Course, 1912-13; Qualified War Staff, 1913; Captain, 1916; Commanding Officer, HMS *Blanche*, 1917-18 and HMS *Dublin*, 1918-19; Director, Royal Naval Staff College, 1919-22; Senior Officers' Technical Course, 1925; Tactical Course, 1925; Flag Captain and Chief Staff Officer in *Marlborough*, 1926-27; Rear Admiral, 1928, Commander, Second Battle Squadron, 1929-30; Director, Manning Department, 1930-32; Vice Admiral, 1932; Commander-in-Chief, North America and West Indies, 1932-34; Commander-in-Chief, Plymouth, 1935-38; Admiral, 1936; Commander-in-Chief, The Nore, 1939-41 and retired, 1941. Recalled for war service.
111 Captain Guy Plantagenet Bigg-Wither (1878-1950). War College, 1912-14; Designated War Staff, 1914; Captain, 1917; Deputy Director, Training and Staff Duties Division, 1918-19; Deputy Director, Royal Naval Staff College, 1919-20; Commanding Officer, HMS *Dartmouth*, 1920-21 and HMS *Caradoc*, 1921-22 and retired, 1922.
112 ADM 196/90 (Bigg-Wither).
113 ADM 196/45 (Drax).

on how higher education and training should proceed in light of his war experience outlining his views in an article shortly after taking office. His overarching desire was to ensure those manning the weapons of the fleet worked to the dictates of the tactician directing operations and not attempt to supplant their respective roles.[114] Given his experience and progressive views on staff education, if any criticism might be leveled of Drax it was that, notwithstanding his quadruple-barreled surname, three Christian names and two titles, he was seen as a rather colourless personality.[115]

Of experience of war, Drax had ample reserves serving as he did in the *Lion* as Flag Commander to Beatty. Always a keen sportsman and avid tennis player, he brought a measure of order and system to a superior imbued with the fire to make things happen, if not with the temperament or gift to record the resulting fallout. It was Drax who thought to record the tactical exercises of the Battle Cruiser Fleet so that they could be analyzed anew.[116] This was not an original idea though it was an original practice. Trained in HMS *Vernon*, a large measure of credit falls to Drax for the improvements in torpedo control made within the Grand Fleet. He stopped the practice of rotating maintenance men from specialization to specialization, which while broadening their knowledge, came at the expense of developing a deeper understanding of any one component.[117]

Drax, however, was a crusader with a sheathed sword and not a zealot in his efforts to change the Service's approach to higher education. Though considered a member of those 'Young Turks', the crusading cabal for naval reform who argued their corner largely in the pages of *The Naval Review*, he made his views known with a measure of tact alien to some of his better-known associates. In common with Richmond, he was a torpedo specialist and very well read. Yet, the two did not always see eye-to-eye. In at least one instance, this conflict played out in the pages of *The Naval Review* for all to see. In this instance, the issue stemmed from Drax's use of historical example to buttress his arguments.[118] Drax was not to be the last officer to cross swords with Richmond over a point of naval history or theory and the merits of the dispute need not concern us here. The episode's value lies in the dogmatic and unbending pose to historical discourse that Richmond was prone to assume. Something considered more fully in due course, a brilliant mind married to a sharp pen, requires a strong measure

114 R. P. Ernle Erle Drax, 'Naval Education,' Earl Brassey, ed. *The Naval Annual 1919* (London: William Clowes and Sons, Limited, 1919), p. 240.

115 Grenfell to Liddell Hart letter of 2 June 1937, Liddell Hart Papers, LH 1/330/9, LHCMA.

116 Chalmers, ed., *Beatty*, pp. 209-10.

117 E. N. Poland, *The Torpedomen: HMS Vernon's Story 1872-1986*. (Privately published, no date, *c.* 1993), pp. 80-81.

118 Drax's original article 'The Influence of an Efficient Home Defence Army on Naval Strategy,' appeared in *The Naval Review* (London: The Naval Society, 1914), pp. 23-38. Richmond's reply 'Home Defence, Some Historical Aspects of. A Reply to R.X.,' pp. 141-59 and Drax's rebuttal 'Home Defence. A Reply by R.X.,' pp. 254-63.

of discretion in a serving officer if the soundness of the argument is not to be lost in the vehemence of its exposition.

Serving as the college's first Assistant Director and coming from the Gunnery and Torpedo Division of the Naval Staff was Captain Wilfrid Egerton,[119] a gunnery specialist and another writer on naval subjects with his *Contraband of War*, a brief monograph surveying the legal aspects of controlling trade, being his foremost work.[120] He was also the principal author of the classified study *Progress in Gunnery, 1914-18* and confirms from an early date that the leading light on tactical discourse at the Staff College was the Deputy Director.[121] Egerton would shortly become the Deputy Director upon Bigg-Wither's return to the fleet and new blood in the form of Henry Thursfield,[122] late of the Training and Staff Duties Division and like Drax a member of the staff serving in the *Lion*, arrived to assume the duties of Assistant Director. Thursfield had attended the first Staff Course in 1912 with Drax and, upon his retirement, was to continue thinking about naval affairs becoming naval editor of *Brassey's*, the respected defence journal while also writing for *The Times* newspaper. A founding member of the reforming circle producing *The Naval Review*, Thursfield's specialty whilst at the Staff College was lecturing and directing the exercises covering the tactics of the Grand Fleet.

The hand of Beatty behind the first appointments to the Staff and War Colleges appears palpable if not provable. Given the several links to *Lion*, his flagship in the Battle Cruiser Fleet, how much he pressed Wemyss as the programme of officer education was being re-established is unknown. Beatty was anything but a shrinking violet and was not afraid to offer his views on matters of concern to the fleet. At some point, he is likely to have expressed strong views on the subject though direct evidence for this conclusion is lacking. Strong links amongst several of the first officers attached to Greenwich are also apparent to *The Naval Review* and whether this was the actual

---

119 Later Rear Admiral Wilfrid Allan Egerton (1881–1931). Qualified for War Staff, 1916; Flag Commander in HMS *Barham*, 1916-18; Captain, 1918; Gunnery and Torpedo Division, 1919; Assistant Director, Royal Naval Staff College, 1919-20; Deputy Director, Royal Naval Staff College, 1920-21; Senior Officers' Technical Course, 1921 and 1930; Deputy Director, Plans Division, 1924-25; Director, Plans Division, 1925-28; Tactical Course, 1928 and 1931; Flag Captain, *Queen Elizabeth*, 1928-30; Rear Admiral, 1930 and Senior Officers' War Course, 1930-31.

120 Wilfrid A. Egerton, *Contraband of War* (London: Gieve, Matthew and Seagrove, 1915).

121 ADM 196/46 (Egerton).

122 Rear Admiral Henry George Thursfield (1882-1963). Five firsts in examinations for Lieutenant; War Staff Course, 1912; War College, 1914; War Staff Duties in HMS *Swiftsure*, 1915 and *Lion*, 1917-19; Training and Staff Duties Division, 1919-20; Captain, 1920; Assistant Director, Royal Naval Staff College, 1920-22; Commanding Officer, HMS *Concord*, 1923 and HMS *Comus*, 1923-24; Naval Mission to Greece, 1924-25; Senior Officers' War Course, 1925; Royal Naval War College, 1925-28; Tactical Course, 1928; Director, Tactical Division, 1928-30; Commanding Officer, *Royal Oak*, 1930-31 and *Renown*, 1931; Rear Admiral and retired, 1932.

source of their selection, likewise, remains undetermined. It may be the best explanation is that both associations worked in harness together. Thoughtful, progressive offices were attracted to *The Naval Review* and Beatty sought such officers on his staff. Knowing of the Admiralty's efforts to reform naval education, he knew a ready cadre of capable officers to help things along.

Unlike its British Army counterpart, the Staff Course followed a one-year and not a two-year curriculum. The shorter course initially adopted allowed a larger cadre of staff officers to be prepared and an ultimate desire to extend the course to the two-year format later after such sufficient numbers had been trained was expressed.[123] This desire, however, to see the course extended was never fulfilled and it remained roughly a single year for the duration of the interwar period. Cost may have been a consideration why this was so, but surely another explanation is that the Service, having other avenues of instruction to aid officer development, simply did not require a second year to be added to the Staff Course. Thus, attending the Tactical School, formed in 1925, or proceeding to either the Senior Officers' Technical Course or the Senior Officers' War Course became possibilities rather than adding a second year to the Staff College curriculum. Further, executive officers of the Navy and RNR officers needed to take the SOTC again at defined intervals. Beyond this, certain officers on the active list—though not RNR officers—took the War Course anew after five years.[124] Moreover, the addition of another yearlong programme of instruction in the guise of the newly formed Imperial Defence College made the arguments for extending the Staff Course more questionable still. This venue beginning in 1927 was also of a year's duration, but the possibility that it might extend to eighteen months or even two year's if a single year proved insufficient to train officers were factors weighing against any extension of the naval course.[125]

As an aside, Lieutenant Colonel Archibald Nye,[126] an officer who rose from the ranks subsequently becoming a barrister, and Wing Commander John Slessor,[127] a future Chief of the Air Staff, argued in a paper prepared for Major General John Dill,[128] Commandant of the Army Staff College, for reducing the current length of the Army Staff Course. This was in response to the attention staff training was then receiving by the Chiefs of Staff Sub-Committee and parliament. Their view stemmed from the realization that the study of large combat formations, which the second year

---

123 *Hansard,* House of Lords Debate, 4 May 1921, v. 45, cc.151-80.
124 Rainbow, 'Red, White and Blue', *The Naval Review*, Vol. XXII, No. 2, May 1934, p. 247.
125 CAB 53/12, Chiefs of Staff Sub-Committee report of 27 May 1926.
126 Later Lieutenant General Sir Archibald Edward Nye (1895-1967). Staff Course Camberley, 1924-25 and Staff College Camberley, 1932-35.
127 Later Marshal of the Royal Air Force Sir John Cotesworth Slessor (1897-1979). Staff Course Andover, 1924-25; Staff Course Camberley, 1931; Staff College Camberley, 1931-34 and Commandant, Imperial Defence College, 1948-50.
128 Later Field Marshal Sir John Greer Dill (1881-1944). Staff Course Camberley, 1913-14; Staff College Camberley, 1919-22; Imperial Defence College, 1926-28 and Commandant, Staff College Camberley, 1931-34.

of the curriculum considered, was of little use to most officers as the likelihood of their ever requiring such knowledge was deemed remote.[129]

Though the subjects covered in both the senior and lower courses offered at Greenwich were never static, the core emphasis remained constant with their emphasis on policy, strategy, operations and tactics. Using current topical issues to highlight the subject at hand, the discussions were sown with relevant historical examples to confirm the presence of the principles of war. Rear Admiral Frederic Dreyer,[130] attending the Senior Officers' War Course in 1924, deprecated the paucity of tactical study in the course. This, though, his course's syllabus contained at least five lectures on the topic.[131] In a reasoned paper presented to his class Dreyer citing the dearth of instruction advocated the establishment of a school committed to tactical instruction and investigation. This was coolly received by his peers. As his session of the War Course included three other officers of flag rank, the thought of attending yet another course may have influenced their reception to his proposal, but it was also the case that Dreyer, though seen as possessing a brilliant mind, was heartedly disliked by some in the Navy.[132] May be it was the messenger more than the message. Dreyer at times came across as pompous and the import of establishing a school was a veiled criticism that tactics were too important to be left to the fleet and the admirals commanding. As it was, Dreyer was not the first officer to call for a dedicated school of tactics. Russell Grenfell, writing in the privately circulated *Naval Review*, trumped Dreyer when he argued the same in 1923.[133]

From his War Course, Dreyer went to the Admiralty to serve as ACNS. Finding a more receptive listener to his arguments in the First Sea Lord, Beatty championed the idea of a school of tactics and quickly established it under Captain Cecil

---

129 John Slessor, *The Central Blue: The Autobiography of Sir John Slessor, Marshal of the Royal Air Force* (New York: Frederick A. Praeger, 1957), p. 86.

130 Later Admiral Sir Frederic Charles Dreyer (1878-1956). Qualified for War Staff, 1912; Captain, 1913; Commanding Officer, *Orion*, 1913-15; Flag Captain, *Iron Duke*, 1915-16; Assistant Director, Anti-Submarine Division, 1916-17; Director, Naval Artillery and Torpedo Division, 1918-19; Director, Gunnery Division, 1920-22; Commanding Officer, *Repulse*, 1922-23; Rear Admiral, 1923; Senior Officers' War Course, 1924 and 1930; ACNS, 1924-27; Tactical Course, 1927 and 1929; Commander, Battle Cruiser Squadron, 1927-29; Vice Admiral, 1929; DCNS, 1930-33; Admiral, 1932; Commander-in-Chief, China, 1933-36 and retired, 1939. Recalled for war service.

131 ADM 1/ 8658/69, 'Precis of Lecture on Tactics -V'.

132 The assessment was rendered by Russell Grenfell in 1937. See Grenfell to Liddell Hart letter of 2 June 1937, Liddell Hart Papers, LH 1/330/9, LHCMA. The flag officers on Dreyer's War Course were Rear Admirals Christopher R. Payne, Vernon H. S. Haggard and Charles R. Beaty-Pownall. See *The Navy List* (London: His Majesty's Stationery Office, April 1924), p. 307.

133 Grenfell, 'Training in Tactics', *Naval Review*, Vol. XI, No. 4, p. 686.

Usborne,[134] another thoughtful and inventive gunnery officer, at Portsmouth.[135] An officer obtaining five firsts in his examinations for Lieutenant, Usborne had attended Camberley briefly before the war and attended the SOWC the year before Dreyer.[136] Credited with a number of naval inventions, between 1919-21, he served as President of the Naval Anti-Aircraft Gunnery Committee whose charter was to investigate the forms of air attack a British fleet could expect to meet and recommending the best means of defeating the threat. The proposals emanating from this review were comprehensive in nature covering not only the means to be applied, but also the training to be pursued.[137] From this appointment, Usborne went to HMS *Dragon* to conduct trials on the Mark M pom-pom battery—perhaps a rare instance where an officer proposing an initiative actually disposed of it too.[138] By all accounts, Usborne was a sound choice to oversee the new school. Indeed, a paper presented during his recent period of study at the War Course and titled 'Naval Tactics, 1924' was fortuitously circulating at the Admiralty at the very time Beatty reached his decision to establish the Tactical School.[139] Still, in praising all that Usborne had achieved in establishing the school on sound lines Dreyer could not refrain from mentioning that the idea for the school resided elsewhere.[140]

Conveying the decision that a Tactical School was being established, a Confidential Admiralty Fleet Order in late 1924 described the anticipated length, purpose, and rationale for the new course which executive officers holding the rank of Commander and above would now attend.[141] The decision to house the new school at Portsmouth

134 Later Vice Admiral Cecil Vivian Usborne (1880-1951). Five firsts in examinations for Lieutenant; Staff Course Camberley, 1908; Captain, 1917; Assistant Director, Naval Ordnance Department, 1919-21; Commanding Officer, *Dragon*, 1921-22; Deputy Director, Gunnery Division, 1922-23; Senior Officers' War Course, 1923 and 1928; Vice President, Chemical Warfare Committee, 1923-25; Director, Tactical School, 1925-27; Commanding Officer, HMS *Malaya*, 1927 and *Resolution*, 1928; Rear Admiral, 1928; Senior Officers' Technical Course, 1928; Director, Naval Intelligence Division, 1930-32; Vice Admiral and retired, 1933.

135 Frederic Dreyer, *The Sea Heritage: A Study of Maritime Warfare* (London: Museum Press, 1955), pp. 278-79.

136 ADM 196/142 (Usborne) and ADM 196/90 (Usborne). Following a period of approximately 28 months, examinations in Seamanship and Navigation concluded a midshipman's period of at-sea instruction leading to promotion to Acting Sub-Lieutenant. Successfully completing examinations in Gunnery, Torpedo, Engineering and Pilotage after an additional period of instruction ashore completed the young officers training with promotion to Lieutenant following.

137 ADM 186/244, *C.B. 1561, Progress in Gunnery Matériel, 1920*, Admiralty, Gunnery Branch, G. 0721/20 of July 1920, p.77.

138 ADM 196/45 (Usborne).

139 ADM 1/8658/69.

140 ADM 196/90 (Usborne).

141 ADM 182/83, Confidential Admiralty Fleet Order, '3168a.—Tactical Courses at Portsmouth for Senior Officers' of 28 November 1924.

and not at Greenwich was deliberate. With several other Service schools in the immediate environ supporting Gunnery, Torpedo, Signals and Submarine training and with an air station at Gosport close-by, the Admiralty sought to emphasize the practical and realistic nature of the intended instruction.[142] Moreover, the announcement coming as a *Confidential* order laid stress upon the restricted nature of the training anticipated.[143]

As the broad study of tactics had been a prominent feature of the War College curriculum during Richmond's tenure—it featuring in no less than ten lectures during the 1921-22 courses—it is worth enquiring whether Dreyer's complaint was a valid criticism of the instruction provided in 1924. In 1922, the year of Richmond's departure, the War College curriculum offered a balanced course of study with lectures covering policy, strategy, operations, and tactics. In the syllabus, tactical analysis was slightly favoured, but caution must be noted in reaching this assessment as surveying the written record of the lectures presented at that time and found in the papers of Richmond and other members of the directing staff cannot measure the tenor of any discussions which followed. Hence, a paper nominally addressing tactics might quickly lead to the question whether operations to support the desired strategy were feasible in light of the tactical conditions prevailing—witness the pre-1914 debate whether mines, submarines, and wireless communications in naval operations made a close blockade a still feasible or necessary evolution in a potential war with Germany.

In 1924, tactics still featured in the War Course's syllabus, but its emphasis was certainly less pronounced. Part of the reason may have been that the current members of the directing staff were not consumed with the topic. Someone, though, who did carry the spark of tactical discourse was Captain James Troup,[144] an officer who would oversee the Tactical School in due course, but for the moment was overseeing the Tactical Section of the Naval Staff who visited Greenwich to speak on the subject. Doubtless, Troup's lectures coming from the font of approved doctrine allowed the students to benefit from his expertise and access. Yet, his primary duties were at the Admiralty and not at Greenwich. For students attending previously, Richmond and Captain Henry Thursfield directed tactical discourses and contributed to the subject's strong emphasis. Richmond's and Thursfield's tactical lectures, though, were

142 *Ibid.*
143 *Ibid,* Confidential Admiralty Fleet Order, '3380.—Tactical Courses—Information as to Institution, etc, To be withheld from Press, Foreign Officers and Public,' of 19 December 1924.
144 Later Vice Admiral Sir James Andrew Gardiner Troup (1883-1975). Navigating Officer and Master of the Fleet in *Queen* Elizabeth, 1920-22; Captain, 1922; Head, Tactical Section, 1923-25; Deputy Director, Tactical School, 1925-26; Commanding Officer, *Cairo,* 1926-28; in Command, HMS *Dryad,* 1928-30; Flag Captain and Chief Staff Officer in *Revenge,* 1930-32; Senior Officers' Technical Course, 1933; Senior Officers' War Course, 1933; Director, Tactical School, 1933-35; Rear Admiral, 1935; Director, Naval Intelligence Division, 1935-38; Vice Admiral and retired, 1938. Recalled for war service.

not studies in the characteristics of specific weapons' performances weighing their utility and use in battle. Rather, they were lectures on general tactical principles such as concentration of fire, the value of speed, styles of command, and the merits of offensive and defensive encounters. To the extent they failed to address the tactical employment of weapons usage, Dreyer's criticisms were valid to a point. The rejoinder to Dreyer's attack on tactical analysis was that the War and Staff Courses aimed to impart knowledge on the higher aspects of the naval profession allowing an officer to appreciate the tactical principles to be followed in battle notwithstanding the specific means to be employed. Accordingly, Richmond argued in one 1920 lecture on tactics:

> We have suffered greatly by subordinating tactics unduly to the demands of the indi-vidual gun. It is not the individual gun we have to provide for, but the collective artillery – using the word artillery to denote every missile weapon – of which a fleet can make employment, large guns, small guns, machine guns, torpedoes, bombs and mines.[145]

The effect of establishing the Tactical School on the War Course, however, was immediate. Tactical instruction lost favour with its discourse downplayed. With its treatment now circumscribed, it became a review of historical practice more than any attempt to codify contemporary execution. In 1926-27, tactical investigation was reduced to two lectures when in Richmond's time up to thirteen lectures had analyzed the subject and its endless possibilities. The higher direction of war—strategy and policy—now formed the heart of the Senior Officers' War Course and the balanced syllabus that Richmond originally developed lapsed. As for its handling at the Tactical School, Dreyer's ambitions culminated in an eight-week course of instruction built around the formal doctrine covering the differing classes of warships, the evolu-tion of tactics in the Navy from 1909 to the present and a review of current tactical problems.[146] The last point is crucial. For just as the War and Staff Courses exam-ined current policy problems for the Naval Staff, so the investigations of the Tactical School were updated at the request of the Admiralty to examine battle against the presumed naval threat of the day.[147]

Predating formation of the Tactical School, however, was the Senior Officers' Technical Course (SOTC), a class conducted at Portsmouth and viewed as a stepping-stone to attending the Senior Officers' War Course, though many officers attended

---

145 'Lectures at Royal Naval War College, Greenwich Spring Session 1920, Volume V, "Tactics,"' Admiral Sir Herbert W. Richmond Papers, National Maritime Museum, Greenwich, NMM/RIC/10/2. Original emphasis.
146 'Course X,' Admiral Sir Frederic Charles Dreyer Papers, Churchill Archives Centre, Cambridge, United Kingdom, DRYR/7/2.
147 Admiral Sir Geoffrey Miles letter to Roskill of 30 August 1964, Captain Stephen Wentworth Roskill Papers, Churchill Archives Centre, Cambridge, United Kingdom, ROSK/7/163.

the SOTC without ultimately proceeding to the War Course.[148] Though a version of the SOTC had existed before the World War, it was recast in content with the first postwar sitting coming in January 1920 and members moved directly from it into the first War Course.[149] Thus, the SOTC as a precursor to War Course instruction was an established tenet but later experience would demonstrate that it was frequently honoured more in the breach.[150] With its studies focusing on 'the capabilities, limitations and methods of use of weapons and material, the course originally lasted for eight weeks.'[151] That the SOTC and the Tactical Course both ran to eight weeks was by design and not by accident. This made officer assignment to the courses and to the SOWC easier to manage. Even still, not all officers were able to begin or complete these courses on time and this owed much to having to meet the manning requirements of the greater Service.

The SOTC and the Tactical Course were primarily intended for officers holding the rank of Captain, though others, including flag officers, were admitted. Royal Marine officers did not sit the two courses which can be seen as something of a missed opportunity. Yet, their professional development did not suffer unduly as they attended the Army's corresponding classes of instruction in the combat arms and also a Combined Operations Course that naval officers did not. Though marine officers were not present at the SOTC, members of the Royal Naval Reserve and Dominion and Commonwealth navies did attend the SOTC and classes of 10 to 14 members were the norm.

Shortly after the establishment of the Tactical School, the Admiralty attempted to obfuscate its existence and the nature of its instruction from any public scrutiny by concealing its existence and operation. Thus, it was initially referred to as the Senior Officers' Technical Course, Part II, a name already known outside of naval circles for the reasons already discussed.[152] In 1930, it was established that an officer should attend his two Technical Courses between his sitting the War Course and officers destined for command afloat were expected to attend at least one of these three-core courses before proceeding to sea.[153]

Roger Backhouse,[154] later an Admiral of the Fleet and First Sea Lord from 1938-39, was one officer who attended the SOTC whilst holding flag rank. Completing the

148 ADM 182/31, Admiralty Fleet Order, '4004.—Senior Officers' Courses, 1922-23.' of 9 December 1921.
149 ADM 196/90 (Collard).
150 'Naval Notes, Officers' Technical Course', *Journal of the Royal United Services Institution*, Vol. LXV, No. 459, August 1920, p. 613.
151 *King's Regulations*, Section 326, p. 326.
152 ADM 182/84, Confidential Admiralty Fleet Order, '613.—Tactical School, Portsmouth— Change of Name' of 6 Match 1925.
153 ADM 182/89, Confidential Admiralty Fleet Order '131.—Captains, R.N.—Technical and War Courses' of 17 January 1930.
154 Admiral of the Fleet Sir Roger Roland Charles Backhouse (1878-1939). Five firsts in examinations for Lieutenant; qualified for War Staff. 1912; Captain, 1914; Director, Naval

course in 1925, Backhouse proceeded to the War Course the following year. Likewise, Andrew Cunningham,[155] another future Admiral of the Fleet and First Sea Lord, attended the Senior Officers' Technical Course whilst a Vice Admiral in 1937, a point the more noteworthy as he had already attended sessions in 1922 and 1933. Indeed, during the period, it was normal to have at least one flag officer attending each session of the SOTC though given the limited number of officers holding such ranks and their patterns of employment, it was practice never elevated to a rule. The typical officer studying was just as apt to hold the rank of Commander as Captain, notwithstanding the guidance of *King's Regulations and Admiralty Instructions*, whilst a significant portion of each class was likely to come from members holding commissions in the Royal Naval Reserve.

Historians, such as Marder and Roskill, have frequently claimed that the study of Jutland became somewhat of an obsession in the interwar Royal Navy as they sought to hone their professional skills. Writing in 1974, Marder noted:

> Naval strategy and tactics were largely conditioned by a determination to make the next Jutland a Trafalgar—when a second Jutland was highly improbable. A reconstruction of Jutland was the *piece de resistance* of the Staff College during most of the interwar period.[156]

Certainly, such study was prominent at Greenwich and at the Tactical School, but it did not become an explicit topic of the Staff and War Colleges until the 1921-22 academic year. Before then, the tactics of the Grand Fleet featured as a generalized topic and Jutland as an unequivocal focus of enquiry remained missing. That this was so must owe something to the ongoing controversy surrounding the battle and the lack of consensus on what its actual lessons were. Until the Official History and the Service's own Staff Appreciation had been prepared, explicitly teaching the battle was fraught with risk. This writer has found no direct evidence that Jutland as a topic was prohibited by Admiralty fiat, but it remains the case that its place in the curriculum only became manifest when Corbett's study and the Dewars' appreciation were well

Ordnance Department, 1920-22 Commanding Officer, *Malaya*, 1923-24; Rear Admiral, 1925; Senior Officers' Technical Course, 1925; Senior Officers' War Course, 1925-26; Controller and Third Sea Lord, 1928-32; Vice Admiral, 1929; Admiral, 1934; Commander-in-Chief, Home Fleet, 1935-37; First Sea Lord, 1938-39 and Admiral of the Fleet, 1939.

155 Admiral of the Fleet Sir Andrew Browne Cunningham, later Viscount Cunningham of Hyndhope (1883-1963). Commanding Officer, HMS *Seafire*, 1919; Captain, 1919; Senior Officers' Technical Course, 1922, 1933 and 1937; Flag Captain in HMS *Calcutta*, 1926-28; Senior Officers' School, Sheerness, 1928; Imperial Defence Course, 1929; Rear Admiral, 1932; Tactical Course, 1933; Commander, Destroyer Flotillas in Mediterranean Fleet, 1933-36; Vice Admiral, 1936; Commander, Battle Cruiser Squadron, 1937-38; DCNS, 1938-39; Admiral, 1941; Admiral of the Fleet, 1943 and First Sea Lord, 1943-46.

156 Marder, *From the Dardanelles to Oran*, p. 48.

advanced. It may be that Richmond and Drax believed the best course was to wait for the Admiralty position on the battle to be resolved before expounding any lessons.

In time, Jutland was to feature prominently in the studies of the Staff College, culminating in a week's instruction by the early 1930s. This, though, hardly appears excessive in a course of instruction running to eleven-months and when weighed against the time devoted to other topics.[157] The truth is that whilst Jutland was examined at some length, equal weight of emphasis was placed on the economic element in war (five lectures), combined operations[158] (five lectures), maritime international law (six lectures), and air operations and strategy (four lectures). If there was a bias in the instructional emphasis, it was in the time devoted to naval and military organization and administration; yet, this is perhaps quite forgivable in what was perforce a *Staff* College education. As to the charge that only a single hour was devoted to submarines, the reality is that five lectures covered submarine operations, anti-submarine warfare, the protection of British shipping and the attack against enemy commerce.

Certainly, one criticism that can be leveled is that in recently fighting a major conflict in which British strategy and operations formed but a part of an allied effort, little discussion has been detected in the Staff and War Colleges' curricula on examining the issues surrounding this problem. After the creation of the IDC, this void is understandable as it could be seen as rightly belonging within the parameters of that course. However, it remains true that the gap existed before the founding of the IDC and surely must have been one of the most obvious lessons of the World War demanding attention. It may be that it was viewed as too difficult a topic to address within the limited War Course curriculum and that it was considered an inappropriate topic for the more junior officers attending the Staff College. It might also be that Richmond and Drax simply missed its saliency, though this observer finds this answer unsatisfying given their outlook and experience. Perhaps if officers from such friendly states as Greece and Chile had attended the higher naval courses, the matter would of necessity arisen. The best answer is probably that the issue of concerting action with allies was tied very much to the higher direction of war and this subject was ever a problem to tackle adequately as later chapters will show.[159]

More remains to be said about the style and type of instruction provided at Greenwich and Portsmouth, but the question remains why the conventional wisdom concerning the education and training provided remains. Several possibilities exist but to this writer the most likely one is that our understanding of the interwar Navy owes much to Marder and Roskill. Neither conducted a systematic review of the curriculum nor

---

157 Holger H. Herwig, 'Innovation Ignored: The submarine problem, Germany, Britain, and the United States, 1919-1939', in Williamson Murray and Allan R. Millett, eds. *Military Innovation in the Interwar Period* (New York: Cambridge University Press, 1996), p. 249.

158 Combined Operations in this case referring to amphibious operations and not the operations of two or more nations.

159 Hankey, 'The Higher Direction of War', Imperial Defence College lecture of 7 November 1927, Alanbrooke Papers, 3/12, LHCMA.

did they place the curriculum within its broader educational context within the Service and the period. Relying mainly on the memories of officers who had attended the Staff College, they concluded that Jutland was the centrepiece of studies. It certainly was a prominent aspect of staff studies, but it was never as central as supposed by these two writers. Others writing in their wake have assumed this to be true, again, without ever systematically reviewing the curricula of the period. The upshot is that it renders a disservice to the interwar Royal Navy as the impression is left of a force more fixated on the past, than on solving the problems of the present and near future.

Memory is a notoriously fickle mistress and how those officers studying at the Staff College came to see Jutland as so central to the course requires an explanation. One reason probably lies in the format and structure of the Staff College course itself, which dedicated a specific block of time to the battle rather than address its issues across several broader subjects. Moreover, it was probably the case that Jutland engendered the liveliest of discussions from those attending, as many had direct experience of the engagement. Additionally, much depended on the questions posed by Marder and Roskill to contemporaries on their recollections of study. As a later chapter will show, Jutland became a major piece of Staff College and Tactical Course study, but it was never the central theme and was but a vehicle for introducing broader topics.

Minus a final qualifying examination—something its prewar counterpart featured—the question remains how challenging was the Staff Course?[160] A considered response is that it was a significant test of one's mental ability, but its format and style left to the individual exactly how much he took away from his period of study. That the Staff Course was conducted in too relaxed an atmosphere cannot be definitively proven, but strong circumstantial evidence exists that this was the case. The late Vice Admiral Leslie Brownfield,[161] a student of the 1934 Staff Course, confirms in a private memoir the extensive periods of leave allowed and the relaxed pace of study pursued, though neglecting to discuss the actual import of instruction pursued.[162] Rear Admiral Ralph Fisher,[163] again, writing in a private memoir, if a little more forthcoming of the instruction provided, noted the light academic atmosphere at times present whilst attending the 1935 course. In this case, members of the naval class joined their Army counterparts from Camberley in a staff ride of the Aisne battlefields of France; a tour conducted with much frequency between the wars focusing especially on the events

<hr>

160   ADM 196/47 (Thursfield).

161   Vice Admiral Leslie Newton Brownfield (1901-1969). Short Course, Cambridge University, 1922; Staff Course, 1934; Squadron Gunnery Officer in HMAS *Canberra*, 1935-37; Squadron Gunnery Officer and Staff Officer Operations in *Royal Oak*, 1937-39; Squadron Gunnery Officer and Naval Attaché, Siam in HMS *Terror II*, 1939; Captain, 1942; Rear Admiral, 1952 and Vice Admiral, 1955.

162   Leslie Brownfield, *Sea Cycle*, no date, http://www.sandylane.plus.com/brownfield/.

163   Later Rear Admiral Ralph Lindsay Fisher (1903-88). Staff Course, 1935, Royal Naval College Greenwich for Junior Officers' War Course, 1936-37; Joint Services Staff College, 1948-50; Rear Admiral, 1954 and retired, 1957.

of 13-15 September 1914. Fisher admitted that an opportunity to consider seriously the battles' imports were largely missed by some of the naval officers present and the impression remains that the trip was viewed as a bit of a lark by the naval members of the party.[164] Finally, it is appropriate for Captain Edward Astley-Rushton[165] to have the last word on the subject as he concluded of one officer's experience of the Staff Course, he allowed 'social activities to prevent him doing full justice to his ability knows his mistake and may be expected not to repeat it.'[166]

With regard to the War Course, the Admiralty assigned to Herbert Richmond the task of regenerating the programme of instruction, an appointment that on the surface should have been both congenial and professionally satisfying to an officer of his talents and sympathies. Yet, this proved to be far from the case. Richmond had little input to the instructors assigned to the Royal Naval College and the Second Sea Lord, Admiral Sir Montague Browning,[167] made it clear to him that for those officers posted to teach, reassignment to seagoing duty was apt if a suitable billet were found for a member of the directing staff.[168] Nor was the issue for Richmond solely the possibility that an instructor was subject to reassignment; he deprecated the background of some of the officers available and questioned their fitness to be instructors in the first instance.[169] This criticism was not wrong, but it was misplaced. The Admiralty had to balance the requirements of the broader Navy and could not consider an officer's appointment to the college as an isolated question. As the service record of Lieutenant Commander Victor Danckwerts[170] makes clear his temporary appointment to the Staff College following completion of his War Staff Course in 1920 was made because

164 *Salt Horse: A Naval Life*, Rear Admiral Ralph Lindsay Fisher Papers, Liddell Hart Centre for Military Archives, King's College, London, United Kingdom.
165 Later Vice Admiral Edward Astley Astley-Rushton (1879-1935). Five firsts in examinations for Lieutenant; War Course, 1913; appointed War Staff Officer, 1913; Captain, 1916; Deputy Director, Training and Staff Duties Division, 1919-21; Director, Royal Naval Staff College, 1922-25; Tactical Course, 1925 and 1931; Commanding Officer, *Malaya*, 1925-27; Rear Admiral, 1927; Director, Manning Department, 1928-30; Commander, Second Cruiser Squadron, 1931-32; Vice Admiral, 1932 and Commander Reserve Fleet, 1934-35.
166 ADM 196/127 (Jackson).
167 Admiral Sir Montague Browning (1863-1947). Captain, 1902; Chief of Staff, Channel Fleet, 1908; Rear Admiral, 1911; Vice Admiral, 1917; Second Sea Lord and Chief of Naval Personnel, 1919-20; and Admiral, 1919.
168 Marder, ed., *Portrait*, p. 362.
169 *Ibid.*
170 Later Vice Admiral Victor Hilary Danckwerts (1890–1944). Five firsts in examinations for Lieutenant; War Staff Course, 1919-20; Royal Naval Staff College, 1920-21; Staff College Camberley, 1923; Royal Naval Staff College, 1923-24; Tactical Course, 1926; Squadron Gunnery Officer and Staff Officer Operations in *Iron Duke*, 1926-27; Commanding Officer, *Windsor*, 1927-28; Imperial Defence Course, 1929; Naval Intelligence Division, 1930; Captain, 1930; Commanding Officer, *Caradoc*, 1930-32; Plans Division, 1932-35; Captain (D) Sixth Flotilla, 1936-38; Director, Plans Division, 1938-39; Rear Admiral and retired, 1940. Recalled for war service,

the make-up of the permanent instructional staff was still an open question.[171] This applied even more in the case of War College appointments where the numbers of qualified officers of suitable background were fewer.

Compounding the issue, officers studying at both the War and Staff Colleges in 1921 had their instruction curtailed when a nationwide coal strike occurred and reserve forces were mobilized as part of the Government's overall response to the industrial action. Lieutenant Stephen King-Hall,[172] a War Staff qualifier, found himself assigned to one of the activated battalions deployed to Gravesend.[173] The irony that some of the ratings in his battalion were themselves striking miners was not lost on King-Hall, but the three-week hiatus in his year long course was accepted with a certain degree of *sang froid* and more easily absorbed than those attending the four and a half month-long War Course.[174] Richmond had to release officers attending the senior course to support the Navy's 'UC' duties where the Service protected key facilities and distribution points.[175] To a man of Richmond's disposition, his letter to the Admiralty noting the effects of lending such support was mild only because naval custom and discipline required such obeisance when addressing superiors:

> I forward herewith a report of the work done by the Senior Officers' Course in the Session March-July, 1921.
>
> The course could not be followed out as designed owing to half of the officers being required for battalion duties during the coal strike. Hence the continuity of the work suffered and the work has been less connected than it was intended to be.[176]

171 ADM 196/144 (Danckwerts).
172 Commander Sir William Stephen Richard King-Hall, later Baron King-Hall of Headley (1893-1966). Five firsts in examinations for Lieutenant; Training and Staff Duties Division, 1919-20; War Staff Course, 1920-21; Staff Course Camberley, 1924; Plans Division, 1924-25; Staff Officer Intelligence, Mediterranean Fleet, 1925-26; Tactical Section, 1928 and retired, 1929. Member of Parliament, 1939-45 and founder of the non-partisan Hansard Society, 1944.
173 Stephen King-Hall, *My Naval Life, 1906-1929* (London: Faber and Faber, 1951), pp. 181-82.
174 The Earl of Cork and Orrery, *My Naval Life. 1886-1941* (London: Hutchinson & Co., Ltd., 1942), p. 154.
175 'UC' was the designation for aid rendered to the civil authorities during periods of unrest. Though the designation may have been adopted from a convenient sleight of hand of the initials formed from the words Civil Unrest, another version has it that UC was used for its easy but not direct association to the Trade Union Congress. For the Navy, 14 different tasks were specified, of which, 'UC 7', the diversion of shipping in the event of a strike by miners and 'UC 13', the protection of 16 London power stations, were germane to the Royal Naval College, Greenwich.
176 ADM 1/8628/120, Admiral-President, Royal Naval College, Greenwich to Admiralty letter of 29 July 1921.

In fairness to the Admiralty, most of the officers sitting the spring 1921 War Course, the one interrupted by the strike, were reassigned to the next available class. Accordingly, their education did not suffer unduly and Richmond's pride remained the most apparent victim.[177]

Edward Altham,[178] a gunnery specialist recently promoted to Captain and lately engaged in supporting Allied naval operations in North Russia was an officer presumably Browning had in mind and not all to Richmond's liking. A member of the first postwar course, Altham was also an inventive officer and soon became a part of the War College's directing staff in June 1920 upon completing his class.[179] Such a turn of mind, though, may have labeled him one from the 'materialist' school and thus far from the type of officer Richmond desired.[180] For Altham, though, a return to active service in Russian waters remained a distinct possibility. Operations in that theatre were still ongoing and given his wide understanding in supporting inshore operations dating from the World War and his recent experience in Russian Waters the possibility of recall was ever present. Possessing an inventive mind, Altham had developed improvements in the fire control arrangements in naval monitors as a result of the shortcomings he witnessed in their performance during the war. Altham's lecture notes have not been located so the precise scope of his responsibilities remains a mystery. Compounding the matter, Altham soon found himself transferred to the Training and Staff Duties Division in January 1921. This was a temporary appointment only, so his move to the Naval Staff after serving at the War College for only six months was most likely made to placate Richmond. Whether the reasons were one of performance, discretion, or just two officers of chalk and cheese temperaments is unknown, but the most likely answer was to be found in Altham's manner, which was frequently brusque and overbearing to both seniors and subordinates alike.[181]

Altham proved to be an able officer when it came to thinking and writing about naval affairs, as later events demonstrated. Though his contributions to *The Naval Review* only began from his time at Greenwich when an article surveying the Navy's role in the Russian intervention appeared, Altham did not shy from controversy.[182] His piece addressing Jutland was tellingly a précis in response to A. C. Dewar, one of the authors of the Staff Appreciation previously discussed. Writing now as a retired officer, his argument was a strident defence of Corbett's treatment of the battle as

177 ADM 203/99, Richmond undated note describing the reassignment of officers affected.
178 Captain Edward Altham (1882-1950). Five firsts in examinations for Lieutenant; Captain, 1918; Senior Officers' War Course, 1920; Royal Naval War College, 1920-21; Training and Staff Duties Division, 1921-22; Senior Officers' School Woking, 1922; Senior Officers' Technical Course, 1922 and retired, 1922. Recalled for war service.
179 ADM 203/99 and *Navy List December 1920*, p. 1865b.
180 ADM 196/142 (Altham).
181 ADM 196/91 (Altham).
182 Edward Altham, 'Some Naval Work in North Russia, 1918-1919,' *The Naval Review* (London: The Naval Society, 1921), pp. 85-155.

presented in the Official History where he refuted the use of divided tactics and the possibility of deploying the Grand Fleet along lines not supported by the doctrine of the day.[183]

Commander Hamilton Hall,[184] a gunnery officer, is another officer whose lecture notes have not survived and a review of *The Naval Review* sheds no further light, as he was not a contributor. Holding the rank of Commander, Hall's lack of seniority relative to a class composed largely of post Captains was a further complication to his tenure, though the absence of uniform and the wearing of *mufti* by qualifiers removed its most visible effects.[185] Hall, in common with a practice Henry Thursfield would follow in 1925, attended an abbreviated War Course before moving to the War College directing staff. In fact, Hall was an officer who Richmond appreciated praising his intellectual attainments and highly accurate mind.[186] Still, his appointment was only a temporary one until a more senior officer was available to serve and Hall soon departed for the *Malaya*. As for his relative inferiority in rank, even an officer such as Richmond could find himself outranked at times, so too much should not be made of his being a Commander. In Richmond's case, amongst the first of those attending the War Course was Vice Admiral Cecil Dampier,[187] late Admiral Superintendent of Dover Dockyard and an officer with experience watching over a brash and headstrong junior flag officer—in this case, Roger Keyes.

A member of the War College's staff whose lecture and exercise notes have survived is Guy Hamilton.[188] Serving in the battleship HMS *Ajax* during the war, he left her upon promotion to Captain and eventually found himself in the Training and Staff Duties Division with responsibility for overseeing the reformation of the War Course.[189] A torpedo specialist and an attendee of the 1913 War Staff Course, he left

---

183 Edward Altham, 'Imaginary Tactics and the Battle of Jutland,' *The Naval Review* (London: The Naval Society, 1924), pp. 608-618.

184 Later Captain Hamilton John Burnett Hall (1884–1949). Five firsts in examinations for Lieutenant; War Staff Course, 1913; War Staff duties in HMS *King Edward VII*, 1915-16; Staff Course Camberley, 1919; Training and Staff Duties Division, 1920; Senior Officers' War Course, 1920; Royal Naval War College, 1920-21; Chief Staff Officer and War Staff Officer in HMS *Columbine*, 1923-24; Second Secretary, Ordnance Committee, 1925-27; Naval Ordnance Department, 1927-30; Captain and retired, 1930. Recalled for war service.

185 'Naval Notes, New War Course Opens', *Journal of the Royal United Services Institution*, Vol. LXV, No. 458, May 1920, p. 412.

186 ADM 196/126 (Hall).

187 Vice Admiral Cecil Frederick Dampier (1868–1950). Captain, 1904; Commanding Officer, HMS *Audacious*, 1913-14; Rear Admiral, 1915; Rear Admiral, Third Battle Squadron, 1916-17; Admiral Superintendent, Dover Dockyard, 1917-18; Rear Admiral, Controlled Minefields, 1918-19; Vice Admiral, 1919 and retired, 1922.

188 Captain Guy Hamilton (1882–1950). War Staff Course, 1913; Captain, 1918; Training and Staff Duties Division, 1919-20; Senior Officers' War Course, 1920; Royal Naval War College, 1920-22; Commanding Officer, *Castor*, 1922 and retired, 1922; Recalled for war service.

189 ADM 196/48 (Hamilton).

early for the Admiralty and special work, yet managed to successfully negotiate the course's qualifying exam.[190] Hamilton lectured and directed exercises on the organization of commands and staffs, the peace distribution of the fleet, and logistics—particularly those governing the wars considered at the War College against the United States or Japan.[191] Having studied at the London School of Economics, Hamilton was responsible, too, for analyzing the economic elements of war and relieved Richmond of much of the administrative burdens of running the War Course.[192]

Of the initial officers assigned to the War College directing staff in 1920, none reached flag rank. Thus, a *prima facie* case that the Admiralty was providing Richmond a team of the second eleven is hard to refute. Still, that is to anticipate events. Altham was a very capable officer who had demonstrated strong leadership skills in Russia with the Admiralty deploying him twice on that front: first, in the light cruiser HMS *Attentive* and, then, in command of HMS *Fox*. Conditions there were difficult, coordination between the Admiralty and the War Office was not always ideal, and, ultimately success proved elusive.[193] Yet, for that, the Admiralty recognized that he had done a credible job under difficult circumstances, desired to retain him in the near term and specially promoted him to Captain.[194] He may have not demonstrated a previous disposition to considering the higher aspects of war, but his time in Russia had certainly shown to him the pitfalls likely to arise in conducting joint operations if cooperation was lacking amongst the Services. Moreover, once at Greenwich he blossomed. Following his naval career, Altham wrote on naval subjects for both the *Encyclopaedia Britannica* and the defence publisher *Brassey's*. He soon joined the staff of the Royal United Services Institution contributing frequently to its journal while also serving as its Secretary and Chief Executive Officer. The son of Lieutenant General Sir Edward Altham, a graduate of Camberley and later a member of its directing staff, Altham *major* did write about the nature of war most conspicuously in 1914 with *The Principles of War Historically Illustrated*. A conclusion that Captain Altham consciously desired to emulate his father would be speculative, but a child frequently follows in the steps of a parent, if only because the way has been shown previously. Only recently promoted to Captain, assignment to Greenwich with its collegial atmosphere could be viewed as reward for a job well done in Russia whilst making the best use of his recent experiences.

190 ADM 196/143 (Hamilton).

191 'Strategy VIII. Naval Lines of Communication,' lecture by Captain Guy Hamilton of 1 April 1921, Richmond Papers, NMM/RIC/10/5a and 'Naval Lines of Communication. The Supply of a Fleet,' lecture by Captain Guy Hamilton 3 November 1921, Richmond Papers, NMM/RIC/10/5b.

192 ADM 196/91 (Hamilton).

193 Clifford Kinvig, *Churchill's Crusade: The British Intervention of Russia 1918-1920* (London: Hambledon Continuum, 2006), p. 182.

194 ADM 196/125 (Altham).

Though never reaching flag rank, Hall subsequently was employed within the Naval Ordnance Department becoming Chief Superintendent of Armament Supply in March 1927. The assignment must have been advantageous to Hall and the Admiralty both, as promotion to Captain followed in 1930 and he continued to oversee these responsibilities upon his retirement from the Navy.[195] Though some viewed employment with the Ordnance Department as a sinecure for surplus officers, that in itself does not detract that competency was still a requirement.[196]

If Richmond questioned the quality of officers made available by the Admiralty as instructors, he could hardly lament those initially assigned to the course itself. Besides Dampier previously mentioned, another flag officer sitting the first course was Rear Admiral Edward Bruen.[197] His presence was curtailed, though, when he was appointed Director of Naval Equipment in May 1920.[198] Thus, Richmond's concern that an officer assigned to the War College as an instructor might see his appointment terminated and returned to other duties applied equally to its students. In this, Captain John Casement[199] was singularly affected. Selected to attend the first War Course, he was unable to proceed to Greenwich remaining in command of gunnery training in the shore establishment HMS *Vivid*. Released to sit the spring 1921 course, this session was much upset by the industrial strike action arising at the time and Casement's SOWC studies suffered again.[200] Casement is also unique in that he attended the War Course, or its later variation, on four separate occasions during his career though not always to a successful result.[201]

As for Richmond's lament, the conclusion of this writer is that an officer of his experience was guilty of over-identifying the specific needs of his present duties at the expense of the Navy's fuller requirements in what was, after all, a *seagoing* enterprise. Thus, comparisons to the running of a public school made by him, while appearing comparable on the surface, discounted that the purpose of naval education

195 *The Navy List* (London: His Majesty's Stationery Officer, January 1932), p. 412 and p. 561.
196 John M. Scott, 'Retirement' *The Naval Review*, Vol. XVIII, No. 1, February 1930, pp. 135-36. As Scott was serving in Singapore in HMS *Tamar* he may have held little sympathy with the Naval Ordnance Department given the tortuous process in constructing the new naval base.
197 Later Admiral Edward Francis Bruen (1866–1952). Captain, 1906; Commanding Officer, HMS *Bellerophon*, 1913-16 and *Resolution*, 1916-17; Rear Admiral, 1917; Commander, Second Cruiser Squadron, 1918; Senior Officers' War Course, 1920; Director, Naval Equipment Department, 1920-22; Vice Admiral, 1922 and retired, 1924.
198 *Navy List* (London: His Majesty's Stationery Office, September 1920), p. 1825.
199 Later Admiral John Moore Casement (1877–1952). War Course, 1909-10; Senior Officers' Torpedo Course, 1910; Senior Officers' Gunnery Course, 1910; Captain, 1915; Senior Officers' Technical Course, 1921 and 1926; Senior Officers' War Course, 1921, 1923-24 and 1927; Flag Captain and Chief Staff Officer in HMS *Courageous*, 1921-23; Commanding Officer, HMS *Benbow*, 1924-26; Rear Admiral, 1926; Commander, Third Battle Squadron, 1928-29; Vice Admiral and retired, 1931 and Admiral, 1936. Recalled for war service.
200 ADM 203/99.
201 ADM 196/90 (Casement).

and training were not ends to themselves but operated in the context of maintaining efficiency in the fleet and furthering individual officer development. Difficult objectives to achieve in the best of times, even more so in the environment immediately following the war as ships were paid-off, officers were reassigned after only nominal time afloat, and others retired whose services were no longer required.[202] The silver lining of such disruptions in directing staff assignment, if one can be claimed, is that instructors had perforce to know the broad themes of the curriculum and not be a narrow specialist lecturing to a pat syllabus.[203]

The assignment of officers to Greenwich was ever a problem for the Navy and even the more so when the surplus of officers available to attend a particular course diminished. Richmond's successors would experience the problem too and accepting the broader issue at play, it was ever a problem for any course. That said, the Admiralty did treat officer education as a weigh-station, at times, and this was surely frustrating to many of the staff at Greenwich with the case of Captain Hugh Shipway[204] being a particular egregious case. That officer passed through the SOWC no less than three times in four years, though his first attempt was the curtailed spring 1921 course interrupted by the national coal strike.[205]

Compounding the question was the matter of officer specialization, as certain billets had to be filled by officers of a given rank and technical ability. The paradox, of course, is that the more training is defined as a requirement before a position can be filled, the fewer will be available to actually hold any vacancy. This applies equally to filling staff and command positions as it does to any other specialist billet. When considerations of age, medical fitness, and rank are added, then the task of drafting becomes that much harder and whilst one can sympathize with Richmond, the issue is never an easy one to balance and his criticism of Admiralty policy needs to be considered in that light.[206] The conundrum goes far to explain why the Admiralty was reluctant to require completion of the Staff Course before posting officers to administrative staffs.

202 Many instances can be cited where officers were reassigned following nominal service in a ship. Amongst such officers were Lieutenant John Harding, commanding *P 39* from 22 August 1918 until transferred to the *Queen Elizabeth* in December 1919, Lieutenant Russell Grenfell, commanding HMS *Renard* from 28 September 1918 until reassigned to *Cumberland* on 1 February 1919 and Lieutenant Eric Brind who served as Gunnery Officer in the monitor HMS *Sir John Moore* from 8 Oct 1918 until January 1919 when placed on half-pay.
203 ADM 196/91 (Laurence).
204 Later Rear Admiral Hugh Skinner Shipway (1880–1961). Captain, 1919; Senior Officers School, Woking, 1920; Senior Officers' Technical Course, 1921; Senior Officers' War Course, March and October, 1921 and 1925; Commanding Officer, *Dublin*, 1922-24; Captain-in-Charge Singapore in HMS *Hawkins*, 1925-27; Commanding Officer, *Malaya*, 1929-31; Captain-in-Charge, Portland Naval Base, 1931-32; Rear Admiral and retired, 1932. Recalled for war service.
205 ADM 196/45 (Shipway).
206 In an age when air conditioning in ships and shore establishment was absent, service in the tropics was a particular limitation for some officers owing to a pre-existing medical condition.

Unlike the Army where most officers attending the Staff College sat a series of examinations, naval officers destined for the Greenwich Staff Course secured admittance by application to the Admiralty with a covering endorsement from their commanding officer.[207] For officers of the Royal Marines, both the path and criteria for acceptance were slightly different. Their applications were forwarded to the Adjutant-General of their Corps and the minimum age of entry was initially set at twenty-five years. Still, whether a naval or marine officer, a preliminary examination was never a bar to entrance.[208]

The examination for Camberley was notoriously difficult and cramming for the papers required was a rite of passage for many an aspiring officer. Even top-flight officers were defeated in their attempts to master it with J. F. C. Fuller,[209] later to be a Chief Instructor at Camberley, even failing at his first try.[210] Competition was keen and though some officers did gain admittance through appointment, the normal path was through examination. General Sir Charles Harington passed his Camberley exam netting one of the six vacancies, but another 180 officers though qualifying did not.[211] In 1908, the future Field Marshal the Earl Wavell,[212] as a junior subaltern, applied enduring eighteen papers in the process. His name appeared top of the list for the year's applicants and the satisfaction of knowing he excelled above all others in such a grueling process can be but imagined.[213]

Such a system of matriculation worked for the British and Indian Armies where competition was keen and regimental duties in the absence of active service allowed one sufficient time to prepare. Indeed, an officer spending a year in rigorous preparation was not unknown and the case of the future Field Marshal Lord Harding[214] is illustrative. After some prompting by his brigade commander, Harding decided to

---

207 Still, the British Army reserved the right to nominate an officer to Camberley without passing the entrance examination and this procedure featured in the appointments of many cavalrymen. See Matthew Hughes, *Allenby in Palestine: The Middle East Correspondence of Field Marshal Viscount Allenby, June 1917–October 1919* (Stroud: Sutton Publishing, Ltd., 2004), p. 3.

208 'Naval Notes, Candidates for Staff Training', *Journal of the Royal United Services Institution*, Vol. LXXII, No. 487, August 1927, p. 657.

209 Major General John Frederick Charles Fuller (1878-1966). Staff Course Camberley, 1913-14 and Staff College Camberley, 1923-26.

210 Anthony John Trythall, *'Boney' Fuller: Soldier Strategist and Writer 1878-1966* (Baltimore: Naval & Aviation Publishing Company, 1977), p 26.

211 B. H. S. "'Tim Harington Looks Back.'" *The Naval Review*, Vol. XXIX, No. 2, May 1941, p. 365.

212 Archibald Percival Wavell, later Field Marshal Earl Wavell (1883-1950). Staff Course Camberley, 1910-1911, Senior Officers' Course Hythe, 1927 and Higher Commanders' Course Versailles, 1934.

213 Adrian Fort, *Archibald Wavell: The Life and Times of an Imperial Servant* (London: Jonathan Cape, 2009), p. 41.

214 Allan Francis John Harding, Field Marshal Lord Harding (1896-1989). Staff Course Camberley, 1928-29.

try for the Staff College and requested a year's home leave whilst serving in India. Returning to England, he took a crammer's course and then sat a fortnight's course provided by the Army that prepared officers for the examination. Still, notwithstanding this level of preparation, he, too, was an officer who only crossed the hurdle at his second attempt.[215]

Indeed, a particular duty of Wavell's whilst posted to Aldershot from 1930-31 was preparing officers for the military history portion of the Staff College examination. If not a gifted speaker, he, nevertheless, possessed a first class intellect and with his previous success in negotiating the exam was an excellent choice to help others do the same.[216] This was all possible because within the Army a culture of examining officers prior to promotion or appointment existed to a degree not present in the Navy where seniority and due regard for officer-like qualities were the norm.[217] Additionally, examining an officer prior to selection to the Staff College in the interwar Royal Navy was thought unworkable given the demands of shipboard routine, the constant absence from the home ports when engaged on foreign deployments, and the total lack of privacy allowing any sort of concentration of effort to be maintained by the would-be-applicant when underway. No doubt, this was all very true, but it does not explain why officers of the accountant branch were not thought similarly disadvantaged when preparing for their examinations leading to Lieutenant Commander.[218] Surely, the most plausible explanation is that the Navy was reluctant to risk selection to the results arising.[219] In the end, the Navy's method may have lacked the academic rigour of the Army, but it was not unique. Shortly after the World War, the Army initially opted to appoint rather than examine officers for entrance to the Staff Colleges and with so many skilled officers having served in a staff capacity during the war, it was recognition perhaps that experience had perforce to account for something.[220] Thus, it was not until February 1921 that the first examinations for applicant screening to Camberley and Quetta occurred. The Royal Air Force, too, elected to enter officer candidates by nomination rather by examination when its Staff College first formed in 1922, though in common with the Army, examinations soon became the norm.[221]

215 Field Marshal Lord Harding, Reel 10, Sound Recording 8736 of 1984, Imperial War Museum, London.

216 Wavell to Liddell Hart letter of 20 October 1930, LH 1/733/23, LHCMA.

217 First Principles, 'Specialists', *The Naval Review*, Vol. XXVII, No. 1, February 1939, p. 18.

218 *King's Regulations, Vol. II*, Part 9, p. 87.

219 The ability to confer patronage may have also been a consideration, but this observer has found no direct evidence that it operated in the selection officers to the Staff Course for many touted as benefiting from attending the Staff College waited years before sitting the course.

220 A. R. Godwin-Austen, *The Staff and the Staff College* (London: Constable and Company, Ltd., 1927), pp. 270-73.

221 Vincent Orange, *Park: The Biography of Air Chief Marshal Sir Keith Park, GCB, KBE, MC, DFC, DCL* (London: Grubb Street, 2001), p. 45.

Of course, one problem with relying on examinations is that a culture of cramming soon becomes endemic, as the fate of an officer's career becomes a zero sum contest against one's professional peers. Unsurprisingly, a cottage industry of firms specializing in preparing officers for these examinations arose. Whilst this no doubt ensured that the quality of papers finally set by the applicant approached a higher standard than if left to one's own efforts and reflected much keenness on the part of the aspirants, it surely must have come at the price of performing one's present duties. That some officers mastered the examinations rather than mastering the subjects examined is doubtlessly true as Fuller's memoirs confirms.[222] This, too, was a reason why the Navy was reluctant to rely on examinations in selecting its officers to attend its Staff Course. As for the entrance examinations employed by the Army and the Air Force, it is hard to escape the conclusion that any officer receiving full marks on the topics posed was in little need of actually attending the Staff College for mastery of the subject material would have already been demonstrated.

In retrospect, the Navy would have done itself little harm if it had required officers to sit an examination prior to attending and probably much good. Officers would have been forced to read and contemplate the issues before arriving, and even for those not selected, the preparation would have led to a better-informed officer. One indication that such a scheme was desirable was the failure of the Admiralty to receive for consideration even a single essay in Naval History when holding its yearly competition in 1931.[223] To win the Naval History prize, an officer needed to submit a 10-12,000-word essay on a topic specified in an Admiralty Fleet Order. The successful officer received a special medal in recognition, £50 in prize money, and preferred consideration for attending the Staff College.[224] As some officers believed the status of the college was declining—of which, more anon—the lack of any entrant for the 1931 competition appears to offer *prima facie* evidence that they were not entirely incorrect. Still, by any measure, the absence of a single essay indicates that the spirit giving rise to the Naval Records Society and *The Naval Review* was now lagging within the Service.[225]

Students assembled at Greenwich somewhat slowly with the first intake of the new War Staff Course arriving in early June 1919 appearing only slightly after the

---

222 J. F. C. Fuller, *Memoirs of an Unconventional Soldier* (London: Ivor Nicholson and Watson Limited, 1936), p. 21.
223 'Naval Notes, Study of Naval History', *Journal of the Royal United Services Institution*, Vol. LXXVII, No. 506, May 1932, p. 414.
224 *King's Regulations, Vol. II*, Appendix II, p. 80.
225 Interestingly, Sub-Lieutenant Neil Kemp did win a Gold Medal in 1931 in the RUSI naval essay competition. Perhaps with a bit of irony, the tag RUSI adopted for that year's theme was 'Business as Usual', set against the forms of attack likely to be faced against British seaborne trade in a war with a European power. As other naval officers, such as Commander Ralph Binney, also submitted essays to the RUSI contest, the failure of any naval officer to offer a paper against the Admiralty's History Prize offers testimony how officers viewed the Navy's internal efforts at fostering critical discourse.

instructors who began arriving the previous January. The initial class was the smallest of the interwar period, not a surprising feature when it is recalled that new territory was being plowed and residual military operations remained to be supported. Affairs began with twenty qualifiers, of whom, fifteen were from the Navy, one coming from the Royal Marines and the Army and Air Force offering the balance.[226] Amongst this last group was Lieutenant Colonel Wilfrid Freeman,[227] destined to become the senior instructor of the Air Force Staff College when soon formed under Air Commodore Robert Brooke-Popham[228] at Andover.[229] With Freeman, who in time became a Commandant of the Air Force Staff College, were two promising naval officers: Victor Danckwerts, a future Director of Plans who would stay on at Greenwich after completing his course, and Commander Robert Fitzgerald, later seconded to oversee the South African Naval Service.[230] Present, too, was Lieutenant Commander Tom Phillips,[231] another future Director of Plans and who later found fame of another sort. It remained, though, for Commander John Tovey to achieve the greatest fame of the course's first qualifiers, eventually becoming Admiral of the Fleet Lord Tovey. Greatest, though not it be must said highest, for Lieutenant Frederic Peters,[232] though missing two months of the course in 1919 upon the death of his father and soon to retire shortly after gaining his *ws* designation because of poor health himself, earned a posthumous Victoria Cross in 1943.[233]

Under ideal circumstances, whether an officer of the interwar Royal Navy entered the Service through the Special Entry programme reserved for applicants of Britain's public school system, came as a Direct Entry applicant of the Royal Naval Reserve, or began his career as a cadet at either the Osborne or Dartmouth Naval Colleges at 13

226 ADM 203/100.
227 Later Air Chief Marshal Sir Wilfrid Rhodes Freeman (1888–1953). Staff Course Greenwich, 1919-20; Air Force Staff College, 1922-25 and Commandant, Air Force Staff College, 1933-36.
228 Later Air Chief Marshal Sir Henry Robert Moore Brooke-Popham (1878-1953). Staff Course Camberley, 1910-12; Commandant, Royal Air Force Staff College, 1922-26 and Commandant, Imperial Defence College, 1931-33.
229 Orange, *Park*, p. 45.
230 *Navy List*, July 1920, p. 1865c.
231 Later Acting Admiral Sir Tom Spencer Vaughan Phillips (1888-1941). Five firsts in examinations for Lieutenant. Staff Course, 1919-20; Plans Division, 1923-24; Staff Officer Operations in *Queen Elizabeth* and *Warspite*, 1925-28; Captain, 1927; Tactical Course, 1928; Assistant Director, Plans Division, 1930-32; Flag Captain in *Effingham* and *Hawkins*, 1932-35; Director, Plans Division, 1935-38; Rear Admiral, 1939; DCNS and Vice Chief of the Naval Staff, 1939-41; Acting Vice Admiral, 1940; Acting Admiral, 1941; Commander-in-Chief, Eastern Fleet and lost in *Prince of Wales*, 1941.
232 Later Acting Captain Frederic Thornton Peters (1889-1942). War Staff Course, 1919-20; Assistant War Staff Officer in *Queen Elizabeth*, 1920 and retired, 1920. Recalled for war service. Acting Captain, 1942; killed on active service, 1942 and subsequently awarded Victoria Cross, 1943.
233 ADM 196/144 (Peters) and ADM 196/52 (Peters).

or 14 years of age should have made little difference on their potential to excel at the Staff College.[234] In the initial period following the war, the minimum age of entering the course was 34 rising to 35 years beginning in September 1923.[235] Thus, notwithstanding the manner of entry, by the time one came to take the Staff Course at least 12 years service as a commissioned officer would have been completed. Those benefiting from a public school education were perhaps better rounded in their outlook, but those officers following the typical path of Dartmouth arrived with more experience of the Service and were academically just as capable. The relative merits of early or late entry was a heated subject for much of the period and the strongest claim to accepting candidates at such a young age was that the field was clear of competition as Army and Air Force officer entrants began their careers typically when eighteen or even later. Engaging their officer cadets at thirteen also posed another set of issues affecting inter-Service relations. Leaving Dartmouth at around seventeen, the typical naval officer could expect to be made a Sub-Lieutenant by twenty and promoted to Lieutenant at about twenty-one or twenty-two. This rankled the other establishments whose officers of similar grade were typically much older and became an acute issue when the Navy appointed Lieutenants as observers who often outranked their RAF pilot but could be several years younger.[236] This anomaly also played its part when it came time for the naval officer to attend the schools of the other Services, as will be later shown.

Of course, an immediate problem for the Navy was that the young naval officer of 1919 had not followed a normal path to commissioning, as they left Dartmouth early with the onset of war to join the fleet. Consequently, their formal education was abbreviated. Their command of English usage was a particular cause for concern, as the War Staff Course dealt, in part, with how to prepare orders, signals, and appreciations.[237] A poorly crafted order might lead to grief in battle, and, sometimes, rather than correct the immediate error, if, for example, occurring during play at the war game table, the directing staff allowed the order to stand thus demonstrating the peril—experience, again, proving to be the best teacher.

Aware of any academic shortfall qualifiers as a group might possess, the Navy could modify instruction at Greenwich as necessary. However, the lack of general academic preparation and a Service schooling emphasizing the technical minutiae of their calling were not the best grounding when selected to attend the Imperial Defence College where their peers from the Army and Air Force began their careers later after completing studies at a public school. Tellingly, most naval and marine officers attending the IDC were graduates of the Staff Course, the Senior Officers'

---

234 In 1921, Osborne closed with its cadets now attending an expanded Royal Naval College Dartmouth.

235 'Naval Notes, Navy War Staff Course', *Journal of the Royal United Services Institution*, Vol. LXVIII, No. 470, May 1923, p. 334.

236 CAB 24/167, Secretary of State for Air to Cabinet memorandum C.P. 4349 of 7 December 1922 and Secretary of State for Air to Cabinet memorandum C.P. 359/24 of 23 June 1924.

237 Anon.,'Royal Naval Staff College', *The Naval Review*, Vol. XX, No. 1, February, 1932, p. 25.

War Course, or both, and an officer studying there who had not, such as Captain Alfred Collinson[238] in 1937, was an exception to this preparatory process.

Indeed, Collinson presents the case of an officer attending the Senior Officers' War Course *after* graduating from the IDC, as he only proceeded to the War Course in 1939. This, though, was unusual as some officers did sit the SOWC multiple times. Admiral Sir Herbert Fitzherbert,[239] demonstrates a case in point, attending the IDC in 1931 sandwiched between sitting War Courses in 1925 and 1937. Meanwhile, Captain William Chalmers,[240] a future Director of the Staff College, took his War Course in 1932 after securing his *idc* in 1928. Admiral Dickens presents a slightly different story. After serving as the first naval member of the directing staff of the Imperial Defence College from 1926-29, he went to Greenwich in 1931 to complete the Senior Officers' War Course. Whether it was a policy of design to guarantee that at least one member of a War Course class had attended the IDC previously is unknown, but the occurrence was common enough and was in keeping with the common practice of the other Services that doctrine be harmonized amongst the several institutions of instruction. The practice also ensured that naval officers not selected to attend that college became familiar with its teachings through contact with students and faculty who had.[241]

Of course, another reason why an officer might sit a course multiple times was that he was adjudged to have failed the requisite standard. As the courses were deemed mandatory for those of Captain's rank, having an officer retake a session made perfect

238 Captain Alfred Creighton Collinson (1893–1970). Commanding Officer, HMS *Torch*, 1924-25, HMS *Stormcloud*, 1925 HMS *Westcott*, 1925-26 and HMS *Wolverine*, 1927-29; Tactical Course, 1929-30; Training and Staff Duties Division, 1933-35; Captain, 1935; Chief Staff Officer in HMS *Resource*, 1936; Senior Officers' Technical Course, 1936; Imperial Defence Course, 1937; Commanding Officer, *Centurion*, 1938-39; Senior Officers' War Course, 1939; Tactical Investigation Section of Tactical School, 1939 and retired, 1945.

239 Admiral Sir Herbert Fitzherbert (1885–1958). Signal School, 1914; Flag Lieutenant in *Iron Duke*, 1914-16; Flag Lieutenant Commander in *Marlborough* and *Revenge*, 1916-17; Captain, 1924; Flag Captain in *Iron Duke*, 1922-23; Senior Officers' Technical Course, 1925, 1929 and 1936; Senior Officers' War Course, 1925 and 1937; Tactical Course, 1925 and 1936; Flag Captain and Chief Staff Officer, HMS *Coventry*, 1926-27; Naval Assistant to First Sea Lord, 1929-31; Imperial Defence Course, 1931; Royal Naval War College, 1931-32; In Command, Signal School, 1932-34; Rear Admiral, 1936; Flag Officer Commanding, Royal Indian Navy, 1937-43; Vice Admiral, 1939; retired and Admiral, 1943.

240 Later Rear Admiral William Scott Chalmers (1888-1971). War Staff Duties in Grand Fleet, 1916-19; Designated for War Staff, 1918; Royal Naval Staff College, 1921-23; Training and Staff Duties Division, 1923-25; Captain, 1927; Imperial Defence Course, 1928; Tactical Course, 1929; Flag Captain and Chief Staff Officer in HMAS *Australia*, 1929-32; Senior Officers' War Course, 1932; Royal Naval War College, 1932-34; Senior Officers' Technical Course, 1934; Commanding Officer and then Flag Captain and Chief Staff Officer in HMS *Delhi*, 1934-36; Director, Royal Naval Staff College 1937-39; Rear Admiral and retired, 1939. Recalled for war service.

241 CAB 53/5, Chiefs of Staff Sub-Committee minutes of 1 November 1934.

sense if he was deemed a capable officer in most other respects. Captain Edward Thornton,[242] a salt horse through and through, demonstrates this side of naval education and training. Attending the SOWC in 1934 he received a poor report on his progress and took the course again the following year.[243]

As the first course of the IDC only began in 1927, it remains that some naval and marine officers earmarked for study did so without previously passing through the Greenwich preparatory courses. This, though, was more a question of timing as the officers in these instances, usually, in the rank of Captain or Colonel, belonged to that cadre whose presumed time of study were affected by the exigencies of war. Captain Andrew Cunningham attending in 1929 is representative of such an entrant to the IDC.

Carp as Richmond may about the manner in which higher level education and training were being initiated in the period immediately following the war, it is this observer's view that the Admiralty deserves recognition for the measures adopted. In selecting two founding members of *The Naval* Review, Rear Admiral Richmond and Captain Drax as President and Director, respectively, of the War and Staff Colleges, it demonstrated a seriousness of purpose in establishing instruction along sound lines. Richmond's reputation for serious naval research was even then a feature of his character, whilst his time at the Admiralty during the war made him cognizant of the requirements of the higher education desired in officers and the training required of those destined for staff assignments, the evolving nature of the staff system itself, and the resulting improvements in administration.[244]

Drax, too, was an extremely intelligent officer well prepared for the duties assumed. Receiving five first class passes in his examinations leading to Lieutenant, he possessed an original and inventive mind.[245] Trained in the theory of staff procedure through his time at Camberley and completing the Navy's prewar Staff Course, he brought direct experience of staff work in the fleet with his time in *Lion* and the Battle Cruiser Fleet during the war. Paired with Bigg-Wither, another officer coming from the Training and Staff Duties Division of the Naval Staff, the Admiralty was laying a serious foundation for the study of war. Furthermore, Altham's presence at the War College

242· Later Captain Edward Chicheley Thornton (1889-1959). Commanding Officer, HMS *Romola*, 1918-19, HMS *Raider*, 1920, HMS *Vancouver*, 1921, HMS *Vivacious*, 1921, HMS *Forres*, 1923-24 and HMS *Carstairs*, 1924-25; Staff Officer Operations and Intelligence in HMS *Pembroke*, 1926-28; Short Anti-Submarine Course, 1928; Commanding Officer, HMS *Vidette*, 1928, *Thruster*, 1928-30, HMS *Wryneck*, 1930-32 and *Duncan*, 1933; Captain, 1933; Tactical Course, 1934; Senior Officers' Technical Course, 1934; Senior Officers' War Course, 1934-35; Senior Officers' School Sheerness, 1936; Commanding Officer, HMS *Milford*, 1936-37 and HMS *Woolwich*, 1938-39 and retired, 1943.

243 ADM 196/92 (Thornton).

244 Diary entry of 25 May 1918, Churchill Archives Centre, the Papers of Lord Hankey of the Chart (Maurice Hankey), HNKY/1/3.

245 Robert L. Davison, 'Striking a Balance between Dissent and Discipline: Admiral Sir Reginald Drax,' *The Northern Mariner*, Vol. XIII, No. 2, April 2003, p. 44.

proved more than a fortuitous choice, bringing direct experience of events in Russia and the Baltic where British naval operations had assumed significant proportions with nearly ninety warships engaged.[246] If the charter to examine contemporary issues featured in the War College's aim, his posting to Greenwich could be little bettered.

Sustaining this foundation was another matter. The short tenures for some holding the positions of Admiral-President or Director of the Staff College were not conducive to establishing a consistent academic regime. To wit, the period of Captains William James[247] and Geoffrey Blake[248] as Directors is notable for the brevity of their appointments. James departed Greenwich in January 1926 and entered a period of unemployment before joining the battleship HMS *Royal Sovereign* as its commanding officer. His early departure saw Blake assume the Directorship of the Staff College, though he had only recently assumed the Deputy Director's chair vacated upon James' recent raising to Director. Moreover, Blake's appearance at the Staff College came after serving only briefly as a member of the War College directing staff. Both officers rose to flag rank, but the time of James coming after the tenure of the very forceful, capable Astley-Rushton, another officer recently of the Training and Staff Duties Division, was not his best appointment lasting but five months before giving way to the period of Blake, a person of which more remains to be said.[249] However, others reaped where Richmond and Drax sowed and those following in their steps often had been students of the Staff Course, the War Course and, increasingly, the Imperial Defence Course.

246 Kinvig, *Churchill's Crusade*, p. 289.
247 Later Admiral Sir William Milbourne James (1881-1973). Five firsts in examinations for Lieutenant; Flag Commander in *Benbow*, 1916 and in *Hercules*, 1916-17; appointed War Staff Officer, 1917; Captain, 1917; Naval Intelligence Division, 1917-20; Senior Officers' War Course, 1923; Deputy Director, Royal Naval Staff College, 1923-25; Director, Royal Naval Staff College, 1925-26; Senior Officers' Technical Course, 1926; Flag Captain and Chief Staff Officer in *Royal Sovereign*, 1926-27; Naval Assistant, First Sea Lord, 1927-29; Rear Admiral, 1929; Chief of Staff, Atlantic Fleet, 1929-30 and Mediterranean Fleet, 1930-31; Tactical Course, 1932; Commander, Battle Cruiser Squadron, 1932-34; Vice Admiral, 1933; DCNS, 1935-38; Admiral, 1938; Commander-in-Chief, Portsmouth, 1939-42 and, retired, 1942. Recalled for war service.
248 Later Vice Admiral Geoffrey Blake (1882-1968). Captain, 1918; Naval Attaché, Washington, DC, 1919-21; Commanding Officer, *Queen Elizabeth*, 1921-23; Royal Naval War College, 1923-25; Deputy Director, Royal Naval Staff College, 1925-26; Director, Royal Naval Staff College, 1926-27; Chief of Staff, Atlantic Fleet, 1927-29; First Naval Member, New Zealand Naval Board, 1929-32; Rear Admiral, 1931; Fourth Sea Lord, 1932-35; Tactical Course, 1935 and 1936; Vice Admiral, 1935; Vice Admiral Commanding, Battle Cruiser Squadron, 1936-38 and retired, 1938. Recalled for war service.
249 ADM 196/90 (James).

William Chalmers was one such officer but so, too, was Robert Raikes[250] and Charles Kennedy-Purvis.[251]

In many respects, Rushton was in a class by himself. Of brilliant intellect and fearless in execution his greatest weaknesses were thought a certain lack of judgment and a lack of sympathy with the lower deck seeing them as but part of a machine. If not the best ship handler, he possessed élan and was thought the ideal seagoing officer who would face any task and see it through to a successful conclusion. Held to be supremely suited to be another Admiral-President of Greenwich, perhaps Rushton would have been save for the motoring accident that claimed his life in July 1935. Maybe the verdict on judgment was not far off the mark after all.[252]

Typically, an officer served as Director of the Staff College for 24-30 months before leaving for his next assignment. Chalmers retiring directly from the post in January 1939 is the exception. Meanwhile, Chalmers' successor, Captain Richard Scott,[253] proved the Director with the briefest tenure for the period under review lasting but nine months. Yet, allowances in his case are permitted. Arriving at his post in January 1939, the school closed with the onset of war in September and Scott became Commodore in the cruiser HMS *Colombo*.[254]

If in employing the facilities of Greenwich rather than a Camberley setting an educational Valhalla was missed, the reasons were mostly fiscal and not fundamental. True, some worried that establishing the War and Staff Courses at a site removed

250 Later Admiral Sir Robert Henry Taunton Raikes (1885–1953). Naval Equipment Department, 1919-20; Captain, 1923; Commanding Officer, HMS *X 1*, 1923-25; Senior Officers' War Course, 1927 and 1932; Imperial Defence Course, 1928; Commanding Officer, HMS *Sussex*, 1929 and HMS *St Vincent*, 1930-32; Director, Royal Naval Staff College, 1932-34; Chief of Staff in *Resolution*, 1934-35; Rear Admiral, 1935; Rear Admiral Alexandria, 1935; Rear Admiral Commanding Submarines, 1936-38; Vice Admiral, 1939; Senior Officers' Technical Course, 1939; Commander-in-Chief, South Atlantic Station, 1940-41; Admiral and retired, 1942. Recalled for war service.
251 Later Admiral Sir Charles Edward Kennedy-Purvis (1884-1946). Five firsts in examinations for Lieutenant; Captain, 1921; Commanding Officer, *Concord*, 1926-27; Director, Signal Department, 1927-30; Senior Officers' War Course, 1930; Tactical Course, 1930; Senior Officers' Technical Course, 1931; Commanding Officer, *Glorious*, 1931-32; Imperial Defence Course, 1933; Rear Admiral, 1933; ACNS, 1934-36; Commander, First Cruiser Squadron, 1936-38; Vice Admiral, 1937; Admiral-President, Royal Naval College Greenwich, 1938-39; Commander-in-Chief, America and West Indies, 1940-42; Admiral, 1942; Deputy First Sea Lord, 1942-45.
252 ADM 196/90 (Astley-Rushton).
253 Later Rear Admiral Richard James Rodney Scott (1887–1967). Commanding Officer, HMS *Lupin*, 1918-19 and HMS *Myrtle*, 1919 including her loss on active service; War Staff Course, 1924-25; Training and Staff Division, 1925-26; Tactical Course, 1929, 1931 and 1938; Captain, 1929; Tactical Division, 1931-33; Senior Officers' Technical Course, 1935; Captain of the Fleet, Mediterranean Fleet, 1936-38; Senior Officers' War Course, 1938-39; Director, Royal Naval Staff College, 1939; Commanding Officer, *Colombo*, 1939-40; Rear Admiral, 1940 and retired, 1940. Recalled for war service.
254 ADM 196/92 (Scott).

from a traditional naval setting might lead to unorthodox teachings, but locale is no guarantee of convention.[255] Learning is more about creating a state of mind given to critical examination and Greenwich, poor physical facility though it may have been, was more than adequate to foster such a temperament. Within Richmond existed a trait never to accept less than perfection. No matter the assignment held, despondency soon arose as the reality of his setting and the limitations of his superiors became apparent. Unfortunately, more such disappointments awaited Richmond. For making every allowance for the limitations of superiors, finance too influenced what was obtainable. Thus, it is to the realities of naval finance and its influence on the period's naval education and instruction that this history now turns.

255 Anon.,'Royal Naval Staff College', *Naval Review*, February 1932, pp. 14 and 16.

2

Naval and Imperial Policies, Finance, and the Limits of Education

I would far rather spend money upon a real Naval Staff College than have another battle cruiser.[1]

Viscount Haldane, 1921

Armed with the best experience a military force can acquire—the experience of war—the Royal Navy might appreciate the ideal to be strived for in preparing and developing its executive officers, but other constraints presented themselves and influenced events. Chief among these were British and Imperial policies defining the objectives of statecraft, the place of the defence establishment, and finance. The Navy, operating within the framework of a British tradition of liberal democracy, was very much not the master of its own house. Others would define the ends of policy and the means afforded to the Service and the Admiralty, though an important agent in defining these ends, was, at best, a junior partner in the firm labeled Cabinet Government.

Ultimately, finance proved the decisive factor in determining the number of naval and marine officers educated to a higher standard and the manner of their training. In claiming this, the writer does not wish to imply that finance dictated the instructional content delivered nor those actually selected to direct the instruction. Yet, the hand of finance influenced the length, setting and numbers taught, and still more so, whether the officer so recently educated found an assignment equal to his training. The necessity to economise in all aspects of its operations was a regular feature of the Royal Navy during the interwar period until rearmament and expansion began anew during the middle part of the 1930s. Lectures by divisional officers on the need for internal economy in ships may have been received with scant enthusiasm, but for all that, they remained a regular part of a ship's routine and formed a part of midshipman's training.[2]

1   *The Hansard*, House of Lords debate, 4 May 1921, v. 45, cc. 151-80.
2   Tennant letter of 3 November 1921, Tennant Papers, NMM/TEN/6.

Economy, whether the need for it, the state of it, or the state's role in it, dominated the counsels of Cabinets throughout the period to a degree never seen before in Britain. This was the case whether the matter was faced by Coalition, Conservative, Labour or National Governments. Conservatives may have favoured a lower rate of taxation, Liberals may have desired higher levels of social spending, and Socialists may have argued for the greater involvement of the state in economic planning and the control of vital industries, but these were only differences in the scope of the economic argument and not in that argument's priority within British political discourse. Defence spending, as a component of total government spending fell, and the Navy would never see the halcyon days of the pre-1914 era repeated when its expenditure would consume over 25% of government outlays.[3] The irony is that as the Service's fortunes receded the level of overall government spending in the interwar period rose as a percentage of Britain's Gross Domestic Product.[4] Thus, it was not a case of electing not to spend. Rather, governments simply elected to spend their funds on other priorities, of which the portion of the budget directed towards paying down the massive debt incurred in waging the recent war was paramount. Thus, by 1925 fully 38% of government outlays were required to meet its debt obligations when in 1912 the figure stood at a more manageable, but, nevertheless, sizable 12.4% of spending.[5]

Britain's relative decline following the World War has been amply surveyed by more skillful writers than the present one and the reader desiring to know more are invited to turn to Paul Kennedy's *The Rise and Fall of the Great Powers* or Correlli Barnett's *The Collapse of British Power* for fuller treatments of a complex subject. Matters, though, striking an appearance of inevitability to the modern reader were not necessarily seen as such by contemporaries. Facing difficult and intractable issues, officials wrestled with their problems with a degree of hope and absent the knowledge of how things actually would turn out. Perhaps more could have been done and sooner to address its many pressing deficiencies, but a nation spending of 5.5% of its Gross National Product on defence, as Britain was in 1937, would not be thought shirking its strategic responsibilities to a modern liberal, democratic electorate.[6]

Liberals may have taken the lead in establishing a new set of priorities for government expenditure with their social programmes begun before the World War, but this reordering in priorities was endorsed by all the major political parties in light of the war. Liberals and Unionists, though disagreeing about reform of the House of Lords, the Welsh Church, or tariffs, could yet agree in 1918 on education, health, housing,

3   Paul M. Kennedy, *The Rise and Fall of British Naval Mastery* (London: The Ashfield Press, 1990), p. 272.

4   G. C. Peden, *The Treasury and British Public Policy 1906-1959* (New York: Oxford University Press, 2000), p. 5.

5   Ross McKibbin, *Parties and People: England 1914-1951* (Oxford: Oxford University Press, 2010), p. 44.

6   Paul Kennedy, *The Rise and Fall of the Great Powers: Economic Change and Military Conflict from 1500-2000* (New York: Random House, 1987), p. 319.

and taxation.[7] Congruency enough allowing a coalition forged by the necessity of war to continue into the troubled waters of peace. Indicative of the changing priorities afoot in the Britain of 1918 was whilst one committee addressed how to reduce the size of the Navy another body was proposing new avenues of social spending and this when the outcome of the war was still in doubt. Extension of the franchise was another new factor and whilst defence of the realm remained a core responsibility of government, few could argue that after four years of war it was a responsibility neglected unlike the many domestic concerns which appeared so pressing.[8]

In the period following the First World War, the need for retrenchment was recognised within the coalition headed by Lloyd George and the Admiralty both, as British financial balances were simply not sustainable in the long-term absent a rebalancing. The shape, form and role of the Royal Navy in the postwar environment was a question much discussed even prior to the war's ending and an internal committee under the Civil Lord of the Admiralty, Ernest Pretyman, and Admiral of the Fleet Sir William May began an examination along such lines early in 1918.[9] By that summer with the German High Sea Fleet no longer a consideration in their plans and with the other major powers either friendly, exhausted or both, the naval power of Britain reached its zenith. Yet, the need for economy was understood and given its apparent supremacy, the arguments following became a question of the risks to be accepted in the near-term and what tasks her forces should meet. The answer came in the form of guidance drafted by Hankey, the Cabinet Secretary, but reflecting the tenor of the Cabinet.[10] Issued in August 1919, this note stressed that in forming Service plans 'It should be assumed for framing revised Estimates, that the British Empire will not be engaged in any great war during the next ten years, and that no Expeditionary Force is required for this purpose.'[11]

In light of all that followed, the pronouncement of a 'Ten Year Rule' has been subject to much censure, but a large dose of this disapproval is merely being wise after the fact.[12] Barry Powers labels the formulation as 'infamous' and whilst it came to assume such a hue, as a general pronouncement on what tasks the forces of the state should align themselves towards, it made sense for its time.[13] Ferris has argued

<hr>

7   R. J. Q. Adams, *Bonar Law* (Stanford: Stanford University Press, 1999), pp. 275-77.

8   As a result of changes in the franchise introduced in 1918, over 20,000,000 people were eligible to vote compared to around 7,000,000 before the war. See A.J. P. Taylor, *English History 1914-145* (New York: Oxford University Press, 1965), pp. 115-16.

9   Roskill, *Naval Policy, Vol. I*, p. 103.

10   Peden, *Treasury and Public Policy* (New York: Oxford University Press, 2000), p. 171.

11   N. H. Gibbs, *Grand Strategy, Volume I, Rearmament Policy* (London: Her Majesty's Stationery Office, 1976), p. 3.

12   Unsurprisingly, the literature covering the Ten Year Rule is vast but worth considering are Christopher M. Bell, 'Winston Churchill and the Ten Year Rule, *The Journal of Military History*, Vol. 74, No. 4, October 2010 and John Robert Ferris, *Men, Money and Diplomacy* (Ithaca: Cornell University Press, 1989).

13   Barry D. Powers, *Strategy Without Slide Rule: British Air Strategy 1914-1939* (London: Croom Helm, 1976), p. 168.

that as a specific construct the 'Ten Year Rule' is more a formulation of those writing after the events in question than a precise piece of law and that as a construct it was rarely used by contemporaries.[14] While correct, politicians and mandarins of the day understood its import even if they rarely gave voice to it in public and here the workings of the Official Secrets Act played its part.[15] Though subsidiary operations in Europe remained, contemplation of another major conflict was not one favoured by the Commonwealth or the man on the Clapham omnibus alike. Certainly, this was the lesson of the Chanak Crisis coming soon after the 'War to End All Wars' had ended with the Coalition Government only appreciating the changed circumstances to its dismay.[16] To that extent, the Ten Year Rule reflected not only the fiscal imperatives of the moment but the general will of the governed.

Criticism of the rule, though, did not have to wait for a second war to arise before being voiced. In one of the first public affirmations of its existence, the Duke of Sutherland, then serving as the Under-Secretary of State for Air, reminded Their Noble Lordships of the rule's guidance when discussing the relative strength of French and British airpower, to which Haldane cautioned that such an axiom was not a measure supported by the recent past noting:

> Anybody who has followed, with any attention, the history of Europe for the last quarter of a century, and more, will see that the situation has varied from year to year. At times we have been on terms of great friendship with the Central Powers. At other times we have been on terms of great antagonism with them. We have gone from one side to the other, and we have done so because we could not help ourselves.[17]

Meeting the demands of the Treasury and supporting the Government's foreign policy concurrently, whilst attempting to treat personnel due for release equitably were not easy matters to achieve. Discontent was always a possibility, and the wonder is that the Navy largely succeeded in meeting its operational commitments with the minimum of disturbance faced. This success must have owed something to the Government's decision to provide a temporary bonus to men of all the armed forces in February 1919 with the promise of a further adjustment in pay.[18] Notwithstanding

14   Ferris, *Men, Money and Diplomacy*, p. 17.
15   Before 1948, Lord Hankey felt constrained about discussing the specific origins and ramifications of the 'rule' for this reason during public debates in the House of Lords. See Hankey to Liddell Hart letter of 23 September 1938, Liddell Hart Papers, LH 1/352/91, LHCMA.
16   The Chanak crisis arose in 1922 out of Turkish hostility to the continuing enforcement of the Treaty of Sèvres, the initial peace agreement reached between Turkey and the Allies. The crisis, residing outside the limited scope of this work is instructive for its bearing on the fall of the Coalition Government, imperial and foreign relations, and British civil-military relations. For background and treatment, the reader is directed to David Walder, *The Chanak Affair* (London: Hutchinson & Co., 1969).
17   *Hansard*, House of Lords debate of 21 March 1923, v. 53, cc470-511.
18   *C.B. 3027(A), Mutiny*, p. 15

the cuts in naval expenditure implemented through early 1921, more retrenchment was yet to come. Thus, an original target of £83,200,000 established by the Chancellor of the Exchequer for the financial year 1921-22 ultimately gave way to a budget of £75,986,141.[19] This change in fortunes owed much to the climate arising in the country for further government economies—a campaign against waste was then underway with manifest political overtones as parliamentary candidates supporting such cuts contested a number of by-elections—and a commission to suggest further reductions soon followed.[20]

Sir Eric Geddes, a former First Lord of the Admiralty, was tipped to chair a committee of the great and good to investigate public spending with an eye to effecting significant economies. Thus, was formed the Committee on National Expenditure in August 1921. Presently, serving as Minister for Transport, Geddes was fundamentally a businessman and to the world of business he would yet return. Even before his latest appointment, he was recommending economies in spending and pushing to relocate portions of departments from central London to Acton.[21] A protégé of Lloyd George and a mentor of the Chancellor, Sir Robert Horne, Geddes selection, sound on its own merits, showed the Government fighting to protect its political flank as much as attempting to find savings not already under consideration.

Whilst the cuts that followed addressed all aspects of expenditure, defence felt its effects disproportionately. Geddes submitted three reports proposing total savings of £87,000,000 against Supply Estimates totaling £528,000,000.[22] Of further interest to this study, Geddes recommended the creation of a Ministry of Defence—which if not accepted immediately—gave rise to a further committee sitting and recommending, *inter alia,* the creation of a joint staff college.[23] Still, that awaited future events and for the Navy, the cuts proposed were particularly severe. As the official recommending these reductions was the former political head of the Service, this perhaps made the pain the greater still. Nevertheless, the proposed cuts could be received with a measure of *sang froid* and in the *Hood* the wardroom dined with an axe, symbolic of the 'Geddes Axe' as the savings were quickly dubbed, above the mess table.[24]

For these officers and those serving elsewhere, the gallows humour displayed would give way to the cold reality of what the cuts portended and those serving in the

19   ADM 1/8598/14, First Lord to Cabinet memorandum F.C. 72 of 28 February 1921 and *Navy Estimates for the Year 1924-25* (London: His Majesty's Stationery Office, 1924), p. 7.
20   Peden, *Treasury and British Public Policy,* pp. 131-69.
21   Austen Chamberlain to Lloyd·George letter of 13 June 1921 cited in Lord Beaverbrook, *The Decline and Fall of Lloyd George* (London: Collins, 1963), p. 271.
22   Andrew McDonald, 'The Geddes Committee and the Formulation of Public Expenditure Policy, 1921-1922', *The Historical Journal,* Vol. 32, No. 2, p. 643.
23   J. F. C. Fuller, *Empire Unity and Defence* (Bristol: Arrowsmith, 1934), pp. 232-33.
24   Midshipman Journal entry of 6 April 1922, Vice Admiral Sir Robert Francis Elkins Papers, National Maritime Museum, Greenwich, United Kingdom, NMM/ELK/1.

ranks of Lieutenant through Commander need fear the most. Surplus officers were paid off and those remaining faced the prospects of spending longer time in rank as promotions were frozen. For Lieutenant Commander Ronald Trevor,[25] a submariner now serving in the battleship HMS *Centurion* with the Mediterranean Fleet, the reasons behind the pending retrenchment were understood. Writing to his parents, he thought the Geddes report entirely sensible, though it would undoubtedly hit some hard as promotions were deferred. Still, he believed a great deal was being wasted on dockyards and the like and only wondered that the effort at securing savings had not been pursued sooner.[26] Trevor, promoted to his rank in 1919, escaped the affects of any freeze in promotions announced in the near term, so perhaps he could afford to be sanguine about matters. However, he spent the next ten years in rank only becoming Commander upon his retirement from the Service in 1932.[27]

For ships serving beyond home waters, extending their commissions was an easy step to adopt and a ship's company could anticipate now finding themselves overseas for three years before returning to their appropriate manning depot.[28] Reducing crew sizes in the heavy ships of the fleet was another measure adopted. Accordingly, *Hood* saw her complement reduced by 18 percent, or 254 men, whilst *Royal Sovereign* saw her crew lessened by 175 men or 15% of its regular establishment.[29] Meanwhile, foregoing some evolutions until necessary, such as in providing naval gunfire support to military forces operating ashore, was an additional measure adopted to reduce the operating costs of the Service. In truth, between the financial years 1921-22 and 1924-25 where savings could be implemented in current operating expenses, reductions in the appropriate accounts within the *Estimates* followed.

The *Navy Estimates* were approved each year by parliament after the Service developed its spending proposals within the guidance provided by the Cabinet and the Treasury on the spending limits to be followed for the coming financial year.[30] If agreement could not be reached between the Treasury and the Admiralty, the First Lord, as the Minister responsible for naval affairs, had the duty to bring the matter to Cabinet and fight his corner. Cabinet decision was final as far as the Chancellor and the First Lord were concerned, but members of parliament frequently introduced amendments to the bill supporting the *Estimates* seeking to reduce a portion of the

25 Later Captain Ronald Adair Trevor (1889-1963). Commanding Officer, *B 1*, 1915, *C 10*, 1915-16, *H 8*, 1916-18, *H 28*, 1918-19, *J 1*, 1919-20; to RAN, 1919-20; Commanding Officer, *K 22*, 1923-25; Army Gas School Porton, 1925; Anti-Gas School in *Pembroke*, 1925-27; Tactical School, 1927-29; retired and Commander, 1932. Recalled for war service and retired with war service rank of Captain, 1946.

26 Lieutenant Commander Ronald Trevor letter to parents of 6 February 1922, Captain Ronald A. Trevor Papers, Imperial War Museum, London, IWM/66/35/1.

27 *The Navy List* (London: His Majesty's Stationery Officer, July 1932), p. 559.

28 ADM 1/8567/250, Admiralty Fleet Order M. 20265/22 of 24 February 1922.

29 ADM 1/8659/74.

30 The British financial year operated from 1 April to 31 March of the following year.

funds being sought for the coming financial year. Typically, the sums involved in such reductions were nominal and the measure was employed more with an eye to making a point on aspects of naval policy than with a desire to change fundamental naval policy. Individual parliamentarians could be extremely knowledgeable about the Service and this is not surprising given the number of members who were former naval officers. In the House of Commons, the interjections of a Kenworthy or a Carlyon Bellairs were frequently at variance with government policy. This was only natural, for though a member such as Stephen King-Hall aspired to neutrality, the notion of a cross-bencher void of partisan views if present in the House of Lords was an example largely alien in the lower house of the Palace of Westminster. Of Kenworthy, in particular, the interjections produced little by way of changes in policy and a solitary figure with a penchant for tilting at naval windmills is an apt conclusion. That he frequently bristled those who he desired to help doubtless did not assist matters, but parliamentary rules also played their role in how Kenworthy was perceived.[31] The primary means for initiating a discussion on naval policy was to raise an amendment against Vote A which supported the manning of the fleet seeking a reduction by a nominal amount. The amendment, thus, made no fundamental change to the Service, but it did provide scope for raising questions of broader policy such as the role and place of the Naval Staff and the Staff College. These actions were disdained by others as it appeared unseemly and ungentlemanly that a former naval officer should work against the apparent interests of his former Service.[32]

The *Navy Estimates* consisted of sixteen separate accounts, or 'Votes', allocating the monies and personnel not to be exceeded during the course of the coming fiscal year.[33] If the Navy found itself running out of financial authority during the year, a supplemental appropriation might be sought, but the normal course followed was to curtail operations such that the authorizations for the account or accounts affected were not exceeded. The Admiralty tracked the level of its obligations and routinely advised the fleet and shore establishments to limit ammunition and fuel expenditure or defer maintenance actions until a more fortuitous moment. Facing a programme of retrenchment following the World War, the *Navy Estimates* faced severe cuts and the Votes supporting new construction and manning were often the target. Conversely,

31   On Kenworthy's repellent personality, see Drage diary entry of 20 October 1927, Drage Papers, IWM/PP/MCR/99, Reel Three.
32   *Hansard*, House of Commons Debates of 24 March 1922, v. 152, cc831-97831.
33   In 1924, the accounts and their associated 'Vote' were: 'A, Average numbers borne'; '1, Wages'; '2, Victualling'; '3, Medical'; '4, Civilians'; '5, Educational Services'; '6, Scientific Services'; '7, Reserves'; '8, Shipbuilding and Repairs'; '9, Naval Armaments'; '10, Works'; '11, Miscellaneous'; '12, Admiralty Office'; '13, Half-Pay'; '14, Naval Pensions'; and '15, Civil Superannuation'. Minor changes in the purpose of some of the 'Votes' occurred during the period and, beginning in 1932, Votes 13 and 14 were referred to collectively under the nomenclature of 'Non-Effective Services' whilst in 1934 the Fleet Air Arm' was now covered by Vote 4.

funds supporting pensions, half-pay, and reserves saw increases, but a notable exception to this pattern were the funds for 'Scientific Services', falling under Vote 6, which rose from £262,886 in 1918-19 to £440,000 for 1924-25 and was primarily attributable to the realignment of chemical research from 'Naval Armaments' under Vote 9.[34] Appendix I provides the actual sums approved by the Government year on year to finance the Navy.

As personnel costs represented a significant portion of the *Estimates*, the savings enacted came primarily by reducing the civilian shore establishment and the officers and ratings borne. In 1921, the Service stood at 127,180 officers and ratings with £480,243 approved to fund the civilians employed in the ashore tail providing fleet support.[35] By 1924, the Service had 99,453 officers and ratings with £190,699 now allocated for civilian fleet support.[36] Such savings in the flesh of the Service were expressly made to protect the funds available for new construction.[37] Staffs afloat and ashore were reduced, ships operated with lesser complements, and funds to support education across the board were pruned.[38] These reductions as applied to staffs were severe, and, by 1926, a Naval Staff reduced to seven division (i.e., Intelligence, Plans, Operations, Trade, Training and Staff Duties, Gunnery and Torpedo) and employing 56 naval and marine officers existed.[39] This was further rationalized in 1929 when the Trade Division formed part of the Plans Division, the Gunnery Division was merged with Training and Staff Duties, and a Tactical Division incorporating the formerly separate Torpedo Division was created. A new Naval Air Division was formed and with the Intelligence Division and the Operations Division remaining unchanged, the Naval Staff was now comprised of six divisions.[40] The above can be contrasted to the time of the Armistice when the Naval Staff formed eleven divisions (i.e., Plans, Operations (Foreign), Operations (Home), Intelligence, Training and Staff Duties, Trade, Mercantile Movements, Troop Movements, Minesweeping, Anti-Submarine, Signals, Air, and Gunnery and Torpedo). Clearly, much was not now covered as before or only being met at a fleet headquarters or subsidiary establishment.[41] The actual numbers employed in the Naval Staff in 1918 are subject to dispute, but even if

34   *Navy Estimates 1924-25*, pp. 6-7 and 79b.
35   *Ibid.*
36   *Ibid.*
37   Admiralty letter No. 460 to Commander-in-Chief East Indies of 12 September 1925, Richmond Papers, NMM/RIC/7/3a.
38   *Ibid.*
39   'Empire Naval Policy and Co-operation,' Admiralty memorandum E. 95 of May 1926, Roskill Papers, ROSK/7/41.
40   ADM 182/55, Admiralty Fleet Order '3251—Naval Staff—Revised Peace Organisation', 28 December 1928.
41   ADM 1/8558/133, 'The Necessity of the Naval Staff under Peace Conditions 1919,' Admiralty memorandum dated 8 May 1919.

limited to the 129 members specified in the Admiralty Technical History discussing its formation and evolution, the reductions implemented were prodigious.[42]

Whilst a lower scale of operational tempo contributed to reducing the size of the Naval Staff, the fundamental reason was the need to secure financial savings. This much was made clear by the First Lord of the Admiralty in his supporting statement covering the 1923-24 *Estimates* advising:

> The actual numerical strength and grouping of the Staff has, however, been exhaustively reviewed by the Board before the commencement of each financial year, with a view to effecting the utmost economy consistent with the requirements of Naval Policy. As a result of the modifications so carried out, the numbers of the Naval Staff now stand at 60 as compared with 87 in 1921, 118 in 1920, 160 in 1919, and 336 in 1918.[43]

Though the *Navy Estimates* were quite detailed on the total personnel allowed by the Service on the active list and their associated allowances, further restrictions governing the types of officers and their respective ranks were regulated through Orders-in-Council.[44] Of course, one's time in rank, the minimum of sea duty as specified in *The King's Regulations and Admiralty Instruction's*, and the favourable report of one's superiors were required before advancement was conferred, but within these limits discretion as to the number of officers of each rank remained for the Navy to determine.[45]

Contemporary readers may sense a strong sense of *déjà vu* in witnessing an Admiralty struggling to find savings elsewhere in its budget to allow new construction to proceed apace whilst staying within the limits of the funds available. In the 1920s, complicating such prioritisations in funding was the need to replace ships owing to the wear and tear of war service and to match the newer designs of foreign powers in light of the treaty system now in place. Moreover, the creation of a third, independent arm, the Royal Air Force, meant that the Navy and Army were no longer the only defence establishments requiring sustainment. Whether the Royal Navy, if allowed to possess an organic air arm, would have developed it more effectively

42   Black, *Naval Staff*, p. 233. Professor Till cites the number employed on the Naval Staff in 1918 as approaching 330. See Geoffrey Till, 'Retrenchment, Rethinking, Revival, 1918-1919', in J. R. Hill, ed., *The Oxford Illustrated History of the Royal Navy* (Oxford: Oxford University Press, 1995), pp. 322-3. This writer's own review of the *Navy List* for January 1919 reaches a staff of nearly 500 officers without counting civil members assigned, nor those officers assigned to research positions nominally within the relevant Division of the Naval Staff.
43   ADM 1/8744/140, 'Command Paper 1818, Navy Estimates, 1923-1924, Statement of the First Lord of the Admiralty Explanatory of the Navy Estimates, 1923-1924', p. 7.
44   *The Navy List* (London: His Majesty's Stationery Officer, January 1920), p. 2293.
45   *King's Regulations*, Chapter VI, pp. 83-136.

than what transpired is unknowable, but given the many competing claims facing the Admiralty for funding, a higher allocation of resources for an air arm was by no means a certainty.[46] This writer believes that they would have done a better job, but cannot prove it for it remains that building more cruisers to meet its trade protection requirements rather than aircraft remained a distinct possibility.

Lest one believe that the 'Geddes Axe' was a one-off measure, it is beneficial to follow the story until the outbreak of the second war. This clearly shows that the proposals of the Committee on National Expenditure were only a foretaste of what was to follow. In the years 1925-27, the Royal Navy regained authority to expand to 101,916 effectives before another wave of retrenchment appeared.[47] Thus, by 1932 when a nadir was reached, the authorized strength for the Royal Navy and Royal Marines was set at 89,667.[48] Slightly more than the 88,000 officers and ratings that Geddes and his committee had recommended.[49] Of more relevance to this study, is the number of officers borne by the Service and, here, a different picture emerges. For the financial year 1924-25, the authorized number of commissioned officers stood at 5,148.[50] By 1927, the number of naval and marine officers had increased to 5,684.[51] These numbers continued to rise such that by 1930, the number of commissioned officers now stood at 6,145.[52] These totals, though, began to recede aided in large measure by a scheme announced by the Admiralty in 1929 allowing Lieutenant Commanders who were within five years of the compulsory retirement age and not actively engaged to retire early from the Service with full pension rights.[53] The effect was immediate. By 1932, there were 5,894 officers carried.[54] Falling to 5,699 in 1934, the number of naval and marine officers carried in the active force from this point increased, albeit slowly, until the rapid expansion evidenced upon the outbreak of hostilities in 1939.[55]

The above figures may be contrasted to those of 1914, the last year of peace when the number of active duty officers stood at 5,293.[56] Expansion with the onset of war was immediate and continuous and in November 1918 there were over 14,000 naval and marine officers, and this not counting those members so recently transferred to the Royal Air Force from the erstwhile Royal Naval Air Service.[57] Thus, once the immediate necessity for contraction had been realized in the 1919-22 period, the officer

46   Schofield, *British Sea Power*, p. 146.

47   *Navy Estimates 1930*, p. 8.

48   *Navy Estimates 1935* (London: His Majesty's Stationery Office, 1935), p. 8.

49   *Hansard*, House of Commons debate of 1 March 1922, v. 151, cc434-6.

50   *Navy Estimates 1924–25*, p. 10.

51   *Navy Estimates for the Year 1927* (London: His Majesty's Stationery Office, 1927), p. 10.

52   *Navy Estimates 1930*, p.12.

53   Scott, 'Retirement' *Naval Review*, p. 135.

54   *Navy Estimates 1935*, p. 12

55   *Ibid.*

56   Viscount Hythe, ed., *The Naval Annual 1913* (Portsmouth: J. Griffin and Co., 1913), p. 464.

57   Brassey, ed. *Naval Annual 1919*, p. 225

corps of the Royal Navy was afforded a degree of protection from the worst effects of financial stringency when compared to the Service's ratings, though promotion may have been longer coming or come not at all. The requirement for spending at least three years afloat in a ship of war in one's present rank of Lieutenant or Lieutenant Commander before promotion to Commander was doubtlessly a consideration in some officers' calculations whether to seek appointment to the Staff Course. With ships being mothballed or scrapped, the risk was ever present that a vacancy would not exist afloat and duty at a shore billet would follow, thus, delaying yet further the prospect for promotion. Commanders, of course, while only having to serve a year afloat in a warship before qualifying for Captain may appear somewhat better provisioned, though the dilemma for them was that fewer billets actually existed on the promotion ladder.[58]

An expedient adopted to mitigate the reduction in ships' companies was to increase the number of officers serving afloat, particularly in the heavy ships of the fleet. This trade-off explains the protection that the officer corps experienced relative to the total numbers carried in Vote A of the *Estimates* between the period 1924-25 and 1930. This practice, though, could not be sustained in the presence of further calls for economy. The 1929 Admiralty Fleet Order offering early retirement to Lieutenant Commanders promised to relieve the periods of unemployment those officers faced by reducing their total numbers, but that was only a part of the matter. The Admiralty had also to find savings in the manning requirements for the active fleet to meet the funding levels available.

A vicious circle had now been reached. The Admiralty and fleet minutes of the period record the surplus of officers afloat compared to previous manning practices. Overlooking that the surplus itself was a measure adopted, *inter alia*, as compensation for the reduced number of ratings borne, the Commanders-in-Chief of fleets were asked to propose further areas where officer reductions could be contemplated.[59] Noting that midshipman and ratings were better educated and trained than was previously the case, the Admiralty decided to eliminate the surplus of Lieutenants and Lieutenant Commanders afloat and sought to develop the leadership potential of junior officers and senior ratings.[60]

In announcing these manning changes, the sense that the Admiralty was trying to make the best of a bad situation is a difficult conclusion to avoid. Yet, frequently it is the case that the pressure of economy forces a change in method or manner which ultimately proves beneficial and may have not been realised otherwise. Benefiting from the improvements in education occurring in Britain as a whole, the ratings engaged

58   *King's Regulations*, Chapter VI, Paragraphs 259-60, p. 84.
59   ADM 1/8762/265, Commander-in-Chief Atlantic Fleet to Admiralty letter No. 1422/A.F.118 of 15 August 1929.
60   *Ibid*, Admiralty Fleet Order '2405.—Commissioned Executive Officers— Revised Complements and Amendment to Seaman Complements—REPORTS. (N.1904/32—28.10.1932)'.

between the wars were better prepared in basic skills than their prewar predecessors.[61] Confirming the soundness of the recent manning changes adopted, the Training and Staff Duties Division recorded in one 1934 report a most favourable example. During a practice shoot held not long after the changes were announced, a very successful demonstration against the *Centurion*, now configured as a fleet target ship, was made with a rating controlling the six-inch fire whilst working in difficult circumstances.[62]

Having a surfeit of officers afloat may have been adopted to protect the careers of many of those engaged during the 1914-1918 war, while also allowing a normal rate of accession to proceed for new entrants. As the number of ships maintained in the active fleet had been reduced though, these officers faced increasing periods of unemployment or half-pay. Promotion, tied as it was to assuming a position of greater responsibility, suffered as the billets to support advancement were simply not available. One consequence of it all was that increasingly officers were thought lacking in zeal. This was no doubt correct though maintaining drive when future prospects were poor, if not outright bleak, was in a sense understandable.[63] Additionally, the morale of officers suffered as their final pension was very much depended on their length of service whilst actively engaged.[64] Bad as all this was, though, another unintended consequence arose. Employing officers in a wider range of duties was seen to stifle the leadership development of senior ratings and resulted in even the most routine duties in a heavy ship now being directed by officers. Previously where a senior rating or junior officer had carried out certain tasks, an officer holding seniority one or two ranks higher might now perform them.[65] Capable as these officers may be, they messed and berthed separately from a ship's deck force, and, therefore, were removed from the men's social setting. Thus, when the forced economies announced in 1931 gave rise to the instances of ill-discipline in the Atlantic Fleet collectively known as the Invergordon Mutiny, senior ratings that might otherwise have influenced events to a more positive outcome eschewed taking action. Their lack of previous training and any opportunity of exercising command had not provided them the skills or temperament to forestall the mutiny. Such was the view of Chatfield serving as Commander-in-Chief, Mediterranean at the time and his view was echoed by other senior officers.[66]

As the cuts applied to all personnel of the Royal Navy and the mutiny was limited almost entirely to the Atlantic Fleet, other factors such as personality, disparities in

61   Fostering the improvement in scholastic achievement, the age of school leaving was raised to fourteen with the Education Act of 1918. See Taylor, *English History*, pp. 170-71.

62   ADM 186/323, *C.B. 3001, Progress in Naval Gunnery, 1934*, Admiralty, Training and Staff Duties Division, April 1934, p. 5.

63   John S. M. Mackenzie-Grieve, 'The Axe,' *The Naval Review*, Vol. XVII, No. 3, August 1929, p. 581.

64   Scott, 'Retirement', *Naval Review*, p. 135.

65   Mackenzie-Grieve, 'The Axe,' *Naval Review*, p. 582.

66   Christopher M. Bell, 'The Royal Navy and the Lessons of the Invergordon Mutiny,' *War in History*, Vol. 12, No. 1, 2005, pp. 80-84.

pay due to the date of one's enlistment, the length of a ship's commissioning, operational tempo, leave, proximity to the national press, and concerns of loved ones doubtlessly played a part. Still, central to events was the poor leadership demonstrated by officers on the scene and the passivity of ratings. Having a surplus of officers in the fleet did not in itself cause the events at Invergordon, but the scope of duties officers performed was germane. This cannot be separated from the numbers and types of specialist officers borne in ships, as well as the number of ships maintained in the active fleet, and, here, the size of the afloat staffs came in for particularly sharp criticism.[67]

The Admiralty formed several committees between the wars to investigate such pressing issues with Admiral Sir William Goodenough chairing one such body in 1925. In a reclama to the drafted report, the Directors of Naval Ordnance and of the Gunnery Division argued that British manning suffered already when set against Japanese ships of comparable classes. This was certainly the case when total shipboard complements were viewed between the two navies, but British ships did carry a greater ratio of officers to men in battleships and battle cruisers: 14.7 to 1 against 17.2 to 1.[68] In 1930, the manning required to support fleet visual and wireless signaling was the subject of another committee, in this instance, led by Rear Admiral Forbes. This writer has seen no evidence that a systematic culling of officers was ever attempted by the Admiralty whereby officers over-supplied in one branch of specialization, such as gunnery, were retrained to another proficiency such as signals, though individual officers did opt to do so.[69] Chatfield's predecessor as Commander-in-Chief of the Atlantic Fleet warned about the over-manning of officers in the fleet's heavy ships shortly before hauling his flag down in August 1929 and the relevant Divisions of the Admiralty already then were proposing a new reduced manning scheme for ships of the *Queen Elizabeth*-class.[70]

One expedient adopted with so many officers facing periods of unemployment was to recycle them through the higher courses available even if one may have only recently attended the class in question. This was a particularly attractive option if the measures of economy were viewed as only temporary and the officer was thought one of promise. Accordingly, some officers attended the Senior Officers' War Course, the Tactical Course or the Senior Officers' Technical Course more than once as a

---

67   R. P. Ernle-Erle-Drax, 'Naval Efficiency and Discipline', *The Naval Review*, Vol. XXXII, No. 4, November 1944, p. 302.
68   ADM 116/2284, Joint minute from Director of Ordnance and Director of Gunnery Division to Manning Committee Report.
69   ADM 1/8740/69, 'Report of W/T and V/S Organisation Committee 1930.'
70   ADM 1/8762/265, Director of Gunnery and Director of Ordnance Divisions minute of 24 November 1928.

means of keeping them employed.[71] Rear Admiral Arthur Bedford[72] first attended the War Course as a Captain in 1924 returning once more to take the course as a Rear Admiral in 1932 before assuming command of the Royal Indian Marine.[73] Captain Cyril Coltart,[74] a submariner, and Rear Admiral Harold Franklin,[75] a salt horse, were two other such officers. Coltart attended the Senior Officers' War Course in 1933 and then again in late 1937,[76] whilst Franklin sat the course as a Captain in both 1926 and 1932.[77] Admiral Sir Lionel Wells[78] affords a more interesting example, as he attended the War Course in 1927 and 1930 and then took the Tactical Course in both 1930 and 1935.[79] As Wells proceeded to the directing staff of the War College in 1930 and became Director of the Tactical School in 1935, attending these courses an additional time were also measures making him current with their present form of

71   Given the obfuscation employed by the Admiralty in describing the 'Tactical Course' some-times as the 'Senior Officers' Technical Course, Part II', it is not always readily apparent in the *Navy Lists* whether one is attending the SOTC or the Tactical Course. The best meas-urement remains an officer's record of service held within the ADM 196 series, but not all records have been released for some officers and for other officers remain closed entirely.

72   Later Vice Admiral Sir Arthur Edward Frederick Bedford (1881-1949). Captain, 1919; Senior Officers' War Course, 1924 and 1932; Rear Admiral, 1931; Chief of Staff Mediterranean Fleet in *Queen* Elizabeth, 1932-34; Commander, Royal Indian Marine, 1934-36; Vice Admiral, 1936 and retired, 1938.

73   *Navy List*, April 1924, p. 307 and *Navy List*, July 1932, p. 307.

74   Captain Cyril George Bucknill Coltart (1889-1964). Commanding Officer, HMS *Alecto*, 1926-27, HMS *Lucia*, 1929 and *Cyclops*, 1929-31; Operations Division, 1931-32; Captain, 1932; Tactical Course, 1932; Senior Officers' Technical Course, 1933 and 1937; Senior Officers' War Course, 1933 and 1937-38; Chief Staff Officer in HMS *Dolphin*, 1933; Commanding Officer, HMS *Glasgow*, 1938-39; Captain (S) Fifth Flotilla, 1939; Commanding Officer, HMS *Diomede*, 1940 and retired, 1942. Recalled for war service.

75   Rear Admiral Harold Gordon Cooper Franklin (1885-1957). Captain, 1925; Senior Officers' Technical Course, 1925, 1929 and 1934; Tactical Course, 1925 and 1929-30; Senior Officers' War Course, 1926 and 1932; Senior Officers' School Sheerness, 1926; Commanding Officer, *Emerald*, 1928; Assistant Director, Naval Equipment Department, 1930-32; Commanding Officer, *Exeter*, 1932-33; to Royal Australian Navy, 1934-36; Rear Admiral and retired, 1936. Recalled for war service.

76   ADM 196/127 (Coltart).

77   ADM 196/49 (Franklin).

78   Admiral Sir Lionel Victor Wells (1884-1965). Five firsts in examinations for Lieutenant; Squadron Gunnery Officer in *Hood*, 1921-23; Senior Officers' Technical Course, 1923; Captain, 1924; Ordnance Committee, 1925-27; Senior Officers' War Course, 1927 and 1930; to New Zealand Division Royal Navy, 1927-30; Tactical Course, 1930 and 1935; Royal Naval War College, 1930-33; Commanding Officer, HMS *Eagle*, 1933-35; Director, Tactical School, 1935-37; Rear Admiral, 1936; Commander, Third Cruiser Squadron, 1937-39; President, Committee to Revise Battle Instructions, 1938; Vice Admiral, 1939; Vice Admiral, Aircraft Carriers, 1939-40; Admiral and retired, 1943. Recalled for war service.

79   ADM 196/49 (Wells).

instruction.[80] Another officer attending the Tactical Course twice was Rear Admiral Ion Tower,[81] sitting the class in 1930 and then again in 1937. As with Lionel Wells, Tower became the Director of the Tactical School in 1937 assuming responsibility from the former.[82] Thus, a reasonable conclusion is that by this point in time, attending the Tactical Course again was becoming a *pro forma* matter before overseeing the school's instruction.

The practice of having officers repeat a particular course did not meet with universal approval, as it was deemed wasteful of public funds when economy was the watchword and funds spent needlessly in one area were funds potentially available to meet unmet obligations in another. Thus, did one author writing in *The Naval Review* deplore the excess of executive officers currently borne in the Service, especially in its heavy ships and shore establishments. Accepting that a surplus of such officers made perfect sense as a matter of normal policy, the practice was not sustainable in what were clearly abnormal times with dire financial circumstances afoot. Indeed, the author's *bete noire* was the Senior Officers' Technical Course and the habit of passing officers through it multiple times.[83]

This charge was not without foundation for when Rear Admiral Bertram Ramsay,[84] only recently promoted to flag rank, found his appointment as Chief of Staff to the Commander-in-Chief, Home Fleet prematurely terminated and without subsequent employment he sought to forestall retirement by seeking to attend another round of senior officers' courses.[85] Thus, whilst fully accepting the above critic's concern

---

80   *The Navy List* (London: His Majesty's Stationery Office, October 1930, p. 444 and *The Navy List October 1935* (London: His Majesty's Stationery Office, 1935), p. 288.

81   Rear Admiral Ion Beauchamp Butler Tower (1889-1940). Five firsts in examinations for Lieutenant; War Staff Course, 1923-24; Air Ministry, 1924-26; Aviation Course, 1925; Naval Air Section, 1926; Captain, 1929; Tactical Course, 1930 and 1937; Senior Officers' Technical Course, 1932-32; Senior Officers' War Course, 1932; Imperial Defence Course, 1933; Flag Captain in HMS *Kent* to Commander-in-Chief China, 1934-36; Assistant Director, Tactical School and Director, Tactical School, 1937-38; Commanding Officer, *Malaya*, 1938-39; Rear Admiral, 1940 and Planning Committee in Imperial Defence College until killed in air raid.

82   *The Navy List October 1937* (London: His Majesty's Stationery Office, 1937), p. 286.

83   Anon., 'The Urgent Need for Economy', *The Naval Review*, Vol. XIX, No. 4, November 1931, p. 663-70.

84   Later Admiral Sir Bertram Home Ramsay (1883-1945). War Staff Course, 1913-14; War Staff duties in HMS *Orion*, 1914; Flag Lieutenant and War Staff Officer in *Marlborough*, 1914-15; Commanding Officer, HMS *Broke*, 1917-19; Flag Commander, HMS *New Zealand*, 1919-20; Captain, 1923; Senior Officers' War Course, 1924 and 1927; Senior Officers' Technical Course, 1924 and 1937; Commanding Officer, *Danae*, 1925-27; Royal Naval War College, 1927-29; Flag Captain, *Kent*, 1929-31; Imperial Defence College, 1931-33; Commanding Officer, *Royal Sovereign*, 1933-35; Rear Admiral, 1935; Chief of Staff, Home Fleet, 1935; Tactical Course, 1938 and retired, 1938. Vice Admiral, 1939. Recalled for war service. Admiral, 1944 and killed in air crash, 1945.

85   ADM 196/48 (Ramsay).

that some passed through classes as a sop to avoiding periods of unemployment, it remained, nevertheless, a reasonable practice. Not every officer was able to attend the War Course within the six-months' window prescribed following completion of Part I of the SOTC.[86] In those instances, the dilemma was to either have the officer take the War Course and perhaps not derive full benefit through their lack of currency in fleet procedures, or take the course again at the nominal expense of attending another sitting. Having officers attend the courses more than once was also a reflection that learning was more than just a rote process of box ticking but a continuous process of professional development if an officer chose to approach it as such.

Of the officers assigned to direct the Senior Officers' War Course and the Staff Course throughout the interwar period, the numbers were largely steady. Beginning with a cadre of four and nine officers for the War College and Staff College, respectively, each school lost a single officer in 1926. The *Navy Estimates* for the year avow that the savings were made in order to secure new furniture for the Greenwich College and whilst this may be the specific case, it remains that the savings were not restored in the years following.[87] Specifically, for the Staff College, the position of Assistant Director ceased. Moreover, the marine billet assigned to instruct the dedicated Intelligence Course no longer featured as a separate line item in the *Estimates*, though the position was retained in practice. The Intelligence Course, was not so fortunate, however, being eliminated with portions of its curriculum now added to the Staff Course to allow those destined to serve as a Staff Officer Intelligence to secure at least a rudimentary exposure to forthcoming duties.[88]

From 1926, the funds allocated to the War and Staff Colleges remained constant with only marginal differences noted from year on year. Appendix II and Appendix III provide the actual sums sought by the Admiralty in each financial year. The sums depicted cover the personnel costs of the naval officers and civil servants supporting the schools and other nominal overhead charges such as insurance for the staff. They do not include the ancillary costs covered through other funds allocated to the hosting station, the Royal Naval College Greenwich, for common services such police and messing nor met under other Votes of the *Navy Estimates* for building and property care and maintenance.

One conclusion is that the savings achieved in 1926 and never restored were necessitated by the creation of the Imperial Defence College. Founded in the middle of the 1926-27 financial year, the Navy simply reduced its Greenwich expenditure and applied it to the IDC to fund its share of the teaching costs. As the first Commandant of the IDC was a naval officer, its share of the initial start-up costs was proportionally greater than those faced by the Army and Air Force as the officers comprising the

---

86   ADM 182/31, Admiralty Fleet Order 4004 of 9 December 1921.
87   *Navy Estimates for the Year 1926* (London: His Majesty's Stationery Office, 1926), p. 68a.
88   ADM 182/42, Admiralty Fleet Order '2612—Intelligence Course—Abolition', 11 September 1925.

directing staff and the Commandant were paid by their parent establishment. The funds assigned to 'Educational Services' within the *Navy Estimates* were rearranged in 1927, as the instructors and schoolmasters serving afloat and overseeing the education of midshipman, cadets and boys were now funded through Vote 1 rather than Vote 5. This accounting change saw the 1926 allocation of £326,800 for 'Educational Services' reduced to £240,700 for 1927.[89] As the funding provided under Vote 1 in 1927 actually decreased upon the sums provided in 1926—even in view of the accounting transfer noted—the shortfall is accounted for by a reduction, again, in the number of effectives allowed.[90] In this case the, educational area was particularly disadvantaged as the number of commissioned instructors and schoolmasters now carried by the Navy fell from 90 and 205 in 1926 to 53 and 54, respectively, in 1927.[91] Of significance to this study, is that faced with making substantial cuts in its educational programme in the 1926-27 period, the Navy largely protected its higher educational efforts at the Staff and War Colleges at the expense of the afloat training provided to its midshipmen, cadets, and boy seamen.

Additional savings were achieved by paring the number of courses conducted and representative of this type of retrenchment is the Senior Officers Technical Course. The SOTC, it will be recalled, was a preparatory class for officers attending the War Course, and typically met four times a year at Portsmouth in classes composed of approximately fifteen students each and lasting two months. In the financial year 1927-28, the SOTC met but twice presumably because of the budgetary pressures prevailing.[92] Consequently, the creation of the IDC, though representing a signal change in the higher education of officers and senior civil servants for Britain and the Empire as a whole, did not come without a price to the Royal Navy and its own educational regime. A price compounded by the Service having also to meet the administrative responsibilities of operating the college in the absence of a Ministry of Defence and providing the school's initial Commandant, Herbert Richmond.[93] Nor was it a case that instruction formerly provided at the SOTC was redundant now with the establishment of the IDC for the two courses were concerned with different aspects and problems in warfare. Certainly, the Admiralty did not think so as many officers attending the IDC had attended both portions of the SOTC and some even managed attending a third session.

Curtailment of courses was not limited to the Senior Officers' Technical Course and to the single year of 1927. In 1930, the Senior Officers' War Course was reduced to a single eighteen-week session from the two classes normally convened in March and October of each year, and again financial stringency is the most likely explanation.

89   *Navy Estimates 1927*, p. 34.
90   *Ibid*, pp. 34a and 6.
91   *Navy Estimates 1926*, p. 78a and *Navy Estimates 1927*, p. 21.
92   ADM 196/49 (Arbuthnot).
93   *Navy Estimates 1927*, p. 44a.

Cancellation of a course, though, was exceptional as the disruption to faculty planning and employment represented real waste. Meanwhile, at the Staff College no course was ever cancelled though interruption to meet an operational commitment was always a possibility. The numbers initially attending the course were quite modest with 17 naval and marine officers sitting the first postwar session rising to 30 for the 1920-21 course. A nominal number of military and air officers augmented these numbers, but the totals paled to those being training by the Army at Camberley and Quetta.[94] By 1930, the Staff College was only accepting 25 naval and marine officers for the annual course and by the middle of the decade until the outbreak of war in 1939 the numbers enrolled were typically around 37 including one or two members from the Royal Air Force and British Army.[95]

Dominion, Commonwealth and Imperial naval officers were eligible to attend the Senior Officers' War Course and the Staff Course at Greenwich and, indeed, given the unified manner in which the Admiralty viewed the separate imperial establishments, attendance in British eyes was tantamount to a necessity. At the time of the Washington Naval Conference in 1921-22 this view was affirmed and,

> So far as the Admiralty are concerned the Naval Forces of the Empire have always been treated as a unit; the sea is all one, and there is in war no dividing line between the area of operations of the British Navy and, for example, the Royal Australian Navy.[96]

That may have been how the Cabinet in London saw matters, but it was not a view that held sway with the Dominions even in 1918 whilst already in the midst of a common war. Train and equip to a common standard they would, but agree to accept unequivocal direction from a single imperial authority they would not.[97]

The Admiralty accepted their point for the moment, but believed that a common navy operating under a common command was a desirable and perhaps still an obtainable goal. Influencing this view were the principles of war, a key tenet of British military teaching instilled at both the Staff and War Courses. Co-operation and concentration, two of these principles, were vital factors of consideration when planning British military operations. The absence of any formal arrangements or obligations between

---

94   ADM 203/100.
95   *Navy List*, October 1930, p. 307, *Navy List*, October 1935, p. 307, *Navy List* (London: His Majesty's Stationery Office, July 1936,) p. 307, *Navy List*, October 1937, p. 306B, *Navy List* (London: His Majesty's Stationery Office, August 1938), p. 306B and *Navy List* (London: His Majesty's Stationery Office, March 1939), pp. 1-66A.
96   CAB 16/61, 'Committee of Imperial Defence, Sub-Committee on the Reduction and Limitation of Armaments, Report, Proceedings and Memoranda, November 1925'.
97   Minutes of the Imperial War Cabinet of 27 June 1918 cited in Nicholas Tracy, ed., *The Collective Naval Defence of the Empire, 1900-1940* (Aldershot: Ashgate Publishing Limited, 1997), pp. 230-32.

Britain and the several Dominions in providing for a common defence offended these principles and goes far to explain why the Admiralty desired a single Imperial Navy. This view was also influenced by a further consideration in war: the importance of time. With a knockout blow, especially from the air, proving more and more to be a concern, the machinery overseeing the higher direction of imperial defence to be adopted in war needed to be the same mechanisms applied during peacetime.[98]

However, two events, one a crisis and the other a political understanding demonstrated that developing such formal mechanisms were a chimera. The crisis was Chanak, a confrontation between Britain and a resurgent Turkey that showed every prospect of leading to war. In the event, a forceful and level-headed general on the scene, Tim Harington, showing that a soldier, too, could display a Nelsonic eye to orders ignored the Cabinet's directive to issue an ultimatum to the Turks now occupying positions within the neutral zone. Harington sought to calm troubled waters and successfully played for time. His actions averted war, but a career that until then offered every prospect of bestowing the fullest honours was harmed and Cabinet censure followed.[99] Yet, a rupture to imperial relations had occurred and the more surprising it is that it was under Lloyd George that the rupture occurred, for he had done much during the war to foster the political influence of the Dominions.[100] Here, Churchill had a hand in matters as the communiqué transmitted to the Empire read as a *fait accompli* of their support and not as a note of consultation. If he failed to appreciate that the days of a British premier acting as Asquith had in 1911 and denying the Dominions a say in the Empire's foreign policy were over, he was not alone in misjudging the tenor of the age.[101]

As for the political understanding, this was the so-called 'spirit of Locarno' that Britain reached with her European rivals absent the explicit support of the Dominions. Beatty, for one, recognized the changed circumstances arising in a post-1918 world. Writing to the Committee of Imperial Defence, the First Sea Lord observed:

> It is, moreover, recognised that the recent Locarno Pact does not throw any obligations on the Dominions, and that, therefore, if there was a war resulting from Great Britain's obligations under the Pact the Naval Forces *actively* employed would not necessarily include the Dominion Forces.[102]

---

98   Central, 'Some Reflections on the Direction of War', *The Naval Review*, Vol. XXVI, No. 3, August 1938, p. 417.

99   B.H.S., 'Harington', *Naval Review*, p. 368.

100   Judith Brown and Wm. Roger Louis, eds. *The Oxford History of the British Empire, Volume IV, The Twentieth Century* (Oxford: Oxford University Press, 1999), pp. 67-68.

101   S.D.S., 'The Canadian View of Naval Policy', *The Naval Review*, Vol. XXII, No. 4, November 1934, p. 629.

102   CAB 16/61. Original emphasis.

Locarno, of course, refers to the series of diplomatic settlements reached in 1925 guaranteeing the territorial status quo of the Versailles agreement between Germany, France and Belgium, of which Britain was a party. Its gestation had arisen when an earlier protocol negotiated at the League of Nations had been rejected by the Dominions and, in turn, Britain in 1924.[103] As Britain had even earlier refused a military pact with France to meet a possible German threat in the distant future absent support coming from the United States she faced a quandary. Not for the last time, the conflicting imperatives of European and imperial responsibilities presented themselves. Geography made Britain a European power, no less than Australia a Pacific nation or Canada an American state. Unable to be indifferent to her European flank and not willing to wreck the Empire by insisting that the Dominions accept European obligations, she adopted the half-measure of agreeing to the Locarno accords. These while ostensibly obligating Britain to the territorial status quo still left the actions required in any breach to the Council of the League of Nations of which she was a permanent member.

Compounding the issue of Dominion integration with the Royal Navy were the several naval arms control agreements of the period. The Washington Naval Agreement and the London Naval Treaties considered the navies of the British Empire as a unified military force. The reality, of course, was something less, as her ability to insist upon Dominion cooperation was nil. This was much appreciated by the Admiralty and Vice Admiral William Fisher,[104] the tall, imposing DCNS who registered the warning:

> The situation up till now with regard to the Dominion Navies and their relation to the Royal Navy has needed no special consideration, for the Washington Treaty only fixed the strength of battleships and aircraft carriers, of which the Dominion Navies did not possess any units.
>
> The extension of the rationing system to all classes of ships except sloops in the London Treaty, and the emphasis which has been laid during the Treaty negotiations on the principle of equal status with the Mother Country, has changed the situation … The Admiralty, however, cannot view this policy without anxiety.[105]

103 Austen Chamberlain, *Down the Years* (London: Cassell and Company Limited, 1935), pp. 151-71.
104 Later Admiral Sir William Wordsworth Fisher (1875-1937). War Course, 1909; Captain, 1912; Director, Anti-Submarine Division, 1917-18; Chief of Staff, Mediterranean Fleet, 1919-22; Rear Admiral, 1922; Chief of Staff, Atlantic Fleet, 1922-24; Rear Admiral, First Battle Squadron, 1924-25; Director, Naval Intelligence Division, 1926-27; Fourth Sea Lord, 1927; Vice Admiral, 1928; DCNS, 1928-30; Vice Admiral, First Battle Squadron and Second-in-Command, Mediterranean, 1930-32; Admiral, 1932; Commander-in-Chief, Mediterranean, 1932-36 and Commander-in-Chief, Portsmouth, 1936-37.
105 ADM 1/8744/125, Deputy Chief of the Naval Staff to First Sea Lord minute of 19 June 1930. Fisher was wrong in his claim that the Dominions were not directly affected by the Washington Naval Agreement, as the battle cruiser *Australia* fell under its remit and was scrapped in accordance with Chapter II, Part 3, Section II of the treaty.

British statesmen understood the dilemma, but it was not one that weighed unduly upon them. Perhaps, it was because to recognize it openly would necessitate greater spending on her part to meet her declared policy of maintaining a 'One Power Standard' in naval armaments. Perhaps, it was the case that agreement with the United States would have been problematic if the greater Empire's naval assets, small as they were, were not counted as part of the British totals. Alternatively, perhaps, British statesmen conditioned to believe and think in terms of empire, could not bring themselves to think otherwise and believed that cometh the hour, cometh the man.[106]

Outside Government circles, the belief that the Empire was more a façade than a fact was noted, too, and *The Times*, the journal of the Establishment, advised its readers that 'the weakness and dangers of the system are manifest.'[107] Yet, too much must not be read into such warnings. For Britons, with a custom of governing themselves with a minimum of formal structure and where traditional practices played no small measure in the evolving nature of government structures, the lack of formal obligations was in keeping with her political traditions. The Admiralty, leaving little to chance and regulating its naval affairs in both *The King's Regulations and Admiralty Instructions* and Fleet Orders, doubtless preferred a more definite arrangement, but adapted itself accordingly.

First muted by the Admiralty in December 1920, the one power standard became the basis for formulating the size of the Navy and the Imperial Conference of 1921 concurred with the yardstick as a basis of future policy.[108] The pronouncement of the standard was acceptable to statesmen and Admiralty alike, because never having truly defined what was implied, all were at liberty to infer a meaning suitable to their purposes. At least Captain Tom Phillips, the Director of Plans, believed as much in 1937.[109] To navalists, the standard represented a minimum naval programme, whilst to politicians it represented a maximum naval effort.[110] The Admiralty sometimes used the one power standard in its arguments with the Treasury when the Exchequer sought reductions based on the guidance of the 'Ten Year Rule'. In this case, the Admiralty could rightly argue that some naval capabilities, whilst not immediately required for a war a decade away, were, nevertheless, necessary to maintain British prestige. Thus, in 1928 Captain Roger Bellairs,[111] the Director of Plans, advised:

106 See especially Foreign Office and Admiralty memorandum of 3 October 1934 cited in W. N. Medlicott *et al*, *Documents on British Foreign Policy 1919-1939, Naval Policy and Defence Requirements July 1934–March 1936* (London: Her Majesty's Stationery Office, 1973), pp. 45.

107 *The Times* cited in Piers Brendon, *The Decline and Fall of the British Empire, 1781-1997* (New York: Alfred A. Knopf, 2008), p. 339.

108 Roskill, *Naval Policy, Vol. I*, p. 21.

109 Christopher M. Bell, *The Royal Navy and Strategy Between the Wars* (Basingstoke: Macmillan Press Ltd., 2000), p. 2.

110 Gibbs, *Grand Strategy*, p. 23.

111 Later Rear Admiral Roger Mowbray Bellairs (1884-1959). Five firsts in examinations for Lieutenant; War Staff Course, 1912; Qualified for War Staff, 1913; War Staff duties in

[T]he fully commissioned fleet is only sufficient now to carry out the tactical and sea training required, to maintain a one-power standard and to uphold our political and commercial interests abroad. So far as the fully commissioned fleet, therefore, is concerned, the reply of the Admiralty to the Chancellor of the Exchequer is not affected by the 10 year decision and no economy in this direction can be suggested.[112]

To the budget conscious, the one power standard was a ready guide to avow that British maritime superiority was being preserved by recourse to highlighting the essential parity in types of naval warships. To the Admiralty, cognizant that naval power was more than merely the size of the active and reserve fleets, but depended on ancillary elements such as industrial potential, bases, oil reserves, and ammunition allowances, having a navy of a given size that could also execute the tasks expected was no less vital. As naval forces exist to protect the interests of the states that raise them, their requirements are bound to be different. The geographic and political contexts of states are never equal. Formulations of parity attractive to diplomats and Treasury mandarins alike are seldom viewed with a similar equanimity by those forced to deal with their practical effects. Even the Washington Naval Agreement, codifying a general parity between the United States and the British Empire in capital ships and aircraft carriers, remained silent on the foundations of power such as naval budgets, manning levels, the provision of reserves, or operational tempo.

This was also the case with respect to the higher education of officers. Direct comparisons are somewhat misleading as the specific manner in which instruction was given were not symmetrical between the two navies. Allowing for such a caveat, the United States Naval War College, the nearest educational equivalent to the Royal Naval War College and the Royal Naval Staff College, had 45 officers in its class of 1922 whilst the British Staff Course mustered only 26 naval qualifiers.[113] By 1934, with the addition of an upper class for the benefit of flag officers and senior captains, the Americans were passing more than 90 officers through Newport.[114] For the Royal Navy, 37 qualifiers, including four Dominion members, were educated in 1937. If the officers sitting the Senior Officers' War Course, attending the Tactical School and taking the Senior Officers' Technical Course are considered then the number of

HMS *Neptune*, 1914; War Staff Officer in Grand Fleet, 1916-19; Naval Assistant to First Sea Lord, 1919-25; Captain, 1920; Chief of Staff and Flag Captain in HMS *Effingham*, 1925-28; Director of Plans, 1928-30; Commanding Officer, HMS *Rodney*, 1930-32; Rear Admiral and retired, 1932. British representative to League of Nations Advisory Committee, 1932-39. Recalled for war service and Head, Historical Section, Tactical and Staff Duties Division, 1948.

112 ADM 1/8737/97, Director of Plans to Chief of the Naval Staff minute 03101/28 of 5 September 1928.

113 Michael Vlahos, *The Blue Sword: The Naval War College and the American Mission, 1919-1941* (Newport: Naval War College Press, 1980), p. 66 and ADM 203/100.

114 Vlahia, *Blue Sword*, p. 66.

officers under instruction approached those studying at Newport. Discounting those officers attending more than one course during the year, for 1934, at least 102 separate British and Commonwealth officers received such training. These numbers do not consider others who studied at the Imperial Defence College, qualified at Andover or Camberley, or passed through the Sheerness Senior Officers' School.

While many senior officers in the Royal Navy, such as William Fisher, lamented that the one power standard was not being maintained and said as much in their official correspondence, such was not the case when it came to the higher education of the Service's officers.[115] However, where matters did diverge from the American example was that Greenwich offered little prospect for further development, as facilities were extremely limited with little to no space available whilst Newport afforded ample room.[116] Additionally, the United States Navy was not averse to training additional officers through correspondence courses, a practice the Royal Navy made no attempt to pursue.[117]

Why no attempt at correspondence courses was made is unknown, but a number of reasons suggest themselves. First, for the Royal Navy, the postwar Staff Course was fundamentally about producing staff officers of a common standard. A correspondence course operated against this precept. Secondly, many commanding officers of ships would not wish to see officers distracted by outside studies operating against the execution of their present duties. Here, the custom of requiring officers desiring to learn a foreign language on their own time to seek Admiralty approval first before pursuing can be cited. This was often given, but it did provide a brake to ensure that an officer was meeting his current responsibilities and resorting to correspondence study would have faced a similar hurdle.

That the Royal Naval Staff College was about producing staff officers may sound axiomatic, but the United States Naval War College had a broader mission to impart doctrine across the broadest reaches of the Service. Secretary of War Henry Stimson, with due allowance for hyperbole, noted in American naval circles that 'Neptune was God, Mahan his prophet, and the United States Navy the only true church' capturing a sense of the zeal American officers felt about their Service and the proselytizing role of Newport.[118] Additionally, British naval officers spent a great deal of time at sea. The adverse environment of shipboard life that made cramming for an examination to the Staff College so problematic applied just as well to attempts at mastering a course through distance learning.

---

115 ADM 1/8737/97, Deputy Chief of the Naval Staff to Chief of the Naval Staff minute of 27 September 1928.
116 'Royal Naval Staff College', *Naval Review*, Vol. XX, No. 1, p. 17.
117 Vlahos, *Blue Sword*, p. 66.
118 Stimson cited in Philip A. Crowl, 'Alfred Thayer Mahan: The Naval Historian' Peter Paret, ed., *Makers of Modern Strategy: From Machiavelli to the Nuclear Age* (Princeton: Princeton University Press, 1986), p. 444.

Though its officers never adopted a post-nominal after their name and rank to reflect attendance at the Naval War College, as was the British custom, Newport was very much a stepping-stone to flag rank in the United States Navy to a degree never realized by Greenwich. Vlahos notes that by 1941, 99% of American flag officers were graduates of Newport;[119] the corresponding figure for Royal Navy, Royal Australian Navy and Royal Canadian Navy flag officers holding the *psc* designation in 1939, the last year of peace, was 24% rising to 44% if graduates of the Imperial Defence College are considered.[120] Finally, the size of the Greenwich staff would not bear it. The limited number of officers on the directing staff was more than fully occupied in overseeing the qualifiers in residence. This and making the necessary changes in the curriculum year on year necessitated by the evolving strategic scene or changes in naval policy and administration left little time for supporting distance learning.

The short-term decision to adopt Greenwich as the home of the Staff and War Colleges rather than establish a new facility in 1919 was entirely logical given the fiscal pressures of the day. The dilemma became that financial stringency never receded as a worry during the period and a decision taken as a stopgap measure, by default, became a permanent solution. In time, however, Greenwich was deemed to offer many tangible benefits. These included its proximity to the Naval Staff, the greater governmental machinery of Whitehall, and the Royal United Services Institution. Moreover, prominent university lecturers had easy access to Greenwich, thus, more than compensating for shortcomings in physical plant.[121]

Operationally and administratively, Britain and her partners understood that only the politicians facing an actual crisis would decide the parameters of accepting any strategic commitment. The dilemma was not new and had faced Britain previously on the cusp of the World War. Colonel Hankey recounting that time cited the words of Herbert Asquith, the premier, who displaying a healthy measure of *sang-froid* said simply, 'leave it to the day.'[122] In spite of the stresses and strains experienced since Asquith's day, a similar view prevailed during most of the interwar period. A point emphasized as late as November 1936 by Sir Archdale Parkhill, the Australian Minister of Defence, when introducing the Defence Estimates in the Commonwealth's parliament advising members that 'subject to sovereign control of its own policy, and without prior commitment in any shape or form, the Government stands for co-operation in Imperial Defence.'[123] Notwithstanding the inherent contradictions in such a policy, training proceeded in the expectation that such support would be mutual and ultimately forthcoming. Brooke-Popham addressing the 1932 IDC class for the last time captured something of this ambivalence remarking:

119 Vlahos, *Blue Sword*, p. 92.
120 *Navy List*, March 1939, pp. 322E and 322G.
121 'Staff College', *Naval Review*, pp. 15-16.
122 CAB 53/6, Chiefs of Staff Sub-Committee minutes of 21 July 1936.
123 Archdale Parkhill, 'Australian Co-operation in Imperial Defence', *The Naval Review*, Vol. XXV, No. 2, May 1937, p. 239.

When I welcomed you here eleven months ago, I made a special reference to the representatives of the Civil Service and of the Dominions, perhaps drawing some distinction in my mind between them and officers of the British Navy, Army and Air Force.

I shall not make any such distinction today because I regard you now as all one and as to one body I wish you Goodbye and Good Luck.[124]

Hankey was more forthright in framing the issue and with his vast experience in the late war knew well of its many problems. Speaking to the IDC in 1927 on the topic of 'The Higher Direction of War' he told his audience:

I have not dealt with the Higher Control from the point of view of the Empire or from that of allies. The former question raises Constitutional issues of a difficult character which are hardly ripe for solution. In the War we solved it, not without consider-able success, by the emergency expedient of an Imperial War Cabinet. I shall not speculate whether that expedient would prove acceptable again to the Dominions.[125]

A later chapter addresses the issue more fully examining how the IDC dealt with its implications when focusing on the threats facing the Empire with their possible responses. That friction arose amongst students with differing points of view as syndicates developed possible courses of action implying imperial unity is a given.

By 1939, ten Australia naval officers on the active list had qualified at the Staff College, whilst the number of Canadian and Royal Indian Navy officers holding the *psc* was sixteen and six, respectively.[126] The ten RAN officers came from an establishment numbering 148 officers in the rank of Lieutenant and above in the Permanent Naval Force. For the RCN, the active component in 1939 consisted of 51 officers in the rank of Lieutenant and above whilst the RIN had 60 such officers. Clearly, distance was one barrier to training more officers from the Dominions; however, the primary constraint was finance. If the actual numbers educated seem few, the reasons had more to do with the fiscal and financial circumstances of the Dominions than any fundamental disagreement in receiving such training. The parent establishments remained responsible for meeting the costs of tuition and maintenance for its officers attending the Staff and War Courses. Thus, ships alone do not calculate the basis of naval power.[127] In 1923, a charge of £300 per officer was levied for attendance at the Royal Naval Staff College and when the Government of India sought to send an

124 Air Marshal Sir Robert Brooke-Popham, 'Final Address' of 15 December 1932, Brooke-Popham Papers, 1/5/13, LHCMA.
125 Hankey, 'The Higher Direction of War', Imperial Defence College lecture of 7 November 1927, Alanbrooke Papers, 3/12, LHCMA.
126 *Navy List March 1939*, pp. 465–66, 483 and 495.
127 *Navy Estimates for the Year 1934* (London: His Majesty's Stationery Office, 1934), pp. 92-93.

officer from the Mahratta Light Infantry a similar amount was specified.[128] In fact, when viewed as a percentage of the officers eligible to attend, the imperial navies made a concerted effort at having their officers educated and trained at the schools and courses. Additionally, many of the Royal Navy officers seconded to the Empire's nascent navies had been graduates of the Staff and War Courses. Thus, if the numbers trained appear small in the aggregate, they are, nevertheless, impressive when viewed against the numbers eligible to attend and the obstacles to be overcome. This was groundwork with a purpose and lent support to the future growth of the separate establishments though exceptions, of course, existed to this policy.

Here other factors were at play. The British Empire though a maritime creation and a commercial trading bloc with many common political traditions was not solely concerned about naval affairs. The military and domestic contexts of the separate Dominions were different for each and different to Britain herself. Being island states, the outlook of Australia and New Zealand in defence approximated closest to England. Canada, India, and the Union of South Africa, though having substantial maritime boundaries, had continental and internal defence considerations that had call on their priorities. Eire was of a class by herself and was treated as such by the United Kingdom. Tradition, geography and recent history may have been partly to blame, though the Statute of Westminster on the surface placed her on an equal footing to the other Dominions. The Dominions Secretary, Sidney Webb, the eminent socialist and now Lord Passfield, had this very much in mind when expressing the view that he could not associate himself with any distinction in the status of the several Dominions. Still, the thought of releasing classified military information to Eire was presented as a barrier by the Chiefs of Staff moving Passfield to remind the Chiefs that Ireland already saw all British diplomatic cables. The more fundamental reason was that the wounds of the recent past were still too fresh and it was asking much to have former combatants attend classes.

Given the size of the Irish Free State's defence establishment, the problem was never an acute one for the Admiralty, as Eire lacked a navy. Still, Admiral Madden was particularly emphatic that her personnel would not receive naval training in any event. Trenchard was of a similar mind when it came to Ireland's officers attending any higher courses, but the Army was less dogmatic.[129] Perhaps this owed something to the large number of Irish officers in the British Army, though the same was also true for the other military establishments.[130] A further complication was that Eire was now sending some of her men to the United States Military Academy at West Point. Though this implied a certain level of mistrust in sharing information, it was also seen as a hindrance in fostering a common pattern in junior officer training. Ultimately, it

128 ADM 1/8640/112, Acting Director of Training and Staff Duties minute of 4 December 1923.
129 CAB 53/3, Chiefs of Staff Sub-Committee minutes of 21 February 1930.
130 CAB 53/28, Joint Planning Sub-Committee Report of 27 July 1936.

was decided that it was better to have Eire inside the tent and not out and access to introductory courses was confirmed. In 1926, a list of schools and courses acceptable to Free State entrants was fashioned of which Greenwich was omitted.[131] Later when Ireland sought to send officers to British military schools, the matter received fresh consideration by the Chiefs of Staff and a new list of courses available was prepared. Significantly, her attendance at the higher schools of military and naval education remained proscribed.[132]

In light of the political and financial circumstances prevailing, the common naval measures adopted by the Dominions, particularly in equipage and regulations, and, more especially, in training, represented no mean achievement.[133] Illustrative of how badly financed the Empire's defenses were, Canada was spending a mere $1.46 per capita on its forces to the United Kingdom's $23.04 at the time of the 'Geddes Axe'.[134] If the number of active ships in the Royal Canadian Navy was few and getting fewer, Naval Headquarters Ottawa at least sought to educate and train its officers for the day when prospects were better.[135]

The opening of the Imperial Defence College in January 1927 was a further step in creating the necessary wherewithal to effectively prosecute war with forces now operating in three physical domains. As a joint headquarters or administrative body to oversee the day-to-day management of the IDC was largely lacking, it fell to the Navy to accept responsibility for the administration of the college. Each Service retained responsibility for funding its officers attending—whether serving as part of the directing staff, attending as students, or acting as its head. The British Army and Royal Air Force shared in meeting the operating costs of the IDC and their contributions appeared as appropriation-in-aid bequests in the *Navy Estimates*. Appendix IV depicts the total funding required to maintain the IDC in each of the years of its operation from 1927-37. These sums, however, do not represent its true cost of operation, as imperial members attending reimbursed the Home Government for their student's

131 ADM 116/2273, Minutes of the 215th Meeting of the Committee of Imperial Defence of 22 July 1926.

132 CAB 53/3, Chiefs of Staff Sub-Committee minutes of 21 February 1930. The formal position taken by the Chiefs Staff along with a recommended way ahead was documented in a report issued in December 1929. See CAB 53/17.

133 For background on the Royal Canadian Navy during the period see Marc Milner, *Canada's Navy: The First Century* (Toronto: University of Toronto Press, 2000), pp. 57-75. Though an otherwise excellent survey of the Australian case, this writer finds Grey's final judgment on the problems facing the RAN at the outset of war too critical, as Australia simply did not possess the means to meet a Japanese threat on her own. Jeffrey Grey, *A Military History of Australia* (Cambridge: Cambridge University Press, 1999), pp. 119-39.

134 Tony German, *The Sea is at Our Gates: The History of the Canadian Navy* (Toronto: McClelland and Stewart, 1990), p. 56.

135 In January 1919, the Royal Canadian Navy numbered 5 ships and 93 auxiliary vessels. By July 1920, the service was reduced to 3 ships in commission. *Navy List*, January 1919, pp. 952-58a and *Navy List*, July 1920, p. 952-58.

tuition. Civil Servants from the non-military departments also attended the IDC, but in this case, no charges were levied against their parent organization, (e.g., Treasury, Foreign Office, and the Post Office) and so, in essence, they attended gratis save for the expenditure of salary and allowances.[136]

Imperial naval members had a distinct advantage in securing admittance to the IDC, for notwithstanding that only one or two spaces would be available in any given year to support their entry, the size of their parent establishments were small. Even when allowance is made that vacancies were competed against a Dominion's civil service, military and air arms, the odds an imperial naval officer faced were certainly better than those of his Royal Navy counterpart.[137] This advantage is reflected in one other important particular: naval officers from Commonwealth naval forces were readily accepted whilst serving in the rank of Commander when their British counterpart could expect attendance at the IDC to follow only after being promoted to Captain. Between 1927 and the outbreak of war in 1939, the Royal Australian Navy sent five officers to the IDC, whilst four representatives of the Royal Canadian Navy attended the course.[138]

Though the funding levels of the IDC remained largely stable throughout the period, the Navy's share of these costs varied and depended on whether the Commandant was a naval member. (It remained that no Commandant for the period came from the Royal Marines.) This senior appointment, of a two-year fixed term, was filled on a rotational basis between the Services with the officer, in the case of the Navy, being either a Vice Admiral or Admiral. Besides Richmond, who served from 1926-28, the other naval officers of the period acting as Commandant were Lionel Preston[139] (1933-35) and Hugh Binney[140] whose appointment in 1939 terminated early with the onset

136 *Navy Estimates 1935*, p. 83.
137 By 1930, one member of a Dominion's civil service had attended the IDC. See *Hansard*, House of Commons debates of 24 Jun 1930, v. 240, cc966-7.
138 The Australian officers attending were Captain Joseph Burnett (1899-1941), Commander, later Acting Captain James Esdaile (1899-1993) Lieutenant Commander, later Rear Admiral Harold Farncomb (1899-1971), Commander, later Captain, Frank Getting (1899-1942) and Commander, later Rear Admiral, Cuthbert Pope (1887-1959). Canadian officers attending the IDC included Captain Victor Brodeur, later Rear Admiral (1892-1976), Captain, later Vice Admiral, George Jones (1895-1946), Captain, later Rear Admiral, Leonard Murray (1896-1971) and Captain, later Admiral, Percy Nelles (1892-1951).
139 Later Admiral Sir Lionel George Preston (1875-1971). War Course, 1907-08 and 1912; Captain, 1914; Director of Minesweeping, 1919; Commanding Officer, *Harebell*, 1919-20; Director, Signal School, 1920-22; Commanding Officer, *Eagle*, 1923-25; Rear Admiral, 1925; Senior Officers' Technical Course, 1925; Rear Admiral, Third Cruiser Squadron, 1926-28; Senior Officers' War Course, 1929-30; Vice Admiral, 1930; Fourth Sea Lord, 1930-32; Commandant, Imperial Defence College, 1933-34; Admiral, 1934 and retired, 1935. Recalled for war service.
140 Admiral Sir Thomas Hugh Binney (1883-1953). Naval Assistant to Fourth Sea Lord, 1919-20 and ACNS, 1920-21; Captain, 1922; Commanding Officer, HMS *Cardiff*, 1922-24; Deputy Director, Plans Division, 1925-27; Senior Officers' Technical Course, 1928;

of war. Of Preston and Binney, more will be offered later, but neither were a remarkable choice if measured by what followed in their career. Each had qualities and experiences, though, that made their appointments sound, if not brilliant choices, and after the experience of Richmond, maybe that is all that was desired by the Admiralty.

How its officers educated at Greenwich, Portsmouth and elsewhere were employed is yet to be examined, but a comment at this point would not be out of place on the effects financial stringency had on those directing and receiving instruction. Richmond whilst commanding the Royal Naval War College lamented that many officers so recently taught or directing the War Course were retired in the wake of the cuts following the Geddes Report; Marder citing 19 of 24 officers.[141] Such wastage on the surface appears to lend credence to the contention that the Service deprecated the value of its higher educational regime until it is realized that the vast majority of officers concerned did not retire until January 1930—long past the time of the 'Geddes Axe'. These officers only became redundant when the Labour Government of Ramsay MacDonald imposed further cuts on the Navy. All of the officers had attended the 'War Course' as a Captain, the vast majority had not advanced beyond that rank, and their prospects of promotion approached nil. The retirements of the officers may have been hard, but it was not brutal. Certainly, they fared far better than their brethren holding ranks of Commander and Lieutenant Commander who were now spending 25% of their time unemployed.

A true measure of the retrenchment facing the Service was the plight of naval aviation. In 1918, some 3,000 aircraft were supporting maritime operations; by the early 1920s, the number had fallen to around 100 aeroplanes.[142] Moreover, unlike a warship scrapped or a station closed, an officer retired was subject to recall to serve again, if the situation required. To be sure, this caveat does not address the complete issue as the question exists whether officers educated *throughout the period* were effectively employed in their future assignments or if the schools were becoming a sinecure of the indifferent—matters awaiting discussion in due course.

In the immediate aftermath of the World War, a surfeit of junior officers of the executive branch who had not completed the normal routine of education for junior officers before commissioning and proceeding to sea existed. Returning these officers to Dartmouth to complete their normal cycle of education was not a realistic option, as the school could not accommodate the numbers required whilst continuing to train cadets being entered for the standard course of instruction. Moreover, the officer of 1919 was no longer the youthful cadet who had departed. An enlightened remedy

Tactical Course, 1928, 1931 and 1936; Flag Captain in *Nelson*, 1928-30; Senior Officers' War Course, 1930-31; Director, Tactical School, 1931-32; Flag Captain and Chief Staff Officer in *Hood*, 1932-33; Rear Admiral, 1934; Commander, First Battle Squadron, 1936-38; Vice Admiral, 1938; Commandant, Imperial Defence College, 1939; Admiral, 1942 and retired, 1943. Recalled for war service and Governor of Tasmania, 1945-51.
141 Marder, ed., *Portrait*, p. 27.
142 Till, 'Retrenchment', in Hill, ed., *Royal Navy*, p. 320.

decided in late 1917 was to send increments of these officers to attend a short course of instruction at Cambridge University lasting for two terms. Captain Eric Fullerton,[143] a respected officer with previous experience of preparing cadets at Osborne, was appointed to oversee this unique naval training effort.[144]

Besides standard courses in mathematics, navigation, engineering and science, these officers also pursued studies in elective courses.[145] Wearing naval uniform and residing in their colleges and with Royal Marine attendants looking after their personal needs, these were undergraduates with a difference. With approximately 150 officers in residence at any time, their conspicuous presence generated a certain amount of tension with the other students, as Cambridge had at this moment a pronounced left-wing fervor. Still, the officers studying there found it to be a welcome change from their normal naval routine and educationally enlightening.[146] Yet, tensions remained and after a number of confrontations, university authorities sanctioned the wearing of cap and gown rather than uniform to lower the presence of the naval contingent.[147]

Nine of the Cambridge's colleges supported the scheme and though the officers attending would not secure a degree in the time permitted, the programme provided a pause whilst the Admiralty grappled at finding a permanent solution to the manning problem.[148] University examinations concluded the period of study, and, though, an officer failing these might not face dismissal from the Service, the results achieved were considered along with the other standard naval examinations taken leading to promotion to Lieutenant.[149] Conversely, the officer doing well in these assessments was able to secure credit towards an accelerated promotion to Lieutenant.[150] Officers were graded as securing First, Second, or Third Class passes, but given the standard applied, few met the typical university criterion for such grades. Instead, the Navy soon adopted the expedient of awarding pass-rates by simply dividing the officers into

143 Later Admiral Sir Eric John Arthur Fullerton. (1878-1962). Royal Naval College Osborne, 1912-14; Captain, 1914; Officer-in-Charge, Cambridge University, 1918-21; Senior Officers' Technical Course, 1921 and 1927; Captain of the Fleet, Atlantic Fleet, 1921-23; Rear Admiral, 1926; Tactical Course, 1926; Naval Secretary to First Lord, 1927-29; Commander-in-Chief, East Indies, 1929-32; Vice Admiral, 1930; Commander-in-Chief, Plymouth, 1932-35; Admiral, 1935 and retired, 1936. Recalled for war service.
144 ADM 196/44 (Fullerton).
145 John Leyland, 'Admiralty Administration and Naval Personnel', *Naval Annual 1919*, p. 217.
146 Captain Robert John Shaw, Sound Recording 9764 of 1987, Imperial War Museum, London.
147 Captain Charles Hawken Drake, Reel 4, Sound Recording 8250 of 2 July 1984, Imperial War Museum, London.
148 The colleges hosting naval officers were Magdalene, Queens, Selwyn, King's, Peterhouse, Gonville and Caius, Sidney Sussex, Trinity Hall, and Emmanuel.
149 For those attending the Cambridge Short Course, the examination following study there replaced the previous exam in Engineering required of all Lieutenants.
150 *Hansard*, House of Commons debates of 4 March 1919, v. 113, cc237-8w.

three equal portions based on their examination results and assigning the appropriate grade.[151]

The decision to send newly minted naval officers to Cambridge, unusual as it was, represented a marriage of opportunity and necessity. Some officers, such as Lieutenant Charles Drage,[152] seeing what was on offer, sought half-pay and took the opportunity to go to university on their own volition. Drage, twenty-one and recently returned from duty in Russian waters where he had witnessed incidents of naval insurrection from the cruiser HMS *Cochrane*, pressed the Admiralty to be released though they initially had other ideas. In the end, they relented and he was granted a year's half-pay. In light of the war and the many men who in the normal course of events would have gone up to Oxford, that university at this moment offered an abbreviated course leading to a degree. Drage took advantage of the scheme going to Christ Church in 1919—his father's former college. Indeed, Geoffrey Drage, the Tory politician, had been the driving force of the younger Drage's entry into the Navy. Drage minor now read Modern History securing a B.A. Honours degree in the Michaelmas term of 1920.[153] Partaking of the rich experience that Oxford had on offer, Drage discussed Turkey and Armenia with Colonel T.E. Lawrence, rode in the hunts and point-to-points, became a freemason and attended his lectures.[154] Drage was blessed that the Dean of Christ Church, Sir Thomas Strong, desired to see the college return to its ante-bellum ways where riding featured so prominently and sought to have the equestrian side of affairs restored by those officers now attending. Strong secured the help of the lecturers in his efforts with extra tuition provided in the evenings allowing the college and officers alike to leave the memories of the recent past behind.[155] Passing his examinations for B.A. but failing to secure a fellowship to All Souls, Drage applied to the Admiralty for a ship on the China Station in order to study Japanese. In this, he was only partially successful—getting his ship but not the China Station.

Drage, no doubt, was an exception and an exceptional officer. Unless one had demonstrated outstanding scholastic achievement at school (and Drage had not), attendance at Christ Church, the *primus inter pares* of Oxford colleges, was an indication more of breeding rather than reading. Others not so gifted or so inclined, left it

151 Onlooker, 'The Navy at Cambridge: 1919', *The Naval Review*, Vol. LV, No. 4, July 1967, p. 231.
152 Later Commander Charles Hardinge Drage (1897-1983). Oxford University, 1919-20; Intelligence Officer in *Cardiff*, 1920; Intelligence Course, 1923; Staff Course, 1927-28; Staff Officer Operations to Commander-in-Chief, East Indies, 1928-31 and retired, 1933. Recalled for war service.
153 Charles Hardinge Drage, Reel 1, Sound Recording 8439 of 1983, Imperial War Museum, London.
154 Lieutenant Charles Drage letter to Secretary of the Admiralty of 18 May 1920, Lieutenant Commander Charles Hardinge Drage Papers, Imperial War Museum, London, PP/MCR/99, Reel 2.
155 Group Captain Frederick William Winterbotham, Reels 3 and 4, Sound Recording 7462 of 1984, Imperial War Museum, London.

to the Navy who recognized that many young officers having left their preparatory college, Dartmouth, early to serve with the fleet had an all too much-abbreviated education before receiving their commissions.[156] Technical courses to qualify in one of the specialist branches, such as gunnery, torpedo, or signals, would typically be in order, but as their initial education had been truncated, the readiness of these officers to receive such instruction was uneven. Moreover, the ability of the naval training establishments to accept the officers given their numbers and uneven preparation were additional factors not to be discounted. The Cambridge education was unique in another respect: naval instructors provided parallel courses, such as in mathematics, to the students to bring them to the standard achieved by those who benefited from the full regime of Dartmouth.[157] However and unsurprisingly, the lectures taught by Dr John Holland Rose, the Vere Harmsworth professor of naval history, proved the most popular with Lieutenant Edward Longsdon[158] believing they were an essential grounding to any officer desiring to pursue staff work and as a preparation for achieving higher command.[159] A conclusion the First Lord would have noted with satisfaction, as identifying those officers with a talent and an inclination for staff work was a primary purpose of Rose's offering.[160]

Longsdon believed that one's time at Cambridge leavened what would otherwise be a highly technical, but limited, education. A view mirrored by Charles Burney, himself a naval officer, now retired, and sitting in parliament as the member for Uxbridge. During consideration of the 1924 *Navy Estimates*, Burney praised the Navy's foresight in sending its young officers to Cambridge commenting:

> [T]here is nothing I have seen introduced which gives me greater pleasure than the fact that naval officers are to be given a course at Cambridge University. The whole tendency of the Navy at the present time is not only to get more and more complicated from the technical and material point of view, but the actual questions become more and more diverse, both from the political and personnel point of view. The education which is given at the University fulfils a want for any officer who is going to fill a higher command.[161]

---

156 Edward Longsdon, 'On the Course at Cambridge University for Naval Officers, *The Naval Review*, Vol. VII, (London: The Naval Society, 1919), p. 229.

157 *Ibid*, p. 230.

158 Acting Captain Edward Henry Longsdon (1896-1984). Short Course, Cambridge University, 1919; to New Zealand Division, 1927-30; Senior Officers' Technical Course, 1932; Naval Ordnance Department, 1932-35; Naval Intelligence Division, 1935; Commanding Officer, HMS *Penzance*, 1935-36; No. 1 Bombing Group, 1937; Tactical Course, 1937; Acting Captain, 1939; Commanding Officer, *Maidstone*, 1939-40, *Cyclops*, 1940 and *Letitia*, 1940 and retired, 1946.

159 Longsdon, 'On the Course', p. 232.

160 ADM 1/8744/140, 'Command Paper 451, Statement of the First Lord of the Admiralty Explanatory of the Navy Estimates, 1919-1920,' of 1 December 1919, p. 16.

161 *Hansard*, House of Commons debates of 12 March 1923, v. 161, cc1197-1239.

If not all naval officers endorsed Burney's plea for the programme's continuance and recalled their experience with more mixed feelings, the Admiralty appreciated its value, nevertheless.[162] Ultimately, though, having attended Cambridge or not, many young officers were separated as the numbers were just too great to carry and those of independent means were especially at risk.

To be sure, not all executive officers leaving Dartmouth early went to Cambridge. A select few, such as Lieutenant Maurice Mansergh,[163] went direct to their specialist course without taking the normal series of examinations for Lieutenant.[164] Still, in the expectation that of those early leavers sent to Cambridge would complete their studies by March 1923, the Admiralty contemplated its next moves. With over 1,600 Sub-Lieutenants and Lieutenants having passed through the programme, it sounded out the University on sending additional officers to study for a full academic year.[165] Though sympathetic to the Admiralty's plan, university officials were less than enthusiastic. Not surprising, perhaps, in an institution where tradition was prized every bit as much as academic excellence for whilst a year's course of study offered more scope for instruction than the present offering, it would yet result in a sizeable number of students not taking a degree. Moreover, as the officer arriving would be older than the typical entrant, and yet not academically on the same level as those occupying the Middle Common Room, they promised to be a class apart. This last concern perhaps begging the question what the university thought of the large number of military officers coming to Cambridge after spending their gap year on active service. Still, their response was positive, but, conditional, insisting that any officers enrolled must be subject to the full regime of university regulation.

Armed with this tentative offer, the Admiralty sought Treasury sanction to expand the programme by allowing up to 25% of Sub Lieutenants not affected by having their Britannia course truncated proceed to Cambridge for a year's study. At a time when economy was the order of the day, the £9,000 requested by the Admiralty proved red meat to the Treasury who argued that if the Navy believed the Dartmouth curriculum was too technical and was producing officers with limited focus the natural measure was to modify their own educational affairs and not to expect the taxpayer to further subsidize officer education. Warming to his task, George Barstow replying for the Treasury curtly concluded:

162 J.M.G.G., 'University in Education', *Naval Review*, p. 150.
163 Later Admiral Sir Maurice James Mansergh (1896-1966). Air Navigation Course, 1925; Navigating Officer in *Glorious*, 1929-31; Navigation Branch, 1931-33; Staff Course, 1933; Staff Officer Operations in *Queen Elizabeth* and *Resolution*, 1933-36; Captain, 1937; Imperial Defence Course, 1938; Assistant Director, Plans Division, 1939; Director, Trade Division, 1939-41; Staff College Camberley, 1941; Rear Admiral, 1946; Senior Course Land-Air Warfare, 1948; Vice Admiral, 1949; Admiral, 1953 and retired, 1954.
164 ADM 196/146 (Mansergh).
165 'Naval Notes, Cambridge Course Ending', *Journal of the Royal United Services Institution*, Vol. LXVIII, No. 469, February 1923, pp. 151-52.

> My Lords feel the gravest doubt whether a year's course at a University would be likely in general to be attended by such beneficial results as are anticipated by the Lords Commissioners of the Admiralty or indeed whether there is any ground at all for postponing the age which a Naval Officer commences to render service to the State in return for his pay and the cost of his education.[166]

This was churlish, but in an office where considering the expense of any problem more than weighing any benefit was the norm, not unpredictable. Compromising the Navy's case was that it failed to consider other options. For example, the British Army at the time allowed an officer time away to pursue a university education, if the soldier was willing to meet the tuition and maintenance fees himself. The Admiralty, aware of this plan, believed that its officers would dismiss any similar scheme if it failed to cover the expenses incurred.

Of course, Drage demonstrates that this was not necessarily so and the greater failure was to not even attempt to test the merits of their beliefs. Here the case of Rear Admiral Walter Ash is instructive. Beginning his career as a civilian electrical apprentice in the Portsmouth Dockyard in 1922, Ash soon secured a scholarship to study for a degree at City and Guilds College. At this point, the Admiralty approached him about entering the Service and continuing his studies at the Royal Naval College, Greenwich. This he accepted and Ash was soon launched on a fruitful naval career. Though it is true that Ash started as a civilian and not as an established naval officer, his case demonstrates that the motivating force in career and educational choices cannot be reduced to a single criterion. In the end, what the Admiralty did attempt was to secure a limited number of university graduates without recourse to public financing by offering commissions to graduates willing to serve as an engineer officer.[167] A similar scheme leading to a commission in the Royal Marines had already been in place since 1935.[168] Tellingly, the executive branch remained forbidden territory, but given the angst already engendered over the entry of public school leavers this was perhaps asking too much for the moment.

As for those currently studying at Cambridge, if the Admiralty thought the dons would make their task of thinning surplus junior officers easier from the *Navy Lists*, they were in for a surprise, as most tutors applied a large dose of sympathy in their marks.[169] Ultimately, though, many officers elected to leave the Service voluntarily and one such officer so recently sent to Cambridge electing to do so was Lieutenant Patrick

---

166 ADM 1/8591/119, H.M. Treasury to Secretary of the Admiralty letter of 31 August 1920.
167 'Naval Notes, Engineers from Universities', *Journal of the Royal United Services Institution*, Vol. LXXXII, No. 527, August 1937, p. 638.
168 'Naval Notes, University Entrants', *Journal of the Royal United Services Institution*, Vol. LXXX, No. 520, November 1935, p. 865.
169 Brownfield, *Sea Cycle*.

Blackett,[170] resigning his commission a month after arriving in the university.[171] Blackett was even at that early date, an officer of much promise having graduated head of his term at Dartmouth and being present at both the Falklands Islands and Jutland encounters.[172] His experiences in these surface gunnery actions were not wasted and soon Blackett was turning his hand to improving British gunnery performance where the margin of hits registered to shells expended had been low—certainly, lower than their German foes. Reaching fame and honour outside the Navy, yet contributing to its success in another war in the field of Operational Research, Blackett may be atypical of the latent talent the Navy risked losing. Still, the risks were real enough for frequently it remains the case that those possessing the greater potential will elect to leave military and naval service when future prospects dim or they are not sufficiently challenged in their daily tasks. Others would have followed Blackett, if outside employment opportunities had been more promising. Even a young midshipmen could appreciate that future career prospects were not assured and Gordon Campbell-Johnston, serving in the Atlantic Fleet battleship HMS *Valiant*, recorded the decision of a close contemporary to chart another course in life much as Blackett had elected to do. In this case, it was his contemporary and fellow member of the *Valiant* gunroom, Charles Cookson, who elected to resign from the Service and pursue studies at Oxford. Campbell-Johnston saw it as a courageous step that others would pursue, including himself, if circumstances only allowed.[173] Campbell-Johnston remained in the Navy, though, unhappy, unsuccessful and unfulfilled retiring in February 1936 as a Lieutenant Commander.[174]

Often, the 'Geddes Axe' and 'Ten Year Rule' are used as a kind of shorthand to ascribe the financial ails besetting the Royal Navy and British defence policy as a whole for the period between the wars. This is incorrect. Both formulae were subject to Cabinet control, and it is to Cabinet where responsibility ultimately falls for the policies adopted. Much to his credit, Lloyd George did not merely endorse the savings proposed by Geddes and ameliorated many of the most drastic recommendations proposed in the initial period.[175] This was especially true of the proposed scheme aiming to reduce the number of executive officers borne by the Service.[176] The Coalition Government was not prepared to reduce the manning of the Service below

170 Patrick Maynard Stuart Blackett, later Baron Blackett (1897-1974). Short Course, Cambridge University, 1919; Director of Operational Research, 1942-45; Nobel Prize in Physics, 1948 and President Royal Society, 1965.
171 ADM 196/146 (Blackett).
172 Lord Zuckerman, *Six Men Out of the Ordinary* (London: Peter Owen Publishers, 1992), p. 15-16.
173 Midshipman Gordon Colin Campbell Campbell-Johnston journal entry of 16 November 1923. (Privately held).
174 *Navy List*, July 1936, p. 623.
175 Peden, *Treasury and British Public Policy*, p. 169.
176 *Hansard*, House of Commons debates of 1 March 1922, v. 151, cc349-50.

the 98,000 figure which the Admiralty believed necessary in the near-term.[177] Still, the opprobrium frequently cast upon Geddes is largely unfair for if Horne had had the final say in the matter, then reductions would have been the greater yet, as the Chancellor was arguing for cuts totaling £130,000,000.[178]

Whilst the forces of the Crown suffered heavy reductions, the levels accepted were sustainable in the near-term given the strategic environment then existing. The problem became one of changing spending patterns when the environment was not so benign and recognizing the need to do so in time adequate to forestall, or if necessary, defeat a new threat. This was understood at the time, but recognition of a problem whilst it may be the first step to ensuring a solution does not guarantee it necessarily happening. John Simon, Home Secretary in 1936 and speaking in the Commons reminded members why foregoing the rule in 1932 had not immediately resulted in increased defence expenditure explaining:

> When the rule had to be given up, there were two considerations which still delayed any provision which might have been contemplated for restoring to a proper condition forces which had fallen into disrepair. One was the financial crisis of 1931 which arose almost at the same time. The National Government was pledged to get the country out of an economic morass. It was then considered that the country's gravest peril lay within its own gates and not without. The risk of national bankruptcy was greater and more immediate than any external risk... The other factor to which I would refer is this: The result of the Disarmament Conference was a terrible disappointment. After years of preparatory work the Disarmament Conference met in February, 1932, and that was the very year in which the 10 year rule had to be abandoned. His Majesty's Government were most unwilling to prejudice the chances of general agreement and any chance which might exist for the limitation of armaments.[179]

Simon, a former aide to Trenchard, is not without knowledge in this case, as he served as Foreign Secretary during the period of the rule's abandonment and was directly conversant with the strategic milieu in play. Accordingly, Simon was not a neutral in these matters and had a past to defend. Indeed, his moral landscape was so averse to contemplating another war that in 1932 he had spoken of Britain being at peace for another fifty years.[180]

If the rule remained subject to Cabinet control, the absence of any reconsideration in the near-term gave the Treasury a strong position when dealing with the Admiralty's

---

177 *Ibid*, cc434-6.
178 John Ramsden, ed., *The Oxford Companion to Twentieth-Century British Politics* (Oxford: Oxford University Press, 2002), p. 264.
179 *Hansard*, House of Commons debate of 10 March 1936, v. 309, cc1973-2099.
180 Uri Bialer, *The Shadow of the Bomber: The Fear of Air Attack and British Politics 1932-1939* (London: Royal Historical Society, 1980), pp. 18-19.

budgetary submissions. Clearly, Richard Hopkins, the Controller of Finance and Supply Services in the Treasury, believed the formulation provided his office a strong-hand in dealing with the defence departments and promoted the idea that the rule should be a floating one: thus, no war was to be considered from ten years onward until changed or cancelled by the Cabinet.[181] Newton's first law applies to bureaucracies as equally to physics. Any proposal for expenditure, which perforce did not conform to the rule, was easily targeted for being at variance to government policy. This was crucial. Most decisions in government never rise to the level of Cabinet consideration; minor officials dispense judgments based on precedent and in conformance to written policy and guidance. The Board of Admiralty waged a constant struggle to retain funds for the Singapore Naval Base, to secure a steady pace of cruiser construction, and to regain control over the Fleet Air Arm. These tactical battles for funding were difficult enough. Far more difficult was it to argue the case for suspending or cance-ling the 'Ten Year Rule', as the issue, though originally a financial formulation soon possessed a greater strategic import.

The fiscal retrenchment experienced led many officers to separate from the Royal Navy who in the natural course of events might have expected a fuller career. King-Hall, the son, grandson, great-grandson, and nephew of flag officers remains a case in point. Having attended the Staff Course in both its naval and military variants and now working as Fleet Intelligence Officer in the Mediterranean Fleet, King-Hall's career was conventional enough to assure him the promise of further promotion and employment. Performing duties and occupying billets typically the providence of more senior officers, he benefited from the patronage and guidance of officers, such as Admiral Sir Roger Keyes. A fuller career remained a distinct possibility, though this conclusion may have not been so evident to him at the time. As it was, King-Hall proved only one of two naval officers from the 1919-39 period attending both naval and military Staff Colleges not to achieve his captaincy.[182]

In his memoirs, King-Hall addressed at some length the perils of being passed-over for promotion to Commander, the keen competition faced in trying to secure another stripe against the limited number of vacancies at hand, and the diminished prospects for the officer missing his chances. In such circumstances, choice assignments were withheld to the passed-over officer and one eked-out the remaining years in back-water duties. Having been denied promotion to Commander twice, King-Hall was flummoxed when at last rewarded. Assignment to the Admiralty for special duties followed, when a period of half-pay was the customary practice following promo-tion. By this point, he had already resolved to leave the Service and rationalized that further promotion to Captain required a minimum of ten years of service whilst an

181 Peden, *Treasury and Public Policy*, p. 215.
182 Commander Cecil Turner remains the only other naval officer not to have been promoted to Captain following attendance at both Staff Courses.

additional decade beyond that would be required if he was ever to reach flag rank.[183] A rational judgment for an officer to make when promotion and assignment flowed from seniority and the Service's prospects for expansion appeared dim. A literate officer with other prospects (he had several volumes published already by this point in his life) he took his leave and stipend, moved to the Royal Institute of International Affairs, stood for parliament, and turned his hand to journalism.

Blackett's case is more difficult to measure. Denied a transfer to the Royal Naval Air Service after Jutland, he was not content to mark time.[184] A man, in time, possessing distinct left-wing political sentiments, how much the conditions prevailing in postwar Britain influenced his decision to leave the Navy is unknown and the suspicion exists that the Admiralty denied his transfer to flying duties because they dare not lose a young officer of such marked talent to the Air Force. Still, with the ongoing British intervention in Russia engaging the attention of his Service widely decried at home and industrial labour disputes arising in Britain following the war Blackett must have thought Cambridge an idyllic refuge. The Cavendish Laboratory and the opportunity to immerse oneself in pure research were tonic to the travails of war and the uncertainties of peace. His departure from the Navy, though, probably owes more to the growth of the man than the receding fortunes of the Service.[185] His case, indeed, highlights the risks of expecting a youth engaged as a cadet at 13 to know his life's path and the dangers of accepting adolescents into the Royal Navy in a social climate far-removed from the ethos of Nelson and Fisher when broader possibilities now abounded.

Blackett and King-Hall, possessing no more the gift of foresight than the politicians of the day, must be measured against the circumstances they faced and not scrutinized in the knowledge of the war to follow. Others could reasonably ask whether the Navy was making the best use of the officers trained at taxpayer expense or if the wastage of members separated was too great when compared to the Army and Air Force officers similarly schooled. This is especially true of the naval and marine officers attending and serving at the Imperial Defence College where direct comparisons to the other forces of the Crown could be made. Such questions featured in parliament and the reply that the Admiralty appreciated the investment made, but other factors weighed in the balance, whilst correct, was received with a large measure of scepticism.[186] In the Royal Navy, the ethos was one of command and not of staff. The flag officers and the staffs of the major fleets were embarked in the very same ships that would face an enemy in battle whether that foe fought at sea, from the land or in the air. Thus, comparisons to British Army and Royal Air Force officers of equivalent rank made by

<hr>

183 King-Hall, *Naval Life*, pp. 258-69.

184 Zuckerman, *Six Men*, p. 15.

185 On Blackett's political views see Brain Cathcart, *The Fly in the Cathedral: How a Small Group of Cambridge Scientists Won the Race to Split the Atom* (London: Viking, 2004), p.121 and Sir Frederick Pile to Liddell Hart letter of 26 November 1948, Liddell Hart Papers, LH 1/575/611, LHCMA.

186 *Hansard*, House of Commons debate 15 December 1937, v. 330, cc144-5.

parliamentarians ignored or discounted the differences in style, performance, risks, and competencies necessary to perform at the most senior levels of command.

Though it was a time of peace, it was far from a peaceful time. Industrial and labour strife were common in Britain, the nation's fiscal problems were endemic, and the relationship of Britain to the Empire was evolving in ways still not settled. Contraction of the Navy was a necessity and even the Service itself was not immune to the social upheaval ongoing. Though it has been said that in every dark cloud there is a silver-lining, the dark cloud of the Invergordon Mutiny proved to be laced with gold or more appropriately, Sterling, as the Royal Navy benefited from an immediate improvement in its financial fortunes. The increased sums that followed the mutiny, however, did not make their way to that portion of the Vote supporting the War and Staff Colleges. If the Admiralty protected the funds set aside to finance the higher education of its executive officers during the years of want, it did not feel a compelling need to make further improvements in such funding when its fiscal climate became brighter. Such funds went to the Fleet Air Arm, the construction of new warships, and the accession of new officers and men.[187] With so many deficiencies in equipment and supply requiring alleviation, this is understandable. Maintaining the operational effectiveness of the fleets in the face of expansion was a difficult proposition, as the necessary priority between the training of entrants and the working up of new ships came at the expense of just the very officers who were eligible for the Staff and War Colleges, the Senior Officer's Technical Course and the Tactical Course. There was logic in the Admiralty's policy of protecting its officer corps at the expense of the lower deck during the retrenchment of 1920s and early 1930s that cannot be distilled by mere financial bookkeeping. The numbers borne may have been excessive to the size of the peacetime fleet, but they represented the nucleus of any future expansion and the reservoir of experience so dearly purchased in war was an asset not to be needlessly discarded by a maritime nation.

During a particularly trying period off the coast of Crete in 1941, Admiral Cunningham could remind the Mediterranean Fleet that losing a ship was easier to bear than sacrificing a reputation. Ever more was this so for the education and development of the Service's middle and higher ranking officers. Many contemporary officers may have held that leaders were born, but strategists and tacticians in the Navy were not. The long and vocal debate in *The Naval Review* regarding specialization, officer entry, and officer education and development was evidence of the strong contemporary feelings on these questions. Thus, this history now turns to Greenwich and to Portsmouth to examine the style and substance of the higher education and training of executive officers within the Service.

187 *Navy Estimates 1936* (London: His Majesty's Stationery Office, 1936), pp. 8-9.

3

Educating a Navy: The Staff Course and the Senior Officers' War Course

---

A fighting officer must never permit tactical training to start from the assumption that we shall fight with superior force. The results of so doing are far reaching; action is cramped, risks are avoided, opportunities missed, and officers imagine they can shelter themselves behind the excuse that their numbers were not adequate ...[1]

Rear Admiral Herbert Richmond, 1920

Though Churchill and others deprecated the preparation afforded to naval officers for higher command in the early part of the century, when at last, the Admiralty created both a War Course and a War Staff Course the Service was never to be the same. Though late to the 'staff game', (the Army having founded its Staff College at Camberley in 1858 following the Crimean War), the Navy came to develop a programme of education and instruction that was broader, more balanced, and infinitely superior to that of the British Army. It was able to accomplish this while simultaneously avoiding the creation of a clique of officers castigated as those 'Bloody Red Tabs' and despised by the greater Service.[2] This is not to say that what evolved was universally admired. On the contrary, criticism aplenty there remained. Yet, the criticisms leveled were often those of the converted desiring that the Navy embrace more of the British Army method and style and not an attack on the staff system itself. This was asking too much. For within the Service, suspicion of staff training and the role of the staff officer remained, but it was to the Navy's credit that the suspicion was of

---

1 'Tactics', Richmond, lecture, Royal Naval War College, Spring Session 1920, Richmond Papers, NMM/RIC/10/2.

2 Taken from the distinctive scarlet facing worn on the uniforms of staff officers in the British Army to distinguish them from non-staff specialists. In 1920, Army Order 539 ordered the removal of the scarlet gorget and head band also worn for staff officers below General Officers rank. A corresponding Admiralty Fleet Order removed the requirement for naval staff officers to wear a distinguishing shoulder strap. See Godwin-Austen, *Staff and the Staff College*, p. 273 and A.R.W., 'Staff Training and the Royal Navy—I', *The Naval Review*, Vol. 64, No. 1, January 1976, p. 11.

the variety of the child looking strangely askance at a new toy rather than a refusal to play with the toy itself.[3]

Amongst those wishing the Service would adopt more of the Army's method was Lieutenant Colonel Sir Rhys Rhys-Williams, late Welsh Guards, who Geddes brought to the Admiralty's Training and Staff Duties Division when becoming First Lord. It says much about Geddes' views of current naval administration that he felt it necessary to bring Rhys-Williams to the Admiralty. Yet, a man willing to terminate Jellicoe's appointment as First Sea Lord in the perfunctory manner accomplished was not a man given to standing on ceremony and the niceties of naval custom. Rhys-Williams, in time also fell out with Geddes, but for the moment writing to Richmond in December 1918, upon the latter leaving the Admiralty and assuming command of the battleship HMS *Erin*, he expressed the wish that:

> I carry away with me the hope, in spite of numerous disappointments, that the ideas which we have put forward will sooner or later be adopted, and that the time we have spent here as "Sowers" will not have been wasted; that the fruits of our sowing will be reaped by a Board of Admiralty of the future, and that we shall live to see a Naval Staff College at Camberley, a scheme of Education in which fitness for war is the ultimate object, under which Examinations are made the servant and not the master, and practice instead of theory rules the early courses of training. I hope the Staff system as we have tried to develop it, on the lines proved by the Army to be correct, may develop in the Navy, and that the close association we have established with the Staff Duties Division will not be allowed to lapse.[4]

More is yet to follow about adopting the Army's methods to naval practice, but the reluctance shown owed as much to the differing needs of the naval service than an inherent distrust of all things military.[5] For some, such as Admiral Sir Douglas Nicholson,[6] the staff system was seen as a Prussian innovation and given Germany's recent failure in the World War, a model not necessarily commending itself to a Service possessing a long, glorious record of success and holding much political influence. He feared adopting a staff system in the Royal Navy—particularly, one modeled on continental lines—would 'obscure the purpose of its existence, and will assuredly tend more and more to reduce its powers of thought and action.'[7]

3   Koe, 'Naval Staff System', p. 667.
4   Marder, ed., *Portrait*, p. 330.
5   Anon., 'Naval Staff College Training', *The Naval Review*, Vol. XX, No. 3, August 1932, p. 490.
6   Admiral Sir Douglas Romilly Lothian Nicholson (1867-1946). Captain, 1904; War Course, 1909; Rear Admiral, 1916; Vice Admiral, 1920, Vice Admiral Commanding, Reserve Fleet, 1922-23; Admiral, 1925 and retired, 1926.
7   Douglas R. Nicholson, 'The Staff System in the Navy: Prussian or British', *Journal of the Royal United Services Institution*, Vol. LXXIII, No. 490, May 1928, p. 246.

As for the proficiency of Army staff methods, it was an article of faith Rhys-Williams shared with Richmond amongst others, but it was not a view, nevertheless, universally held. Cuthbert Headlam, a British Army staff officer, but not a graduate of its Staff College, found much to lament and believed that the methods of the French Army were superior yet writing to his wife:

> This is a heretical thing to say, but I am quite prepared to back it up. They have method in what they do. Their staff work is infinitely superior to ours. The work is coordinated and logical. We are groping in the dark. Our staff people our infinitely pleased with themselves – but they are not masters of strategy that they imagine and compared to the French I cannot help feeling that we are amateurs.[8]

Headlam, in time, would lecture to qualifiers of the Navy's War Staff Course on intelligence procedures in the field, so his views, if sharp, do not detract that others held him gifted and knowledgeable. That officers, such as Rhys-Williams and Headlam, voiced misgivings in their private moments on the frustrations of staff work should not prove surprising and care must be exercised in taking their comments too literally. Still, the audit of war was not all to the Army's credit. The proficiency the Navy showed following the Battle of Jutland in effecting repairs, making good its losses, and caring for its men stands in mark contrast to the administrative skills displayed by the Army at Kut, the Dardanelles, and at Third Ypres. Of the last named battle, caution is required, as the very name of Passchendaele has become a byword for military incompetence which clouds perspective. This is especially the case when commenting on General Sir Hubert Gough. Still, as Nicholas Perry has commented 'Gough and his Fifth Army, directing its opening phases, earned an unenviable reputation for poor operational planning, sloppy staffwork and an uncaring attitude to those under their command.'[9] Such a poor performance would have been understandable in 1914 when the British Army went to war, in the words of Hew Strachan 'with a command structure in which nobody was trained and for which no staff establishment had been laid down' yet that was now long ago.[10]

Of course, comparing a battle lasting but a day to campaigns extending over several months is not a fair comparison, but the point remains that at the operational and strategic levels the Navy was highly effective. It suffered flaws, as the Army did, and no more so than in its response to the submarine threat. It may be that its greatest

---

8   Headlam letter to his wife of 7 February 1917 cited in Jim Beach, ed., *The Military Papers of Lieutenant-Colonel Cuthbert Headlam, 1910-1942* (Stroud: The History Press, 2010), p. 157.

9   Nicholas Perry, ed., *Major General Oliver Nugent and the Ulster Division 1915-1918* (Stroud: Sutton Publishing, 2007), p. 102.

10   Hew Strachan, 'Operational Art and Britain, 1909-2009', John Andreas Olsen and Martin van Creveld, eds., *The Evolution of Operational Art: From Napoleon to the Present* (Oxford: Oxford University Press, 2011), p. 110.

failing was that it did not effectively articulate its case and, if so, the blame resides squarely with its leadership and its staff. Ethos had much to do with how the Service viewed its staff and the concept of a staff system. A desire to fashion staff officers without creating division was key. Key also was the belief that the hallmark of a naval officer's career was command and not coordination. This is not to avow that military officers did not desire to command their regiment or higher parent formation, but a full career in staff assignments whilst rising through the gradations of General Staff Officer—GSO3, GSO2 and GSO1—was possible for those pursuing a military career in a manner distinctly eschewed by the Royal Navy. If the path to command in the British Army was through being first a good regimental officer and then demonstrating ability as an able staff officer, the naval path to command was to prove oneself, above all else, as a competent seagoing officer.[11] Naval officer equivalents to Field Marshal Sir Henry Wilson,[12] Field Marshal Sir William Robertson, or Major General Sir Edward Spears—prominent for achieving general officer rank with little or no command experience—are noticeably lacking. Moreover, the military officer failing to hold a *psc* from one of the Staff Colleges could feel that the path to the apex of his profession was blocked by this gap in his experience despite proving to be very capable in all other respects. Certainly, this was the view of General Philip Chetwode[13] telling Lieutenant General Edmund Ironside[14] that it was 'very improbable' that he would serve as CIGS, as he lacked the *psc* leitmotif and this from an officer who had held several staff appointments and was presently serving as Commander-in-Chief, India.[15]

The experience of the Earl of Cavan and Field Marshal Sir John French demonstrates that securing appointment as Chief of the Imperial General Staff without being a graduate of the Staff College was possible for an exceptional officer, but this ran counter to form. Importantly, the Navy established that executive officers once trained were to alternate between staff appointments and service afloat in duties commensurate with their rank.[16] Though an ideal and often honoured in the breach—witness the career of Rear Admiral Roger Bellairs, Beatty's long-serving Naval Assistant—the intent was clear. The Service desired its officers to remember that the ultimate purpose of their calling was the sea and leading fleets and men in battle. This requirement may be thought a distinction without a difference, as a primary purpose of military and air staff training was to prepare its officers for command. Yet, many of their

11   Bond, *Military Policy*, p. 52.
12   Field Marshal Sir Henry Hughes Wilson (1864-1922). Staff Course Camberley, 1891-92 and Commandant, Staff College Camberley, 1907-10.
13   Later Field Marshal Philip Walhouse Chetwode, Baron Chetwode (1869-1950).
14   Sir William Edmund Ironside, later Field Marshal Lord Ironside (1880-1959). Staff Course Camberley, 1913-14 and Commandant, Staff College Camberley, 1922-26.
15   Ironside to Liddell Hart letter of 13 June 1932, Liddell Hart Papers, LH 1/401/87, LHCMA.
16   *King's Regulations*, Article 233, p. 75.

officers remained in staff assignments while continuing to advance, a norm that the Navy eschewed and the more appropriate analogy is that such military and air officers approached in kind officers of the accountant branch in the Navy who forewent the chance of command and served in a supporting role.

Moreover, the Army looked askance at an officer serving on the staff who was not a product of the Staff College and this aspect of 'trade unionism' was commented upon by those not a member of the elect such as Headlam, who though a GSO1 was not a formally trained staff officer.[17] Nor was this view simply the thoughts of a bitter outsider with Ironside confirming this general prejudice and avowing the:

> Army placed too much belief in the fact that only a professional could be a Staff Officer. That is where the non-professional should have risen as the war went on. Instead of that all the brilliant young Staff Officers were retained on the Staff and the <u>Commanders</u> became more and more mediocre. The Staff Officers eventually had to remain there to carry these commanders. Result, we have few professionals who have exercised command. There is room for the brilliant brain on the Staff and that is its place.[18]

For the Navy, this desire to command, at times, gave rise to confusion over the actual mission of the Staff College. Whether to prepare officers merely for staff duties or to accomplish that task whilst also preparing one for higher command in the Service were opinions responsible officers could hold contrasting views and did.

During his period commanding the Royal Naval War College, William Boyle[19] recommended that the Admiral-President assume responsibility for coordinating all courses touching upon strategy and tactics, and not merely those taught at Greenwich. The Fourth Sea Lord, Vice Admiral Lionel Preston, with his responsibilities for logistics endorsed this proposal adding that consideration should also be given to relocating the Accountant Officers' Technical Course to Greenwich from Portsmouth 'as good relations between the Staff and Secretariat are essential in all Staff work both ashore and afloat.'[20] To this, Sir Vincent Baddeley, a principal secretary of the

17   Beach, ed., *Cuthbert Headlam*, pp. 117-18.
18   Ironside to Liddell Hart letter of 29 March 1937, Liddell Hart Papers, LH 1/401/178, LHCMA. Original emphasis.
19   William Henry Dudley Boyle, later Admiral of the Fleet the Earl of Cork and Orrery (1873-1967). Naval Intelligence Department, 1909-11; War Course, 1912; Captain, 1913; Naval Attaché Rome, 1913-14; Naval Attaché Rio de Janeiro, 1914-15; Commanding Officer, *Repulse*, 1917-18; Staff Duties in *Lion*, 1918-19; Commanding Officer, HMS *Tiger*, 1919-20; Rear Admiral, 1923; Senior Officers' Technical Course, 1925; Tactical Course, 1925; Senior Officers' War Course, 1925-26; Vice Admiral, 1928; President, Royal Naval College, Greenwich, 1929-32; Admiral, 1932; Commander-in-Chief, Home Fleet, 1933-35; Commander-in-Chief, Plymouth, 1937-39; Admiral of the Fleet, 1938 and retired, 1939. Recalled for war service.
20   ADM 1/9041, Fourth Sea Lord minute of 22 October 1932.

Admiralty strongly dissented. Writing to the ACNS, Rear Admiral John im Thurn,[21] against the initiative, Baddeley stressed that the primary aim of the Staff College was to develop staff officers and that any other aspects of its training were tangential to this purpose.[22]

Vice Admiral William James, an officer with much experience of the Staff College to his credit and chairing a committee in 1935 examining how officers should be trained for war, also felt the heat of opinion on this question. One school of thought believed the Staff College should prepare officers for higher command and not just educate them in the higher aspects of war—an important difference—and that attendance should be mandatory. Certainly, Rear Admiral Humphrey Smith[23] was one of those advocating such an approach though, importantly, not the only proponent.[24] That the Staff College did not aim to prepare naval officers for command was unique amongst the Services and this omission in task was further reflected in the relative rank the Director of the college enjoyed. He was but a Captain and much inferior in rank to his military and air counterpart holding general or air officer rank.

Given the ongoing changes in officer higher education made since the war and the disquiet existing about its variety, duplication, conflicts, and overall usefulness, it was natural for the problem to face fresh consideration. The terms of reference provided to the James Committee avow that the recent expansion in the scope of the Junior Officers' War Course (JOWC), an introductory class providing Sub Lieutenants an early grounding in the study of war, was the ostensible reason for examining the lines of executive officer education. Initially begun at Cambridge to introduce the higher aspects of war to officers whilst relatively new to their profession, the course migrated to Greenwich when the Cambridge scheme concluded and operated under the general direction of the Captain of the College.[25] The recent war demonstrated the need for a course to inform relatively young officers on the broader aspects of their profession as the responsibilities such officers had assumed whilst carrying out inspections,

21   Later Vice Admiral John Knowles im Thurn (1881-1956). Appointed War Staff Officer, 1917; Captain, 1918; Torpedo and Mining Department, 1918-20; Director, Signal Department, 1920-21; Commanding Officer, HMS *Ceres*, 1921-23; Flag Captain and Chief Staff Officer in *Hood*, 1923-25; in Command, Signal School, 1925-28; Tactical Course, 1928; Chief of Staff to Commander-in-Chief, Mediterranean, 1928-30; Rear Admiral, 1929; Senior Officers' War Course, 1930-31; ACNS, 1931-33; Commander, First Cruiser Squadron, 1933-35; Vice Admiral and retired, 1935. Recalled for war service.

22   ADM 1/9041.

23   Later Vice Admiral Humphrey Hugh Smith (1875-1940). War Course, 1908-09; Captain, 1915; Commanding Officer, HMS *Argus*, 1918-20; Senior Officers' Technical Course, 1925; Senior Officers' War Course, 1925-26, Rear Admiral and retired, 1926 and Vice Admiral, 1931. Recalled for war service and lost with ship, 1940.

24   H. H. Smith, 'The Education of the Naval Officer to Fit Him for Eventual Higher Command', *The Naval Review*, Vol. XVI, No. 4, November 1928, p. 722.

25   ADM 1/8744/140, 'Command Paper 451. Statement of the First Lord Explanatory of the Navy Estimates, 1919-1920' of 1 December 1919, p. 16 and ADM 196/91 (Davenport).

boardings, and the like could only be executed if they appreciated the intent of their orders. Moreover, such a course would assist officers in their self-education until they were of age and experience to attend a Staff or War Course.[26] With staff duties now added to the mix of the JOWC curriculum, all new officers were to secure training on the role and place of the staff and staff work.[27] The growing plethora of courses now available—witness, Staff, War, Tactical, and Technical training within the Navy and the expanded JOWC—the need to assess the effectiveness of this seemingly hodge-podge of training was becoming paramount.

This was all true enough, but the actual genesis of the committee's work arose out of a parliamentary enquiry in July 1934 into the apparent contradictions and duplication of officer education amongst the Service Staff Colleges and the Imperial Defence College. This prompted Stanley Baldwin, the premier, to refer the matter to the COS Sub-Committee where Wing Commander John Hodsoll,[28] an Assistant Secretary, sent letters directly to the Commandants and Directors of the several schools for their thoughts and, thus, bypassed the Service Ministries in the case of the Staff Colleges in the process.[29] The responses returned by the colleges, in the main, concluded that any duplication was slight at best. However, they also raised the possibility of restructuring the existing scale of Staff College training to have qualifiers from the three Services, or at least a portion of them, attend a further joint course separate from the IDC.[30] At this point, the Admiralty intervened and established the James Committee in March 1935 to arrive at a considered Service position in view of the initial findings presented to the COS. Whether Admiral James and the other committee members meeting between March and May were aware of this origin is unknown.[31] As the committee's terms included key points from the initial findings provided to the COS, and the letter from the COS Sub-Committee to the colleges setting in train the review discussed the political origins of the questions, the likelihood is that they did.

The four courses the James Committee examined in detail were the Staff Course, the War Course, the Tactical Course, and the Imperial Defence Course. A fifth matter of concern was the Navy's cooperation with the other forces of the Crown.[32] If the terms of reference do not mention the Senior Officers' Technical Course or ancillary courses, such as the Accountant Officers' Technical Course, the supplemental questions drafted and framing the committee's remit provided ample scope for their consideration. James was no novice in serving on an *ad hoc* committee to address a pressing issue for

26   Anon., 'Royal Naval Staff College', *Naval Review*, February 1932, p. 12.
27   CAB 53/5, Chiefs of Staff Sub-Committee minutes of 9 October 1934.
28   Later Sir Eric John Hodsoll (1894-1971). Staff Course Camberley, 1923.
29   CAB 53/24, Assistant Secretary, Chiefs of Staff Sub-Committee to Director, Royal Naval Staff College letter of 18 October 1934.
30   *Ibid*, Acting Secretary note of 8 November 1934.
31   ADM 196/90 (James).
32   ADM 1/9041, 'Committee to Investigate the Future Training of Naval Officers for War Terms of Reference'.

he had been a member of the 1920 team investigating the status and position of Petty Officers.[33] Now, James worked with Captain Patrick Macnamara[34] and Commander Geoffrey Norman,[35] as the body's additional members. Macnamara and Norman were gunnery specialists (as was James) and trained staff officers who possessed practical and theoretical knowledge of staff work and staff training. Macnamara's formal staff training had terminated early when war arose in 1914, though he subsequently completed the Imperial Defence Course.[36] Both officers achieved flag rank in due course though in Macnamara's case, as the courtesy title granted to a retiring officer. Whether the recommendations following from the committee would have been materially different if officers of other technical backgrounds had served is impossible to say. It may be the presence of Macnamara and Norman owed a large measure to the fact that they were available to serve. The former only finishing his War Course the month the committee first met where Vice Admiral Ragnar Colvin[37] rated his performance as outstanding in every measure.[38] Additionally, Macnamara had previously served with James on the staff of the *Excellent* before the World War so James would have been knowledgeable of his overall abilities. Norman came from the *Drake* where he had been Staff Officer Operations for the last three years to the Commander-in-Chief, Plymouth and was now marking time in the Admiralty pending appointment to a ship.[39]

33   ADM 196/46 (James).
34   Later Rear Admiral Sir Patrick Macnamara (1886-1957). War Staff Course, 1914; War Staff duties in *King Edward VII*, 1914-15; Liaison Officer to Rear Admiral Hugh Rodman, 1917-18; Plans Division, 1919-20; War Staff duties in Atlantic Fleet, 1921-23; *Excellent*, 1923-24; Captain, 1925; Assistant Director, Plans Division, 1926-27; Imperial Defence Course, 1930; Naval Intelligence Division, 1930-31; Senior Officers' Technical Course, 1931; Naval Attaché, Washington, DC, 1931-33; Flag Captain in *Nelson*, 1933-34; Senior Officers' War Course, 1934; Chief of Staff to Commander-in-Chief, Plymouth, 1935-36; Rear Admiral and retired, 1936. Elder Brother Trinity House, 1936; Recalled for war service.
35   Later Vice Admiral Sir Horace Geoffrey Norman (1896–1992). Short Course, Cambridge University, 1919; Staff Course, 1929; Staff Officer Operations and Squadron Gunnery Officer in *Barham* and *Warspite*, 1930-31; Staff Officer Operations in *Drake*, 1932-35; Captain, 1938; Imperial Defence Course, 1939; Commanding Officer, *Queen Elizabeth*, 1943-45; Rear Admiral, 1947; Vice Admiral and retired, 1950.
36   ADM 196/50 (Macnamara).
37   Later Admiral Sir Ragnar Musgrave Colvin (1882–1954). Captain, 1917; Assistant Director, Plans Division, 1918-19; Commanding Officer, *Caradoc*, 1919-21; Naval Attaché Tokyo, 1922-24; Flag Captain, *Revenge*, 1924-26; Director, Tactical School, 1927-29; Senior Officers' School Sheerness, 1929; Rear Admiral, 1929; Chief of Staff to Commander-in-Chief, Atlantic Fleet, 1930-32; Vice Admiral, 1934; Admiral-President, Royal Naval College, Greenwich, 1934-37; First Member, Naval Board Australia, 1937-41; Admiral, 1939 and retired, 1942.
38   ADM 196/91 (Macnamara).
39   ADM 196/122 (Norman).

Of greater significance was the disposition of the Naval Staff to entertain the suggestions proposed. Here, practical considerations intervened. Recommending major changes to the several curricula had to account for the type and seniority of officers able to attend. This was especially true of the Staff Course given the length of its instruction. The margin of officers once available to attend any course had fallen as the surplus previously borne of Commanders and Lieutenant Commanders was now absent. Moreover, the James Committee heard evidence that the quality of officer now attending the Staff Course was less than before. Such was the view of the present Director of the Staff College, Captain Bertram Watson,[40] who also thought that the status of the course had slipped.[41] As the Committee recorded that upwards of 75% of executive officers were applying to attend the Staff College at some point in their career, it was a loss not evidently visible to those desiring to attend.

Still, one can accept Watson's view regarding the apparent loss of prestige in staff training at face value. The consensus that existed at war's end on the necessity of staff training and education was probably less at the moment the James Committee was sitting than in 1918, as the shortcomings demonstrated in battle receded. Certainly, the Committee's report believed so noting:

> It is perhaps inevitable that when the compelling influence of war is removed, tech-
> nical efficiency and practical seamanship tend to force into the background the
> importance of other factors that go to make up a fully efficient fighting machine.[42]

Here, the actual personnel accepted for Staff College training unwittingly played a part. Desirous not to harm an officer's chance for promotion, naval qualifiers typically accepted to the Staff Course were officers recently promoted. Lieutenant Commanders eligible to attend, but still in the zone for promotion to Commander, were usually not accepted because of the fear that if an officer failed to secure promotion, his time at Greenwich could be seen as the cause.[43] The corollary to such a policy was that an officer recognized that his promotion once gained owed nothing to the Staff College. Lieutenants accepted to the Staff Course did not face a similar risk as their half-stripe followed automatically at eight years seniority if their performance was otherwise thought satisfactory. Still, they, too, had to mind that they spent sufficient time at

---

40   Later Vice Admiral Bertram Chalmers Watson (1887–1976). Plans Division, 1919-21 and 1924-26; Commanding Officer, *Hollyhock*, 1921-23; Captain, 1925; Senior Officers' War Course, 1926; Royal Naval War College, 1926-28; Captain (D) in *Broke*, 1928-30; Assistant Director, Tactical School, 1930-32; Commanding Officer, *Curlew*, 1932-33 and *Valiant*, 1933-34; Director, Royal Naval Staff College, 1934-37; Rear Admiral, 1936; Rear Admiral, Submarines, 1938-39; Vice Admiral and retired, 1940. Recalled for war service.

41   ADM 1/9041, James Committee Report of 21 May 1935.

42   *Ibid*.

43   ADM 203/69, Captain Edward Astley-Rushton, 'The Staff College', lecture to Senior Officers' War Course, *c.* 1924, p. 5.

sea. Thus, Lieutenant Desmond Tufnell,[44] following time in Japan learning Japanese, sought in January 1923 to attend the War Staff Course for 1924. A natural enough follow-on appointment and though promoted to Lieutenant Commander the next month, having spent three years ashore already operated against his desire. Rather, a return to sea to command a destroyer was deemed the more pressing matter in his professional development.[45] The policy if sound on its own merits went far to ensuring that a staff clique would not develop in the Navy in the manner seen in the British Army. It also meant that officers who probably should have aspired to the Staff College as a means to further their professional knowledge did not as they sought other avenues of employment.

It is difficult to generalize on a subject such as the perceived value of staff education and training when the reasons an officer attended were as much personal as professional and the factors influencing the matter appear infinite. To many, staff training was but another form of specialization and the issue went beyond the merits of the Staff College and touched upon the whole question of what sort of officers were now required. If the value of the Staff College was now thought less, then it did not follow that staff training was not viewed as essential by the wider Navy—else why expand the scope of the JOWC. Chatfield was a proponent of reducing the length of the Staff Course and most definitely did not want to add a second year of study to further develop jointness amongst the Services. On the surface, this view appears set against staff training and confirming the conclusions of the James Committee on the relative place of staff education within the Navy. Yet, Chatfield was also the principal proponent of providing staff training to the Service's newest officers in the form of a broadened JOWC. Faced with also sending officers to the IDC, manning the fleet, and finding personnel for an expanded FAA, it is possible to conclude that the increased importance of the staff and staff training was recognized at the highest levels of the Service, but this did not translate into an enhanced status and role for the Staff College.

For Royal Marine officers, the situation was somewhat different. Following the World War, they were eligible to apply for staff training upon reaching the age of twenty-five. In 1927, the age requirement was dropped, and in its place, it was established that an officer now had to have five years seniority in the rank of Lieutenant before being eligible to attend.[46] Still, able to attend was not the same thing as actu-

---

44   Captain Desmond Nevill Cooper Tufnell (1892–1965). In Command, *TB* 4, 1917-18 and HMS *Ferret*, 1918-19; School of Oriental Studies, 1920-21 and *Hawkins* for language study in Japan, 1921-24; qualified interpreter in Japanese, 1922; Government Code and Cipher School, 1929; Commanding Officer, HMS *Bryony*, 1929-31; Staff Course, 1932; Staff Officer Intelligence in *Kent*, 1933-34; Captain, 1936; Commanding Officer, *Coventry*, 1937-38; Naval Intelligence Division, 1938; Naval Attaché Tokyo, 1939-41 and retired, 1946.

45   ADM 196/145 (Tufnell).

46   'Naval Notes, Candidates for Staff Training', *Journal of the Royal United Services Institution*, August 1927, p. 657.

ally attending, and it is instructive that no Royal Marine officer holding a rank below Captain ever attended the Staff Course during the period. Rather, qualifiers usually held their captaincy or majority, instead, at the time of their acceptance.

As to Watson's claim that the quality of officers attending had suffered, his views must carry strong consideration given his previous experience serving on the staffs of both the War College and the Tactical School. Yet, it cannot be the only explanation. Officers having significant war experience were now largely beyond the window of eligibility for attending the Staff College. Of those attending, the quality of ability may have been equal, but the quality of experience was not. This pool of experience was diminishing and not to be replaced. One remedy would have been to broaden the pool of those eligible to attend by allowing non-executive officers to take the Staff Course. Though the terms of reference raised the issue as a matter for investigation, the James Committee failed to offer any recommendations to the question and here fault with its conclusions is ripe. This is not being wise after the event. That engineer, accountant, and medical officers offered unique expertise in their areas of responsibility was accepted, and, as will be shown, the War and Staff Courses were not afraid to make use of their talents. Here, one must conclude that an unhealthy element of orthodoxy now governed matters within the naval service. It had not been that long ago that George Trewin, an Assistant Paymaster, had served as the Naval Observer in the only British air reconnaissance conducted at Jutland.[47] More recently, Paymaster Commander John Webster had served in the postwar Intelligence Division before becoming, first, District Intelligence Officer and then Staff Officer Intelligence in New Zealand.[48] Then, the Navy was willing to accept candidates from all sources, as professional prospects were not so bleak. Tellingly, at roughly the same moment the Service was narrowing its avenues of instruction for non-executive officers, the Air Force Staff College had taken the opposite tack and was moving to allow officers of its stores branch—their equivalent to the naval accountant branch—to qualify for the *psa* at Andover.[49]

Another factor unwittingly affecting who could attend the Staff Course was the timeframe in which the Admiralty made its selections. Those aspiring to attend a course beginning in September needed to have their application forwarded through channels to arrive at the Admiralty by the preceding May.[50] Officers stationed abroad were singularly disadvantaged from applying, as the time available for securing release from current duties and making one's passage to Britain if accepted was severe. Accordingly, officers

47   Arthur Longmore, *From Sea to Sky: 1910–1945* (London: Geoffrey Bles, 1946), p. 55.

48   *Navy List* (London: His Majesty's Stationery Office, September 1920), p. 1815, *Navy List*, April 1924, p. 260 and *Navy List* (London: His Majesty's Stationery Office, July 1925), p. 260.

49   Wing Commander Ephraim W. Havers and Flight Lieutenant Frank N. Trinder were awarded their *psa* in 1929. See *The Hawk: The Journal of the Royal Air Force Staff College*, No. 9, December 1936, p. 136.

50   'Naval Notes, Naval Staff College', *Journal of the Royal United Services Institution*, Vol. LXXI, No. 481, February 1926, p. 169.

able to attend the Staff Course were overwhelmingly those having present appointments in ships serving in home waters or ensconced in shore billets in Britain. Such a policy was sensible given the expense of posting officers to distant stations and the desire to see personnel complete their time during the period of a ship's foreign commission. Yet, with upwards of 40% of naval personnel serving abroad at any one moment, the policy operated against having the best officer attend the course rather than those merely available to attend.[51] Lieutenant Gerald Muirhead-Gould,[52] having good French and Russian and with a working knowledge of Turkish and Hindustani, had been set to attend the first postwar War Staff Course, but the Admiralty rescinded the appointment at the request of the Mediterranean Station, as he could not be spared from supporting ongoing operations in the Black Sea.[53] Consequently, the officer serving abroad typically had to bide his time until reassigned to a domestic appointment before his application for the Staff Course or any course, for that matter, came to fruition.

The Naval Staff, too, recognized the value of non-executive officers—particularly the Naval Intelligence Division—where employing such men in key roles had been a frequent custom. The case of Surgeon Lieutenant Oscar Parkes is perhaps one of the most noteworthy moving to the editorship of *Jane's Fighting Ships* upon leaving the Intelligence Division and the Service.[54] Meanwhile, in 1921, the first officer from the accountant branch was appointed to be an Assistant Naval Attaché. Paymaster Lieutenant Commander Lloyd Hirst, accredited to both Argentina and Brazil, divided time between Buenos Aires in the summer and Rio de Janeiro for the winter months.[55] Why the James Committee missed the opportunity to address the question is unknown, but the suspicion that the three officers considering the future

51   ADM 1/9181, Director of Personal Services minute M/03289/37 of May 1937.

52   Later Acting Rear Admiral Gerald Charles Muirhead-Gould (1889-1945). Flag Lieutenant in HMS *Emperor of India*, 1915-16, in *Southampton*, 1916-17, in HMS *Birmingham*, 1917-18 and in HMS *Lord Nelson*, 1918; Qualified for War Staff, 1919; War Staff Officer in *Emperor of India*, 1919-20; Intelligence Duties in *Iron Duke*, 1920-22; Naval Intelligence Division, 1922-23, 1929-30, 1931-32 and 1936; Staff Course Camberley, 1925; Commanding Officer, *Bluebell*, 1926-27; Staff Officer Operations in *Hawkins*, 1928-29; Senior Officers' Technical Course, 1930; Tactical Course, 1930 and 1936; Captain, 1931; Commanding Officer, HMS *Active*, 1932-33; Naval Attaché Berlin, 1933-36; Commanding Officer, HMS *Devonshire*, 1936-39; Tactical School, 1939-40; to RAN, 1940-44; retired, 1941. Recalled for war service. Acting Rear Admiral, 1942 and Senior Naval Officer Germany, 1945.

53   ADM 196/144 (Muirhead-Gould). Muirhead-Gould's service record reflects that he attended the Staff Course for two weeks in 1930. As the time noted was during the course's third term, the probable explanation is that following his recent service on the China Station as Naval Liaison Officer to the Shanghai Defence Force, he lectured and supported Staff College exercises considering Japan before proceeding to Portsmouth to perform the same task for the Tactical School.

54   Obituary, 'Oscar Parkes, O.B.E., M.B., Ch. B.', *British Medical Journal*, Vol. 2, 5 July 1958, p. 52.

55   'Naval Notes, Paymasters as Naval Attaché', *Journal of the Royal United Services Institution*, Vol. LXVI, No. 464, November 1921, p. 729.

training of naval officers for war sought to protect the prerogatives of the executive branch is a strong possibility. After all, James had been the Assistant Director of Naval Intelligence in 1919 and served as Deputy Director, Royal Naval Staff College in 1923 when Astley-Rushton moved to eliminate the dedicated Intelligence Course, transferring its curriculum into the Staff Course proper. James, presumably, more than anyone could vouch for the value of non-executive officers and remind others that the attendance at the Intelligence Course had not been the preserve solely of the 'X' branch. Alternatively, James, knowing of this history, believed any reconsideration was not worth the resulting discord. The final report suggests the latter reasoning as its introductory sections, tempered with cautionary words about the perils of educational reform, sought to evolve and not fundamentally restyle, officer education.[56]

If the James Committee thought its cautionary words would forestall criticism to its findings, then it was sorely mistaken. Though the report was endorsed by the current Admiral-President, Ragnar Colvin, dissent arose from other quarters including Gerald Dickens, the Director of Naval Intelligence (DNI), and Captain Watson.[57] No immediate tangible changes followed from the deliberations of the James Committee though the Junior Officers' War Course, the immediate cause for the committee's creation, moved from Greenwich to Portsmouth in 1938.[58] Exactly why the venue was changed is unknown and the move at first glance appears to have been a retrograde step removing it from the oversight of the Admiral-President of Greenwich. The most likely explanation is that it was done to circumscribe the perceived authority of the Admiral-President who had immediate oversight of the War Course and substantial influence over the Staff Course. His oversight of the JOWC may have, on further reflection, been thought rather too much grip for one officer to have on the higher education of the Navy. Boyle was a firm advocate for the JOWC and his move to consolidate all higher educational courses under the auspices of the Admiral-President may have prompted the change by the Admiralty as a mean to trim the sails of Greenwich influence.[59]

Thus, a measure of uncertainty greeted any proposal for changes in officer education and while many reasons flourished specifying what was deplored, harder was it to find consensus on how to proceed. A root cause of this uncertainty was that the Service was reluctant to define a single path to senior command. Rather, more valuable than any staff training with its theoretical underpinnings was the importance of judging an officer's potential and fitness for higher command in the cauldron of fleet operations and seeing how one actually performed from the bridge of a ship. For during the period under consideration, an officer commanding a fleet did so afloat

56   ADM 1/9041, Report of Committee to Investigate the Future Training of Naval Officers for War of 21 May 1935. See, especially, paragraph 5.
57   ADM 1/9041.
58   ADM 1/19839.
59   Anon., 'The Study of Strategy by Junior Officers', *The Naval Review*, Vol. XIX, No. 4, November 1931, p. 605.

and shared the same risks as the men in his trust. This may have not been conducive to supporting joint operations and maintaining situational awareness of events when wireless communications were not always possible, but that is to anticipate events. A flag officer's place remained afloat and most definitely not sheltered in a bunker, or worse, yet, a chateau, and divorced from the fleet entrusted to his care.

In early 1926, the dedicated six-week Intelligence Course ceased instruction. Now, officers destined to act as Staff Officer Intelligence, Supervising Intelligence Officer, District Intelligence Officer, Assistant District Intelligence Officer, or as an Intelligence Officer on the staff of an ashore command attended the Staff College instead.[60] Captain Rushton, the Director of the Staff College, proposed this change and, unlike the growing number of rationalizations occurring within the Service, budgetary considerations do not appear as the prime motivating factor. In this case, administrative efficiency was the mainspring. Though endorsed by the Admiralty and receiving the general approval of the DNI and the Director of Training and Staff Duties (DTSD), a closer examination of their reasoning shows that if the change was welcomed, then the arguments of the principals were hardly congruent.

To the DNI, Rear Admiral Maurice FitzMaurice,[61] the change promised to provide a better specimen of intelligence professional, as the Intelligence Course, though concentrated and believed satisfactory, was limited in its objectives and failed to address the broader areas covered by Staff Course instruction.[62] Recognizing that not all new officers destined for intelligence work would attend the Staff College, the DNI suggested that closer cooperation between District Intelligence Officers and War Staff Officers serving in flagships who had attended the course could mitigate shortcomings in career preparation.[63] This was an important concession because some of those taking the Intelligence Course were junior officers only recently commissioned and not eligible to attend the Staff Course.[64] Joining these officers were those in the twilight of their career who would shortly retire from naval service.

Commander Sidney Hill[65] was one such officer, but far from a solitary example. Additionally, Royal Marine officers, with their frequent posting to intelligence

---

60   ADM 182/40, Admiralty Fleet Order '2172.—Intelligence Duties—Instruction' of 15 August 1924 and ADM 182/42 Admiralty Fleet Order '2612.—Intelligence Course—Abolition' of 11 September 1925.

61   Later Vice Admiral Sir Maurice Swynfen FitzMaurice (1870-1927). War Course, 1910; Captain, 1910; Chief of Staff, Eastern Mediterranean, 1916-17; Rear Admiral, 1920; Director, Naval Intelligence Division, 1921-24; Commander-in-Chief, Africa Station, 1924-27, and Vice Admiral, 1926.

62   ADM 1/8668/172, Director of Naval Intelligence minute of 14 January 1924.

63   *Ibid.*

64   Lieutenant James Ashby was one such officer. With seniority dating from June 1923, he sat the Intelligence Course from November 1923 to January 1924 and would not have been eligible for the Staff Course.

65   Later Captain Sidney Arthur Geary Hill (1881-1953). Commanding Officer, HMS *Daisy*, 1912-14 and HMS *Buttercup*, 1917; Hydrographic Department, 1919-20; Commanding

duties, appeared in the Intelligence Course in numbers which the Staff Course would not support unless the Service fundamentally altered its approach to staff education and training. Consequently, elimination of the Intelligence Course would either alter the style of officer employed in such work or guarantee that most officers would not be sufficiently trained to task. In fact, the latter is what largely occurred as will be shown.

To be sure, there was a sound argument for eliminating the course. Many of those attending never applied the skills required or used them but briefly. For Royal Navy officers, the conclusion that the course was but a finishing school in a junior officer's training is a hard one to avoid. Their attendance marked not so much a prelude to intelligence duties as another basic topic of instruction, much like gunnery, torpedo, and navigation, pursued on the path to commissioning. Some would serve as Intelligence Officers but others soon specialized along other lines. To Captain Hugh Tweedie,[66] the DTSD, the change proposed was simply a more efficient means of training Intelligence Officers and such work, of necessity, was best performed by executive officers.[67] It may be only coincidental that this debate was occurring at roughly the same moment when the status of engineer officers was being changed, too, and Tweedie posited his views on who should take the Intelligence Course. Still, both decisions can be seen as enhancing the 'X' branch officer at the expense of those not wearing the executive curl.[68]

It is this writer's view that the change whilst easing the burdens of the Staff College Director by placing the Royal Marine Officer leading the Intelligence Course in a more congenial relationship to him was an ill-considered move. The Intelligence Course originally initiated by the DNI during the war to prepare officers for duties within that Division of the Admiralty, usually met four times a year lasting approximately six weeks with ten to fifteen officers attending. The Staff College curriculum varied from ten to fourteen months during the interwar period and could accommodate a maximum of 40 officers, though initially it was limited to only 16 naval officers.[69] The amalgamation of the two courses represented a compromise and a compromise where the objectives of intelligence training received much the lesser consideration. The five new lectures covering intelligence responsibilities added to the Staff Course came from an original syllabus of 34 lectures. Well could the DNI

Officer, *Endeavour*, 1921 and 1925-27; Intelligence Course, 1924; Captain and retired, 1927. Subsequently, Naval Assistant to the Hydrographer.

66   Later Admiral Sir Hugh Justin Tweedie (1877-1951). Captain, 1914; Director, Training and Staff Duties Division, 1923-26; Rear Admiral, 1926; Vice Admiral, 1930; Senior Officers' Technical Course, 1930-31; Commander-in-Chief, Africa Station, 1931-33; Commander-in-Chief, The Nore, 1933-35 and Admiral, 1935.

67   ADM 1/8668/172, Director of Training and Staff Duties minute of 16 February 1924.

68   K. G. F. 'New Schemes', *The Naval Review*, Vol. XXXIII, No. 2, May 1945, p. 132.

69   ADM 1/8549/20, Admiralty letter to Admiral-President, Royal Naval College, Greenwich of 2 February 1919.

insist that any new graduate would need to spend a month within the Intelligence Division before assuming his duties in the face of such truncated instruction.[70] In truth, in stating such a need the DNI was on safe ground as it was common practice for recently qualified staff officers to spend a period at the Admiralty whilst waiting for their next assignment.

The risk now existed that an officer destined for an intelligence assignment would not secure the training required or, at best, partially trained personnel would fill billets. Examinations to assess an officer's mastery of the curriculum were integral to the Intelligence Course unlike the manner officers were measured at the interwar Staff College where continuous but not final assessment was employed. Additionally, familiarizing officers on British intelligence procedures, and, more importantly, requirements before assuming command of one of the many lesser vessels employed abroad was a benefit of the existing arrangement. Certainly, Admiral Sir Henry Oliver,[71] Second Sea Lord and Chief of Naval Personnel, thought so, telling colleagues:

> The existing course covers a total period of 52 days, and a good many Officers have been able to take the course before going abroad to command small ships where they are often detached for long periods on their own resources....
>
> If the Intelligence courses are practically restricted to Staff Officers, the number of Officers in the Service acquainted with Intelligence work will become more restricted, as the Staff College takes a year.[72]

Perhaps as the first commander of the Navigation School and later to become the *Dryad*, Oliver was more than unusually sympathetic to the training issue. Certainly, as a prior DNI, he was not without knowledge of the skills required of those working in that discipline. Still, the Intelligence Division had personnel requirements difficult to square with a Staff College that considered applicants only from the executive branch. Engineer officers with their knowledge of marine propulsion frequently served within the Division and had attended the Intelligence Course as a precursor to so doing. So, too, was this the case of accountant officers. Finally, intelligence work placed a premium on having officers qualified in foreign languages, and though some Staff College qualifiers possessed such skills, not all who did were destined for intelligence work.

<hr>

70   ADM 1/8668/172, Director of Naval Intelligence minute of 14 January 1924.
71   Admiral of the Fleet Sir Henry Francis Oliver (1865-1965). Captain, 1903; Director, Navigation School, 1903-05; Naval Assistant to First Sea Lord, 1908-12; Rear Admiral, 1913; Director of Naval Intelligence, 1913-14; Chief of the Admiralty War Staff, 1914-17; DCNS, 1917-18; Commander, Battle Cruiser Squadron, 1918-19; Vice Admiral, 1919; Commander, Second Battle Squadron, 1919-20; Second Sea Lord and Chief of Naval Personnel, 1920-24; Admiral, 1923; Commander-in-Chief, Atlantic Fleet, 1924-27; Admiral of the Fleet, 1928 and retired, 1933.
72   ADM 1/8668/172, Second Sea Lord minute of 8 April 1924.

The officers who qualified at the Royal Naval Staff College were overwhelmingly full-time executive officers of the Royal Navy, the Royal Marines, or their equivalent counterparts from the other forces of the Crown. Though the *Navy Lists* of the early 1920s at times reflected the attendance of officers from the accountant branch at the Staff College, this was never the case. These officers were actually attending the co-located Intelligence Course and the practice of listing them for the Staff College was done for reasons of operational security. Paymaster Commander Edward Jones,[73] a Japanese interpreter later moving to the Government Code and Cipher School, is a case in point. Listed as studying at the Staff College, it is telling that his period of study began approximately two months after the commencement of the War Staff Course joining Paymaster Lieutenant Commander Gordon Franklin[74] who arrived in late September 1920.[75] Franklin was an officer with much staff experience working, first, in the Anti-Submarine Division of the Admiralty, and then, subsequently, within the Naval Intelligence Division after the Anti-Submarine Division ceased operations.[76] Neither of these two officers appears in the log maintained by the successive Directors recording the qualifiers, lectures, and expenses of the Staff College and their attendance at the Intelligence Course is confirmed by their service records.[77]

With the abolition of the Intelligence Course, the limited number of accountant officers going to Greenwich ceased. Instead, they now attended the Portsmouth-based Secretaries' Course or the Accountant Officers' Technical Course where aspects of their intelligence responsibilities were covered. Indeed, their four-month Technical Course mirrored portions of the Staff Course covering as it did record and file keeping, codes and ciphering, and seaborne trade. However, the crux of instruction was in surveying all aspects of the law applying to the Navy and lectures on Naval, Military, Air Force, Criminal, Commercial, Prize, and International Law were prominent.[78]

In turn, both the Staff Course and the Senior Officers' War Course received lectures from accountant officers on the syllabus of the Secretaries' Course and the relationship of a secretary to his senior officer. Thus, Rear Admiral FitzMaurice was frustrated in his desire to see non-executive officers taking their place at the Staff College. The then DTSD, Tweedie, was against the concept arguing vehemently that intelligence work was essentially the preserve of executive officers as it supported planning and operations. He deemed the training of other officers as wasteful and inefficient,

---

73 Paymaster Captain Edward Percy Jones (1897-1949). Intelligence Course, 1920-21; Government Code and Cipher School, 1924-26; Captain and retired, 1927.
74 Paymaster Commander Gordon Franklin (1887-1969). Anti-Submarine Division (1917-19); Intelligence Division, 1919-20 and Intelligence Course, 1920.
75 *Navy List*, December 1920, p. 1865c.
76 *Navy List*, July 1920, p. 1815.
77 ADM 196/171 (Franklin) and ADM 196/171 (Jones).
78 *King's Regulations and Admiralty Instructions, Volume II, Appendices* (London: His Majesty's Stationery Office, 1937), p. 88.

as the curriculum was composed of topics largely beyond the scope of their duties.[79] Such a conclusion was correct if one subscribed to the narrow view that the purpose of the college was to train and not educate officers and that logistics support was not a controlling factor in operational planning.

Oliver saw a peril in eliminating the Intelligence Course and relying solely on the Staff College, as the number of officers familiar with intelligence matters would be less than if the current arrangements survived.[80] Commander Tom Phillips of the Plans Division, whilst recognizing that existing practice provided for more trained Intelligence Officers than were necessary for the moment, thought an important side-benefit was its keeping officers employed who otherwise would be on the beach.[81] In the end, the elimination of the Intelligence Course owes much to the argument that the proponents of consolidation were more explicit in their endorsement than those holding any reservations ever were.

The elimination of the Intelligence Course, then, owed much to Rushton and Tweedie. If it made the potential Staff College qualifier a better commodity, then it did so at the expense of the greater needs of an Intelligence Officer. Still, the change proved that in course content the Navy was not simply following the lines of military staff training and that the views of the Service on the role, position, and style of the staff officer remained very different from the British Army. Of course, a major reason why its view of the Staff College was fundamentally different from the position Camberley held within the Army was that it had only recently formed—dating from 1912 and then lying idle during the war years. Consequently, few officers had actually passed through its regime of study. Some might argue that it represented a valuable asset, but that value was largely unproven. Those promulgating the gospel of staff training recognized that the Staff College remained in its infancy. Speaking to the officers of the War Course, Rushton called attention to the fact that:

> There is another closely allied aspect of the differences between the Naval and Military Staff Colleges. Though not entirely true, it is not far from the mark to say that in the Army an officer must have passed through the Staff College in order to obtain General's rank, and he obtains that rank to a great extent because he is p.s.c. That idea will not commend itself to our Service.[82]

Warming to his subject, the Director surveyed the number of officers who could expect to be promoted to Captain that year—approximately twenty—and suggested that the day may yet come when those twenty would be Staff College graduates. However, they would be promoted not because they had taken the Staff Course, but

79   ADM 1/8668/172, DTSD minute of 16 February 1924.
80   *Ibid*, Second Sea Lord minute of 8 April 1924.
81   *Ibid*, T. S. V. Phillips minute of 1 April 1924.
82   ADM 203/69, 'The Staff College', Astley-Rushton, p. 7.

that they attended the Staff Course because they were noted as the officers most likely to rise in the Service.

That the Royal Navy ultimately developed a strong regime of education and training was understood by contemporary observers, if not always by those in the Service itself. The Staff College motto of *Respice, Aspice, Prospecte* (Reflect the past; do not be blind to the present; search well into the future) may only have been a pastiche of Latin to most. Still, it encapsulated the desire of its most fervent supporters for it to be a force for progress and not simply a bastion of tradition.[83] If what followed did not achieve all that was expected, then it still yielded more than is usually credited. Colonel Pierse 'Pat' Mackesy,[84] a sapper and experienced staff officer, even went so far as to adjoin British Army officers:

> When you are stationed at a naval base, when you have opportunity to go for a cruise in one of His Majesty's ships, if you have the opportunity of going to the Naval Staff College or later on of attending what, to my mind, is the most valuable of all courses, the Senior Officers' War Course at Greenwich, take your opportunity and make the most of it.[85]

If Mackesy's effusive view owed more to his association with Quetta, where he had been a member of its directing staff, and not Camberley, his verdict on the Senior Officers' War Course remains telling because, in substance, its instruction was so unlike what the British or Indian Armies attempted at their Staff Colleges with their focus on operations.[86] True, the second year at Camberley focused on corps level formations, but these were merely operations of a larger kind. How such formations came to be in the theatre in the first instance—the strategic rationale of their employment—was largely absent from consideration. This conclusion appears at odds with Professor David French's assessment that Camberley and Quetta focused on making officers effective strategists and not successful practitioners of the operational art.[87] This writer accepts that military staff training did address aspects of policy and strategy. Indeed, a regular scheme of Camberley saw a politician invited to play the role of prime minister chairing a cabinet session with qualifiers rounding out his team of ministers.[88] Additionally, one of the principle reasons why the Aisne battle featured so prominently in the Army's teachings with its frequent tours to France was that

83   C. M. Dawson, 'Royal Navy College Greenwich Centenary 1873-1973', *The Naval Review*, Vol. LXI, No. 4, October 1973, p. 335.
84   Later Major General Pierse Joseph Mackesy (1883-1956). Staff College Quetta, 1927-30.
85   'Imperial Strategy II', Mackesy lecture delivered 1933 in Field Marshal Sir John Dill Papers, 2/3, Liddell Hart Centre for Military Archives, King's College, London, United Kingdom.
86   Philip Warner, *Auchinleck: The Lonely Soldier* (London: Cassell & Co., 2001), p. 51.
87   David French, *Raising Churchill's Armies: The British Army and the War against Germany 1919-1945* (Oxford: Oxford University Press, 2000), p. 164.
88   Godwin-Austen, *Staff College*, pp. 289-90.

taking place during the war's opening phase, the door for strategic decision-making still existed for how best to employ a British Expeditionary Force.[89] Yet, the impression remains that these were tangential to the central focus of instruction and removed from the Navy's approach of focusing on the larger issue of where to employ British force in the first instance and what purpose that force was to achieve once committed. More typical of the training provided at Camberley and Quetta were the lines of conducting a campaign on the Northwest Frontier or how to integrate tanks into operations. Important lessons, to be sure, but not of a variety designed to shed light on the place of strategy in war. Moreover, that the Army actually removed 'object' as a principle of war from its doctrinal foundation, the *Field Service Regulations* (*FSR*) in 1929 indicates that considerations of the higher direction of war were not necessarily at the core of its education and training.[90]

Besides officers of the Royal Navy and Royal Marines attending, officers from the Dominions also studied at Greenwich along with two members of the Army and Air Force. In this regard, the experience of Wing Commander Alec Coryton[91] is illustrative. Serving as a staff officer in Coastal Area Headquarters for the past four years, the Air Ministry simply informed him he was to attend the Greenwich Staff Course set to convene in January 1937. By this point in his career, Coryton had already done a period of service at the Army School of Co-operation at Old Sarum and had served a four-year stint in the Air Ministry's Operations and Intelligence Branch where he focused on the air threat of Britain's erstwhile enemies of the World War.[92] Officers of the Indian Army did not as a rule attend, though Colonel Eric de Burgh[93] did sit the first term of the 1926 course before proceeding to the inaugural IDC course.[94] De Burgh's brief time at the Staff College, senior officer that he was, indicates that the Government of India evidenced a desire to improve military-naval cooperation on the sub-continent. Perhaps if the Admiralty had shown a similar interest by sending an officer to study or instruct at Quetta, this pattern would have continued. They did not and with the opening of the IDC, the Government of India no longer felt the same need to send its military officers to Greenwich.

Until the founding of the IDC, the SOWC was the one Service school dedicated to studying the higher aspects of war and was little concerned about the preparation of

89   Brooke lecture for Aisne tour of 1935, Alanbrooke Papers, 3/17/1, LHCMA.
90   On the importance of the 'object' at the higher echelons of war see Russell Grenfell, *The Art of the Admiral* (London: Faber and Faber Limited, 1937), p. 166.
91   Later Air Chief Marshal Sir William Alec Coryton (1895-1981). Staff Officer Coastal Area Headquarters, 1933-37 and Staff Course Greenwich, 1937.
92   Air Chief Marshal Sir William Alec Coryton Sound Recording 3190 of 18 April 1978, Imperial War Museum, London.
93   Later General Eric de Burgh (1881-1973). Staff Course Quetta, 1913-14; Senior Officers' School Belgaum, 1924-26; Staff Course Greenwich, 1926-27; Imperial Defence Course, 1927 and Staff College Quetta, 1928-30
94   ADM 1/8714/169, C.W. minute of 14 July 1926.

officers in staff duties. This was the case in 1920 and only became more pronounced as time passed and the Staff College and Tactical School assumed greater responsibilities in their respective areas. Within the Service, though, the Staff College was seen as the more important of the two Greenwich schools. It will be recalled that Richmond was emphatic that the reinstitution of the Staff Course was the more pressing concern in 1918, but, importantly, he was far from the only DTSD holding such views. Rear Admiral Vernon Haggard,[95] occupying the office in 1923, thought so as well when he tried to secure the library of the late Sir Julian Corbett for the Staff College rather than see it reside with the War College.[96] Haggard's intervention was prompted in the knowledge that whilst the War and Staff Colleges were currently co-located and shared common administrative means, including a library, this arrangement was not something to be taken for granted. That the junior class was academically the more challenging, had better prospects of survival, and had the larger number of qualifiers were further considerations influencing his wish for securing a widow's bequest.[97]

The possibility that the War and Staff Colleges might cease to exist in close proximity was a risk if the junior course moved to a venue closer to Camberley at the future date so often discussed; it was also a possibility if the War College were to return to Portsmouth. However, another possibility now presented itself. Discussions were ongoing about establishing a new tri-Service college. Discussions that Haggard, as the DTSD, was very much a part of when he met with his Naval Staff counterparts in his office to reach a common naval position on the purposes and aims of such a school.[98] The Admiralty's views of establishing a school to train officers for a common general staff, a possible Ministry of Defence, or to attach officers to the existing CID were at best lukewarm. Here, it shared common ground with the other Services. For the creation of the IDC promised to diminish the stature of the Services and their own body of staff officers and with it the ability to define strategy and the best means of exercising it.[99]

95   Later Admiral Sir Vernon Harry Stuart Haggard (1874-1960). Captain, 1913; Admiralty Liaison Officer to War Office, 1913; Assistant Director, Paravanes, 1917-18; Commanding Officer, *Ajax*, 1919-21; Director, Training and Staff Duties Division, 1921-23; Rear Admiral, 1923; Senior Officers' War Course, 1924; Senior Officers' Technical Course, 1924; Tactical Course, 1925 and 1928; Flag Officer, Submarines, 1925-27; Vice Admiral, 1928, Fourth Sea Lord, 1928-30; Commander-in-Chief, America and West Indies, 1930-32; Admiral and retired, 1932.

96   ADM 1/8650/244, Director Training and Staff Duties Division minute of 29 June 1923 to Vice Admiral Commanding Royal Naval War College.

97   *Ibid.*

98   ADM 1/8644/169, Record of Informal Conference of 27 October 1922. Attending this meeting besides Haggard were Captain Dudley Pound (Director of Plans), Captain Edward Astley-Rushton (Director of the Staff College), Captain Charles Forbes (Deputy Director of the Staff College) and Captain Charles Scott (Deputy Director of Training and Staff Duties).

99   Geoffrey Till, 'Richmond and the Faith Reaffirmed: British Naval Thinking Between the Wars,' Geoffrey Till, ed., *The Development of British Naval Thinking: Essays in Memory of Bryan Ranft* (London: Routledge, 2006), p. 130.

Rather than create new structures and institutions, they desired that the existing courses be located nearer to each other to foster the necessary cooperation and efficiency required in preparing joint plans and appreciations. Staff training, it was held, existed to serve the ends of the separate Services and they each possessed differing requirements. Moreover, an officer, no matter how well trained at a common institution, could not be expected to offer advice to a joint body on affairs separate from his own parent arm and must certainly not proffer advice regarding another establishment.[100]

In fairness to those arguing for a joint school of training, as good as the present Staff Colleges were, they did not train civil servants—even their own. The problems they wrestled with in their schemes and war games were always the ones most germane to the Service setting them and were not necessarily the ones of most pressing importance to the nation. Criticism might be leveled that establishing a joint school before the issue of the higher organization for war had been settled was premature, but a start had to be made somewhere. Questions of defence in the absence of war rarely are a politician's first concern. Forming a joint school now would at least allow the many issues surrounding defence cooperation to be framed and served up in a fashion which the Services could influence.

Root and branch, the officers forming the Service's view on the question sought to protect the existing training regime and most particularly, the Staff and War Courses. Perhaps recognizing that collocation of courses was not likely, as at least two of the present schools would have to move, and knowing that any initiative adopted had perforce not to increase the total cost of staff training, the officers meeting with Haggard proposed a middle way. Thus, they suggested:

> The present War Course for Senior Officers, which can be taken by the majority of Captains, is too valuable to be reduced or modified to form a Joint War Course. But no objection is seen to the consideration of a further joint course for any senior officers who may benefit by it and are available provided the War Course is left intact.[101]

What followed from the Navy's proposals is discussed elsewhere, but the Wood Committee, the body formed by the CID to make concrete proposals on future joint training largely accepted the Admiralty's views that the existing methods should be retained. Thus, the committee's report, *inter alia*, concluded:

> We do not recommend the amalgamation of the existing Staff Colleges of the three Services. The primary object of those Colleges is to train officers in staff duties. They do not claim to combine with that object higher training for command or the investigation of major strategic problems. The existing Staff Colleges suffice only to train enough Staff Officers to provide for the requirements of their respective Services, and

100 ADM 1/8644/169.
101 *Ibid.*

they must be retained for that purpose. The creation of the proposed Joint College would not therefore permit of any reduction of the number of officers now trained at the Staff Colleges or of the curtailment of instruction now given at them.[102]

If the language is redolent of Richmond, then it should prove unsurprising that he was the principal naval member serving on the Wood Committee and his hand in drafting a large measure of the report is discernible.[103]

The claim that the primary mission of the Staff Colleges was to train officers for staff duties but not to prepare qualifiers for higher command was correct only in so far as that was the purpose common to all four schools. Only the Royal Navy truly held to this view and then only in the case of its Staff Course, but not importantly to its War Course. Meanwhile, for both the Army and the Air Force a corollary mission of their staff education was to prepare officers for higher command. Additionally, the Wood Committee was incorrect in its finding that the Staff Colleges did not investigate strategic problems. The directing staff and qualifiers at the Royal Naval Staff College were alive to the major issues of the day with the Admiralty tasking them to investigate and prepare specific appreciations. The evidence supporting this assertion following anon. Richmond knew this to be so, as a former Admiral-President where correspondence covering such requests originating at the Admiralty came to him. So, too, Commander Henry Moore,[104] the committee's Assistant Secretary, was likely aware that such was the case having served at the Staff College from 1919-21 before moving to the CID. Moore's position at Greenwich surely made him privy in the knowledge that some of the schemes pursued had their origins at the Admiralty. Exactly why the committee offered misleading evidence is unknown, but two explanations suggest themselves. First, though amenable to seeing a joint school created, they did not want the existing colleges to suffer through its creation. Secondly, acknowledging that the Staff Colleges conducted planning at the behest of their respective Ministries would have implied that economies were possible in the Plans Divisions

102 *Ibid.*
103 The committee was comprised of Edward F. L. Wood (President of the Board of Education) acting as chairman, Richmond (Admiralty), Major General Cecil F. Romer (War Office), Air Vice Marshal Sir Geoffrey G. H. Salmond (Air Ministry), and E. W. H. Millar (Treasury). Lieutenant Colonel Sir John Robert Chancellor and Commander Henry R. Moore served as Secretary and Assistant Secretary, respectively, to the committee.
104 Later Admiral Sir Henry Ruthven Moore (1886–1978). Designated War Staff Officer, 1919; Royal Naval Staff College, 1919-21; Assistant Secretary, Committee of Imperial Defence, 1921-24; Captain, 1926; Imperial Defence Course, 1927; Commanding Officer, *Caradoc*, 1928-30; Deputy Director and Director, Plans Division, 1930-33; Commanding Officer, *Neptune*, 1934-36; Chief of Staff to Commander-in-Chief, Home Fleet, 1936-38; Rear Admiral, 1938; Chief of Staff to Commander-in-Chief, Portsmouth, 1938-39; Commander, Third Cruiser Squadron, 1939-40; ACNS (Trade), 1940-41; Vice Admiral, 1941; Vice Chief of the Naval Staff, 1941-43; Commander-in-Chief, Home Fleet, 1944-45; Admiral, 1945; Commander-in-Chief, The Nore, 1948-50 and retired, 1951.

of those bodies. With the Treasury represented on the Wood Committee in the form of Mr Eric Millar, the costings of any recommendations made and their sources of funding were matters of no small interest. Significantly, the final report issued by the Wood Committee included Millar's reservations on the savings possible if a joint school was established.[105]

The last point having further credence in light of an order issued shortly after the course was reformed expressly stating that the Staff College had no planning or executive function. Beyond preparing appreciations specifically requested by the Admiralty, the Staff and War Colleges also forwarded papers written by officers during the course of their studies it believed worthy of consideration and perhaps even of a broader distribution. This reluctance to have the Staff College investigate specific topics was partly attributable to the uneven talent of the qualifiers when first arriving. Until they were familiar with the concepts taught, using them in any studies was largely premature. The same was not the case of the directing staff. Yet, if employed on such investigations, their primary duties of facilitating course instruction would suffer. Further, with the development of a proper staff system at the Admiralty during the war, the need for the Staff College to perform such a function conflicted with those parts of the Naval Staff now established and available to so act.

Such then was the background to the Admiralty order issued in August 1919 advising that the Staff College was:

> "Solely an instructional establishment engaged in training officers for War Staff duties, and working under the general superintendence of the Director of Training and Staff Duties.

> "It has no planning, executive, or administrative functions, nor is it responsible for advice to the Board on these matters."[106]

The order issued owed more to ensuring that the prerogatives of the Naval Staff were protected than an absolute statement of naval practice. For as demonstrated below, Greenwich was consulted and advice was tendered on a regular basis to the Naval Staff. Responsibility, though, rested at the Admiralty and the Staff College had no *ex officio* role in policy formulation. The relationship between the Director of the Staff College and the DTSD, the immediate Naval Staff official responsible for the college, was not always an easy one. A contributing factor was that the DTSD was typically the junior officer in seniority and his posting to the Staff College as its Director in a subsequent assignment was not unusual. Moreover, other members of the Naval Staff had important equities in how the course operated.

105 ADM 1/8644/169.
106 Admiralty Weekly Order 2678 of 9 August 1919 cited in 'Naval Staff College', *Naval Review*, p. 18.

Foremost was the First Lord, but practice demonstrated that unless a politician of strong proclivities held the office, such as a Churchill or a Geddes, his power was more latent than realized. Such was not the case of the First Sea Lord and, since 1917, when he assumed the additional role of Chief of the Naval Staff, the officer responsible for the operations and movements of the fleets and staff work in general. Assisting him were two senior subordinate officers, the Deputy Chief of the Naval Staff and the Assistant Chief of the Naval Staff, having strong interests in the work of the Staff College.[107] The Permanent Secretary of the Admiralty, a civilian member responsible for overall staff organization and procedure, was another official of the Admiralty Board having direct interests in the college, its relationship to the Admiralty in general, and the training it provided to qualifiers.[108] The Admiralty Adviser on Education and, of course, the Directors of the major Naval Staff Divisions had interests when matters arising from the Staff College touched their immediate areas. The Commanders-in-Chief of the several fleets and the major shore establishments, possessors of the staffs where qualifiers ultimately were destined to serve, also exerted no little influence in how the Staff College operated. Yet, it remained for the Admiral-President, Royal Naval College, Greenwich, due to conditions of rank and presence exerted the greatest influence over the workings of the Staff College.

When the War Staff and Senior Officers' War Courses were reformed at Greenwich following the war, the Admiral-President's relationship to the courses was one of a host to his tenants. Upon the termination of Admiral Sir Frederick Tudor Tudor's[109] appointment in 1922, and with Richmond overseeing the War Course already present at Greenwich as another flag officer, the two offices merged. The President of Royal Naval College, Greenwich now assuming the additional title of Vice Admiral Commanding the War College and becoming responsible for not only the college in general, but the War Course specifically. The Director of the Staff College, an officer holding Captain's rank, oversaw the common facilities shared between the Staff and War Courses including the civilian support staff, the chartroom, and the library.[110] The Director of the Staff College reported to the DTSD, an officer perhaps of flag rank, but not always, and definitely junior to the officer holding the command of the Royal Naval College and reporting to the First Sea Lord. Lastly, the Director of the Staff College had, until early 1926, the additional obligation of overseeing the Intelligence Course also residing at Greenwich and meeting the requirements of the DNI.

---

107 Oswyn A. R. Murray, 'The Administration of a Fighting Service', *The Naval Review*, Vol. XXV, No. 2, May 1937, p. 248.

108 *Ibid.*

109 Admiral Sir Frederick Charles Tudor Tudor (1863-1946). Captain, 1902; Rear Admiral, 1913; Third Sea Lord and Controller, 1914-1917; Vice Admiral, 1917; Commander-in-Chief, China, 1917-19; President, Royal Naval College, Greenwich, 1920-22; Admiral, 1921 and retired, 1922.

110 'Naval Staff College', *Naval Review*, p. 17.

If the previously discussed arrangements were less than Teutonic in their efficiency, then it merely reflects that each of the courses created and now residing at Greenwich and Portsmouth appeared at separate times to address separate needs and did not arise from the master design of a single mind. Each course arose to solve a specific set of problems. If they overlapped in portions of their content, especially in tactical instruction, then this was seen as not a fault but as an essential requirement. Not all executive officers were destined to go to the Staff College, and though it was desirable to have officers complete the SOTC before proceeding to the War Course, the need to man the fleet and shore establishments meant that officer education at times was secondary to the requirement of actually operating a navy. One flag officer recognizing the perils of overlap, promised to coordinate the two lectures on contemporary tactical problems then offered at Greenwich with the Director of the Tactical School, if that establishment could not send its own officer to lecture on the subject. This approach proving amenable to Captain Ragnar Colvin, the newly installed Director of the Tactical School, and his immediate predecessor, Captain Usborne, and the War Course continued to survey tactics from both a historical and contemporary perspective.[111]

The tact demonstrated by Vice Admiral Richard Webb[112] and found in the above approach was not a trait present in his successor, 'Ginger' Boyle. A capable if rather impetuous officer, in 1925 he had pleaded for the development of a specially modified cruiser to face the growing air threat—an idea that would wait many years before seeing adoption.[113] Now, Boyle sought to have all naval training touching upon war made subject to the coordinating efforts of the Admiral-President. This proved to be a proposal too far. His initiative, if sound on its own merits, was difficult to square with the equities of the diverse parties concerned. The Admiralty was having none of it and the fear that if adopted it would lead in time to the Admiral-President accruing a status similar to that enjoyed by the Commandant of the Army Staff College is hard to escape.[114]

Though the Navy eschewed embracing doctrine—the word noticeably absent from its 1925 edition of the *Naval War Manual*—doctrine ever increasingly was imparted in its education and training as precepts codified in a series of manuals, instructions,

111   ADM 1/8710/138, Vice Admiral, Royal Naval War College to Secretary of the Admiralty letter of 22 February 1927.

112   Later Admiral Sir Richard Webb (1870-1950). Captain, 1907; War Course, 1907-08; Royal Naval War College, 1908-09; Director, Trade Division, 1913-17; Rear Admiral, 1918; Vice Admiral, 1924; British Naval Mission to Greece, 1924-26; Admiral-President, Royal Naval College, Greenwich, 1926-29; Admiral, 1928 and retired, 1929. Editor, *The Naval Review*, 1931-1950.

113   ADM 1/8685/151, Commander-in-Chief, Atlantic Fleet to Secretary of the Admiralty letter No. 682/A.F.1 of 15 April 1925.

114   ADM 1/9041 and Cork and Orrery, *Naval Life*, pp. 155-56.

orders, and practices.[115] In time, an Admiral-President would seek to propose some rationalization to a structure that had grown like topsy-turvy and the interests involved and having a stake in naval education and training would demonstrate just how strong their equities were. To this, it is emphasized that the training and education pursued were not ends in themselves, but operated within the context of an establishment struggling to ensure that what followed was proper for the Navy and not merely the grafting of military means to a naval setting.

Though the order quoted previously by the Admiralty left little doubt that the Staff College had no formal role in providing advice and preparing plans, evidence demonstrates that the Naval Staff sought and accepted its assistance nonetheless. Moreover, a major purpose of the annual joint exercise conducted at Camberley amongst the Staff Colleges was to evaluate problems forwarded by the CID or posed by the War Office, the Air Ministry, and the Admiralty working in common. The writer will leave it to another chapter to address this part of the equation. However, Anthony Wells is mistaken to avow that 'At no time during the inter-war period did the RN Staff College conduct a scheme or war game which was directly sponsored by a Naval Staff department or maintain a system of direct feedback other than for the content of lectures given by Admiralty Staff.'[116]

Reviewing the exercises conducted by the main fleets and drawing attention to possible lines of further investigation was one particular manner how the Naval Staff employed the Staff College. After assessing the results of exercise 'K.E.', an Atlantic Fleet serial of March 1927 which demonstrated new screening and attack dispositions at night, and 'L.B.2', an Atlantic Fleet night-time exercise held 24 January 1928 which saw light forces attacking a screened battlefleet, the Staff College had much to say.[117] It advised the Admiralty that anti-submarine equipment, in this case, the Type 114 Asdic might aid a fleet in locating an opposing surface force.[118] This was an important possibility, as the primary available means for detecting enemy forces at night—searchlight and star shell—though effective in their own spheres came at the price of announcing one's own location to an opposing force.[119] In exercise 'L.B.2', the British force of heavy ships known as Red Fleet and consisting of the Battle Cruiser Squadron in company with the battleships *Nelson* and *Iron Duke* was attacked by the cruiser and destroyer forces of Blue Fleet. Blue Fleet operating without heavy ships was able to launch torpedo attacks against the Red Fleet. These attacks were thought

115 ADM 1/8710/138, Colonel L. H. R. Pope-Hennessy, 'The Services and a Common Doctrine of War', lecture of 17 January 1927, Royal Naval War College.

116 Wells, 'Staff Training', in Bond and Roy, eds., *War and Society*, p. 103.

117 ADM 186/146, *C.B. 1769/29(2). Exercises & Operations, 1929, Volume II.* Admiralty, Naval Staff, Tactical Division, June 1930.

118 ADM 1/8735/71, Admiral-President, Royal Naval College, Greenwich to Secretary of the Admiralty letter S.C. 4515 of 10 December 1928.

119 In 1937, an improved flash-less star shell was introduced minimizing the defects of the type available in 1928.

successful as two of the three torpedoes employed displayed evidence of striking either *Hood* or *Repulse*.[120] Further trails were carried out along the lines discussed by the Staff College, and in 1931, the Tactical Division was advising the fleet that such use of Asdic and known as Radio Acoustic Position Finding could be successfully employed to fix an opposing vessel up to 200 miles with an average error of 6 miles.[121] Rare is it, though, that any tactical method does not come without a trade-off. In this case, wireless communications were required when employing Radio Acoustic Position Finding and so announcing one's own presence, if not precise location, was a risk for the British force employing.

Of course, the recent war had highlighted many shortfalls in fleet tactical procedures; shortfalls that provided ready topics for investigation to the qualifiers. Before a battle was possible, an opposing force had to be located and the Royal Navy defined several methods of conducting searches to fix an enemy fleet in *C.B. 920: The Cruiser Manual*. A revised manual drafted by King-Hall under Astley-Rushton's sponsorship when serving in the Training and Staff Duties Division and adopted in 1920, was ripe for review.[122] A gunnery specialist who once went aloft in an experimental kite to observe the fall of shot and now serving at the Staff College as its Director, Astley-Rushton led a series of investigations demonstrating that the searches specified in *The Cruiser Manual* were difficult to implement without transmitting excessively long signals.[123] If the signals were revised and abbreviated, then the appropriate search for a given situation might be used rather than adopting the search procedure that was simply the easiest to order.[124] The college's thoughts, in this case, were valuable as conducting searches formed a regular part of the many schemes conducted during the course and were regularly repeated each year by successive qualifiers. Any recommendations, thus made, by the Staff College benefited from seeing the common errors repeated as officers worked to define solutions to a similar problem set each year.[125] Confirming that the problem identified by the college was a matter of no small import

120 Midshipman Hugh A.V. Haggard journal entry of 24 January 1928, Haggard Papers, IWM/85/21/3.

121 *C.B. 3016/30, Progress in Tactics, 1930*, Admiralty, Naval Staff, Tactical Division, July 1931, British Sources, Box 12, Naval Historical Center, Washington, DC.

122 King-Hall, *Naval Life*, pp. 170-71.

123 Robert Travers Young, *The House that Jack Built: The Story of H. M. S. Excellent* (Aldershot: Gale and Polden Limited, 1955), p. 106.

124 ADM 116/2471, Director, Royal Naval Staff College to Admiral-President, Royal Naval College, Greenwich letter S.C. 2064 of 10 July 1925.

125 ADM 203/100.

are the actions of Admiral Sir Dudley Pound.[126] He was still working to abbreviate the signals covering tactical evolutions during his time in the Mediterranean Fleet.[127]

In a similar vein, the Staff College called attention to the lack of training in naval gunfire support and the neglected opportunities this represented. Writing to the Admiralty in late 1929, Admiral Boyle, recalling the 1923 decision to only train in such measures when events actually required it, noted the absence of personnel presently familiar with such shoots and the lack of appropriate targets on-hand if the requirement to support an opposed landing was to arise.[128] As a key feature of the interwar curriculum was instructing qualifiers in the Navy's role in supporting combined, or what are typically now labeled amphibious operations, Boyle and Rear Admiral Charles Little,[129] the Director of the Staff College, were highlighting a serious shortfall in capability.

Moreover, the Admiralty was not adverse to tasking the Admiral-President of Greenwich to investigate a particular question of concern. Thus, in 1927 Richard Webb chaired a committee to review the influence of international law on naval strategy and tactics including how it specifically applied to ships of the fleet.[130] A later discussion will revisit the contemporary debate about international law and its place in officer education, but Richmond probably was aware of Webb's work when he contemporaneously raised the matter in association with the Imperial Defence College to the CID.

As qualifiers gained proficiency in the subjects taught, they researched topics set by the directing staff, prepared appreciations, and presented their findings in their

126 Later Admiral of the Fleet Sir Alfred Dudley Pickman Rogers Pound (1877-1943). Royal Naval War College, 1913-14; Captain, 1914; Qualified War Staff, 1914; Assistant Director, Plans Division, 1917-18; Director, Operations Division (Home), 1918-19; Director, Plans Division, 1922-25; Tactical Course, 1925; Chief of Staff to Commander-in-Chief, Mediterranean, 1925-27; Rear Admiral, 1926; ACNS, 1927-29; Vice Admiral, 1930; Second Sea Lord, 1932-35; Admiral, 1935; Chief of Staff to Commander-in-Chief, Mediterranean, 1935-36; Commander-in-Chief, Mediterranean, 1936-39; First Sea Lord, 1939-43 and Admiral of the Fleet, 1939.

127 Anon., 'Admiral of the Fleet Sir Dudley Pound, G.C.B., O.M., G.C.V.O.' *The Naval Review*, Vol. XXXI, No. 4, November 1943, p. 285.

128 Director, Royal Naval Staff College to Secretary of the Admiralty letter S.C. 5123 of 13 December 1929, Roskill Papers, ROSK/7/153.

129 Later Admiral Sir Charles James Colebrooke Little (1882-1973). Five firsts in examinations for Lieutenant; Captain, 1917; Assistant Director, Trade Division, 1919-20; Director, Trade Division, 1920-22; Chief of Staff to Commander-in-Chief, Mediterranean, 1922-24; Royal Naval War College, 1924-26; Commanding Officer, *Iron Duke*, 1926-27; Senior Officers' War Course, 1927; Director, Royal Naval Staff College, 1927-30; Rear Admiral, 1929; Commander, Second Battle Squadron, 1930-31; Tactical Course, 1931; Commander, Submarine Flotilla, 1931-32; Vice Admiral, 1933; DCNS, 1933-35; Commander-in-Chief, China, 1936-37; Admiral, 1937; Second Sea Lord and Chief of Naval Personnel, 1938-42; Commander-in-Chief, Portsmouth, 1942-45 and retired, 1945.

130 ADM 196/43 (Webb).

own lectures. Frequently, qualifiers presented papers of which they had direct knowledge. Others, though, wrote to common topics that appeared regularly from course to course. Though not written to today's academic conventions, the papers were, nevertheless, thoughtful discourses that supplemented the many lectures provided by the instructors or presented by the many guest lecturers appearing at Greenwich. Charles Drage, for his paper on 'Some Modern Naval Mutinies', visited the Historical Section of the French Ministry of Marine to secure background material in preparation for his lecture in the course's final term. On his honeymoon at the same time, what the new Mrs Drage thought of his dedication is not recorded, but as a daughter duly followed in the normal course of events it was not all work for Drage. It was also whilst in France that Drage met Admiral Sir Roger Keyes, so recently of the Mediterranean Station. Keyes, much interested in Drage's research, recounted the tale of a flag officer who had faced a mutiny as a cadet (the crew of his cutter threw their oars overboard), midshipman (his ratings tried to abandon him) and sub-lieutenant (the boys in an old sailing ship refused to come down from the rigging and topmast).[131] Upon returning home, Drage visited the champion of the lower deck, James Wood, but more commonly known as Lionel Yexley, editor of *The Fleet* and a regular speaker at the Staff College on the affairs of the lower deck to enquire about the events of October 1918.[132] Such thoroughness in preparing a paper was outside the norm, but Drage now with an Oxon MA in Modern History and winning naval history prizes in 1922 and 1923 had a reputation to uphold.[133]

Beyond the sharing of common administrative means, an important benefit of having the Staff Course co-located with its more senior partner was that qualifiers oft times attended talks sponsored by the higher course, an important consideration when the funds supporting outside lecturers to the Staff College were significantly reduced in 1932.[134] Still, a great deal of the course's instruction was provided by the directing staff, limited in size though it was, augmented by student efforts such as Drage's. More than furthering a qualifier's skill in one particular subject area, the habit of having qualifiers present their work to their peers developed presentation skills and allowed the directing staff to identify those officers possessing the requisite aptitude who might return and serve as a future instructor. Accordingly, during his time in attendance, the qualifier could expect to hear between 160-215 lectures of which the directing staff presented, at best, only half of them.[135]

131 Diary entry 2 August 1928, Drage Papers, IWM/PP/MCR, Reel 3.
132 *Ibid*, entries 29 August and 17 September 1928.
133 ADM 196/146 (Drage).
134 In 1931, the funds available for outside lectures were £450 for both courses. In the following year, the War College received a similar amount but they were reduced to £200 for the Staff College. As the typical guest lecturer received a stipend of roughly £5, a major reduction in the number outside speakers was introduced. See *Navy Estimates for the Year 1932* (London: His Majesty's Stationery Office, 1932), p. 73.
135 Anon., 'Royal Naval Staff College', p. 28 and ADM 203/100.

'The Fleet Submarine', 'The Influence of the Panama Canal on Strategy' and 'Divided Tactics' are typical of the papers presented by qualifiers in recurring years appearing successively in the 1919-20 and 1920-21 sessions. Still, 'Fleet Communications' offers the case of a topic originally presented by a qualifier later featuring as a core competency in future War Staff Courses. Lieutenant Commander Stewart Spicer[136] originally covering the matter as a qualifier and then presenting it anew when retained as a member of the directing staff following completion of his 1919-20 course. An experience similarly shared by Danckwerts who remained at Greenwich in 1920 after securing his War Staff designation. As a new member of the directing staff, Danckwerts continued to lecture on 'Divided Tactics' whilst specializing on the defence of trade, convoy, and blockade.[137] Finally, matters that came to be presented routinely by the directing staff, at times, had their origin in a paper presented by a qualifier. Thus, 'Esprit de Corps and Morale', a regular subject of discourse, was first presented by Lieutenant Colonel William Barber of the King's Royal Rifle Corps and a qualifier at the inaugural postwar session and continued to be presented by subsequent directing staffs. Meanwhile, Captain John Wickham, Royal Engineers, another qualifier from the 1919-20 session, presented two papers in his area of expertise. 'Topography in War' and 'The Effects of Topography on Tactics', whilst attending the War Staff Course and returned to Greenwich in succeeding years to speak on these same themes.[138]

When first arriving at the Staff Course, qualifiers were offered a strong measure of instruction in the minutiae of naval administration. This is unsurprising as the primary purpose of the course was preparing qualifiers to assume the role of a staff officer. Lectures in naval administration, file and record keeping, Admiralty organization, plotting, and the drafting of orders and instructions combined with early schemes covering the same featured. In order not to overwhelm the qualifier in pure tedium, other lectures also appeared of a historical nature covering naval warfare, naval strategy, and the Seven Years War. Combined operations were always a strong suit of the Staff Course and here the recent experiences of the Dardanelles campaign, the German Navy assault on Oesel in October 1917, and the Zeebrugge raid conducted on St. George's Day, 1918 were at the fore. Initially, Lieutenant Colonel William Godfrey,[139] Royal Marine Light Infantry, who had served as staff officer throughout

---

136   Later Captain Sir Stewart Dykes Spicer, Bart. (1888-1968). Flag Lieutenant in *Orion*, 1913 and HMS *Antrim*, 1913-16; War Staff Course, 1919-20; Royal Naval Staff College, 1920-21; War Staff Officer to Commander-in-Chief, North America and West Indies, 1921-23; Staff Course Camberley, 1926; Senior Officers' Technical Course, 1928; Tactical Course, 1929; Operations Division, 1929-30; retired and Captain, 1933. Recalled for war service.
137   ADM 203/100.
138   *Ibid.*
139   Later General Sir William Wellington Godfrey (1880–1952). War Staff Course, 1912-13; Temporary Lieutenant Colonel, 1917; GSO1, 1917; GSO1 in HMS *Arrogant*, 1917-19; Staff College, 1919-24; Plans Division, 1924-27; Lieutenant Colonel, 1927; Senior Officers'

the Dardanelles campaign, surveyed the subject in a series of five lectures expanding his treatment in later years by adding lectures specific to the Gallipoli operation itself.[140] Godfrey's knowledge on the Dardanelles was of a special type as he was one of two officers who had devised the initial plan of attack against the Turkish forts.[141] Admiral of the Fleet Lord Keyes valued Godfrey's abilities—especially his clarity of expression—and as he would later serve as Adjutant General Royal Marines, Keyes' judgment in this instance was one shared by others.

One purpose in surveying recent amphibious operations was to derive lessons for future British practice, but a corollary reason was to inform British thinking on the defence of Singapore.[142] Whilst some officers drew as the primary lesson of the Dardanelles campaign that such assaults were always difficult and only becoming increasingly so, others noted that effective execution required proper planning, the allocation of sufficient force, and the will to prevail pointing to the highly successful and aforementioned German operation at Oesel, the largest of the two islands governing access to the Baltic Sea from the Gulf of Riga. Thus, it was not enough to say that Singapore was far removed from Japan and the threat was slight, as others might not subscribe to such a view.

Consequently, from the beginning of its reforming, the Staff College covered Japan at great length in lectures, schemes, and war games and the analyses pursued were not limited merely to an examination of a possible Anglo-Japanese war. Frequently, Greenwich tackled the question from the perspective of a possible Japanese-American conflict given the known friction existing between the two powers. Accordingly, Lieutenant Frederic Peters, a Canadian serving in the Royal Navy and qualifying in 1919-20, presented a paper surveying the place of the Philippine Islands in a possible Japanese-American war. In 1920-21, Lieutenant Alfred Phillips,[143] who won the 1919 Admiralty prize in naval history, wrote as his essay, 'The Pacific Ocean and Japan-United States Relations', presenting his conclusions to peers in an accompanying

School Sheerness, 1927; Brevet Colonel, 1930; Colonel Commandant, 1933; Major General, 1935; Adjutant General, 1936; Lieutenant General, 1937; General and retired, 1939. Recalled for war service.

140 ADM 203/100.

141 Cecil Aspinall-Oglander, *Roger Keyes: Being the Biography of Admiral of the Fleet Lord Keyes of Zeebrugge and Dover* (London: The Hogarth Press, 1951), pp. 128 and 221.

142 CAB 53/14, Chief of Naval Staff note of 15 February 1928.

143 Later Rear Admiral Alfred Jerome Lucian Phillips (1893-1979). Five firsts in examinations for Lieutenant; Staff Course, 1920-21; Naval Ordnance Department, 1928-30; Tactical Course, 1930 and 1937; Staff Officer Operations and Squadron Gunnery Officer in HMS *Centaur* and *Cairo*, 1932-34; Training and Staff Duties Division, 1934; Captain, 1934; Senior Officers' School, Sheerness, 1935; Imperial Defence Course, 1936; Senior Officers' Technical Course, 1937; Tactical Investigation Section in Tactical School, 1939; Deputy Director, Royal Naval Staff College, 1939; Commanding Officer and Chief Staff Officer in HMS *Norfolk*, 1941; retired and Admiral, 1944. Recalled for war service.

lecture.[144] Indeed, the qualifiers of the Staff College were introduced to the Japanese question from any number of angles whether historical (Russo-Japanese War), contemporary (United States-Japan or British Empire-Japan), or economic. In the last instance, Ernest Fayle offered his thoughts on Japanese economic vulnerabilities and their susceptibility to interruption by any British naval action.[145]

The pattern of training though varying from year-to-year, invariably concluded with the playing of a major strategical game where the skills acquired during the year's study were fully drawn upon in preparing orders and instructions, drafting supporting appreciations, assigning forces to tasks, employing the *Battle Instructions* and *Manoeuvring Orders*, and conducting play.[146] With qualifiers detailed to play the roles of specific force commanders for both Red and Blue, the directing staff acted as umpires overseeing the game and ruling on exchange results and signals. Other members of the course served as recorders and plotters capturing the progression of play and the results arising. At the conclusion, the Director of the Staff College submitted to the Naval Staff a full summary of the strategical game noting any significant aspects, but especially reporting on those points requested by the Admiralty for investigation.

A recurring theme for the strategical game was Britain against Japan in a future Pacific war and for the 1920-21 session, the game occurred over five afternoons beginning on 24 June and ending on 1 July 1921. During the 1921-22 course, a presumed Anglo-Japanese war in the Pacific lasted eighteen afternoons and the suspicion remains that the recent termination of the Anglo-Japanese alliance arising from the Washington Naval Conference placed a premium on understanding the new strategic setting; hence, the game's growing complexity and duration.[147] Again, in 1924, the theme of the major game conducted as the course was approaching its end was a possible war between Japan and the British Empire. In this case, the Admiralty tasked the Staff College to use the format to investigate the role of a British China Squadron in any operations before the arrival of the British main fleet.[148]

Lieutenant Commander Robert Stewart,[149] a Signals Officer and a member of the 1922-23 War Staff Course, visited both the Admiralty and War Office as he researched his paper on how to defend Hong Kong in a war with Japan. With the

144 ADM 196/55 (Phillips) and ADM 203/100.
145 ADM 203/100.
146 ADM 186/78, *C.B. 3011. War Game Rules, 1929.* Admiralty, Naval Staff, Tactical Division, April 1929, p. 26.
147 ADM 203/100.
148 ADM 203/69, Astley-Rushton lecture 'The Staff College' p. 12.
149 Later Captain Robert Ross Stewart (1893-1988). Five firsts in examinations for Lieutenant; Flag Lieutenant in *Hawkins*, 1919-22; War Staff Course, 1922-23; Squadron Signal Officer in *Barham*, 1923-24; Fleet Signal Officer in *Iron Duke* and *Queen Elizabeth*, 1924-26; Naval Air Division, 1931; Staff Course Andover, 1931-32; Tactical Course, 1935; Naval Intelligence Division, 1935-36; Captain, 1935; to RAN, 1938-41; Plans Division, 1941 and retired, 1945.

year 1926 set as the backdrop, his concern was to examine the means available of defending the colony in light of the prohibition on fortifications accepted by Britain in the Washington Naval Treaty. The Admiralty, Stewart found, planned to keep mines, boom defence equipment, indicator loops, and guns at Singapore only moving them forward to China when events proved necessary. Such a plan rested in having adequate military forces present at the colony to buy the necessary time for the defensive aids to be transported at the day of reckoning. Unfortunately, the level of defence required to wait out any investment by the Japanese was beyond the means of the War Office to support, as they were loathe to immobilize troops they did not have for a contingency, in their view, unlikely in any case to arise.[150]

Of course, before the strategical game was played or qualifiers presented their own contributions to a presumed Pacific war, much background material had been provided by the directing staff and the outside lecturers engaged as subject matter experts. Between 1919-22 Captain Wilfrid Egerton, the Assistant Director, provided lectures covering 'Notes on the World Situation in 1914' and 'War Plans Pacific 1914'. He offered a further three papers on 'Operations in the Pacific in 1914' and concluded his review with a summary of 'Japanese Naval Manoeuvres 1919'. Supporting Egerton's review of Japan was Instructor Commander Guy Rayment of the Naval Intelligence Division who provided the necessary political context and summarized the current information held against that power.[151] Not without a measure of irony were Rayment's lectures offered, as that officer was soon to receive the Order of the Sacred Treasure from Japan for his work during the recently concluded war.[152] Still, the Japanese situation was but one topic of many surveyed in the Staff Course and others such as Russia, the Near East, and the United States possessed saliency too. What is abundantly clear is that the Battle of Jutland far from being an obsession with the Navy, was but one venue amongst many for instilling the critical skills necessary to serve as a staff officer.

Though an Anglo-Japanese war never lost resonance as a topic between 1919-39, its emphasis waxed and waned as other contemporary matters were addressed. The rise of other naval threats closer to home by the early 1930s played a role in this shift and schemes considering Italy and Germany as adversaries made their appearance as the spirit of Locarno became a distant memory. By way of example, the maintenance of a submarine patrol in Mediterranean waters where Italy was the presumed Blue force featured as an exercise during the 1936 Staff Course. In Short Scheme 518, qualifiers prepared an appreciation for operating British submarines on ten-day patrols in the Gulf of Taranto during a period of strained relations.[153] In its essentials, the scheme is

---

150 ADM 203/47, 'Examination of the Possibility of Losing Hong Kong in a War with Japan in 1926', Qualifier's lecture by Lieutenant Commander Robert R. Stewart delivered 9 May 1923.
151 ADM 203/100.
152 Supplement to the *London Gazette* of 8 March 1920.
153 Captain Richard Oliver-Bellasis Papers, National Maritime Museum, Greenwich, United Kingdom, NMM/BEL/151.

but an updated version of those featuring in the early 1920s when qualifiers weighed the moves facing Admiral Sir Berkeley Milne and Rear Admiral Ernest Troubridge when dealing with SMS *Goeben* and SMS *Breslau* in August 1914. Though the enemy, time, and forces available were different, the objectives and the necessary skills for assessing and advising the possible courses of action were one.

As for Jutland, its first appearance as a specific Staff College topic came during the 1921-22 session when Kenneth Dewar, co-author of the pending Staff Appreciation, visited and delivered four lectures on the battle.[154] His treatment was hotly received by qualifiers and directing staff alike, so much so, that he was not asked to return and speak on the topic again.[155] Thus, the following year witnessed the topic becoming a regular feature of the curriculum with Captain Charles Turle,[156] the Assistant Director, continuing the four-lecture format inaugurated by Dewar. Eventually, though, lecturing on Jutland became the responsibility of the Deputy Director, the officer responsible for tactical studies at the Staff College.[157] Though full treatment was given to recounting the lines of the battle, its primary purpose was to introduce the principles of war and to highlight failures in staff duties. Indeed, the latter reason increasingly assumed the greater prominence and by the time of the 1931-32 session, the battle was appearing in seven lectures provided by Commander William Tennant[158] with Captain Robert Raikes, the Director, offering a summary of the encounter's most important points at the conclusion of Tennant's survey.

Thus, the battle was a central feature of the curriculum, but not, unduly so, given the breadth of other topics covered and previously discussed. Captains John Godfrey[159] and

154 ADM 203/100.

155 Barry D. Hunt, *Sailor-Scholar: Admiral Sir Herbert Richmond, 1871-1946* (Waterloo: Wilfrid Laurier University Press, 1982), p. 117.

156 Later Rear Admiral Charles Edward Turle (1883-1966). War Staff Course, 1914; Captain, 1921; Assistant Director, Royal Naval Staff College, 1922-24; Flag Captain and Chief Staff Officer in HMS *Frobisher*, 1924-26; British Naval Mission to Greece, 1927-29; Senior Officers' War Course, 1929-30; Director, Naval Air Division, 1930-32; Tactical Course, 1932; Commanding Officer, *Resolution*, 1932-33; Rear Admiral and retired, 1934. Recalled for war service and Naval Attaché Greece and Bulgaria, 1940-41.

157 Drage Papers, IWM/PP/MCR/99, Reel 3.

158 Later Admiral Sir William George Tennant (1890-1963). Operations Division, 1926-27; Staff Course, 1927-28; Tactical Course, 1930; Royal Naval Staff College, 1930-32; Captain, 1932; Imperial Defence Course, 1934; Flag Captain in HMS *Arethusa*, 1935-37; Imperial Defence College, 1937-39; Commanding Officer, *Repulse*, 1940-41; Rear Admiral, 1942; Vice Admiral, 1945; Commander-in-Chief, America and West Indies Station, 1946-49; Admiral, 1948; and retired, 1949.

159 Later Admiral John Henry Godfrey (1888-1971). Designated War Staff Officer, 1918; War Staff duties in Mediterranean Fleet, 1918-19 and Atlantic Fleet, 1919-20; Plans Division, 1921-23; Royal Naval Staff College, 1923-25; to New Zealand Division, 1925-28; Captain, 1928; Tactical Course, 1928 and 1933; Senior Officers' War Course, 1928-29; Deputy Director, Royal Naval Staff College, 1929-31; Senior Officers' Technical Course, 1931; Commanding Officer, *Kent* and HMS *Suffolk*, 1931-33; Deputy Director, Plans Division,

Lancelot Holland[160] were the principal agents in writing the lectures used at the Staff College when serving respectively as Deputy Directors and subsequent members of the directing staff penned revisions to reflect the changing perspectives of the engagement.[161] Both of these officers were to have fruitful careers and the quality of Godfrey's talents was such that in 1946 his was short-listed for consideration as the Chichele Professor of the History of War in the University of Oxford. In the event, the appointment went to Cyril Falls, a former officer having a strong record of published historical writing as a member of the team preparing the Official History of the World War before becoming a journalist for *The Times*. Godfrey, if personally dismissive of his own academic talents, was hindered only in that he wrote primarily for an internal, private audience and lacked the public recognition of a Falls.[162] He, moreover, was instrumental in bringing questions of logistics and administration to the fore in the curriculum of the Staff Course.[163]

As the received wisdom of the interwar period depicts a Navy consumed with the battle to the detriment of other matters, the question arises does the documentary evidence support such a conclusion. The short answer is yes, but not for the reasons typically given and consequently not to the ill-effects supposed. Jutland does feature prominently at the Staff College, the Tactical School, and at other venues where naval officers studies and lectured. This frequency of appearance as a subject, though, owed more to its very familiarity with contemporary officers than any other reason remaining the most recent example of a fleet action.[164] In the War Course, it featured as a topic of discussion during the late 1920s under the tutelage of Bertram Ramsay. Yet, it was never a core subject as it comprised but four lectures.[165] This is only natural when one recalls that Jutland as a battle lasted but a day and the shortcomings present were of the operational and tactical type and not of the strategic and political, the foci

1933-36; Commanding Officer, *Repulse*, 1936-38; Director, Naval Intelligence Division, 1939-42; Rear Admiral, 1939; Commander, Royal Indian Navy, 1943-46; Vice Admiral, 1942; Admiral and retired, 1945.

160 Later Vice Admiral Lancelot Ernest Holland (1887-1941). Five firsts in examinations for Lieutenant; Fleet Gunnery Officer in *Iron Duke*, 1922-24; Captain, 1925; Senior Officers' War Course, 1926; Royal Naval Staff College, 1926-27; Tactical Course, 1929 and 1934; Flag Captain in *Hawkins*, 1929-31, Head British Naval Mission to Greece, 1931-33; Flag Captain in *Revenge*, 1934-35; Imperial Defence Course, 1933; Chief Staff Officer in *Valiant*, 1936; Rear Admiral, 1938; ACNS, 1937-39; Commander, Second Battle Squadron, 1939; Vice Admiral, 1940; Vice Admiral, Battle Cruiser Squadron in *Hood* and killed in action, 1941.

161 Tennant, 'Jutland I', lecture of 1932, Tennant Papers, NMM/TEN/41/1.

162 Danchev, *Alchemist of War*, p. 222.

163 ADM 196/127 (Godfrey).

164 Commander Charles S. Daniel attending the Air Force Staff College in 1932 presented at least two papers on Jutland as did Captain Henry J. Egerton to the British Army Senior Officers' School in 1938.

165 Arthur Marder, *From the Dreadnought to Scapa Flow, Volume III, Jutland and After: May 1916–December 1916* (London: Oxford University Press, 1966), p. viii.

of the War Course after the founding of the Tactical School. Even at the Staff College, however, Jutland rarely appeared in the schemes and war games. This may owe much to the battle's very scale. Most schemes worked addressed a limited problem, such as the *Breslau* and *Goeben* search. Those schemes conducted of a comparable scale to the Battle of Jutland usually were of the nature of an Anglo-Japanese conflict, which always had more relevance to classes of the period. Nevertheless, when members of the Air Force Staff College visited Greenwich, Jutland did appear in the joint schemes worked, but with a twist. The British and German fleets now sailed with a full complement of aircraft carriers and doing battle were '"FURIOUS", "HERMES", "EAGLE" and "ARGUS" on the British side, and the "BURIOUS", "BERMES", "BEAGLE" and "BARGUS" on the German side'.[166] As Germany was disarmed and not a naval threat, the nature of the exercise offered the chance to retest the lines of the encounter by changing an important variable now relevant in any future naval engagement. Thus, the 1916 battle served as the control in the contemporary experiments used to measure a future naval encounter.

Of course, another reason for studying Jutland was to gain insight into the failings of staff procedures evidenced during the engagement. The battle was deconstructed in a series of lectures arranged chronologically and by the time of the 1931-32 course, the topic was lasting over a week. Highlighting the multiple errors of commission and omission, the aim was to impress upon future staff officers the vital importance of paying attention to detail and understanding the necessity for showing individual initiative in operations. Another use of Jutland was to introduce the principles of war in a tactical setting. This element shines through in Captain Geoffrey Blake's summing up lecture following a week's study of Jutland in 1927. Here, he noted how the principle of concentration was demonstrated by the deployment of the Grand Fleet and was suggested by the tentative efforts of integrating the gunfire of multiple ships to a single target being considered at the time. Again, Blake called attention to the lamentable lack of initiative shown by officers, particularly, their failure to report the movements of the High Sea Fleet to the British Commander-in-Chief, Admiral Sir John Jellicoe.[167]

As with its senior counterpart, however, Staff Course lectures covering tactics in the initial period following the war eschewed an emphasis on Jutland. This changed in 1921-22 at the time of the third postwar session and became only more pronounced as the years passed. Exactly why this proved so is difficult to gauge, but one consideration must be the practice of successive members of the directing staff enhancing the Jutland lectures by adding newer material not considered previously. Tennant's

166 ADM 203/69, Astley-Rushton lecture, 'The Staff College', p. 14.
167 'Royal Naval Staff College, Greenwich (Session 1926-1927) Summing Up by Director Royal Naval Staff College on Jutland Lectures 1927', Captain Allan Thomas George Cumberland Peachey Papers, National Maritime Museum, Greenwich, United Kingdom, NMM/PCY/1.

homage to the debt of Holland and Godfrey before proceeding with his thoughts on the battle confirms such a pattern at work. Another possible reason is that the officers now attending the Staff Course had been junior officers or even midshipmen and their experience of the battle was not of a nature to highlight its major lessons. Tennant's lectures of 1931-33 surveying the battle in seven lectures suggested a host of lessons to be drawn. Providing a synopsis of each lecture with the relevant lessons indicated, the qualifiers were left in little doubt by Tennant where matters stood and exactly how British performance was deemed wanting.

Again, the battle was analyzed from the perspective of the principles of war and these certainly featured in many of the suggested lessons. Addressing the principle of cooperation, Tennant called attention to the work of the Fifth Battle Squadron operating under the command of Rear Admiral Hugh Evan-Thomas[168] and now temporarily attached to Beatty's Battle Cruiser Fleet. Here, he argued that squadrons seeking to cooperate were more effective when working from the same harbour.[169] The principle of concentration in both it tactical and strategic variations was another element highlighted by Tennant. The former shown by the Fifth Battle Squadron in company with the Battle Cruiser Fleet and the latter depicted by the rendezvous of the Grand Fleet to the Battle Cruiser Fleet itself.[170] In his third lecture, Tennant stressed the principles of initiative and surprise in British operations, though it is worth mentioning that initiative was not an explicit principle of war in British naval thinking. Noting the initiative of Commodore William Goodenough,[171] flying his broad pennant in the light cruiser *Southampton*, for signaling the presence of enemy battleships to the Battle Cruiser Fleet whilst the British heavy ships were already engaged to ill-effect against their German opposite number received Tennant's praises. Praise, too, was bestowed on the actions of the destroyers HMS *Moresby* and *Onslow* in attacking the First Scouting Group of the High Sea Fleet, a much superior force. This foray conducted at night and in the absence of explicit orders stood in contrast to the actions of others. *Onslow's* efforts if not better served in their results were nevertheless all that could be asked for given the damage that she had already received. Promotion to Commander, a mention in dispatches and a postwar award of a well-deserved Distinguished Service Order to Tovey, her commanding officer, confirms that higher authority shared Tennant's views. Finally, the arrival of the Third Battle Cruiser Squadron during a critical moment of the action was employed to good effect confirming the value of surprise as a principle. Yet, as *Invincible*, the squadron's flagship, came to grief in a

---

168 Later Admiral Sir Hugh Evan-Thomas (1862-1928). Captain, 1902; Rear Admiral, 1912; Vice Admiral, 1917, Admiral, 1920 and retired, 1924.
169 'Jutland I', lecture of 1931, Tennant Papers, NMM/TEN/41/3.
170 'Jutland II', lecture of 1931, Tennant Papers, NMM/TEN/41/3.
171 Later Admiral Sir William Edmund Goodenough (1867-1945). Captain, 1905; Rear Admiral, 1916; Vice Admiral, 1920; Commander-in-Chief, Africa, 1920-22; Commander, Reserve Fleet, 1923-24; Commander-in-Chief, The Nore, 1924-27; Admiral, 1925 and retired, 1930.

cataclysmic inferno at Jutland, Rear Admiral the Honourable Horace Hood and the qualifiers listening might have shared a different twist on the principle's application than the one intended by Tennant's lecture.[172]

Of the many lessons expounded in the lectures, foremost was the importance of initiative. Calling attention to this consideration, Tennant stressed, 'It is suggested that our greatest need of all at JUTLAND was INITIATIVE. Are we better in this respect today?'[173] If the question posed was more rhetorical than Socratic in its intention, its import was clear enough given the tenor of all the previous lectures. That the *Naval War Manual* did not ascribe initiative specifically as a principle of war the question remains did lecturers and qualifiers actually *read* the *War Manual*? The answer, of course, is yes but the confusion must owe something to adopting the *FSR* as the source document when preparing the *War Manual*. If the Naval Staff had written their treatise from scratch and not borrowed from the *FSR*, the importance of initiative as a principle in contemporary naval thinking would have become manifest and probably appeared in the *Manual*.

It remained for one officer writing over a decade after their naval adoption to argue that the principles of war as expounded in the *Naval War Manual* were far from the accepted principles so frequently claimed. The anonymous author (probably Russell Grenfell and then serving at the Staff College) noted how Drax, after seeking the opinions of other naval officers on the principles as then espoused by the Army and their applicability to the Service, compromised in their acceptance by adopting a slightly different phraseology.[174] Grenfell further noted that of the naval experts originally questioned by Drax, one had postulated twenty tenets whilst Drax himself believed the case for ten principles existed. Believing that if principles were required they needed to be few and truly applicable to naval warfare, and not merely the grafting of military thinking on to a maritime dimension, Grenfell resolved on three principles: The Object, Concentration at the Decisive Point, and Initiative.[175]

Drax's compromise was probably made in the hope of keeping doctrinal differences amongst the Services to a minimum. Not an insignificant consideration with officers attending each other's training establishments where joint exercises were conducted. Having different principles made the foundations upon which they rested weak, for could it be claimed with authority that they were universal and applicable to all manner of warfare if not commonly held? The above highlights, too, that in a military and naval setting education is not an end in itself. Rather, it is pursued with the practical purpose of supporting the organizations conducting it. After all the aim

172 'Jutland III', lecture of 1931, Tennant Papers, NMM/TEN/41/3.
173 'Jutland VI', lecture of 1931, Tennant Papers, NMM/TEN/41/3. Original emphasis.
174 R.G., 'The Principles of War', *The Naval Review*, Vol. XXII, No. 2, May 1934, p. 227. The conclusion that Grenfell is the author is based on the pseudonym of the article's author and a textual comparison between the article and Grenfell's other writings particularly 'Training in Tactics' appearing in *The Naval Review* in 1923.
175 *Ibid*, pp. 232-33.

is not to produce dispassionate philosophers, but to develop the necessary reasoning and judgmental skills in officers allowing them to face any new problem where a large measure of uncertainty is apt to be present.

Time was probably a further complicating factor for Drax and the need to publish the *Manual* as quickly as possible cannot be discounted. Without some doctrinal footing, the risk that the directing staffs of the Staff and War Colleges would work to separate lines was ever present. This would explain borrowing the format of the *FSR* and lifting passages from it and their appearing in its naval version. In any event, Grenfell's article demonstrates that the *Naval War Manual* was read at the Staff College and its conflicts with what was being taught in the Staff Course, at least as far as the principles of war were concerned, had not gone unnoticed.

Where initiative did feature prominently in the naval doctrine of the period was in the *Battle Instructions*, and it is to here that one can find the primary reason why as a lesson and a tenet it appears so regularly in the lectures of the Staff Course. The emphasis of initiative within the *Instructions* was unmistakable, appearing within its opening clauses and in boldface typescript cautioning officers that:

> The varying and unforeseen circumstances which always arise in action demand the display of initiative and co-operation by all concerned in carrying out the general plan. The squadrons of a fleet vary in mobility and striking power. The intensity of the enemy gun and torpedo fire will also vary in different parts of the line and the situation outside the battle line or in different parts of the line may not be apparent to the Admiral. Commanders of all squadrons must take full advantage of every incident of the action to press the enemy, always remembering that the destruction of the enemy, particularly his battlefleet, is the object that they must relentlessly pursue.[176]

Such is why the *War Manual* cannot be read in isolation, as it is only the melody to the harmony provided by other doctrinal publications of which the *Battle Instructions* supplemented by the *Manoeuvring Orders, the Cruiser* and *Signal Manuals,* and the *Destroyer Attack Instructions* were at the fore.[177] It would have been better and clearer to historians coming later, if the Navy had written its doctrine afresh following the late war, taking the time to edit each in turn to ensure a consistency of reading. Yet, that is asking too much. Marder failed to appreciate the trinity of the Service's doctrinal foundations, but in this he was not alone. Gretton too failed to see beyond the *War Manual* to the role played by subsidiary publications.[178] As it was, the *Tactical Manual,*

---

176 ADM 186/106, *C.B. 01821, Battle Instructions* as amended, 15 April 1931, p. 1. Original emphasis.

177 The relationship and dependence of these publications to each other was summarized for officers in a contemporary publication, ADM 186/75, *C.B. 01720. Chronological Summary of Fighting Instructions* of 1 June 1928.

178 See Petter Gretton, 'Why Don't' We Learn from History?' *Naval Review,* Vol. XLVI, No. 1, January 1958, p. 23.

the initial heir to the *Grand Fleet Battle Orders*, appeared in 1921 at the same moment of the *Naval War Manual* whilst a second edition promulgated in 1925 coincided with the release of the first postwar *Battle Instructions*.[179]

With its genesis in the *FSR*, the *Naval War Manual* in its initial guise was focused on operations and propagating the principles of war. Consequently, grand strategy, strategy, logistics and administration, were missing and the treatise assumed a level of knowledge by the reader in the broader aspects of war that proved unwarranted. These issues would be addressed in time, but their absence highlighted that the *Manual* as it appeared in the early 1920s was only a first step in the formulation of doctrine and not the last.[180] As for Jutland, though it provided ample scope for teaching most of the principles of war enunciated in the *War Manual*, it was deficient in serving as a vehicle for conveying the utility of one: the principle of economy of force. As the Staff and War Courses both covered economy of force in their curricula, its omission from the Jutland lectures series indicates that as a principle of war it was of more relevance when course discussions were addressing strategy and not reviewing tactics. This is not surprising. The allocation of forces to separate theatres of operations is a major strategic concern and while assigning forces to tasks is a problem at the tactical level, the Royal Navy's tactical doctrine in the wake of Jutland was to concentrate *all* forces available upon the primary objective: the destruction of the enemy. To this end, Tennant quoted from the post-Jutland Battle Cruiser Fleet Orders:

> It becomes the duty of subordinate leaders to act in the spirit of the Commander-in-Chief's requirements. These are only two and they are very simple. So long as the enemy ships remain afloat <u>we must locate and report, attack and destroy</u>: but to perform either duty without the other is to fall short of the co-ordination which ensures success, nor should it be thought that to perform one duty efficiently it is necessary to abstain from the other.[181]

Still, if Jutland was not the ideal vehicle for understanding the concept of economy of force, officers were taught its value in other schemes. In a very real sense economy of force would only be logically understood when the intellectual discipline of operational research came to fore in the next war. For all that, the principle was appreciated and it may be that just as it took Frederick Lanchester to enunciate the mathematical underpinnings of the N-square law used by Nelson at Trafalgar it would take Patrick Blackett and others to define the mathematical foundations of economy of force.

---

179 *C.B. 1601, The Tactical Manual* and *C.B. 934, The Naval War Manual*, both from 1921, are not on file at the National Archives.

180 *Naval War Manual*, 1946 (Draft), Vice Admiral Sir Ballin Robertshaw Papers, IWM/72/75/1, Imperial War Museum, London.

181 'Jutland V', lecture of 1931, Tennant Papers, NMM/TEN/41/3. Original emphasis.

The value of the Jutland lectures served another purpose beyond confirming the principles of war: they provided lessons for the budding staff officer on the likely problems he could expect to face in battle. Emphasizing to qualifiers that tactical communications be redundant and signals were repeated was a strong theme of the lectures. Additionally, officers needed to be mindful of the visibility conditions prevailing, as this posed problems in estimating the exact location of the enemy when making contact reports and hindered the receipt of any visual signals made. Tennant recommended standardizing signals and deprecated the practice of making frequent changes in call signs. The marginal increase in security provided came with a corresponding loss of efficiency, as it became more difficult for all executive officers to comprehend their meaning. Meanwhile, care in executing any challenge and reply when approaching unknown ships was cautioned by Tennant, as the Germans (and presumably any future foe) had had a most efficient organization for capturing British naval procedures.[182] As a trained Navigating Officer, it was natural for Tennant to pay close attention to any failings in this aspect of the battle and, thus, he reproached the qualifiers to guarantee the accuracy of their current position when making contact reports. Moreover, the primary purpose of the fleet's scouting and reconnaissance forces was to secure information about enemy disposition and numbers to allow the British Admiral to form his tactical plan. Officers must not assume that the enemy's location was known throughout the fleet and must provide reports. As staff officers were responsible for maintaining the plot necessary for making torpedo attacks, he stressed the need for Navigating Officers, staff officers, and torpedo specialists to be insistent on that weapon's use when circumstances permitted.[183]

In his final lecture on Jutland, Tennant sought to summarize the major findings previously surveyed. The pitfalls of centralized command were exposed, and, if he did not use the formulation himself, his message was that subordinate leaders needed to be conversant with any plan and exercise full initiative in meeting the commander's intent. Signal communications had fostered an attitude of waiting for orders and failing to take initiative when opportunities presented themselves. Yet, it was just not that signals transmitted were lost. It was also the case that many signals made were unnecessary, whilst others, which should have been sent, never made it to the signal pad for transmission.[184]

As Jutland represented the sole case of a fleet action during the Great War, its study in detail was only natural. It was natural, again, that others would study the battle for these very same reasons. The United States Navy, both at Newport and in the fleet re-fought Jutland, for much the same reasons that the British had examined the battles of the Russo-Japanese War—to understand better the probable lines of a modern naval encounter. The official despatches published by the British in 1920 and

<hr>

182 'Jutland II', lecture of 1931, Tennant Papers, NMM/TEN/41/3.
183 *Ibid*, 'Jutland III'.
184 *Ibid.* 'Jutland VI'.

including copies of the battle's signals supported such study. Indeed, notwithstanding the merits of its several controversies, thanks in large measure to the electronic communications revolution that brought about the engagement, no previous naval battle offered such a wealth of source material documenting the actions of both adversaries. Plots, signals, and battle damage reports combined with the official despatches and histories allowed an examination of the battle to proceed in a manner not previously possible. Thus, whilst commanding the America and West Indies Station in 1931, Vernon Haggard recalled:

> A Lieut. Lovette U.S.N. was attached to my staff as liaison officer during our stay... He told me a war game was being played in the Fleet. Jutland was being refought and officers had been detailed to take the parts of all the principal commanders on both sides. Each officer had to make a complete study of the character and career of the man he represented. In the game only such information was to be acted upon as was available in the battle itself.[185]

Captain William Tait commanding *Capetown* and serving earlier on the same station confirms the centrality of Jutland to American naval education. In this case, he forwarded the substance of a conversation with another friendly American officer noting:

> As regards their battle tactics, they have a War College at Newport where current problems are worked out with models on a tactical board. Rear Admiral Brumby told me that at every session they have, as a stock piece, the Battle of Jutland. The game commences five minutes before the British fleet deploy. Up to that moment both sides have been manoeuvred as in the real battle. After that the commander of each side is allowed to do what he likes, but he commences only with such information as was available to the British and German C. in C's. He added, as an interesting point, that if ever the British Fleet were deployed to starboard in the game, the British battlefleet always got severe punishment and the Germans managed to get away comparatively undamaged to the eastward.[186]

Whether Tait's comments reached Kenneth Dewar is unknown, but a criticism of Jellicoe leveled in the Staff Appreciation he co-authored (and never officially released), and presumably highlighted in his Staff College lectures, was that a deployment to starboard by the Grand Fleet was preferred to the deployment to port actually

---

185 'Narrative of Command of the America and West Indies', Volume VII, p. 18, Haggard Papers, IWM/85/21/2.

186 *Capetown* Letter of Proceedings No. 476/102 of 7 September 1929 to Commander-in-Chief, America and West Indies Station, Admiral Sir William Eric Campbell Tait Papers, National Maritime Museum, Greenwich, United Kingdom, NMM/TAI/4.

made.[187] (This because a starboard deployment brought the two fleets closer towards each other.) The results of a neutral power testing the battle in postwar games remains but an interesting conjecture, but does serve to demonstrate that displaying a regular interest in Jutland was far from being the concern of a single navy.

Fundamentally, the postwar critics who argued that the battle consumed the attention of the Service to the detriment of other topics have assumed that the primary reason for its study was tactical and not administrative. This was certainly a reason for its analysis at the Tactical School, of which more is to follow, but became less the reason for its study by the Staff College where it remained a major fixture of the curriculum but never the primary topic. That was reserved for the college's emphasis on combined operations, which pursued in conjunction with the other Service colleges, was the centrepiece of Greenwich study. This was a complex topic which afforded many avenues of instruction touching as it did upon joint operations, the higher direction of war to say nothing of the administrative and technical issues arising. The Staff College addressed each of these facets but beyond these sent observers to witness actual evolutions conducted by the fleet such as the Mediterranean landings of July-August 1926.[188] Additionally, just how central this topic was can be measured by a complaint registered by the Air Staff. Writing to the Deputy Secretary of the CID in response to an Admiralty proposal to establish a special striking force in support of amphibious operations, Air Vice Marshal Richard Peirse,[189] Deputy Chief of the Air Staff, complained that:

> The Staff Colleges now spend over a month every year – in our case about ⅛th of the whole course – studying amphibious operations. We have long felt that this represents a higher proportion of our available time than we are really justified in devoting to this subject, in spite of the undoubted value of the combined operations exercise in promoting liaison and mutual understanding between the three Services.[190]

Still, a corollary mission of the Staff College was to provide officers with a higher education in war and here an introduction to the *Naval War Manual* served as an important point of departure. If the *Manual* failed to address doctrine explicitly, members of the directing staff were not so reticent and this, too, was a reason for examining Jutland. Studying the evolving nature of the Grand Fleet's Battle Orders was central to the Staff College's survey of not only Jutland, but of the World War also. These surveys showed that sensible as the *Orders* appeared when drafted, their effectiveness as instruments of a Commander's will was only demonstrated by the

187 *C.B. 0938. Naval Staff Appreciation of Jutland*, p. 85, Jellicoe Papers, Add MSS 49,042.
188 ADM 196/62 (Foster).
189 Later Air Chief Marshal Sir Richard Edmund Charles Peirse (1892-1970). Staff Course Andover, 1922-23 and Imperial Defence Course, 1927.
190 CAB 54/13, Deputy Chief of the Air Staff to Deputy Secretary, Committee of Imperial Defence letter of 8 February 1938.

test of battle and here they proved lacking. This was a major reason why qualifiers were adjoined to accept nothing at face value at Greenwich, as how much remained of British naval orthodoxy to be proved wanting? Moreover, co-operation and maintenance of the object required that all officers follow a common doctrine. Those who advocated divided tactics had first to address this doctrinal vacuum, as Thursfield reminded his audience:

> The fact of the matter appears to be that a C-in-C. cannot adopt and apply a real principle of decentralisation of command in a fleet that does not possess a common tactical doctrine. With no common doctrine in the fleet, full initiative cannot be given to subordinate commanders of units: they must be bound by the orders they receive, and full advantage cannot therefore be taken of the rapidly changing situations that occur in a fleet action.[191]

Thursfield's analysis is perceptive. Yet, the principles of war, as already discussed, omitted the principle of initiative and divided tactics was *prima facie* a violation of the principle of concentration. The rejoinder by the advocates of divided tactics was that concentration of fire, effort, and time, were the essentials and not of space, whilst initiative ('courage, energy, determination, and the bold offensive spirit') were enshrined in the *Naval War Manual*'s preamble if not specifically in its catalogue of the principles of war.[192] All this was true, but their public writings made little mention of these caveats. That the *Naval War Manual* was a confidential publication was partially to blame, as reference to it by writers had to be circumscribed. Even in the privately circulated *The Naval Review*, reference to the *Manual* was uncommon and discussion of its contents circumscribed.[193]

Where the Staff College made a unique and vital contribution in investigating contemporary problems was in supporting the exercise programmes of the combined Atlantic and Mediterranean Fleets. Typically, these exercises were conducted annually between the months of January and March in the Atlantic waters near Gibraltar or in the Mediterranean Sea proper. They provided the fleets an opportunity to test the unified tactical doctrine of the Royal Navy—slight variations existed between the two principal fleets in their use of the *Battle Instructions*—and game potential

---

191 Captain H. G. Thursfield 'Development of Tactics in the Grand Fleet, I', Royal Naval Staff College lecture of 2 February 1922, Rear Admiral Henry George Thursfield Papers, National Maritime Museum, Greenwich, United Kingdom, NMM/THU/107.

192 ADM 186/66, *C.B. 973*, p.1.

193 Open reference to the *Naval War Manual* was infrequent and the only appearances detected in contemporary literature found are *The Naval Review* articles 'Advice to Young Officers, I' published in November 1926, 'Advice to Young Officers. A Reply.' appearing in February 1927, 'Functions of Destroyers' from November 1927, 'The Seeds of Morale' featuring in the May 1928 issue, 'The Royal Naval Staff College' of February 1932, and 'Morale, Efficiency and Organisation' published May 1933.

war scenarios. Before the exercises commenced, however, the Staff College was made aware of the general lines of play including the objectives to be achieved. The directing staff used the information to set schemes for the qualifiers to consider. In turn, the instructors developed school solutions to the schemes posed and reviewed the qualifiers' submissions returning their solutions with critical comments. Additionally, the school solution was provided to the qualifiers as a model for what should have been developed by them. The school solution was not *the* answer, but representative of what a correct appreciation should consider. As war was deemed an art and not a science, no one answer could be deemed correct, as many paths were possible to achieve the objectives set. The purpose of the school solution was to highlight to the qualifier where his work fell short or to demonstrate other possibilities to achieving the scheme's objectives.[194]

These solutions were also forwarded to the fleet commander playing either the Red or the Blue forces before the combined exercises commenced. Thus, before the exercises started the Atlantic and Mediterranean Fleet commanders had a strong sense how things might proceed in the evolutions they were about to conduct. Of course, a Staff College appreciation was but one possibility and factors such as weather, chance, and the vagaries of the war game rules ensured that what was possible on paper was subject to validation in practice. The best example of the proceeding description is the 1928 combined fleet exercises demonstrating the possible lines of an Anglo-Japanese war. In the 'Relief of Alboran Island', Gibraltar representing Hong Kong and Alboran, a small island off the coast of Algeria and portraying Singapore were the critical positions at risk.[195] The exercise testing the concepts for the defence and relief of the Empire's Pacific territories, 'M.U.2', was set to begin 15 March 1928 with the Commander-in-Chief, Mediterranean playing Red, the British fleet, and the Commander-in-Chief, Atlantic Fleet directing the Blue force and portraying the Japanese fleet. Importantly for our consideration is that the Staff College had forwarded its appreciation of the courses of action open to Red and Blue on 2 March 1928 to the fleet commanders.[196]

Five sections appreciated exercise 'M.U.2' with three syndicates playing Blue and two groups considering the question from the perspective of the Red Fleet.[197] The general line of play was that Singapore was under investment from Japanese naval forces and that the main British fleet was racing to relieve the garrison before it succumbed. As the British force had been at sea for some weeks already, the exercise was assuming the necessity of deploying a fleet from Home and Mediterranean waters rather than having a fleet in the local environ already. Time was of the essence, as if

194 Anon., 'Royal Naval Staff College', p. 27.
195 Robert Jackson, *Strike from the Sea: A Survey of British Naval Air Operations, 1909–1969* (London: Arthur Barker, Limited, 1970), p. 49.
196 ADM 203/86, Director, Royal Naval Staff College to Chief of Staff, Atlantic Fleet letter S.C. 4328 of 2 March 1928 with Royal Naval Staff College, Scheme 159, Staff Solution.
197 *Ibid.*

Singapore fell, then further British operations against the Japanese would be difficult in the extreme. The appreciation for Blue aimed to attrite British forces as they negotiated the constricted waters surrounding Singapore, which perforce the Red Fleet would have to pass to reach the fortress and dockyard. If successful in his efforts, the Blue Fleet would then be able to meet the British force on more equal terms and deny him the ability to relieve the fortress.[198]

The Staff College solution for the Red Fleet was to negotiate the restricted channel necessary to reach Singapore under cover of night. Screening forces would precede the Red Fleet and sweep the channel for naval mines whilst protecting the main force from any Blue Fleet torpedo attack. Two different cruising dispositions were specified in the appreciation for the Red Fleet when entering the constricted waters near Singapore. Proceeding at ten knots, it recommended cruising order 'R' when operating in the widest part of the channel representing the Lahore Straits. However, when facing the narrowest part of the approach, the fleet would form into cruising order 'Q' to negate the threat from any mines the Blue force might have sewn in the vicinity where Red Fleet faced its greatest risk: Red Fleet's aircraft carrier would provide reconnaissance for the main fleet, strike at Blue's aircraft carrier if found, and attack any submarines detected in that order of priority. If the Red Fleet found itself under the guns of Blue's heavy ships and unable to effect deployment, then Red Fleet's carrier aircraft were to eschew all other tasks and attack Blue's heavy ships with torpedoes.[199]

In Scheme 158, the same 1927-28 Staff Course addressing the defence of Singapore and Hong Kong, further examined a possible British-Japanese war occurring in 1945 in which both the United States and the Netherlands remained neutral. The subject of the scheme was an investigation into British cruiser policy at the request of the Admiralty. This investigation, though, was an examination of cruiser policy in its broadest meaning and not merely a narrow review of a particular type of warship. Here, the aim was to so disrupt Japanese commercial trade as a possible means of forcing a fleet action. Such operations it was believed would compel the Imperial Japanese Navy to defend her maritime trade or see it lost to the predations of the British fleet. These actions would serve the three-fold purpose of forcing Japanese cruisers to protect their trade, and, thus, in turn, protect British commercial shipping. Additionally, they promised to attack Japanese morale and economic strength, and, ultimately, lead to a fleet action as the Japanese battlefleet would sally to support its cruisers. Naturally, the working assumption of the scheme was that the presumed superior British battlefleet, now operating in the Pacific, would prevail in the foregoing encounter.[200]

198 *Ibid.*
199 ADM 203/86, Royal Naval Staff College appreciation playing Red Fleet.
200 ADM 203/87, Royal Naval Staff College, Session 1927-28, Scheme 158, Cruiser Policy.

Hindsight being perfect, it is all too easy for a modern reader to see the gross assumptions that underlay the appreciations developed in projecting the possible course of a Japanese war 18 years into the future. Qualifiers were being asked to assume more the skill of one familiar with a crystal ball than display the talents normally acquired in reading an Admiralty Monthly Intelligence Report. A lot would have to transpire before a British battlefleet could operate in the waters of the Yellow Sea as Scheme 158 anticipated. Moreover, the changes in technology and the political underpinnings upon which any appreciation rests become harder to measure the further one operates from the present. Yet, their value for the contemporary Royal Navy was not that the appreciation prepared was correct and would be put to *operational* use. Rather, they were the necessary rigor in the analysis supporting the arguments employed by the Admiralty in presenting their case for funding to the Treasury. Thus, Scheme 158 forwarded by the Staff College concludes by indicating what types of cruisers Britain should build and how many of each type year-by-year to meet the stated Admiralty goal of having 70 cruisers by 1945. With a budget of £85,000,000 posited in the Director of Plans tasking letter to the Staff College, the appreciation prepared identified the mix of cruisers required and concluded that savings of millions in Sterling were possible whilst still meeting the Naval Staff's underlying requirement.

As the Royal Naval Staff College routinely held an annual exercise with its military and air counterparts at Camberley with its emphasis on combined operations, it should not be surprising to discover that when the time came to consider the types of landing craft to support such evolutions the Admiralty would seek its guidance. Thus, in 1926 the DTSD requested the views of the Staff College on the necessary craft to conduct an opposed landing in light of their unique knowledge.[201] Meanwhile, in 1929, the college prepared a consolidated appreciation of the major Combined Fleet exercise held in March of that year between the Atlantic and Mediterranean Fleets. Having secured the general lines of exercise 'M.Z.' which '*was a strategical exercise embodying an attempt by a weaker fleet to intercept and destroy a portion of a superior fleet which is detached from the main fleet for convoy protection*', the Staff College went to work. It considered several alternatives for achieving the exercise's objectives from the governing memorandum that specified the serial's aims and the forces available.[202] The general lines of the exercise were that Red Fleet (Britain) had to pass two slow moving and separate convoys to their destinations in the face of a single, superior Blue Fleet.[203] An appreciation prepared based on the strategical games set as Schemes 178

---

201 ADM 203/73, Director of Training and Staff Duties letter to Director, Royal Naval Staff College of 9 October 1926.

202 ADM 186/145, *C.B. 1769/29(1). Exercises & Operations, 1929, Volume I.* Admiralty, Naval Staff, Tactical Division, December 1929.

203 Joseph Moretz, *The Royal Navy and the Capital Ship in the Interwar Period* (London: Frank Cass, 2002), pp. 237 and 249.

and 178A at the Staff Course provided its findings to contrast against the results actually achieved in exercise 'M.Z.'[204]

Apart from demonstrating that the fleets must have believed that the college enhanced their understanding of the exercises they conducted—else why continue to provide their synopses of current planning problems to Greenwich—Scheme 178A retains value for the historian as it remains the one case in extant documenting the qualifier assignments made. From a review of these assignments, it can be seen that they in large measure followed the unique skills and experiences of the officers attending. Though the billets of Staff Officer Intelligence and Captain of the Fleet were not employed (presumably because data collection and analysis and fleet logistical issues were not elements of the exercise), those focused on operational control were. Thus, Lieutenant Commander William Beech,[205] RCN, an experienced submariner, played the part of Staff Officer Submarines in Red Fleet while Commander Arthur Bridge,[206] having previously served in cruisers such as HMS *Royalist* during the war, acted as Rear Admiral Commanding Red Fleet Cruisers.

The 1929 Staff Course besides having two members of the RAF in attendance also included Squadron Leader Ambrose Shearer of the Royal Canadian Air Force. Ambrose's experience included duty with No. 3 Wing Royal Naval Air Service during the World War and for purposes of the scheme he served as one of the two officers controlling HMS *Furious*, a Blue Fleet carrier.[207] Meanwhile, Wing Commander André Walser,[208] RAF, fulfilled a similar capacity within Red Fleet, serving as one of the two officers detailed to oversee the carrier *Courageous* in the scheme's play. As for the two military officers participating, Major Francis Grant-Suttie, Royal Artillery, acted as Staff Officer Operations in Red Fleet whilst Major Daly served as that fleet's Chief Staff Officer.[209] Other qualifying officers served as umpires, plotters, and historians leaving the directing staff to control the movements of the two convoys of Red Fleet. The Deputy Director, in keeping with his general responsibilities for tactical studies, acted as the Chief Umpire of the war game allowing the Director to

204 ADM 203/90, Director, Royal Naval Staff College letter S.C. 4736 of 12 April 1929 to President, Royal Naval College, Greenwich.

205 Later Captain William James Robert Beech (1895- ?). Staff Course, 1929; National Defence Headquarters, 1930-31; Staff Officer Operations, First Submarine Flotilla, 1936; Captain, 1941; and retired, 1945.

206 Later Admiral Sir Arthur Robin Moore Bridge (1894–1971). Service with RAN, 1920-22; Navigating Officer in *Furious*, 1927-28; Staff Course, 1929; Fleet Navigating Officer and Staff Officer Operations in *Despatch* and *Delhi*, 1929-31; Royal Naval Staff College, 1932-34; Captain, 1936; in Command, Navigating School, 1937-39; Commanding Officer, *Eagle*, 1939-41; Director, Naval Air Division, 1941-43; Director, Naval Air Organisation, 1943; Rear Admiral, 1945; Vice Admiral, 1948, retired, 1951 and Admiral, 1952.

207 ADM 203/90.

208 Later Air Commodore André Adolphus Walser (1889-1966). Staff Course Andover, 1923-24; Staff Course Greenwich, 1929 and Air Force Staff College, 1932-36.

209 ADM 203/90.

have no set responsibilities but to watch the play unfold. Fourteen qualifiers oversaw the military forces of Red Fleet leaving ten officers to serve the same function for Blue Fleet. Why the imbalance existed is not clear, but as the four additional officers filled key staff officer roles in Red Fleet (e.g., Chief of Staff, Staff Officer Operations, Chief Staff Officer, and Staff Officer Submarines), the likelihood is that arriving at a sound appreciation for the Commanders-in-Chief of the Atlantic and Mediterranean Fleets was but one purpose of the exercise. The primary role of the college remained preparing qualifiers to serve as staff officers and staff officers in the British Navy (Red Fleet) at that. The strategical game was a means to that end and though having greater utility, remained for the Staff College, primarily a tool of learning.[210]

For the Staff Course, the benefit of pursuing such examinations was that they fostered the skills required of a future staff officer whilst dealing with real issues of relevance to the Admiralty. This made the work interesting to the qualifiers and valuable to the Naval Staff—not a trivial issue when budgets were austere and the relevance of the Service to Britain no longer assumed the proportions officers had previously known when entering the Navy. To the Naval Staff, the analysis supplemented their own evaluation of problems and given the press of day-to-day naval administration, allowed matters to be examined at greater length than otherwise could be achieved. In short, the Staff College was acting much as a 'Red Team' does today in defence planning and intelligence assessment—attempting to think outside of the box, trying to avoid pat answers and developing a range of solutions to a set problem.

Given their seniority and experience, the officers directing and attending the Senior Officers' War Course were in a better position to review current strategic issues and have their analyses received with a measure of interest. If their investigations remained like their Staff College counterparts unofficial, they at least served the very useful purpose of making officers familiar with the very same problems that they would have to consider when holding positions of real authority in the very near future. In 1921, Captain Guy Hamilton lectured and directed a seminar on the topic of the 'Peace Distribution of the Fleet' for the War Course. His paper, in time, was forwarded to the Admiralty where it received a favourable reading in at least one quarter with the Director of Training and Staff Duties, Captain Vernon Haggard, endorsing Hamilton's finding of keeping the battlefleet concentrated in a single operational theatre to facilitate training as his note to the docket covering the paper makes clear.[211] Meanwhile, during the session that Hamilton was surveying the proper composition of the fleets, the senior course also considered a possible war against the United States occurring in 1925. This was played out in scheme 'R'. Given the geographic positions and respective responsibilities of the British Empire and the

---

210 Anon., 'Royal Naval Staff College', p. 18.
211 ADM 1/8628/120, Director of Training and Staff Duties minute of 26 November 1921.

United States of America, such a war would likely be a global affair and the appreciation forwarded to the Admiralty documented the objectives and possible courses of action open to both parties.[212]

As was the case in the lower division, the War Course spent much time examining the lines of a future war against Japan. When the 1926-27 sitting came to consider the case, they benefited from having a recent military attaché to Tokyo amongst the course's members in Colonel Francis Piggott.[213] This officer had much experience of Japan beginning from childhood when his father the jurist, Francis Piggott, served as a constitutional adviser to the Japanese Government, and would return in the following decade to serve in Tokyo yet again. Piggott's attendance, as with the later use made of Muirhead-Gould, demonstrates the lengths pursued to ensure the training provided by the Greenwich courses was anchored in contemporary intelligence and analysis. Consequently, the quality of papers written by officers attending was frequently high. In this regard, Godfrey Orde,[214] a marine officer and a Fellow of the Royal Historical Society is representative. His paper 'Propaganda in War' replete with examples from the recent conflict and containing thoughts for its possible future employment written while attending the SOWC in 1924 made a strong impression on Vice Admiral Sir George Hope,[215] the Admiral-President. He submitted Orde's essay to the Admiralty recommending that it receive a wider reading even to the point of including it in a forthcoming 'Monthly Intelligence Report', the Admiralty series covering current intelligence and analysis.[216] In a similar vein, Vice Admiral Webb noted the value of three essays prepared during a 1927 sitting of the War Course. These were monographs written by Rear Admiral Herbert Hope,[217] Colonel Ladislaus

212 *Ibid*, Royal Naval College, Greenwich submission of Scheme 'R' of May 1921.
213 Later Major General Francis Stewart Gilderoy Piggott (1883-1966). Military Attaché Tokyo, 1921-26 and 1936-39 and Senior Officers' War Course, 1926-27.
214 Lieutenant Colonel Godfrey Philip Oppenheim, surname changed by deed poll to Orde in 1915 (1884-1947). Plans Division, 1917-19; designated War Staff Officer, 1918; General Staff Officer, 2nd Grade, 1918, Local Defence Division, 1919-20; Brevet Lieutenant Colonel, 1919; Naval Staff Officer to Paris Peace Conference, 1919; Senior Officers' War Course, 1924; Naval Intelligence Division, 1925-29; Senior Officer's Course School of Land Artillery, 1929; Battery Commanders' Course, 1929 and retired, 1931. Recalled for war service.
215 Later Admiral Sir George Price Webley Hope (1869-1959). Five firsts in examinations for Lieutenant; Captain, 1905; War College, 1908-09; Commanding Officer, *Queen Elizabeth*, 1915-16; Rear Admiral, 1917; Second Director, Operations Division, 1916-18; Deputy First Sea Lord, 1918-19; Vice Admiral, 1920; Commander, Third Light Cruiser Squadron, 1919-21; President, Royal Naval College, Greenwich, 1923-26; Admiral, 1925 and retired, 1926.
216 ADM 1/8657/44, 'Propaganda in War' by Lieutenant Colonel G. P. Orde, RM.
217 Later Admiral Herbert Willes Webley Hope (1878-1968). Five firsts in examinations for Lieutenant; War Course, 1910; War College, 1913-14; Appointed War Staff Officer, 1914; Naval Intelligence Division, 1914-17; Captain, 1915; Ordnance Committee, 1920-23; Commanding Officer, *Repulse*, 1924-26; Rear Admiral, 1926; Senior Officers' War

Pope-Hennessy,[218] late Oxfordshire and Buckinghamshire Light Infantry, and Major Henry Inman, Royal Marines.[219]

Herbert Hope, an early member of the Admiralty's code breaking unit colloquially known as 'Room 40', presented a reasoned a paper on 'The Case For and Against a Ministry of Defence'. This topic, ever featuring in contemporary debates, was sure to find interest at the Admiralty whilst Inman's 'The Future Functions and Training of the Royal Marines' spoke to an issue that the Navy was apt to take for granted. This essay came with Webb's strong endorsement of Inman's capabilities noting not only the clarity of thinking displayed but also the moderation of views demonstrated. Meanwhile, Pope-Hennessy's review of the doctrinal issues facing the Crown argued in 'The Services and a Common Doctrine of War' retains value even today for those facing a more contemporary variance of the issues at hand. Written by a talented military officer given to reflection, Webb advised its forwarding to the War Office for consideration.[220]

The number of instances where papers presented by qualifiers of the Staff College and officers of the War Course and appearing in Admiralty records is *prima facie* evidence that if Greenwich was not formally charged in regulation to investigate matters, the Naval Staff continued to seek their input. Tasking both with identifying possible solutions to specific problems, it employed their findings with alacrity in its own internal discussions. To that extent, the War and Staff Courses served as adjuncts to the Naval Staff. If the practice was less formal than occurring in Sir John Fisher's day when he sought to have the War Course enlighten the development of plans and strategy, it offered confirmation that the Naval Staff itself had matured and was now a permanent fixture in British naval administration.[221] The Greenwich establishments now approached the use that the Admiralty had formerly made of the American Naval Planning Section at Grosvenor Gardens during the latter part of the recent war—a vehicle for considering issues and providing recommendations void of the pressures of

Course, 1926-27; Tactical Course, 1927, Senior Officers' Technical Course, 1927; President, Ordnance Committee, 1928-32; Vice Admiral and retired, 1931 and Admiral, 1936.

218   Later Major General Ladislaus Herbert Richard Pope-Hennessy (1875-1942). Senior Officers' War Course, 1926-27.

219   Later Colonel Henry Basil Inman (1881–1947). Senior Officers' Course Hythe, 1921 and 1926; Combined Operations Course, 1926; Senior Officers' War Course, 1926-27; Senior Officer's Course Netheravon, 1927; Senior Officers' School Sheerness, 1928; Lieutenant Colonel, 1929; Higher Commanders' Course Netheravon, 1930; Colonel and retired, 1933. Recalled for war service.

220   ADM 1/8710/138, Vice Admiral Commanding Royal Naval War College to Secretary of the Admiralty letter of 23 February 1927.

221   Andrew Lambert, 'The Naval War Course, *Some Principles of Maritime Strategy* and the Origins of "The British Way in Warfare"', Keith Neilson and Greg Kennedy, eds., *The British Way in Warfare: Power and the International System, Essays in Honour of David French* (Farnham: Ashgate, 2010), pp. 228-29.

meeting day-to-day naval responsibilities and removed from the bureaucratic infighting of the Naval Staff.[222]

Still, papers forwarded to the Admiralty from Greenwich appeared for another less benign reason: to ensure the gospel propagated was void of heresy. In this, the Naval Staff acted as the defender of the faith. The need to guarantee that the instruction imparted was consistent with current naval doctrine was essential, for a strand of opinion within the Service still looked askance at the whole staff system and staff training. For some holding such views, an element of reactionary orthodoxy to anything new was the font of such beliefs, but, more importantly, such thinking owed much to idea that the staff system attacked the very foundations of naval hierarchy by lessening the prerogatives of rank and reducing the pillars of individual responsibility and accountability.[223] The need to ensure conformity whilst investigating the current problems of strategic and tactical import was essential given the responsibilities others had in these areas. Foremost, here, were the Commanders-in-Chief of the Mediterranean and Atlantic Fleets.[224]

By the time Vice Admiral Boyle assumed responsibility for the War Course from Webb in April 1929, the use of the strategical game had lapsed. Exactly why this was so is unknown, as surviving documentary evidence is fragmentary. The most likely explanation is that the Admiral-Presidents of the College immediately preceding Boyle believed this portion of the course rightly belonged with the Tactical School now that it was a going concern. The number of lectures dedicated to tactics was reduced and the opportunity may have been taken at the same moment to eliminate the strategical game and concentrate more on the higher direction of war and its economic underpinnings. A comparison of the overall subjects presented at the SOWC from Richmond's time to Boyle's arrival offers little light on the matter, as there is scant appreciable difference save for a reduced emphasis on tactics. Be that as it may, Boyle, a firm proponent of the offensive in naval action and a highly gifted lecturer, soon re-established the game as part of the SOWC curriculum and conducting three games became the norm during each course.[225]

Meanwhile, as future staff officers, the question of administration was a core subject covered in the curriculum of the Staff Course. The emphasis, though, proved more than a survey of British naval administration covering as it did the British Army, the Royal Air Force and the Committee of Imperial Defence with the differences of each noted. Guiding the Staff College in its training of administration were *C.B. 3013, The Naval Staff: Its Work and Development, 1929* and a revised edition, *B.R. 1984,*

---

222   See Michael Simpson, ed., *Anglo-American Naval Relations 1917–1919* (Aldershot: Scolar Press, 1991), pp. 145-59.
223   Koe, 'Naval Staff System', pp. 668-69.
224   The Atlantic Fleet was restyled the Home Fleet in 1932 following the mutiny at Invergordon.
225   Cork and Orrery, *Naval Life*, p. 155.

Naval Staff Handbook.[226] These reference works tracing the responsibilities and roles of staffs, supplemented by lectures provided by the directing staff and key officers from the Admiralty, were augmented with those covering the sources and uses of intelligence. Here, the Staff College benefited in having a resident Royal Marine overseeing the Intelligence Course co-located at Greenwich. These subjects are largely absent from the curriculum of the War Course and running to over twenty separate lectures in 1919-20 represented a substantial portion of the foundational work of the junior course.[227] Additional lectures covering technical subjects, such as wireless telegraphy and tactical plotting, featured in the Staff Course whilst being also present to some degree in the War Course.

Thus, a degree of unavoidable overlap existed amongst the courses. Necessary though this was, it probably resulted in some officers thinking that their time was better employed pursuing other endeavours—even golf. When Lieutenant Commander Ridley Waymouth,[228] the officer responsible for wireless instruction in the Signal School, came to lecture one War Course he found officers had done just so giving themselves the afternoon off, or in the vernacular of the service a 'make and mend'. Doubtless, Waymouth's inferior rank allowed them to consider such a course of action, but that many had probably already heard the substance of his talk elsewhere only added to their sense of immunity. This left Waymouth to present his instruction to a class of one: the Admiral-President, 'Ginger' Boyle.[229]

The education provided in the Staff Course was documented and promulgated to the Service in an official use manual, *O.U. 5255, Syllabus of the Royal Naval Staff College, 1921* and a later version also of the same title issued in 1936 as *O.U. 5338*. As Admiralty publications, they possessed the cachet of official endorsement, but the truth is that no manual could do more than outline the objectives and topics of the course, as a key variable was the background of the officers attending. So much of the course depended on the work of the qualifiers and the subjects they covered in their own papers while the arguments they made in the discussions following the lectures was itself a vital component of the course's value. A typical means to ensure a robust discussion at lecture's end was to provide a précis of the lecture at the end of the class

226 Neither *C.B. 3013* nor *B.R. 1984* have been located. The existence of the two works is confirmed by their reference in other Admiralty publications.

227 ADM 203/100.

228 Later Captain Gilbert Ridley Waymouth (1901-1975). Five firsts in examinations for Lieutenant and Goodenough Prize in Gunnery. Short Course, Cambridge University, 1922-23; Assistant Fleet Wireless Officer in *Revenge* and *Nelson*, 1927-29; Flag Lieutenant in *Royal Oak* and *Barham*, 1929-30; Staff Course, 1931; Flag Lieutenant Commander in *Dolphin*, 1932; Signal School, 1932-34; Fleet Wireless Officer in *Nelson*, 1935-36; Signal School, 1936-38; Commanding Officer, HMS *Leith*, 1938-40; Captain, 1942; Commanding Officer, *Delhi*, 1943-44; Imperial Defence Course, 1950 and retired, 1952.

229 Unpublished manuscript, p. 56, Captain Gilbert Ridley Waymouth Papers, Imperial War Museum, London, IWM/87/16/1.

so that note taking was avoided, but warning the qualifiers that one would be asked to lead the discussion at the talk's conclusion. This ensured that minds did not wander unduly and forced all to master the topic at hand. Richard Webb writing of the War Course, but providing an assessment that applied in large measure also to the Staff Course, observed:

> 4. As in former Sessions, the Lectures have been followed by discussions, and these have not only induced to clear and logical thinking of the various problems set before the Course, but have brought out a considerable amount of useful and instructive information which has added materially to the value of the Course.

> 5. All the officers undergoing the War Course have shown great keenness, combined with a very high standard of clear thinking and a good grasp of the theory and principles of War. The discussions have been on a high level, and the Course generally has, in consequence, achieved very valuable results.[230]

If others might hold that Webb was a little too generous in letting officers voice their views when a stronger guiding hand was appropriate, the value of argument and discourse in furthering the aims of both courses was understood and one can accept Webb's praise on its own merits.[231] To the above, it is noteworthy that the Naval Staff endorsed the work of both colleges making their work available to a greater audience by issuing the more considered lectures to the service as a whole. Thus, *C.B. 904, Greenwich Introductory Lectures, 1919* and *O.U. 5233, Royal Naval War College Address by Rear Admiral Richmond, 1920* were collated and distributed to the fleets and shore establishments.

Officers desiring to attend the Staff College were enjoined to prepare themselves by reading preliminary works focusing on staff work within the Navy, military strategy and history, and combined operations. Of the last subject, it was desired that officers appreciate the uniqueness of what are now thought of as amphibious and joint operations. Works recommended for consideration included those covering the Russo-Japanese War and Gallipoli, especially, Cecil Aspinall-Oglander's two-volume Official History. Commander John Creswell's *Naval Warfare,* only recently published, was endorsed as a work worth reading by any applicant and coming from an officer who would later oversee the Tactical School during the next war was testimony that the Service did not always have to look outside its own ranks for a thoughtful reflection on naval war.

---

230 ADM 1/8710/138, Vice Admiral Commanding, Royal Naval War College to Secretary of the Admiralty letter of 22 February 1927.
231 H. G. T. 'Admiral Sir Richard Webb', *The Naval Review*, Vol. XXXVIII, No. 1, February 1950, p. 3.

Yet, outside writers remained at the fore of recommended reading and here C. R. M. F. Cruttwell's *History of the Great War* along with Corbett's *England in the Seven Year's War* were prominent. Both the German and British Official Histories of the First World War featured in the Admiralty's suggested reading lists, whilst works of military history believed worthy included those by Brigadier General Edward Spears, Field Marshal Sir William Robertson and Sir Frederick Maurice.[232] In all that they read, officers were reminded to do so not with the idea of collecting a body of facts and dates, but to appreciate the underlying principles suggested in the wars and campaigns reviewed.[233] This was crucial, as a vital aspect of British naval higher education of the period was securing an appreciation of the principles of war as codified in the Service's primary doctrinal publication, the *Naval War Manual*.

Central to Staff College instruction was to prepare officers in writing an appreciation, the formal assessment developed by a staff guiding likely courses of actions open to the Royal Navy in a war, a campaign, or a battle. The components of the appreciation were defined in Appendix I of the *War Manual* and consisted of a series of logical summaries developed into a formal paper providing a top-down approach to address nearly any problem at hand and the measures available to secure British objectives.[234] Accordingly, a typical naval appreciation ran to the following format:

    a)   Review of the situation.
    b)   Our object.
    c)   Course of action open to the enemy which may affect the attainment of our object.
    d)   Enemy's most probable course of action.
    e)   Our possible courses of action in order to attain our object.
    f)   Our proposed course of action.
    g)   Our proposed plan.[235]

The above format was the one apt to be followed when preparing an appreciation where the initiative rested with an adversary, but if the British held the upper hand the paper might follow the template of (a), (b), (e), (c), (d), (f) and (g).

For the SOWC, the bedrock of its instruction was informed by the *Naval War Manual* with the lectures and schemes of the course closely following the themes

---

232 ADM 182/72, Admiralty Fleet Order, '1751—Preliminary Study by Candidates for the Royal Naval Staff Course' of 16 July 1936.
233 ADM 182/66, Admiralty Fleet Order, '1168—R.N. Staff Course—Revised Instruction' of 25 May 1934.
234 The first postwar edition of the *Naval War Manual* was issued in 1921 as *C.B. 934*. Revised editions were released in 1925 as *C.B. 973* and *c.* 1931 as *O.U. 5394*.
235 'The Preparation of Combined Appreciations for a Major War, Appendix A, Notes on Service Appreciations,' Imperial Defence College lecture of 1933, Alanbrooke Papers, 3/12, LHCMA.

raised in that work.[236] Hamilton's paper previously mentioned, 'Peace Distribution of the Fleet', was written against paragraph 101 of the *Manual* which covered the 'Strategic Distribution of the Fleet', but its influence can be seen in many of the other lectures presented. Richmond's first lecture covering tactics covered the principles of war and corresponded to paragraph 3 of the *Naval War Manual*,[237] whilst his lectures covering the role and function of battleships closely followed paragraphs 37 and 38, which discussed the rationale of battleships and battle cruisers.[238]

If the *War Manual* remained the foundation of the SOWC curriculum, then it also served as the medium for introducing new concepts to officers. This is understandable as the *Manual*, though a keystone of naval doctrine and representing the received wisdom of the Service, was by its nature a work of compromise. Roger Bellairs, a founding member of *The Naval Review*, believed it leaned a little too much on the *FSR*. Writing to Richmond in early 1925, the Captain complained:

> The War Manual has come up to the 1st Sea Lord for approval to issue to the fleet. I know it has received consideration by the staff & afloat, but nevertheless I am not quite happy about one or two points in Chapters 1 & 2 concerned with the principles of war. There is perhaps too much adherence to the Army Field Service Regulations manual & also perhaps battle is allowed to sink into the background rather too much.[239]

Bellairs' criticism that, as drafted, the revised *Manual* deprecated battle may have been cant. We have already seen where some critics lamented the pre-1914 War Course's treatment of its place in war and this owing to Corbett's influence. Still, that did not make his argument any less telling and the knowledge that the *Manual* was the primary catechism for the current War Course, of which Richmond was the postwar father, was not a point to be lost on the letter's reader.

Bellairs' view that the *FSR* was influencing naval thinking perhaps a little too much in 1925 remains the more interesting argument. There can be no doubt that the Navy adopted much of military practice and formal doctrine when it established its own staff and war training as recounted in the first chapter. Drax, having studied at Camberley, saw much to admire in the British Army's method of staff training and seeing a vacuum in the Royal Navy's manner could well appreciate the difference. Richmond, too, thought the Army during the war had demonstrated a better grasp of how to

---

236 ADM 186/66, *C.B. 973, Naval War Manual, 1925*. Admiralty, Naval Staff, Training and Staff Duties Division, October 1925.
237 'Tactics I, Fundamental Principles', Richmond lecture, Autumn Session 1921, Richmond Papers, NMM/RIC/10/6.
238 'The Battleship, I' and 'The Battleship, II', Richmond lectures, Autumn Session 1922, Richmond Papers, NMM/RIC/11/3.
239 Roger M. Bellairs letter to Richmond of 17 February 1925, Richmond Papers, NMM/RIC/7/4/B.

employ a staff effectively. That his deputy in the Training and Staff Duties Division even in 1918 was a military officer is ample confirmation that the severest critics of the Service believed it still had far to go in settling its affairs. This belief, though, predated the war. In 1913, Richmond argued for the adoption of something akin to the *FSR* for use by naval officers. A point Bellairs may have forgotten, but assuredly, Richmond had not.[240] Though the 1921 edition of the *Naval War Manual* has not been traced, portions of its contents can be inferred by reviewing Drax's surviving lecture notes on 'Tactics' for the 1921-22 Staff College session. These show that in discussing the phases of battle, the Navy framed the matter in a slightly different hue than the Army by specifying four distinct stages: '(i) Minor strategy and gaining contact. (ii) The approach, manoeuvring for position, deployment and opening fire. (iii) Close action. (iv) The pursuit.'[241] A handwritten note covering his lecture records that the *FSR* merged phases (i) and (ii). When the 1925 edition of the *Naval War Manual* appeared this, too, was now the case for the Service's doctrinal publication with the further change that 'Close action' was now styled 'The Action'.[242]

The Earl Wavell, moreover, provides indirect confirmation that the Navy acceded to the British Army's formulation of the principles of war. Revising the *FSR* in 1935, Wavell informed Liddell Hart that he was not free to make the changes recommended by the military critic and journalist answering 'to alter the principles of war, even had I wanted to or been capable of drafting a better set, was not, I think, "on". The War Office, after a great deal of correspondence, I believe, induced the Navy and Air Force to accept the same set of principles, and could not have changed them without reference to the other services.'[243] Thus, though the *Naval War Manual* began as a maritime version of the *FSR*, the British Army doctrinal publication covering operations, it was never a slavish copy. This, notwithstanding, that passages were lifted from the *FSR* and appeared in its naval counterpart. An example of such constructive borrowing and the nuance of interpretation presented includes:

> The Army will be trained in peace and led in war in accordance with the doctrine contained in Field Service Regulations.[244]

contrasted to the Navy's:

240 Herbert Richmond, 'Orders and Instructions', *Naval Review*, p. 172.
241 'Tactics', Royal Naval Staff College, Greenwich, Session 1921-22, Admiral Sir Reginald Aylmer Ranfurly Plunkett-Ernle-Erle Drax Papers, Churchill College, Cambridge, DRAX/7/2.
242 *Ibid.*
243 Wavell to Liddell Hart letter of 10 February 1935, Liddell Hart Papers, LH 1/733/103, LHCMA.
244 *Field Service Regulations, Volume II, Operations, 1929* (London: His Majesty's Stationery Officer, 1929), p. 2.

> The Navy will be trained in peace and led in war in accordance with the accepted
> principles of war on which this book is based.[245]

From the previous, it will be seen that the Navy avoided any explicit pronouncement of doctrine. This was intentional as a tenet of Royal Navy ethos was that it was for the fleet commander to specify how battle would be fought, whilst the Admiralty concerned itself with raising and providing the forces required for that battle. Successive fleet commanders were very jealous of protecting this prerogative and any publication or order issued by the Naval Staff thought to infringe on this right was viewed with prejudice.

That the *Manual* quickly diverged from the *FSR* is not surprising as they were Service and not joint publications. Both documents evolved over time to account for the changing nature of war, the experiences of the Services drafting them, and the inherent differences in the nature of land warfare and battle at sea. The influence of the *FSR* on the *War Manual* and portions of the *Battle Instructions* is unmistakable. How the greater Navy would have reacted had the degree of borrowing from the *FSR* been known can well be imagined given the views of some who believed a little too much of military thinking was already evident. The documents also evolved owing to the foibles of their respective contributors. Perhaps this latter aspect is no better illustrated than in the writings of the mercurial and brilliant J. F. C. Fuller. Believing in the workings of the trinity—the idea that much of life and of thought were shaped by patterns necessarily reduced to groupings of three including the principles of war— Fuller's habits of thought demonstrate the weakness of only reading any work at a surface level.[246] Still, that is to anticipate events. The edition of the *FSR* available in 1921, the year the *Naval War Manual* first appeared, specified eight principles and the 1929 edition listed seven as the principle, 'object', was dropped.[247] Fuller credited Brigadier John Dill, then serving as a member of the Camberley directing staff, and Major Bernard Paget,[248] then of the War Office, with securing their inclusion in the *FSR*.[249] Perhaps a lesser individual would have been content to rest knowing that his work had found official favour, but Fuller was soon espousing a ninth principle which duly appeared when his *The Foundations of the Science of War* appeared in book form.[250]

---

245 ADM 186/66, *C.B. 973*, p. 1.
246 Brian Holden Reid, *J. F. C. Fuller: Military Thinker* (Houndmills: Macmillan Press, Limited, 1987), pp. 83-85.
247 'Co-operation', Brooke-Popham lecture of 29 March 1932 to Royal Naval War College, Brooke-Popham Papers, 9/4/5, LHCMA.
248 Later General Sir Bernard Charles Tolver Paget (1887-1961).Staff Course Camberley, 1920; Staff College Camberley, 1926-29; Imperial Defence Course, 1929; Quetta Staff College, 1932-34 and Commandant, Staff College Camberley, 1938-39
249 Fuller to Liddell Hart letter of 11 April 1949, Liddell Hart Papers, LH 1/302/388, LHCMA.
250 Reid, *Fuller*, p. 95.

Fuller's first appearance at the Royal Naval Staff College was during the inaugural postwar course when he took as his initial theme 'Tanks'. Recently assigned to the War Office with responsibility for establishing the nascent Tank Corps on permanent lines, Fuller was as much proselytizer of the new arm as reasoned staff officer. He soon returned in March 1920—again during the course's first sitting—discoursing this time on 'The Science of War' where he laid out the principles of war—tenets he had been formulating since 1912. Publicly, Fuller tempered his comments by conceding that it was for the Navy and the qualifiers to confirm whether his principles were applicable to the maritime environment,[251] though privately he voiced no doubt cautioning his new acquaintance Liddell Hart at roughly the same moment:

> I cannot help feeling that your views as regards the principles of war are based more on a special case and particularly a tactical case than on the general conditions of war itself. A principle should aim at being a fundamental truth applicable under all circumstances.[252]

As a serving officer then ensconced at the War Office, Fuller spoke with a necessary measure of reticence that would not always be present. What Fuller did avow was that the eight principles were of equal importance and were the edifice upon which any understanding of war followed. Continuing, he said:

> The pure science of war has really nothing to do with this application for this belongs to the art of war. We, therefore, may definitely say there is a science of war common to all types of war and that this science forms the foundation of the whole edifice of the art of war on land or sea or in the air.[253]

Fuller was not the first writer to argue that principles existed in war and he did not claim their application was common to the Services. His introductory comment that it was for the Royal Navy to determine their application in the maritime environment reads as a caveat introduced after the body of the lecture had been prepared. In the event, the Service quickly accepted his general formulation and they appeared in the *Naval War Manual,* which began with a review of the eight principles. Citing maintenance of the object, offensive action, surprise, concentration, economy of force, security, mobility, and cooperation, it eschewed any discussion on the roles of infantry, artillery, and cavalry, substituting rather discourses on the functions of the differing classes of warships, the distribution of the fleet, and the purpose of battle.[254]

251 J. F. C. Fuller, 'The Science of War', lecture to Royal Naval College, Greenwich of 31 March 1920, Liddell Hart Papers, LH 1/302/673, LHCMA.

252 Fuller to Hart letter of 10 June 1920, Liddell Hart Papers, LH 1/302/2, LHCMA.

253 Fuller, 'Science of War' lecture of 31 March 1920, Liddell Hart Papers, LH 1/302/2, LHCMA.

254 ADM 186/86, *C.B. 973, Naval War Manual, 1925,* p. 1.

In accepting Fuller's general thesis, some in the Navy did not accept his argument that the principles were of equal importance. Richmond, for one, held that the principle of 'object' dominated all others. This was classic Richmond—establishing the overarching concept from which subsidiary arguments flowed. As the Army removed 'object' as a principle in a later edition of the *FSR*, but kept it as an introductory precept preceding the discussion of the other seven, it was a view shared by others. Of course, the Royal Navy's appreciation of the principles of war predated their espousal in the *War Manual* as Drax's 1919 lectures at the Staff College show. That the Director himself addressed the principles and did not leave it to another member of the directing staff to review speaks to the importance of the topic and successive Directors continued this practice.[255] Though both the Staff and War Courses covered the principles of war, their reasons for doing so were subtly different. For the War Course, the principles were the essential grounding a senior officer received to prepare him for battle. Decisions in war were to follow the principles and the Navy would wage war in accordance with their precepts—so avowed *The Naval War Manual*.[256] For the officer qualifying at the Staff Course, it followed that mastery of the principles of war was essential if he were to serve his commander well. Astley-Rushton expressed the rationale of providing the staff officer a background in the principles of war noting:

> There may be those who, not having closely considered the matter, wonder why the Staff Officer should require knowledge of this type, but the matter becomes clear when we ask ourselves what is the first thing we demand of an assistant. Is it not that he should be really able to assist us; to relieve us of a great burden of detail, and leave us free to give decisions upon main issues? If that is what we demand, it is evident that we require him to be well acquainted with our own function and our own duties. We do not want to train him up ourselves – time does not permit of it these days. We like to know that he comes to us trained, and understands what we are talking about from the word "Go".[257]

Meanwhile, when initially lecturing at the War Course, Richmond eschewed any direct exposition on the principles of war preferring to introduce them as broad themes in other lectures. Accordingly, his discourses on 'Tactics' and 'Strategy' are replete with references to principles. Where the Army in 1921 listed eight principles: objective, offensive, mobility, security, surprise, cooperation, concentration, and economy of force, Richmond's 1920 SOWC lectures established twelve: object, offensive,

255 ADM 203/100.
256 ADM 186/66, *C.B 973*, p. 1.
257 ADM 203/69, Astley-Rushton lecture, 'The Staff College', pp. 2-3.

surprise, cooperation, economy of force, initiative, deception, speed of execution, momentum, power, concentration, and secrecy.[258]

Richmond's enunciation of power, or will, as a moral and physical factor, anticipates Fuller's advocacy of the principle of determination when he came to write *The Foundations of the Science of War*, a work first published in 1926 but summarizing much of what that brilliant officer previously had covered in lectures to the qualifiers at Camberley and Greenwich. Further, Richmond's survey of the principle of economy of force accords with the distribution of the fleet as codified in the *Naval War Manual* and amplified by Hamilton's lectures.

Yet, in the 1921-22 academic year Richmond provided an explicit lecture on the principles of war. Now, he closely followed the eight principles identified in both the *War Manual* and the *FSR* allowing for the Navy's variance in how it defined economy of force. Exactly why Richmond adopted this change omitting the principles of will, initiative, speed of execution, deception, and momentum in his second year is not known, but ensuring conformity to what was being presented at the Staff College and the need to track with the now published *War Manual* are certainly strong considerations.[259] As the principles were held universal with only their application changing, conformity was a *sine qua non*. Viscount Haldane, no average intellect when addressing defence, endorsed this view arguing:

> The principles of war are just what they were in the time of Hannibal but, though they are the same principles, the modifications of their application are infinite and it requires a highly trained mind to bring those principles into modern life and to apply them in the way they should be applied.[260]

It was also the view of the *War Manual* for it stressed:

> Although the principles of war are not in themselves abstruse, their application is difficult, and must vary according to circumstances. No two situations are identical; consequently the application of principles cannot be made subject to rules.[261]

To this even Richmond himself subscribed, closing a War Course lecture with the tag, 'In principles be an unchanging conservative; in their application be a red hot radical.'[262]

258 'Tactics' and 'Tactics, II', Royal Naval War College lectures Spring Session 1920, Richmond Papers, NMM/RIC/10/2.
259 'Tactics I, Fundamental Principles', lecture, Autumn Session 1921, Richmond Papers, NMM/RIC/10/6.
260 *Hansard*, House of Lords debate, 16 June 1926, v. 64, cc. 415-46.
261 ADM 186/66, *C.B. 973*, p. 2.
262 'Tactics VI', no date provided, Richmond Papers, NMM/RIC/10/3.

Though the *Manual* with its distillation of the principles of war traces its roots to the *FSR*, its emphasis and structure were of a slightly different construction than the Army's doctrinal interpretation and even more so when set against the Air Force's view of the matter. Where the Army specified the 'objective', the Royal Navy spoke of 'maintenance of the object'.[263] In both instances, the object or objective was apt to be a physical element such as striking at an opponent's military forces. For the Air Force, their elucidation of the principles had evolved from that originally espoused in the *Manual of Air Operations* issued in 1922 and by 1928 was now encapsulated in the *Royal Air Force War Manual* which stressed striking at the enemy's morale.[264] This was more nebulous and the ramifications of the Services having distinct interpretations to the principles of war remains for later consideration.

This, though, is not to claim that the Navy deprecated the importance of morale. The *Naval War Manual* emphasized that success in war depended more on moral factors than on any physical consideration.[265] To this end, it claimed that:

> Success in war depends more on moral than on physical qualities. Neither numbers, armament, resources, nor skill can compensate for lack of courage, energy, determination, and the bold offensive spirit which springs from a national determination to conquer. The development of the necessary moral qualities is, therefore, the first object to be attained in the training of the Navy. Next in importance are organization and discipline, the training of the mind and body, and the proper use of weapons. The final essential is skilful, resolute and understanding leadership.[266]

This emphasis on moral factors, in moral qualities, indeed, in officer-like qualities, explains why education and training alone were never the sole indicators of officer promise and why within the Service the Staff Course never became the requisite for higher command a few officers desired. The Staff Course might produce competent officers. It might even produce clever officers. Yet, if the necessary moral qualities were absent in the man, then it would not produce a great officer. Comments by his superiors on the presence of these moral qualities were an essential part of an officer's evaluation and their absence was the death knell to advancement.[267] The specific attributes an executive officer was assessed against included:

<hr>

263 *Ibid.*
264 Allan D. English, 'The RAF Staff College and the Evolution of British Strategic Bombing Policy, 1922-1929', *Journal of Strategic Studies*, Vol. 16, No. 3, pp. 411-12.
265 ADM 186/66, *C.B. 973*, p. 1.
266 *Ibid.*
267 Though the criteria did not change, following the World War officers also received a numeric rating to their written assessment allowing their performance to be more easily weighed against their peers. See Lord Chatfield, *The Navy and Defence: The Autobiography of Admiral of the Fleet Lord Chatfield* (London: William Heinemann Ltd., 1942), p. 201.

a)   General Conduct;
b)   If **Not** of temperate habits;
c)   Professional Ability;
d)   Zeal and Energy;
e)   Power of Command;
f)   Judgment;
g)   Initiative; and
h)   General Opinion of Officer.[268]

Though these qualities, doubtlessly, meant differing things to different officers there was a general understanding of what the essentials were. Here, the Staff College extensively surveyed the place of *esprit de corps*, discipline, leadership, organization, psychology, and the differing types of naval officers. A work by the American navalist, Alfred Mahan, was instructive as he had written on the last named subject drawing as his material from the officers of the Royal Navy during the age of sail. Drax employed Mahan to shape his lecture and calling on the expertise of the Navy's medical officers, additional talks on disease and its effects on fighting efficiency were covered by Surgeon Commander Arthur Gaskell. Meanwhile, Yexley, the advocate of social welfare for marines and ratings, spoke yearly on the discipline and welfare of the lower deck calling attention to an officer's particular responsibility of care.[269]

Such softer subjects as discipline and organization featured in the War College curriculum, but as the senior course lasted only slightly longer than four months, compared to the Staff College's typical length of 10-12 months, brevity was the watchword. The same applied to the use of historical example and though lectures on past experience make their appearance, the balance is strongly in the favour of the Staff College when employing the model as a means of instruction. In 1920-21, the historical examples employed in the War Course were mostly the examples of 1914-18. True, Richmond did speak to the war plans of the eighteenth century and coverage of the Russo-Japanese War featured, but the emphasis remained surveying the rich experience of the most recent conflict.

The Admiralty Fleet Order calling attention to Corbett's treatment of the Seven Years War (1756-1763) matched the stress that conflict enjoyed at the Staff College. Originally surveyed in five lectures during the first session, by the time of the second postwar course, this early global war was analyzed in seven lectures and this format was retained for the next several years. In time, the American War of Independence and the wars of the Napoleonic period supplanted the primacy of the Seven Years War. Still, as a topic it never truly lost favour.[270] As a subject, the Seven Years War had much

268 ADM 196/91 (Shipway). Original emphasis from Standard Form, S.206 and to be completed by the reviewing officer.
269 ADM 203/100.
270 Anon., 'Royal Naval Staff College', p. 29.

to commend. The argument that the elder Pitt's control of the Army and the Navy approached the truest demonstration of the value of a unified Ministry of Defence, a subject of much contemporary angst, was not the least, or so avowed Fuller.[271] His belief though was far from unique or original and to this view Richmond, Ellison, and, of course, Corbett subscribed.[272] Moreover, far removed from the late war and the direct experiences of officers attending, the Seven Years War offered a vehicle of learning divorced from recent controversies.[273]

Though the first War Staff Course began in June 1919 and ran until the following June, this twelve-month format was highly unusual. Subsequent courses normally ran from early September until July of the following year, though the second course suffered a two-month interruption when classes were suspended from 11 April through 11 June 1921 due to the previously mentioned miners' strike and qualifiers supported duties in aid of the civil power. In this case, the course ran slightly longer finishing on 28 July when Captain Drax gave the closing lecture.[274] From 1929, the Staff Course moved to a January starting date running to as early as August or to as late as December. Exactly why the January matriculation date was adopted has not been determined, as direct evidence noting the rationale for the change is lacking. Funding for the Navy was no more acute than usual, as Appendices I and III amplify. Nor, does the need to support emergent operations appear as the likely explanation as 1928 and 1929 were relatively quiescent on that score. The most likely explanation is that the Navy sought to arrange its schedule to coincide with those of the other Staff Colleges and the recently created IDC, which typically began their proceedings in January. This allowed the other Services to send their officers to the Royal Naval Staff College in a pattern matching their own practices for making officer assignments. Thus, an Army or Air Force officer would not need to extend or shorten a tour to meet a September matriculation date and the Navy could now assign qualifiers or members of the directing staff to the joint course without any undue problems. Commander James Esdaile,[275] RAN, followed completion of his Staff Course by attending the IDC in January 1936 and Lieutenant Commander Frank Getting,[276] also of the Commonwealth's Permanent Naval Force, managed the same in 1934. Adjusting the schedule was a salutary benefit for Commonwealth officers, as it alleviated the need to return officers home unnecessarily or find temporary employment in the Royal Navy whilst waiting commencement of the IDC.

271 Fuller, *Empire Unity and Defence*, p. 230.
272 Ellison, *Amateur Strategy*, p. 131.
273 ADM 100/69, Astley-Rushton lecture 'The Staff College', p. 11.
274 ADM 203/100.
275 Later Acting Captain James Claude Durie Esdaile (1899- ?). Staff Course, 1935; Imperial Defence Course, 1936; Acting Captain, 1940 and retired, 1959.
276 Later Captain Frank Edmund Getting (1899-1942). Staff Course, 1933; Imperial Defence Course, 1934; Commanding Officer, *Canberra*, 1942 and died of wounds on loss of ship.

The January start date of the IDC also had the potential for affecting officers attending the War Course's October session, as this class ended in the following February. Robert Raikes currently enrolled in the October 1927 SOWC was selected to attend the 1928 IDC. Consequently, he left his SOWC a month early to begin his new studies in the joint course. Perhaps mindful that he had not completed the SOWC, when the moment came for Raikes to replace Thomas Calvert[277] as Director of the Staff College, he was obliged to sit a month of the March – July 1932 War Course before assuming his duties, something that Calvert himself had done upon completion of his Imperial Defence Course.[278]

One benefit of the new schedule, however, was that it allowed Staff Course qualifiers and members of the directing staff to be attached to military units during the Greenwich summer leave period when the Army typically conducted its manoeuvres in the field.[279] Charles Drage, last seen at Christ Church, Oxford, and a member of the last course convening under the old format arrived at Greenwich in October 1927. His extended course had their matriculation date deferred, as the course now completed in December 1928. This allowed Drage to secure attachment for a week with the Second Light Infantry Brigade during its period of manoeuvres. Joining on 9 September 1928, Drage was much impressed by two things: the general standard of the Grenadier Guards and the manner in which the battalion commander issued his operations orders before a night attack commenced.[280] Demonstrating the problem of a September matriculation date, Captain Lancelot Holland, Deputy Director of the Staff College, had traveled to Germany in the previous year to observe the British Army of the Rhine in its manoeuvres with the French Army.[281] As the Staff Course for 1927 was set to begin on 27 September, Holland had little time to prepare for his primary responsibilities upon returning and a January opening date would have been more conducive to his travel. Whatever the exact reason for the change, the Staff Course for the balance of the period operated with a January opening.

Typically, the morning period ran from 0930–1300 hours when the day's lecture was given and resumed following the midday meal at 1400 when schemes and individual work were accomplished. Formal work ceased at 1600 hours, but officers still had much to do in the evenings with visits to Chatham House, RUSI, and Woolwich

277 Later Rear Admiral Thomas Frederick Parker Calvert (1883–1938). In Command, *K 11*, 1916-18; Captain, 1922; Training and Staff Duties Division, 1920-21 and 1923-24; Commanding Officer, HMS *Pegasus*, 1921-23; Naval Air Section, 1924-26; Flag Captain in *Frobisher*, 1926-28; Imperial Defence Course, 1929; Senior Officers' War Course, 1929-30; Director, Royal Naval Staff College, 1930-32; Commanding Officer, *Renown*, 1932-33; Chief of Staff to Commander Second Battle Squadron, 1933-34; Rear Admiral, 1934; Tactical Course, 1936 and Commander, Second Cruiser Squadron, 1936-38.

278 ADM 196/50 (Raikes) and ADM 196/91 (Calvert).

279 Anon., 'Royal Naval Staff College', p. 31.

280 Diary entries 9 and 10 September 1928, Drage Papers, IWM/PP/MCR/99, Reel 3.

281 ADM 203/85, Captain L. E. Holland letter to Admiral-President, Royal Naval College, Greenwich of 7 October 1927.

frequently the norm to hear topical lectures. Drage, now an Oxford MA, during his time on the Staff Course, supplemented his Greenwich instruction by attending a series of lectures on the 'Crisis of Communism' at King's College, London. Meanwhile, qualifiers drew lots to attend evening debates in the House of Commons as the college received a quota of tickets for the Distinguished Strangers' Gallery.[282] When not attending such talks, reading required material, revising papers, sports predominated.[283] If Greenwich failed to maintain the horses and hounds of Camberley or the Moths and Harts aircraft of Andover for fostering pilot proficiency, sports, nevertheless, remained a central feature. Inclined officers could even visit nearby Woolwich secure a mount from the Gunners' Riding School before listening to one of the regular lectures often held there.[284]

Regular competition amongst the Staff Colleges, most notably, during the week of the joint exercise held at Camberley was testimony that course work was just not class work. Indeed, given the regularity that the Royal Naval Staff College bested its military and air counterparts in tennis, cricket, squash, and golf, it is hard to escape the conclusion that a deciding factor in an officer's acceptance was his athletic prowess.[285] Even for those whose best days on the playing field were but a fond memory got caught up in the enthusiasm of Epsom and the Derby when the college shut down for the day and all and sundry made a day of the races. Frivolous? Perhaps, but it also fostered cohesion, teamwork, and a sense of camaraderie across the ranks in a profession that was more of a calling than a career.[286] Likewise, the qualifier could expect to dine with the Director and Assistant Director of the Staff College at some point during the year and regular dances were held emphasizing that the qualities required of a staff officer were broader than a thorough understanding of the principles of war and *King's Regulations*.

Every two weeks a debate occurred in the evening hours on a matter of current interest to foster skills in logical reasoning, repartee, and social presence. Officers

282 Diary entries 20 November, 1 December 1927 and 1 March 1928, Drage Papers, IWM/PP/MCR/99, Reel 3.
283 ADM 203/69, Astley-Rushton, 'The Staff College', p. 8.
284 Diary entry 17 October 1927, Drage Papers, IWM/PP/MCR/99, Reel 3.
285 In 1936, the Staff College defeated its military and air counterparts in golf by the score of 30 points to 24 and 18 points, respectively. In squash, the Air Force lost to the Navy 2 matches to three whilst in tennis, the Navy took five matches to three against the Air Force Staff College. In 1937, the Royal Naval Staff College heavily defeated the Hawks of Andover who, in turn, defeated the Owls of Camberley. In tennis, the Air Force lost to Greenwich four matches to five during their week of visiting. Again, in squash, the Greenwich Pelicans defeated the Air Force Staff College. In 1938, the Air Force Staff College beat Greenwich in a closely fought match whilst losing to the Navy in tennis nine matches to seven and losing also in golf. Squash favoured the Air Force that year, as they defeated both Camberley and Greenwich.
286 Humphrey Hugh Smith, *An Admiral Never Forgets: Reminiscences of Thirty-Seven Years on the Active List of the Royal Navy* (London: Seeley, Service & Co. Limited, 1936), pp. 308-09.

when assigned to the Naval Staff or seconded to the CID would have to work with civil servants most likely possessing a university education. The Navy was alive that the education of its officers to that point in their career whilst technically severe came at the expense of the more liberal arts or classical grounding others possessed and sought to rebalance any deficiency present. More often than not, the topics had little to do with naval affairs. Thus, debates resolving whether a central Department of Labour was desirable and foxhunting was inherently cruel took their place alongside those arguing that a married naval officer was necessarily more inefficient than his counterpart who remained single.[287]

In common with those studying at Camberley, the syndicate featured as a primary means of overseeing instructions. Composed of five qualifiers with one acting as lead, the syndicate was monitored and mentored by a member of the directing staff. The format allowed the several minor schemes and exercises to be considered as a group, fostered teamwork, and allowed specialist knowledge to be shared amongst members. At Greenwich, the name syndicate was deprecated preferring the term sections, though the function served was much the same as at Camberley. Each section possessed at least one non-naval officer—whether from the British Army, the Royal Air Force, or the Royal Marines—and assigning officers of different specializations amongst the sections ensured a divergence of perspective to any problem weighed. Periodically, it became the practice to rotate the members of sections so that all qualifiers had experience working with each other and the further members of the directing staff.[288] Though the instructor overseeing his section was likely to be senior to his qualifiers, this was difficult to guarantee in the first sessions conducted following the war. By 1924, though, the practice of overseeing a section by an officer of Lieutenant Commander's rank ceased and with it the problem of having a qualifier outranking the section's instructor.

That problems arose at Greenwich when the anomaly occurred is unknown, but indirect evidence exists that it probably did as the Air Force noted an ill-effect in the practice. It moved to provide Squadron Leaders assigned to the Andover directing staff the temporary rank of Wing Commander—albeit, unpaid—for the duration of their tour.[289] Such practice was the norm already for equivalent Army officers serving at Camberley and Quetta. The Air Force move ensured that debating points arising at its Staff College from naval qualifiers did not end up being settled merely on the number of stripes worn by the protagonists as recourse to rank was a tendency air officers noted in their naval counterparts.[290] Still, the problem may have been no more than a reflection that a common ethos amongst the Services was largely absent

---

287 ADM 203/100 and diary entry 24 October 1927, Drage Papers, IWM/PP/MCR/99, Reel 3. When Drage's course debated 'Public School versus Dartmouth Entry', the motion carried for the latter by a margin of 14-25.

288 Anon., 'Royal Naval Staff College', p. 25.

289 *The London Gazette*, No. 34064, 26 June 1934, p. 4063.

290 Air 2/526, Director of Staff Duties minute of 21 November 1932.

when having to consider the views of nominal superiors. One naval qualifier noted a pronounced tendency of Army and Air Force members to defer too easily to members of the directing staff during arguments and further comments regarding the matter are offered in a subsequent chapter.[291]

Reference has already been made to several of the schemes conducted by the Staff College and a qualifier could expect to participate in upwards of fifty such exercises as part of his coursework. Writing an appreciation for the opening moves faced by British Mediterranean naval forces seeking the German ships *Goeben* and *Breslau* in 1914 was a common early scheme faced by qualifiers with necessary background to these events being covered in multiple lectures offered by Wilfrid Egerton and successive members of the directing staff. Another scenario frequently appearing were the events surrounding the outbreak of a popular uprising in Portugal and, of course, the means of conducting a search as previously discussed. In 1926-27, a scheme addressing the characteristics of a future capital ship was conducted for the Admiralty whilst the 1928-29 session set Schemes178 and 178a to coincide with the combined fleet exercise 'M.Z.' held in March 1929, as previously mentioned.[292]

Beyond the work pursued at Greenwich proper, officers attending visited Camberley each year for a week to participate in a common exercise with their military and air counterparts whilst members of the Air Force Staff College came to Greenwich for a week's work leading to a common joint exercise. Beginning with the 1921-22 session, excursions further afield became the norm and supplemented the Staff Course. In May, day trips to survey the commercial docks of London and briefings on the experimental work in chemical warfare occurring at Porton Down became regular diversions from the daily grind of Greenwich.[293] Whilst in the following month, qualifiers visited the shore installations residing at Portsmouth and Portland for a week's lectures and demonstrations covering current fleet work. This offered the chance to observe air work from carriers, watch Coastal Motor Boats go through their paces, and see an Asdic anti-submarine exercise executed by flotilla forces before visiting a nearby air station of the RAF. A stop at the Tank School located at Wool, Dorset and a demonstration of British armour in the field completed the June tours. When he returned to Greenwich, only final lectures and the tackling of the major strategical game remained of the course.[294]

Not quite, for following the strategical game, there remained one last assignment before the Director rounded off the year with a summing up lecture. Qualifiers prepared a paper discussing the duties of a War Staff Officer. The paper, allowing the qualifier to draw upon all that he had experienced, learnt, and argued previously, provided

291 Godfrey Brewer, 'The Melody Lingers On – II', *The Naval Review*, Vol. LXII, No. 1. January 1974, p. 57.
292 ADM 203/90, Director, Royal Naval Staff College to President, Royal Naval College, Greenwich letter S.C.4736 of 12 April 1929.
293 ADM 203/100.
294 ADM 203/69, Astley-Rushton, 'The Staff College', p. 16.

him free range to offer his thoughts on presumably the duties that awaited him as a newly minted, designated War Staff Officer, or, after 1924, Staff Officer and *psc*.[295] No examinations followed unlike prewar practice, but achieving the designation was not automatic. The Director was responsible for reporting on the officers qualifying, and, if the judgment rendered was often in the nature of a canned assessment, it still served the useful purpose of bringing to the attention of higher authorities the latent potential of an individual. Typical of the reports rendered was Drax's assessment of Commander Ralph Binney,[296] Vice Admiral Hugh Binney's younger brother, made in June 1922 as the course was approaching its culmination. Drax reported that his progress had been steady and that the officer had applied himself diligently.[297] If not a ringing endorsement, it was marginally better than Drax's final evaluation which concluded 'that he may possess other valuable qualities, but has no aptitude for Staff work & only just succeeded in qualifying.'[298] Ralph Binney passed perhaps demonstrating in the process that if the minimum was not good enough there would not be a minimum. He secured the War Staff designation avoiding the fate of more than one qualifier who failed.

By the end of the course, the qualifier learnt where his next posting would be. Some might move on to another course, but the more typical assignment was to a staff position afloat employing his skills and learning the practical aspects of the staff officer's world. This need not occur immediately and time with the Admiralty whilst receiving the necessary regime of inoculations for those destined for duty abroad was the Navy's reward for successfully negotiating his time at Greenwich. In Ralph Binney's case, he spent two months in the Intelligence Division before serving an exchange period with the Royal Australian Navy as its Director of Naval Intelligence. Given his rather poor report upon completing his course, the posting appears curious and it may owe much to Binney's stated desire to serve with the New Zealand Division and this was the closest assignment that could be found.[299]

Exactly why the Service failed to re-adopt a final examination for the postwar Staff Course is not known, but the most likely explanation is that it was thought unnecessary when so many officers had direct experience of war.[300] Still, if this was the reason,

---

295 ADM 1/8668/172, Admiralty Fleet Order, '2038.—Substitution of the term Staff Officer for the term War Staff Officer, and the letters p.s.c. for the letters W.S.' of 1 August 1924.

296 Later Captain Ralph Douglas Binney (1888-1944). Six firsts in examinations for Lieutenant; Naval Ordnance Department, 1921; War Staff Course, 1921-22; Naval Intelligence Division, 1923; Director Naval Intelligence, Royal Australian Navy, 1923-25; Senior Officers' Technical Course, 1926; Staff Officer Operations in *Tamar*, 1928-30; Naval Ordnance Department, 1931-34; Captain and retired, 1934. Recalled for war service.

297 Report on Commander Ralph D. Binney dated 5 June 1922, Captain Ralph Douglas Binney Papers, Imperial War Museum, London, 75/98/1. Original emphasis.

298 ADM 196/127 (Binney).

299 *Ibid.*

300 Confirmation of the *ws* designation being awarded following the passing of an examination is found in Bertram Ramsay's service record, ADM 196/143 and the distinctions found in *King's Regulations* for use of *qs* and *ws* designations.

it fails to explain why it was never subsequently adopted and here the conclusion that the designations *qs* and *ws* had fostered a split in the value of the staff officer remains strong. Though the *qs* designation was retained for a time when the newer *psc* distinction replaced the *ws* post-nominal, the Training and Staff Duties Division probably did not want to see a similar divide arise over those awarded the *psc* without examination and those subsequently awarded the distinction who had. Even in the absence of a final examination, qualifiers who were successful were rated as passing as either Staff Officer Grade Two or Staff Officer Grade One.[301]

Officers attending the War Course were not graded as staff officers, but a standard assessment was employed to render a judgment on their performance during the course. This rated their general capacity against their peers, their powers of expression in writing and in oral debate, an officer's ability to buttress his arguments with appropriate facts or principles and their application of the principles taught.[302] If a numeric standing was not applied, it was still possible to discern how an officer did against his brethren when his report was compared to others.

It is vital to note that the Staff College was the first of its higher educational institutions reestablished by the Royal Navy following the war and, as such, the curriculum adopted influenced the shape of the other courses following. This is particularly true of the Senior Officers' Technical Course, Part I, which borrowed heavily from the Staff College syllabus. Additionally, as the Staff College predated the reforming of the War Course, it provided Rear Admiral Richmond a venue to lecture to the first postwar qualifiers whilst he awaited the arrival of the rest of his staff and the officers slated to form the first intake of the SOWC meeting on 1 March 1920. The topics covered included a single presentation on policy, four lectures on strategy, a further four papers on war plans, three discourses on trade covering both its offensive and defensive aspects, three discussions addressed tactics, and, finally, a single paper surveyed invasions and raids. Drax had previously covered much of this ground in the 21 lectures he delivered in the twelve months of the course. If Richmond's offering of 16 lectures in the course's final months was overkill, it was overkill with a purpose. Four of the sixteen naval and marine officers sitting the first course later returned at some point to serve as an instructor at the Staff College.

In the following years, Richmond continued to offer lectures to the Staff College whilst presiding over the War Course though his contributions now became more circumspect. Still, the Admiral took pains to ensure that the two courses' efforts worked to similar lines and he maintained the closest touch with the junior course.[303] His successor as Vice Admiral Commanding, Royal Naval War College, Sir George Hope, avoided the practice of giving lectures to the Staff College, and, if the two

301 ADM 196/52 (Willis).
302 ADM 196/91 (Custance).
303 H. G. T, 'Admiral Sir Herbert Richmond', *Naval Review*, Vol. XXXV, No. 1, February 1947, p. 5.

courses still complemented each other, the conclusion remains that the regime of Hope and Astley-Rushton lacked the previous intimacy of Richmond and Drax.[304] This, though, should not be taken to mean that the tandem of Hope and Rushton was a poor one. Rather, it reflected that the Staff Course was now on solid lines and Hope concentrated on developing the War Course with the Their Lordships acknowledging the efforts achieved at placing it on a sound footing upon his departure from Greenwich.[305]

Clarity of thought and brevity of expression were hallmarks of the trained staff officer. To this end, instruction in English composition featured in the course all with the intent of ensuring that orders, signals and appreciations expressed succinctly the action required. Admiral Sir Henry Jackson made the point clear in his opening address to qualifiers in June 1919 and it was a tenet of instruction retained throughout the period.[306] Demonstrating the centrality of clarity of expression during tactical and strategical war games all orders drafted by qualifiers passed through Signal Umpires who judged their content according to this standard.

In 1919-20, there were at least sixteen separate lectures and papers presented by the directing staff, outside speakers, or qualifiers touching on some aspect of tactics. Whether it was a survey of the 'Battle Tactics in the Sailing Era' or 'Smoke Tactics' offered by Henry Moore or the expositions of Admiral Sir Doveton Sturdee,[307] tactical discourse was a strong feature of the curriculum in the initial sittings of the War Staff Course. With the founding of the Tactical School at Portsmouth, the need to examine tactics at length lessened and the topic was now surveyed primarily from an historical perspective thanks mainly to the efforts of Captain Brian Egerton,[308] the Deputy Director between 1926-27.[309] However, qualifiers need not worry that current problems were being shortchanged for members of the Tactical School visited Greenwich usually for a week during the spring term providing lectures on the *Battle*

---

304 ADM 203/100.
305 ADM 196/43 (Hope).
306 Anon., 'Royal Naval Staff College', p. 13.
307 Later Admiral of the Fleet Sir Frederick Charles Doveton Sturdee (1859-1925). Captain, 1899; Assistant Director, Naval Intelligence, 1900-02; Chief Staff Officer, Mediterranean Fleet, 1905-07; Chief of Staff to Commander-in-Chief, Channel Fleet, 1907-08; Rear Admiral, 1908; War Course, 1909; Vice Admiral, 1913; Chief of War Staff, 1914; Commander-in-Chief South Atlantic and South Pacific, 1914-15; Admiral, 1917; Commander-in-Chief, The Nore, 1918-21 and Admiral of the Fleet, 1921.
308 Later Rear Admiral Brian Egerton (1886–1973). Six firsts in examinations for Lieutenant; Squadron Torpedo Officer in *Coventry*, 1921-23; Captain, 1923; Assistant Director, Royal Naval Staff College, 1924-25; Deputy Director, Royal Naval Staff College, 1926-27 and 1928; Commanding Officer, HMS *Calypso*, 1927; Senior Officers' Technical Course, 1928; Director, Torpedo and Mining Department, 1928-31; Commanding Officer, *Cumberland*, 1931-32; Flag Captain and Chief Staff Officer in *Courageous*, 1933-34; Rear Admiral and retired, 1935. Recalled for war service.
309 ADM 196/91 (Egerton).

Instructions and running several war games to demonstrate precepts.[310] All of this was conducted with the purpose that what was taught at the Staff Course remained consistent and did not deviate from what was being attempted elsewhere.[311] Of course, tactics can never really be divorced from the technology of the weapons themselves so what Greenwich offered was primarily a survey of tactical theory based on historical example and not tactical employment based on the means then available. Thus, it was this distinction that separated how Greenwich approached the matter differently from the Tactical School.

By the late 1920s and early 1930s, the Staff Course focused a greater portion of its time on joint operations and operations of a purely military nature than had been the case in 1919. In addition, a pronounced emphasis on the role of airpower in supporting military and naval operations formed major portions of the curriculum. Thus, lectures such as 'The Development of the Flying Boat', 'The Effect of Air Power Upon the War at Sea', and 'The Exercise of Air Control in Iraq' were features of the 1927 course. In 1931, Arthur Tedder,[312] a future Marshal of the Royal Air Force, surveyed 'The Air Aspect of Combined Operations', whilst other air officers addressed 'Air Power in Relation to Imperial Defence', 'The Employment of Air Forces in War', 'The Organisation of the Royal Air Force', 'Coastal Defence and Reconnaissance', and 'Air Raid Precautions'. Significantly, Tedder had attended the 1923-24 Greenwich Staff Course. Though he was to earn strong praise for his role in supporting military operations in the Western Desert during a subsequent war, his running arguments with Andrew Cunningham at the same moment over the air support rendered to the fleet indicate that education and familiarity of the naval role had their limits.

Historical perspective was matched with current thinking as papers covering air operations in the World War took their place beside those analyzing the French Air Force and air co-operation with the Army. By 1931, the Staff College was surveying the military experience in Belgium and France in two lectures and the Mesopotamia campaign was also suitably treated. General Edmund Allenby's Palestine operations were not neglected and the Army perspective in combined operations featured as topics for the qualifiers. From these lectures, the importance of the principles of war and sound administrative planning in operations emerged. Never neglected as a subject, the Battle of Jutland remained only one of the many topics that qualifiers were expected to understand and weigh.[313]

It is difficult to generalize about a course operating over a twenty-year period given the ever-changing officers both instructing and attending. Foremost, it can be claimed that the Staff Course sought to impart the skills required of one to serve as an effective

---

310 ADM 116/6364, 'Tactical School, Annual Report, 1937'.
311 ADM 1/9041.
312 Later Marshal of the Royal Air Force Arthur William Tedder, Baron Tedder (1890-1967). Staff Course Greenwich, 1923-24; Imperial Defence Course, 1928 and Air Force Staff College, 1929-32.
313 This summary is based on the Tennant Papers, especially, NMM/TEN/41/6.

member of a staff system, whether serving ashore, afloat, or in joint body such as the CID. The higher education in war provided, resting on the principles of war, attempted to enlighten officers on what worked best for Britain and her Empire by recourse to historical example and assessment. Tactical discourse featured, but its prominence waned over time as other venues, particularly, the Tactical School, absorbed the topic. At the strategic level, the qualifier appreciated the many challenges facing the Royal Navy, and war with Japan was the major problem consistently weighed. Operationally, the officer came to understand the competing tasks that a navy faced whether it was maintenance of her lines of communication, projecting the British Army to a distant shore, attacking enemy trade, and, yes, fleet action. The study of Jutland was a means to an end. The battle was never the centre of staff education, nor was it the culminating point of the course—that honour belonged to combined operations. Here, the Dardanelles campaign was key and given the growing dependency of the Services on each other became ever the more so.[314]

Where the War Course diverged from the lower course was in its securing the views of others outside of the British naval establishment to complement the lectures Richmond and the other members of the directing staff provided. This allowed officers to hear the views of those not subject to any regime of naval orthodoxy, facilitated contacts with other departments in Whitehall having an interest in naval affairs, and allowed officers to learn the thoughts of those outside the Royal Navy. They also served to provide officers with the necessary insight to the strands of British policy, the legal framework British naval operations were expected to follow, and the likely views from the other side of the hill—those nations that Britain may face in conflict. Finally, the War Course was structured to provide officers with not only the political and strategical underpinnings necessary to draft appreciations, but also an understanding of the technical capabilities of opposing naval forces.

Thus, Sir Percy Bates lectured to a joint sitting of the War and Staff College directing staffs on 'British Commerce in War' in 1921. With firsthand experience of his subject from time spent with the Ministry of Shipping during the war, Sir Percy and two other civilian officials discussed at length the effects of Royal Navy practices and procedures on her trade during the recent war.[315] The knock-on effects of British contraband control soon became apparent to the officers gathered, and, anticipating the role he was to play in another venue, Richmond saw that the subject was not addressed solely from the perspective of his own Service. Group Captain Charles Samson,[316] late Royal Navy and one of the Naval Wing's first pilots having gained his

---

314 How much Gallipoli weighed on contemporary officers may be judged by the author's copy of Grenfell's, *The Art of the Admiral*, inscribed by Grenfell *'and in memory of the Dardanelles'*.
315 ADM 1/8628/120, Admiral-President, Royal Naval College, Greenwich to Admiralty letter of 29 July 1921.
316 Later Air Commodore Charles Rumney Samson (1883-1931). Chief Staff Officer, Coastal Area, 1919-21 and Senior Officers' War Course, 1921.

wings in 1911, attended the lecture from the Royal Air Force. Introducing Sir Percy to the assembled officers and guests, Richmond explained his desire:

> What I want to do is open the channels a little, which would help us get together and get into better touch with people in the shipping world and the Board of Trade. That seems to me to be the vital thing. I do not pretend today that we are going to solve any problems. But I do want to hear the views on this question of control. And what is possibly more important, to establish a permanent connection between this college and the shipping world.[317]

Richmond's ambition was of an order removed from what the lower course attempted and owed something to his rank, the status of the War Course, and the seniority, experience and knowledge of the officers attending.

In inviting the American naval attaché Rear Admiral Albert Niblack to speak on the subject of policy, Richmond offered the chance for members of the War Course to hear the perspectives of a friendly, but rival, naval power on a foundational subject of the curriculum. Niblack, having served at Gibraltar and in European waters during the war, was an officer possessing close ties to the Royal Navy and is the only known case of a non-British officer lecturing at the War College in the interwar period. He is emblematic of the informal links maintained between the two navies during the period between the wars and comes closest to approaching for the Royal Navy the pattern of association that the British Army maintained with the French Army. That Niblack was the only non-British officer to appear probably reflected the Admiralty's desire to keep the content, extent, and any limitations of the War Course secret mirroring their initial attitude towards the Tactical School and its associated course. The Admiralty was not pleased with Richmond inviting Niblack to Greenwich and their ostensible reason was his failure to acquire Their Lordships prior approval.[318] This was the proximate reason, but the indiscretion would have mattered little if the larger point about secrecy had been properly weighed by Richmond.

Given the high-level of its curriculum and the broad sweep of war that was central to Richmond's own surveys of history, it is unsurprising that the SOWC drew from such a range of lecturers. Economics and its role in war featured in numerous lectures and, with the War College close to London, it benefited from drawing on members of academia and the greater Whitehall establishment. Ernest Fayle, the official historian writing on the commerce war, drew the officers' attention to the economic element in war in a series of five lectures in 1920-21. Professor A. J. Sargent of the University of London considered the economic importance of the Great Lakes shared between the United States and Canada and the naval architect Westcott Abell, of Lloyds,

---

317 ADM 1/8628/120, Richmond introduction to lecture by Sir Percy Bates 'British Commerce in War'.
318 Hunt, *Sailor–Scholar*, p. 128.

surveyed the American merchant marine. Topics of some relevance when the schemes addressing a possible war between the United States and the British Empire were considered later in the course. As blockade was a key aspect of British naval practice and the Navy was the principal guardian of a maritime empire, such a regime of talks is largely unremarkable. More remarkable was the quality of discourse Richmond established in the War Course as Fayle would in due course provide similar foundational instruction at the Imperial Defence College becoming that school's Adviser on Economics.[319]

Group Captain Charles Samson returned and provided officers two lectures on air strategy whilst Dr Pearce Higgins, the Whewell Professor of International Law in Cambridge University, covered his specialty in two talks. Higgins is especially noteworthy. Having previously lectured at the Portsmouth-based War College, he wrote a treatise on the legal aspects of Defensively Equipped Merchant Ships in 1917 and was a recognized authority on the Law of Armed Conflict. He attended the conferences held at The Hague before the World War and sat on the committee investigating war crimes that had been formed during the war by the Government. Even today, his work retains value as it remains a standard in print notwithstanding he succumbed more than 70 years ago.

Meanwhile, Dr James Headlam-Morley, a former member of the Political Intelligence Department of the Foreign Office and a technical adviser at the Paris peace settlement with responsibility for German issues was another earlier lecturer at the revised War Course. Possessing a strong resume of academic writings, a founding member of the Royal Institute of International Affairs, and now serving as the Historical Adviser to the Foreign Office, Headlam-Morley addressed foreign affairs in two lectures at the War College during 1920-21. Whether Headlam-Morley touched upon exercising propaganda in enemy countries, which had been a staple of his duties during the 1914-18 war is unknown, but the possibility must exist.[320] Richmond, though, did not neglect the expertise available within the Service itself turning to the Naval Intelligence Division. Here, Instructor Commander Guy Rayment, a long-serving officer in that office, discussed 'Japan and her Navy' in two lectures. Meanwhile, Sir Eustace Tennyson-d'Eyncourt, the Director of Naval Construction, provided an assessment of the technical capabilities of the German High Sea Fleet from the knowledge gleaned during the period of its internment at Scapa Flow.

With the broad framework of the political, legal, and economic foundations of the contemporary world setting thus provided, the directing staff was free to address the specific problems facing the Royal Navy employing historical examples from the recent past as signposts to indicate what was feasible, desirable, and administratively sound. During the War Course's initial period, besides covering appreciations and

<hr>

319 *Navy List*, July 1936, p. 454.
320 Campbell Stuart, *Secrets of Crewe House: The Story of a Famous Campaign* (London: Hodder and Stoughton, 1920), p. 60.

the nature of war plans, Richmond typically handled those subjects covering policy, strategy and tactics with Hamilton focusing on the organization of staffs and the higher direction of war. This left Captain Ernest Denison[321] to review the preparation of orders and instructions. With Richmond covering the overarching principles, the other members of the staff were responsible for lecturing on British naval operations drawing heavily on the historical example provided from the recent war. Captain George Tomlin covered the use of blockade and shared lecturing on the war plans of 1914 with Captain George Lewin.[322] Denison also covered the role of Scandinavian convoys during the recent war and shared combined operations with Captain Adrian Smyth.[323] Of the latter topic, Richmond rated him one of the foremost experts on the subject and on the strength of his knowledge was recommended for appointment as librarian to the Staff College in order to prepare detailed literature on the question.[324] Still beyond Denison's undoubted expertise, Richmond sought to make the course more than just theoretical. To this end, he employed officers of the Portsmouth Division of the Royal Marines having experience of amphibious operations to anchor proceedings in actual practice.[325] Meanwhile, Smyth addressed the further subjects of shipping and convoy organization; topics where he had ample reserves of practical knowledge stemming from the war.[326]

The practice of having outside lecturers supplement those offered by Richmond and his cohorts continued long after his departure and in 1937 and 1938 Air Marshal Sir Arthur Longmore,[327] late RNAS and now Commandant of the IDC, lectured to the SOWC on 'British Strategy' whilst Brigadier Richard Dewing,[328] also from the same institution, addressed 'Military Strategy'.[329] In his survey, Longmore introduced the Chinese philosopher Sun Tzu. This was a notable departure as most presentations by naval authorities of the period eschewed any direct reference to theoretical writers such as Clausewitz or Jomini and remained more firmly grounded in the formal doctrine of

321 Captain Ernest William Denison, later Baron Londesborough (1876–1963). War Course, 1907; Captain, 1916; Senior Officers' War Course, 1920-21; Royal Naval War College, 1921-23 and retired, 1923. Librarian to Royal Naval Staff College, 1923-37.
322 Captain George Edgar Lewin (1881–1934). Captain, 1918; Royal Naval War College, 1920-21 and retired, 1921.
323 Captain Adrian Holt Smyth (1878–1951). Hydrographic Department, 1909; Commanding Officer, HMS *Teutonic*, 1915-16, HMS *Moladvia*, 1916-18 and HMS *London*, 1918-19; Captain, 1918; Senior Officers' War Course, 1920; Royal Naval War College, 1920-22 and retired, 1922. Recalled for war service.
324 ADM 196/90 (Denison).
325 ADM 196/64 (Simpson).
326 ADM 196/91 (Smyth).
327 Later Air Chief Marshal Sir Arthur Murray Longmore (1885-1970). Staff Course Camberley, 1922 and Commandant, Imperial Defence College, 1936-38.
328 Later Major General Richard Henry Dewing (1891–1981). Staff Course Camberley, 1923-24; Imperial Defence Course, 1934 and Imperial Defence College, 1937-39.
329 CAB 53/43, Report by the Commandant of the 1938 Course of the Imperial Defence College, 22 December 1938.

the *Naval War Manual* and the *FSR*. Both officers lecturing brought a joint perspective to their subjects and ensured that those bodies addressing similar questions were acquainted with the views of the other Services.[330]

Just as the Staff College conducted schemes and ran war games at the request of the Admiralty so, too, was it a common practice of employing the War College to investigate contemporary issues. In 1938, the Naval Air Division turned to the War College for just such assistance. The requirement was for two separate games to be played using the same personnel in each event to gauge the proper ratio of carriers to cruisers in a defence of trade scenario. The presumed enemy was Germany featuring a pocket battleship of the *Deutschland*-class and demonstrates that the Naval Staff continued to employ games to inform its decisions on material acquisition.[331] For these games, the Admiralty defined special rules governing probable casualty rates. Here, the lack of recent battle experience intruded and the results realised were accepted as no more than provisional given the degree of uncertainty over how closely the games reflected current trends. As with the previous scheme conducted by the Staff College in 1927 investigating cruiser policy, a prime purpose of these games were to assist the Naval Staff in reaching a decision on the proper mix of forces to a given budget.[332]

However, the War College did not always wait to be prompted before offering its thoughts on a current problem. In 1937, Vice Admiral Sidney Bailey[333] wrote to the Admiralty recommending that a specialized force trained to conduct opposed amphibious landings be raised. It may be thought that given the number of combined exercises the Staff Colleges had held since the war a suggestion of such an obvious deficiency would have been made earlier. That it was not owed much to the conclusions reached during these exercises that such operations were no longer tenable. What prompted Bailey was not that the results were now different, but rather the knowledge that United States and France had such forces and the Japanese had actually demonstrated that such operations remained feasible. The Admiralty acted with promptness and the DCNS soon brought the matter to the attention of the Committee of Imperial Defence.[334]

330 Longmore, *From Sea to Sky*, pp. 184-85.
331 ADM 1/9466, Director, Plans Division to Director, Naval Air Division Division minute P.D. T.016250 of 10 February 1938.
332 *Ibid.*
333 Admiral Sir Sidney Robert Bailey (1882-1942). Flag Commander and War Staff Officer in *Lion* and *Queen Elizabeth*, 1916-19; Captain, 1918; Deputy Director, Operations Division, 1920-21; Naval Intelligence Division, 1921 and 1922; Naval Attaché Washington, 1921-22; Captain (D), in HMS *Mackay*, 1923-25; Senior Officers' War Course, 1925 and 1936-37; Naval Assistant, First Sea Lord, 1925-27; Tactical Course, 1927; Commanding Officer, *Renown*, 1927-29; Director, Tactical School, 1929-31; Rear Admiral, 1931; Chief of Staff to Commander-in-Chief, Mediterranean, 1931-33; ACNS, 1933-34; Commander, Battle Cruiser Squadron, 1934-36; Vice Admiral, 1935; Admiral-President, Royal Naval College Greenwich, 1937-38; Admiral and retired, 1939. Recalled for war service.
334 CAB 54/13, DCNS to Assistant Secretary, CID letter of 13 January 1938. It is an open

As the preparation of war plans was considered at some length, it should prove unsurprising that Richmond had much to offer on this topic, too. He offered historical perspective with the plans Britain followed in the eighteenth century and those featuring in the Russo-Japanese War. Of the relevance of the latter plans, the Admiral claimed:

> There is good reason for us to examine this question, for it is conceivable that an analogous problem may face ourselves on a far larger scale. If a Far Eastern question is likely to arise, are we to divide our fleet into two parts, one in Europe and one in the Far East, or to concentrate the fleet – bearing in mind the small force probably in continuous commission for some years – and if so where?[335]

Meanwhile, Tomlin reminded officers that any plan prepared was subject to the changing vagaries of their political context. Drafters needed to consider the assumptions governing their use and he emphasized his point by drawing attention to the situation facing the Royal Navy in 1914:

> Previous to the late War the Admiralty had drawn up war orders and instructions giving to all Commanders in Chief and Admirals commanding Squadrons &c. instructions as to action to be taken and as to the dispositions of the Squadrons in the event of war.
>
> War Orders No. 1 dealt with plans in the event of war between Great Britain and Germany.
>
> War Orders No. 2 dealt with plans in the event of Great Britain and France being allied against Germany …
>
> It will be noted that no plans were made for a war in which Germany had an Ally or Allies.[336]

question whether the idea for a trained striking force to conduct amphibious operations originated with Bailey, Ragnar Colvin or, rather, was the brainchild of Captain Bertram Watson, the Director, Royal Naval Staff College. The most likely scenario is that Colvin made the proposal as part of his endorsement to Watson's 1936 proposal that an inter-service committee be formed to further the research of conducting opposed landings. Whether Colvin originated the idea of a special striking arm or merely endorsed the proposal does not detract from the argument that the War College was free to bring fresh questions to the attention of authorities. The record is clearer that it was Watson who originally suggested the establishment of a committee to investigate the training and material required to actually conduct such operations resulting in the eventual establishment of the Inter-Service Training and Development Centre. See CAB 53/33, Report of the Chiefs of Staff Sub-Committee, 14 Oct 1937 and CAB 54/3 for his actual memorandum

335 Richmond, 'War Plans IV, Russo-Japanese War' lecture, no date provided but *c.* Autumn Session 1921, Richmond Papers, NMM/RIC/10/7.

336 Tomlin, 'War Plans, 1914', lecture, Autumn Session 1921, Richmond Papers, NMM/RIC/10/7.

Though the *War Manual* included an appendix covering orders, instructions, and reports, officers attending the SOWC secured but an overview of these matters for it was at the Staff College where such subjects received their fullest treatment.[337] Indeed, such administrative matters were the lifeblood of the 'staffie's' lot. The exact duties a staff officer performed, of course, depended on whether he served afloat or ashore, and, further, if he was designated as a Staff Officer Operations or Staff Officer Intelligence. With the elimination of the Intelligence Course in 1926 and its curriculum subsumed into the Staff Course all staff officers were trained to a common standard. Additional lectures reviewing the correct handling of naval signals, correspondence, and intelligence followed those reviewing Admiralty organization and the role of the Naval Staff in the administration of the Service.[338] Instruction in map reading, charting, and plotting were covered at length, as a primary responsibility for the Staff Officer Operations serving afloat was to maintain either the Strategical Plot or the Tactical Plot depending on whether he was carried in a flagship or a private ship.[339]

Much as was the case for those attending the senior course, qualifiers of the Staff College benefited from having guests lecture on topics where their insights could shed light on issues being raised in the curriculum. Still, the practice though common enough never approached the levels attempted during the Richmond regime of leading the SOWC where that officer cast a wide and deep net to get the best speakers available. Moreover, though qualifiers at the Staff College were charged to accept nothing at face value, when listening to an outside speaker, particularly those from outside the Service, an exception was made. Qualifiers needed to show a healthy measure of respect and courtesy and to leave it to another day before embarrassing themselves and the college.[340]

Engineer Commander George Ross,[341] direct from serving as the Assistant Naval Attaché in Tokyo, spoke to the Royal Naval Staff College *c.* 1936 on his impressions of the current state of the Imperial Japanese Navy. Unusual in his case, Ross believed himself to be the first engineer officer invited to address the Staff College, though it must be remarked that lectures by non-executive officers were not out of the norm as evidence has shown.[342] Meanwhile, Rear Admiral Gerald Dickens, late Director of Naval Intelligence, provided similar background material for the 1935 War Course in his review of 'Japan and Sea Power'. Dickens emphasized the ongoing modernization

337 ADM 186/66, *C.B. 973*, p. 50.
338 ADM 1/8668/172, Major Walter Sinclair report of December 1923.
339 ADM 182/85, Admiralty Fleet Order, '1611—Strategical and Tactical Plotting' of 11 June 1926. A private ship was a vessel which did not fly the standard of an embarked flag officer.
340 ADM 203/69, Astley-Rushton, 'The Staff College', p. 9.
341 Later Rear Admiral George Campbell Ross (1900-1993). Assistant Naval Attaché, Tokyo, 1933-35 and retired, 1953.
342 '*Memoirs*', p. 238, Engineer Rear Admiral George Campbell Ross Papers, Imperial War Museum, London, United Kingdom, IWM/86/60/1.

of Japanese heavy ships, her increases in personnel, the rapid development of her naval air arm, and the scale and frequency of her fleet manoeuvres.[343]

If instruction at the War Course lacked the broad historical grounding imparted at the Staff College, it does not follow that reference to history did not shape its teachings. Given Richmond's interest in history, the discourses he delivered benefited very much from historical perspective drawing as he did lessons and conclusions from the past to understand the present. Indeed, in accepting that principles governed military affairs whether discussing matters of strategy, operations, or tactics, the War Course, no less than the Staff Course, placed a value on historical perspective to a degree that was not always beneficial. After all, that one could effectively argue that the past demonstrated a particular lesson was no predictor of future events, as battle cared little for vehemence of conviction, power of advocacy, and clarity of exposition. Codifying the principles of war in the *Naval War Manual* was the easiest means of ensuring their understanding, consideration, and use by naval and marine officers, but such did not guarantee their accuracy or continued relevance. Whilst officers were adjoined that their application was ever changing, was that anything more than the confidence trick of the knave or the semantic gymnastics of the egg toying with Alice? In arguing for both a principle of concentration and of an economy of force, the *Naval War Manual* was attempting to square a circle. The application of one operates explicitly at the expense of the other. The principles of security and cooperation have a similar conflict if not so openly apparent, and if Richmond and Fuller both could see the necessity for a principle of determination or will, then it remained that the *Naval War Manual* stayed silent.

The danger of employing historical example in military education is that the past may yet become an anchor and not a signpost. Richmond's lectures indicate this peril. From his second lecture on tactics, his view where the fulcrum of naval power existed was clear claiming:

> Fleets, as constituted to-day, consist of powerful gun armed ships, cruisers and a flotilla, variously armed, but the marrow of the Fleet resides in the battleships. If they are destroyed, the remainder have no power of resistance; they depend wholly on ships of the line, and without them are as gossamer to be swept away by the brooms of the heavier ships.[344]

Even if it was ever thus in 1720 and 1820, did the experience of the recent war count for nothing? It is not that Richmond just discounts the rise of the newer weapons of war; he frequently ignored them. Historical example may prove that the first-rate ship

---

343 Gerald Dickens, 'Japan & Sea Power', lecture of 15 May 1935, Admiral Sir Gerald Charles Dickens Papers, Liddell Hart Centre for Military Archives, King's College, London.
344 'Tactics II', p. 1, Spring Session 1920, Richmond Papers, NMM/RIC/10/3.

had always been preeminent, but the risk was that where the Admiral sought recourse to scientific method in his argument only a tautology followed in its place.

The risk of over-employing historical example applies also to his consideration of the principle of concentration. In this case, when examined as a tactical precept and its importance in fire distribution, he argued that the enemy's battlefleet was the primary object telling his charges:

> It is a principle to which reference has been made earlier that the attack should be made upon the vital part of the enemy. Which of these bodies is the most important? This question admits of one answer only – the battleships. They carry the heavy artillery, the dominant weapon of naval force: without their protection the smaller craft cannot offer combat. They are the support of all vessels in all the theatres of operations – at any rate for the present. This admits no argument. It then follows that the attack should be made upon the battleships, and the efforts of our <u>fleet</u> should be concentrated upon their destruction.[345]

Richmond, a man not given to half measures, took his argument to its logical conclusion and continued by lamenting the experience of the recent past commenting in a 1920 seminar:

> We have suffered greatly by subordinating tactics unduly to the demands of the <u>individual</u> gun. It is not the individual gun we have to provide for, but the collective artillery – using the word artillery to denote every missile weapon – of which a fleet can make employment, large guns, small guns, machine guns, torpedoes, bombs and mines.[346]

Richmond's advocacy anticipated paragraph 197 of the 1925 *Naval War Manual* with its charge that, 'The simultaneous blow with all weapons available is essential to rapid decision. All vessels should co-operate with this aim in view.'[347]

This theme was developed further in a 1922 lecture when he came to discuss the role of battle cruisers in any fleet action. Addressing the principle of cooperation, but drawing on several of the other principles Richmond emphasized:

> If the doctrine of the offensive, of concentration, and of co-operation have any meaning at all, do they not lead to the conclusion that the objective of the battle-cruisers is the enemy's battlefleet, and that they should be used in accordance with their capacity, that is, to enfilade or support other attacks. If superior numbers are

345 'Tactics', p. 22, Spring Session 1920, Richmond Papers, NMM/RIC/10/2. Original emphasis.
346 *Ibid*, p. 32. Original emphasis.
347 ADM 186/66, *C.B. 973*, p. 30.

available, we might expect to bring nine or ten into action against no more than five or six of the enemy-we ask ourselves how is that superior force to be used? It is possible that if battlecruisers set out to attack the enemy's van or enfilade it, the enemy, to save himself from the consequences of such action, will use his battle-cruisers defensively. Can five or six ships stand up to nine or ten? What must be the result? – Destruction or retreat.[348]

Richmond's logic if impeccable demonstrates a weakness of the Staff Course and more, especially, of the War Course. The principles of war, if they were to rise above an incantation had at some point to be applied. What theory may suggest, experience could disprove. At a certain level, both directing staff and qualifiers recognized this caveat as a common admonition maintained that the principles if universal in their validity, varied in their application. Here, the strongest case presented itself for the creation of a school devoted to tactical analysis. Precepts introduced at Greenwich and appropriate for the higher direction of war, or perhaps correct in theoretical tactical discourses, might not serve the needs of the fleet during an actual encounter against a defined enemy. It is to the Service's credit, and largely to Admiral Beatty, that it accepted the case for the Tactical School. Others might argue that this was a false path and that better would it have been to extend the War Course and eliminate the Senior Officers' Technical Course and the Tactical Course.[349] Perhaps, though this writer is not in a position to judge. What is clear is that the Service recognized that theory and understanding must be tempered with practice, trial, and, yes, even error. The fleet needed Greenwich, just as Greenwich needed the fleet, and both benefited from the founding of the Tactical School.

It is not without interest and supreme irony that the first Director of the Tactical School, Cecil Usborne, later crossed swords with Richmond over the principles of war in the pages of the *Journal of the Royal United Services Institution*. Usborne questioned their universality and composition in a Socratic exposition held between a budding neophyte of war and a learned Philosopher. Richmond took Usborne to task for seeking to apply the principles in areas where their usage was not appropriate and reminding him that the principles were those 'which govern the *employment* of armed force' and were not appropriate for applying to discussions regarding force structure.[350] Richmond knowing of the origins of the Service's adoption of the principles of war in the *Naval War Manual* and of their culling from the *Field Service Regulations II, Operations* had little time for Usborne's efforts to recast the principles. Usborne held his ground, though. In his reply, he called attention to Richmond's habit of piling

---

348 'Tactics IV, Co-operation (1)', p. 17, lecture delivered 2 May 1922, Richmond Papers, NMM/RIC/11/2.
349 Smith, 'Education of the Naval Officer', *Naval Review*, p. 723.
350 H. W. Richmond, 'Principles of War: A Criticism', *Journal of the Royal United Services Institution*, Vol. LXXIV, No. 496, November 1929, p. 714. Original emphasis.

on historical example atop historical example in argument with a degree of light mockery writing, 'At last arose a learned man renowned for his knowledge of past wars who buried our young principles under a mountain of criticism, and on it planted crosses inscribed with names of great campaigns.'[351] The differing views espoused by Richmond and Usborne may reflect nothing more than contrasting perspectives of the strategist to the tactician. Each has their place and each will always be required in war.

The aim of the Senior Officers' War Course was to foster in officers an appreciation for the higher aspects of his calling and the linkages of naval strategy to policy—the vital concern of those reaching flag rank. The Staff Course sought to prepare officers to act as a flag officer's primary assistants, as the complexity of war made it impossible for a single leader—no matter how gifted—to direct its affairs without recourse to help. Where Camberley and Quetta aimed to prepare one as a Brigade-Major, the goal of the Greenwich Staff Course became one of primarily fitting one to act as a Staff Officer Operations or Staff Officer Intelligence. Those naval officers attending the SOWC or the Staff Course required an understanding of strategy, but for differing reasons. Creation of the Tactical School followed because an understanding of strategy was insufficient and nor was it enough to receive a cursory instruction in tactical discourse. Ever a problem in any course was how to make academic theory relevant to the practical problems that an officer would face at sea.[352] Captain Drax, as committed to the staff system concept as any naval officer, understood the centrality of tactics warning qualifiers, 'Bad strategy may be redeemed by successful tactics, but there is no remedy for defeat in Battle.'[353] The courses taught at Greenwich were an important grounding in a naval officer's career, but by themselves they were insufficient. Thus, it is to Portsmouth and the role of the Tactical School and the Senior Officers' Technical Course as agents in preparing officers for war that this history now turns.

351 C. V. Usborne, 'The Principles of War: Another Dialogue', *Journal of the Royal United Services Institution*, Vol. LXXV, No. 497, February 1930, p. 60.
352 ADM 196/92 (Harwood).
353 'Tactics', Royal Naval Staff College lecture 1921-22 Session, Drax Papers, DRAX/7/2. A variation of this quote is present in Dreyer's 1924 paper arguing for the creation of a Tactical School. See ADM 1/8658/69, 'The Study of War and the Conduct of Naval Operations'.

4

Educating a Navy: The Senior Officers' Technical Course and the Tactical Course

It is in fact a matter of surprise that, notwithstanding the recent increase in educational and technical courses, a tactical course should still be non-existent. The first business of a Navy is to fight and it is indisputably the purpose of peace training to prepare the Navy for battle.[1]

Lieutenant Commander Russell Grenfell, 1923

Though the overwhelming majority of executive officers entering the Service were trained to a common standard, the pattern was soon broken once an officer reached his Lieutenancy. Whether to specialize in one of the technical branches was soon a decision facing the officer and though most officers in the period opted for specialist training, others did not with no ill-effect on their future prospects.[2] Godfrey Brewer,[3] opting to specialize in gunnery, was serving in the battle cruiser *Tiger* and waiting to begin his course at Whale Island where the *Excellent* was housed only to be forestalled in his attempt when it was learned that he was soon to marry.[4] A full career in small ships was his reward along with a happy wife. Many were happy to have a career in small ships where responsibility came quickly and the relations between those commanding and those commanded were less formal than in heavy ships.[5]

1   Grenfell, 'Training in Tactics', *Naval Review*, p. 686.
2   By 1935, 88% of executive branch officers Lieutenants were specializing. See ADM 1/9041, James Committee, *Report*.
3   Captain Godfrey Noel Brewer (1901-1988). Short Course, Cambridge University, 1921-22; Commanding Officer, HMS *Wrestler*, 1931-32; Tactical Course, 1933; Commanding Officer, HMS *Whitehall*, 1934 and HMS *Diamond*, 1934-35; Staff Course, 1937; Tactical Course, 1938; Commanding Officer, HMS *Maori*, 1938-40 and HMS *Stork*, 1942-43; Captain, 1943; Captain (D), Sixth Flotilla, 1944-46 and retired, 1952.
4   Brewer, 'Melody Lingers On – II', p. 52.
5   Eric Grove, ed., *The Battle and the Breeze: The Naval Reminiscences of Admiral of the Fleet Sir Edward Ashmore* (Stroud: Sutton Publishing Limited, 1997), p. 21.

Increasingly, the Navy required specialists and soon sub-specialists. Thus, the decision to specialize no longer became a question simply of individual officer preference. In order to achieve the number of Naval Observers required as part of its efforts to expand the Fleet Air Arm, the Service took the step of assigning officers to observer training just before the Second World War.[6] Likewise, some decided to remain a salt horse (i.e., an executive officer without a specialist skill) after being ordered to a destroyer as a Sub Lieutenant and found that the environment was congenial to them.[7] The issue of specialization or not is important to this story, as it explains in large measure why the courses the Navy adopted for officer education were constructed as they were with their repetitive themes and emphasis on first principles. Though officers might take the courses at similar points in their career, the paths that brought them there were very dissimilar. The issue was even more acute when joint training was attempted, of which more is yet to be said.

Unlike the British Army where promotion to Major required one to pass examinations and wait for a vacancy to appear in his regiment or corps, promotion to Lieutenant Commander was a right of seniority and not by examination.[8] Here, the assessment of his superiors was critical in establishing whether the necessary traits of ability, professional knowledge, tact, of temperate habits, powers of command, and zeal were present. In short, did one exhibit officer-like qualities? From the beginning of an officer's career demonstrating such qualities was key and only became the more so as one progressed in rank.[9] Promotion above Lieutenant Commander was based on selection where the observations of superiors noted in his yearly confidential report weighed heavy in determining one's prospects. It was also upon reaching Lieutenant Commander that an officer first came to experience the effects of the 'promotion zone', the window when he would be eligible and considered for promotion to Commander. The actual 'zone' was a period of varying length set to meet the needs of the Service against the number of officers of like rank available for consideration. This was the first true test of one's professional ability relative to his peers. Secure promotion and one waited to enter the zone again for the next rank; fail to secure promotion and an officer bided one's final years in the Navy until the age of retirement and the minimum time of service were completed.

Amongst the qualities considered essential in an officer was the demonstration of temperate habits and if one had shown proficiency in a second language, so much the better. Professional knowledge might be demonstrated through specialization and this no doubt was a consideration why some elected to do so, as officers were aware of the general criteria employed in assessing performance and fitness for future promotion.

6   Beagle, 'Specialization and Promotion', *The Naval Review*, Vol. XXXVI, No. 2, May 1948, p. 169.
7   *Ibid.* pp. 169-70.
8   *King's Regulations, Vol. I*, Article 323, p. 109.
9   *Ibid, Vol. II*, p. 63.

As even the objective elements used had elements of subjectivity about them (i.e., had the officer chosen French as a second language or elected for one viewed more difficult to master such as Japanese) a single report on an officer said little. However, officers competing for promotion to Commander would have been observed by multiple superiors in differing assignments during a period of some fifteen years. This collective assessment was deemed to provide a truer picture of one's fitness for accepting greater responsibility, and, in consequence, promotion.

It is difficult to generalize why some officers did not attend the Staff Course. That many more applied than were accepted indicates that those securing the naval *psc* had passed a hurdle of selection, but the need to balance the Staff Course with officers of varying technical specialties resulted in some not studying there who in an ideal world would have.[10] Disappointment might follow, but not disfavour for the path to promotion was not a single one. Even as late as 1939, the majority of officers on the flag list were not Greenwich *psc's* with only 14 of the 66 admirals of the executive branch holding the distinction.[11] Accordingly, failure to attend the Staff College was not a career-killer for the period under review though increasingly the tide was changing.

What was not changing was the need to acquire technical knowledge and though one could miss the 8:05 to Greenwich or Blackheath, a naval officer desiring to command at sea and eventually reach flag rank had to continue his education by attending the Senior Officers' Technical Course and the Tactical Course. The influence of these courses on the development of naval officers before the Second World War is a much-neglected field of investigation. This is a serious indictment given that more officers attended these courses than ever passed through the Senior Officers' War Course or qualified at the Staff College between the wars. That more officers attended the two courses is hardly surprising as they typically met four times a year with each session lasting two months. For the SOTC, its sessions normally fell in January, March, August and October whilst the Tactical Course usually met in January, April, June and October. As the War Course met in March and October, a lucky officer could take his SOTC or Tactical Course in January and proceed immediately to his War Course beginning in March, the very month his first course was ending. Because of the close dependency of the three courses an important consideration was to keep the length of the Tactical and Technical Courses approximately equal to assist in further scheduling.

Comment has been made previously that a single path to command in the Royal Navy did not exist and confirmation is offered by the officers who prospered in the absence of attending either of the two Greenwich courses. Admiral of the Fleet Sir James Somerville offers the case of an officer who failed to attend the Greenwich courses, but did take his complement of Technical and Tactical Courses and managed

10   In 1924, 95 applicants applied for the 28 vacancies available in that year's course. See ADM 203/69, 'The Staff College', Astley-Rushton lecture, p. 20.
11   *Navy List*, March 1939, p. 322e.

a full career nevertheless.[12] As a Signals Officer, Somerville attended his share of technical courses related to his area of expertise, though this alone does not account for his failing to sit the Staff or War Courses. Other officers specialising in his field did so with Admiral of the Fleet Earl Mountbatten of Burma perhaps the most notable example.[13]

Though not a graduate of the Staff Course, Lord Louis attended the SOTC, the Tactical Course, and the War Course before assuming command of his destroyer flotilla and HMS *Kelly* in 1939. Of course, Mountbatten's career benefited in other ways not the least his being the son of a previous First Sea Lord. Some contemporaries, such as General Sir Alan Brooke and Admiral Sir Andrew Cunningham, attributed the patronage of Winston Churchill as playing a large measure in his success.[14] This was no doubt the case, but connections to the Royal Family did not hurt his career prospects either. Moreover, he had enjoyed the interest of Chatfield, Beatty, and Keyes at differing points during his career. The interest of others might help (and protect) one along the way, but one still had to demonstrate ability and Lord Louis did have ability with even his critics conceding as much. In fairness, Mountbatten was the type of officer apt to prosper no matter the climate. The same cannot be said of Admiral Sir Angus Cunnighame-Graham whose path to responsibility and promotion owed more to his own talents.[15] Also a Signals officer, Cunninghame-Graham

12   Admiral of the Fleet Sir James Fownes Somerville (1882-1949). Captain, 1921; Commanding Officer and Flag Captain in *Benbow*, 1923-24; Director, Signal Department, 1925-27; Tactical Course, 1927, 1931 and 1936; Commanding Officer and Flag Captain, *Barham*, 1927-29; Imperial Defence College, 1929-31; Commanding Officer, *Norfolk*, 1931-32; Rear Admiral, 1933; Director, Personal Services Department, 1934-35; Rear Admiral, (D) Mediterranean, 1936-38; Vice Admiral, 1937; Commander-in-Chief, East Indies, 1938-39; retired, 1939 and recalled for War Service, 1940; Admiral, 1942; Admiral of the Fleet and retired, 1945.

13   Lord Louis Francis Albert Victor Nicholas, later Admiral of the Fleet Earl Mountbatten of Burma (1900-1979). Short Course, Cambridge University, 1919-20; Flag Lieutenant and Fleet Wireless Telegraphy Officer Reserve Fleet in *Centurion*, 1926; Tactical Course, 1934; Naval Air Division, 1936-37; Captain, 1937; Senior Officers' Technical Course, 1938; Senior Officers' War Course, 1938-39; Tactical Investigation Section, Tactical School, 1939; Captain (D), Fifth Destroyer Flotilla, 1939-40; Commander, Combined Operations, 1942–1943; Supreme Allied Commander, South East Asia, 1943-46; Rear Admiral, 1946; Vice Admiral, 1949, Fourth Sea Lord, 1950-52; Commander-in-Chief, Mediterranean, 1952-54; Admiral, 1953; First Sea Lord, 1955-59; Admiral of the Fleet, 1956 and Chief of Defence Staff, 1959-65.

14   Adrian Smith, *Mountbatten: Apprentice War Lord* (London: I. B. Tauris & Co Ltd, 2010), p. 18.

15   Admiral Sir Angus Edward Malise Bontine Cunninghame-Graham (1892–1981). Flag Lieutenant in HMS *Colossus*, 1917-20; Fleet Signal Officer in *Iron Duke*, 1922-24; Tactical Section, 1924-25; Tactical School, 1925-26 and 1928-29; Senior Officers' Technical Course, 1928; Signal Department, 1928-29; Staff Course, 1929-30; Staff Course Camberley, 1930-31; Staff Course Andover, 1934; Staff Officer Operations and Intelligence in *Pembroke*, 1935-36; Captain, 1935; Commanding Officer, HMS *Tarantula*,

managed to complete not only the Navy's Staff Course, but managed to sit the courses at Camberley and Andover too. Clearly, even for those specializing in signals a single path did not operate, but common to all were that each had completed their quota of Portsmouth-based advanced courses. This is testimony to the overall importance of these classes, and notwithstanding the divergent path an officer's development might follow, certain strands yet remained common.

Though the SOTC and the Tactical Course were essential courses in preparing officers for future command, attending them came at a cost for some as they lost the supplemental allowances in pay associated with serving in a specialist billet. This was certainly the case for the pilots and observers of the Fleet Air Arm. Only in 1928 did Observers, and not until 1932 for qualified pilots, was a stipend paid in lieu of their specialist allowances when attending the Tactical Course.[16] This was a change for the better for before 1932 not a single designated naval pilot had completed the Tactical Course though Captain Cosmo Graham,[17] a graduate of the Aviation Course, had sat the 1931 course before proceeding to the New Zealand Division.[18] Nor were officers from the RAF found in the Tactical Course with any regularity. Wing Commander Harold Collins completed the course in 1937 having served at the maritime air station at RAF Gosport in 1933. Subsequently, Collins served in *Courageous* as a member of her assigned Air Staff and later was employed at Headquarters, Coastal Command at Lee-on-Solent in 1936 before taking his Tactical Course. Thus, as with Graham, though an officer possessing considerable experience in naval cooperation, by the time of his Tactical Course his active flying days were largely behind him. No doubt, Collins' talents were useful in the subsequent staff positions held, but the value of sending such officers to the course was less than for those yet remaining in a flight status or actually assigned to a tactical unit.

That historians have failed to examine the place of the SOTC and the Tactical School is also surprising as conventional wisdom suggests correcting the shortcomings displayed during the Battle of Jutland was a cardinal feature of British naval training between the wars. Professor Holger Herwig has gone so far as to claim that remedying the operational ills of Jutland was an obsession of the interwar Staff College.[19] This

1937-38; in Charge, Signal School, 1939-41; Rear Admiral, 1945, Vice Admiral, 1948, retired, 1951 and Admiral, 1952.

16   *London Gazette*, No. 33849, 26 July 1932, p. 4859.

17   Later Rear Admiral Cosmo Moray Graham (1887–1946). Commanding Officer, HMS *Douglas*, 1921 and HMS *Vansittart*, 1921-23; Torpedo and Mining Department, 1924; Aviation Course, 1924-25; Staff Course, 1925-26; Deputy Director, Naval Air Division, 1928-30; Captain, 1930; Tactical Course, 1931; Flag Captain in *Curacoa*, 1931-32; Senior Officers' War Course, 1932; Senior Officers' Technical Course, 1932; Commanding Officer, *Diomede*, 1933-36; Deputy Director and Director, Naval Air Division, 1936-39; Rear Admiral and retired, 1941. Recalled for war service.

18   ADM 196/50 (Graham).

19   Holger H. Herwig, 'Innovation Ignored: The submarine problem, Germany, Britain, and the United States, 1919-1939' Williamson Murray and Allan R. Millet, eds., *Military Innovation in the Interwar Period* (New York: Cambridge University, Press, 1996), p. 249.

writer cannot agree that Jutland was ever that central to the Staff Course as has been discussed already, yet, if the argument does have resonance, then one would expect to see it confirmed at the Tactical School where tactical discourse was its *raison d'être*. Though the SOTC was not so clearly focused on tactics, but on tactical means, here, too, one would expect to see the influence of Jutland felt. The discussion following addresses the place of these two courses in the higher education of naval officers and the complementary nature of the entire educational regime. Whether Jutland featured so prominently in the two courses is weighed, but more importantly, the strengths and weaknesses of the courses will be considered.

Ideally, a senior officer destined for command attended the Senior Officers' Technical Course followed by the Senior Officers' War Course, and then, finally, sat the Tactical Course in that order. Attendance at one's first course was normally to occur six months following promotion to Captain.[20] This was a rule of thumb only and attendance earlier was possible, if required to meet the exigencies of the Service. In all events, by 1930, an officer now was expected to complete at least one of these foundational courses before assuming command afloat for the first time.[21] The Admiralty Fleet Order specifying the pattern for taking the higher-level courses applied to naval officers only, for Royal Marine officers did not attend either the SOTC or the Tactical Course. This was a significant omission indicating that both courses were narrowly scripted to address naval problems or those focused on naval battle. Supporting tasks did receive attention in the courses, such as combined operations; yet, the record is clear that these were covered only in passing. The systematic analysis of such evolutions largely escaped attention at the tactical table or failed to feature in the instruction of the SOTC. It may be that the Navy considered that these issues were addressed sufficiently at the Royal Naval Staff College and in the joint exercises held each year at Camberley and Quetta. Yet, the establishment of an Inter-Service Training and Development Centre at Portsmouth in September 1938 confirms that a serious gap in examining these types of operations existed which denying Royal Marine officers from attending did nothing to alleviate.[22]

Certainly, it is the case that some Royal Marine officers believed that attending Camberley or the Army Senior Officers' School were more appropriate to their professional needs. Whilst this may have been true for some, it surely was not the case for all, and, in particular, those officers who elected to pursue flying duties such as Godfrey Wildman-Lushington.[23] Serving as a Probationary Second Lieutenant, Wildman-

---

20   ADM 182/89, Confidential Admiralty Fleet Order '131.—Captain, R.N.—Technical and War Courses' of 17 January 1930.
21   *Ibid.*
22   *Hansard*, House of Commons Debates of 9 May 1940, v. 360, cc1396-97.
23   General Godfrey Edward Wildman-Lushington (1897-1970). RAF Leuchars, 1925; Pilot in HMS *Hermes*, 1926; *Eagle*, 1926-27 and *Argus*, 1927-28; Staff Course Andover, 1930; Royal Naval Staff College, 1931-32; Squadron Royal Marine Officer in *Hood*, 1933-34; Staff Officer Intelligence, Bermuda in HMS *York*, 1934-36; Senior Officers' School Sheerness,

Lushington soon volunteered for the RNAS and flying duties with the understanding that he would complete the usual round of marine preparatory courses following the cessation of hostilities. Unlike most officers, he did not transfer to the RAF in 1918, but remained with his Corps though he was soon in Air Force guise again when he resumed flying duties once more in 1925. Following time with the FAA, including service in Shanghai in command of No. 441 Flight FAA, he went to Andover to complete its Staff Course before serving on the directing staff at Greenwich. What he failed to do as a marine officer was attend either the SOTC or, more especially, the Tactical Course and this was an opportunity not realized.[24] A failure to return to flying duties cannot be the sole explanation why this officer was not considered for these courses, as it has been shown with Collins that RAF officers in similar circumstances did attend. Nor was it the case that only Royal Marine officers on flying duties were being neglected by the Admiralty. Officers who had made a name for themselves as gunnery experts were similarly ignored, as candidates for attendance at the courses or as instructors. Though it is true such officers would not command a ship, never mind a fleet, the thought how to employ their knowledge and experience to best advantage was an opportunity missed.

Unlike the Staff Course that was largely restricted to members of the active force, members of the Royal Naval Reserve were eligible to attend both the SOTC and the Tactical Course and did so in substantial numbers. Commander Joseph Heenan,[25] RNR, may have been atypical of the Merchant Navy officers attending the course because of the path that his career subsequently followed. Attending the SOTC in 1935, the SOWC, and the Tactical Course in the following year, he married and moved to Canada shortly thereafter. Beginning a new career as a civil servant, when war broke out a second time, Heenan soon found himself back in naval service. Serving first in staff positions, he transferred to a seagoing billet and assumed command of a ship. Heenan's attendance at the courses demonstrates the value of preparing officers in peacetime to face the uncertainties of the future. As a transplanted Englishman, Hennan would find a new life in Canada, but the value of his earlier training remained proving fortuitous for the two nations he considered home.

As with what was taught at Greenwich, the need to balance conformity against change applied also to the Tactical Course and the Senior Officers' Technical Course. Changes in emphasis within the curricula will be weighed and here it should be

----

1938; Senior Officers' War Course, 1938-39; Lieutenant Colonel, 1939; Royal Naval Staff College, 1939; Acting Colonel, 1941; Acting Colonel Commandant, 1942; Chief of Staff Combined Operations Command, 1942-43; Chief of Staff to Supreme Allied Commander Southeast Asia Command, 1944-45; Commandant, School of Combined Operations, 1945-47; Major General, 1946 and retired, 1950.

24   ADM 196/64 (Wildman-Lushington).

25   Later Captain Joseph Alfred Heenan, RCNR, formerly, RNR (1892-1985). Senior Officers' Technical Course, 1935; Senior Officers' War Course, 1935-36; Tactical Course, 1936; Commanding Officer, HMCS *Provider*, 1942-43 and retired, 1945.

remarked that if the *Naval War Manual* was the mainspring of the Senior Officers' War Course, then for the Tactical School its catechism was the *Battle Instructions*.[26] Both, however, were complementary to each other with the *Instructions* serving the purpose of showing how the principles enunciated in the *War Manual* were to be applied in battle.[27] Moreover, it was the case that the War Course complimented and was not a substitute for the Tactical Course. Both courses addressed strategy and tactics with the essential difference being the centrality that either strategy or tactics assumed in the actual course attended.[28] This was an essential feature of interwar education and training, as the Navy sought to ensure a commonality of doctrine across the Service.

In addition to these foundational publications, an amplifying reader was the two-volume *Naval Tactical Notes*. Volume One was a historical survey of the evolution of naval tactics and under Richmond's influence had been revised to reflect the significance of the principles of war.[29] Volume Two, initially prepared by Commander Alfred Taylor,[30] addressed contemporary naval tactics in light of the capabilities and characteristics of current British and foreign warships. Meanwhile, subsidiary Admiralty publications such as *The Tactical Manual* also informed the curriculum of the course.[31] Dating from 1921, *The Tactical Manual* was a revised version of the *Grand Fleet Battle Orders* compiled by the recently created Tactical Section of the Naval Staff to acquaint all commands with the level of tactical development at war's end.[32] These instructions along with the *Manoeuvring Orders*[33] issued by the respective fleet commanders—and

26   The *Battle Instructions* were issued in 1925 and 1930 as *C.B. 01715* and *C.B. 01821*, respectively. They are retained at the National Archives as ADM 186/72 and ADM186/106. The *Instructions* were restyled the *Fighting Instructions* and issued in 1939 appearing as *C.B. 04027* and are retained as ADM 239/261.

27   ADM 186/72, Atlantic Fleet covering memorandum of 1 October 1925 to *C.B. 01715*. *Battle Instructions*.

28   ADM 196/91 (Watson).

29   Secretary of the Admiralty to Richmond letter M. 03390/27 of 7 January 1928, Richmond Papers, NMM/RIC/7/2.

30   Later Rear Admiral Alfred Hugh Taylor (1886-1972). Six firsts in promotion to Lieutenant; Plans Division, 1918-19 and 1924-25; War Staff Officer to Commander-in-Chief, Atlantic/ Home Fleet, 1919-21; Naval Assistant to ACNS, 1922-23; Captain, 1924; Head, Tactical Section, 1925-27; Tactical Course, 1927 and 1929; Senior Officers' School Sheerness, 1929; Imperial Defence Course, 1930; Short Course in *Osprey*, 1931; in Command, *Vernon*, 1932-34; Commanding Officer, *Valiant* and *Ramillies*, 1935-36; Rear Admiral and retired, 1936. Recalled, 1938 and retired, 1945.

31   Admiral Sir Geoffrey Miles to Captain Stephen Roskill letter of 30 August 1964, Roskill Papers, ROSK/7/163.

32   ADM 1/8585/64, Commander-in-Chief, Atlantic Fleet letter No. 567/A.H.966 of 28 April 1920 and ADM 182/86, Confidential Admiralty Fleet Order 447 of 18 February 1927.

33   Appearing in 1925 and 1936, two separate editions of the *Manoeuvring Orders* were issued between the World Wars and appearing as *C.B. 1716* and *C.B. 1822*. Their TNA references are ADM 186/636 and ADM 186/637, respectively.

not by the Admiralty—were the doctrinal foundations of the Tactical School's work.[34] As the keystone of tactical thinking, their coverage by the school was central to instilling a common doctrine across the Navy. Additionally, though, they served as the point of departure for any new investigations pursued. Where deviations occurred to test an idea or practice, these were made known to the officers undergoing instruction for the *Battle Instructions* in their unexpurgated form remained the operational coda of the fleet to guide the British force and any allies fighting in her company during the naval encounter.[35]

Here it may remarked that variations existed between the *Battle Instructions* used within the fleets and what was taught at the Tactical School, as the Fleet Commanders issued local supplements. These supplements addressed how newer weapons were to be employed where definite guidance had not yet been formulated or might specify how a fleet was to fight against a specific enemy.[36] The force of these edicts was only local, and, if the main fleets were to unite and fight, the approved *Battle Instructions* minus these supplements would hold sway.

Where the Staff College was answerable to the Director of Training and Staff Duties in the first instance, and thence to the ACNS, and the War College reported directly to the ACNS, the Tactical School operated under the guidance of the Tactical Division (or as it was styled for a while, the Tactical Section of the Naval Staff) reporting to the ACNS. Physically residing within *Dryad*, the Navigation School,[37] the Commander-in-Chief, Portsmouth was responsible for its administration as he remained for the operations of the SOTC housed in the *Vernon* and he reported on the fitness of officers commanding both schools.[38] Meanwhile, proximity allowed successive Commanders-in-Chief, Atlantic and Home Fleets to view the work of the Tactical School with an eye not available to their Mediterranean Fleet opposite numbers. Its officers were encouraged to visit the school as time permitted to further their knowledge of its capabilities and the topics undergoing investigation there.

The need for such a school, it will be recalled, was not universally acknowledged when Dryer argued as much during his War Course. Dudley Pound who was slated to attend the first session of the new course following his time in charge of the Plans Division admitted to Richmond that the new school was the subject of severe criticism in some quarters. This criticism Pound stressed was mistaken and stemmed, in part, by the fear that the school would endeavor to dictate new tactics to the fleet with the consequence that initiative would become circumscribed. Pound, writing to

34   ADM 182/86, Confidential Admiralty Fleet Order '447—Tactical Publications' of 18 February 1927.
35   ADM 186/106, *Battle Instructions*, p. 1.
36   ADM 182/86.
37   Officers assigned to the Tactical School as part of its staff were carried in the *Victory* in the period's *Navy Lists*. The school proper was supported by the *Dryad*, as the distribution list for the *Battle Instructions* makes clear.
38   Porthole, 'Staff Courses', *The Naval Review*, Vol. XLI, No. 3, August 1953, p. 303.

Richmond, thought this view was entirely wrong and likened the quality of play by the All Blacks in rugby to their understanding of the game's principles. Armed with such knowledge that they could impose initiative against an opponent based on their understanding of the rules at work.[39] Yet, this fear of ceding tactical responsibility to a shore establishment was losing suasion as the natural numerical superiority of the Royal Navy receded. It was not that the Royal Navy did not exercise at length testing new tactical precepts, but this was no longer sufficient. These exercises, of every value to the senior officers participating, were of less utility to the junior officers involved as they lacked the necessary grounding in tactical theory. As presently conducted, the exercise might foster skill in execution, but not skill in comprehension, and if the benefits of these evolutions were understood by the Admiral commanding, it was not always the case that those participating recognized their ultimate import.[40] Certainly, this was Grenfell's view and Dreyer, too, would have solidly endorsed the argument.[41]

Where the major elements of strategy and policy were amenable to independent study and rarely required speed in thought and execution, tactics was of a class by itself. Rapidity of thought and promptitude in action were the hallmark of proper tactical execution and cried out for a school ashore where the necessary foundation could be laid.[42] It says much about the supposed conservatism of the Service that the Tactical School was formed in the face of the resistance encountered by Dreyer to impart and test the limits of doctrine. This decision may be contrasted with the reluctance of the British Army to adopt a similar institution as part of its establishment during the same period.[43] Moreover, that such a school was often directed by a senior Captain and staffed by Commanders and Lieutenant Commanders deigning to instruct flag officers in the ultimate calling of their profession indicates that the Navy's ethos was changing and changing for the better. A strong dose of tact was no doubt mandatory when applying any criticism to a senior and if an officer evidenced a slight nervousness when correcting a superior, then it was certainly understandable.[44] Still, in a Service more often remembered for the courts-martial arising from the *Royal Oak* affair, when an Admiral was ordered to haul down his flag and two senior officers were dismissed their ship, it does indicate that relations amongst officers were not the caricature often portrayed, as long as the setting was private and Chatham House rules applied.[45]

39   Pound to Richmond letter of 19 January 1925, Richmond Papers, NMM RIC/7/3a.
40   Kay, 'Tactics and the Two-Striper', *The Naval Review*, Vol. XXVII, No. 2, May 1939, p. 240.
41   Grenfell, 'Training in Tactics', *Naval Review*, p. 682.
42   *Ibid*, pp. 684-85.
43   David French, 'Officer Education and Training in the British Regular Army, 1919-39', in Gregory C. Kennedy and Keith Neilson, eds., *Military Education: Past, Present, and Future* (Santa Barbara: Praeger, 2002), p. 117.
44   ADM 196/127 (Wake-Walker) and ADM 196/126 (Schurr).
45   In March 1928, following a series of incidents minor in themselves transpiring in the

In truth, any organization heavily depended on employing the implements of modern technology for its survival can never truly be called conservative or it will not long be of use. Admiral Beatty deserves much credit for acceding to the need for a Tactical School, for his actions at Jutland were not as Caesar's wife. That the proposal came from Rear Admiral Dreyer, an officer very much associated with Admiral Jellicoe, speaks even more to his credit. This too says something about the cliques and camps dividing the Service after that battle. To be sure, they were present with even one Director of the Tactical School pleading for it all to end—albeit anonymously.[46] Still, progress was possible even in their presence and vital is it to recall that the officers of the Royal Navy always held more in common with each other than those elements potentially separating them.

The creation of the Tactical School is unique for another reason. Of all the courses created for the higher education and training of naval and marine officers, its existence was only tangentially confirmed in the period's *Navy Estimates* and appointments to it never appeared in *The London Gazette*, the official newspaper communicating events of record. Public reference was eschewed as to its purpose, the size of its establishment, and the frequency of its meeting and except for the charges associated with employing a civilian typist, it fails to appear even in the *Estimates*.[47] Even the Intelligence School operating at Greenwich appeared openly in the *Estimates* and in unclassified Admiralty Fleet Orders. Not so the Tactical School and its associated course. From its inception, the Tactical School operated *sub rosa* appearing initially under the rubric of the Senior Officers' Technical Course, Part II. As the personnel assigned to the school's staff were covered under Vote 1 of the *Estimate*'s, a distinct sub-head for the school was avoided. In time, this cover was dispensed with, as it only fostered confusion within the Navy and even more so in the Air Force when discussions concerning the SOTC occurred.

The initial secrecy covering the Tactical School and the ready reception Dreyer received from Beatty, the First Sea Lord, in his proposal probably owed much to the recently ratified Washington Naval Treaty with its quantitative and qualitative limitations governing the size and composition of Royal Navy heavy ships and aircraft carriers. Tactical proficiency in the face of such restrictions became more important than ever and forming a school to foster these skills was one means of dealing with the fallout of the naval agreement. Contemporary naval literature emphasized this point in the face of the parity now operating and securing tactical proficiency was seen as a

Mediterranean Fleet battleship *Royal Oak*, a Court of Enquiry was quickly assembled by Admiral Sir Roger Keyes to investigate the actions of the three principal officers involved: Rear Admiral Bernard St. George Collard, Captain Kenneth Dewar, and Commander Henry Daniel. Collard was ordered to strike his flag and Dewar and Daniel to return forthwith to Britain.

46   Ragnar Colvin, 'Jutland', *The Naval Review*, Vol. XV, No.4, pp. 861-63.
47   *Navy Estimates 1935*, p. 31.

necessary premium if the Navy were to prevail in battle.[48] When, at last, the Tactical School was acknowledged it was only because the fiction was no longer tenable for the American naval attaché soon knew of its presence and purpose, if not the exact lines of its workings.

Fleet tactics were the heart and soul of the Tactical School and whilst the initial focus of training was on developing tactical prowess in senior officers and an appreciation in subordinate officers how his commander was likely to act in a given situation, the school soon evolved.[49] In time, a separate section was formed within the establishment to specifically address tactical investigations and running parallel to the courses preparing officers for command. This innovative measure came only in January 1939 so the fruits of such endeavours was of necessity a sparse harvest. Operating under Captain Hubert Acland,[50] a gunnery specialist, the Tactical Investigation Section was composed of thirteen other officers of senior rank out-numbering those officers directing the instruction of the Tactical School proper who were limited to nine.[51] The Admiralty's original intention was that such investigations would be conducted in one session every year, and for 1939, five strategical and six tactical problems were considered by Acland's group. Time did not allow even this limited problem set to receive full treatment and only two of the six tactical issues were sufficiently addressed:

'357. What is the best cruising disposition to adopt by day and night with a force comprising two battle cruisers or battleships, an aircraft carrier, four "Town" class cruisers and one flotilla of destroyers, in circumstances when there is expectation of air and submarine attacks by day and torpedo attack by light craft at night?'

and

'358. How best to use the high speed of *King George V* class battleships when accompanied by *Warspites* and *Nelsons* in a fleet action.'[52]

---

48   Kenneth H. S. Cohen, 'Functions of Destroyers', *The Naval Review*, Vol. XV, No. 4, November 1927, p. 812.

49   ADM 182/83, Confidential Admiralty Fleet Order '3168a.—Tactical Courses at Portsmouth for Senior Officers' of 28 November 1924.

50   Later Captain Sir Hubert Guy Dyke Acland, Baronet Acland (1890-1976). Six firsts in examinations for Lieutenant; Naval Intelligence Division, 1925-27; Fleet Gunnery Officer in *Kent,* 1927-29; Captain, 1932; Tactical Course, 1932; Senior Officers' Technical Course, 1933; Senior Officers' War Course, 1933; Senior Officers' School Sheerness, 1933; Commanding Officer, HMS *Halcyon,* 1934; *Harebell,* 1935-36, *Centurion,* 1936, *Australia,* 1937-38 and HMS *Albatross,* 1938; Tactical Investigation Section in Tactical School, 1939; Commanding Officer, HMS *Vindictive,* 1941-42 and retired, 1942. Recalled for war service.

51   *Navy List,* March 1939, p. 298.

52   ADM 239/142, *C.B. 03016/39, Progress in Tactics*, Training and Staff Duties Division, Naval Staff Admiralty, June 1939, pp. 79-80.

Additionally, short courses, typically lasting a week and forming part of their broader specialist training, were provided by the school to officers undergoing instruction in signals or later serving as Naval Observers with the Fleet Air Arm.[53] In 1939, instruction was expanded to include air officers assigned to Coastal Command and in a reciprocal agreement the Commander-in-Chief, Coastal Command agreed to lecture each future session on the tactical employment of shore-based aircraft.[54] Finally, from 1938 the Junior Officers' War Course became an additional responsibility of the Tactical School.[55] The emphasis on fleet tactics is important for what the school did not attempt in the period under review was to make an officer a better ship driver. Simulation of ship handling was then in its infancy and though other aids, such as the attack teachers employed in the training of submarine officers could be found at Portsmouth and Chatham or the spotting-tables used in gunnery instruction and used in the *Excellent* were now available, such aids did not form part of the Tactical Course's instruction.[56]

Predating the Tactical Course, though, was the Senior Officers' Technical Course established to provide officers with the latest information on the capabilities and limitations of the weapons employed by the Royal Navy. Using the facilities of the technical schools housed in the Portsmouth environ, the course was seen as necessary grounding before proceeding to the War Course though many never actually went to the latter.[57] Before the World War, the SOTC had been organized along slightly different lines as separate senior courses lasting about a month for navigation, gunnery and torpedo instruction were conducted.[58] Now, the course was both expanded and contracted. Expanded in the sense that it sat for a single two-month period and covered other topics, but contracted as the time allotted for each subject was less than previously attempted. Meeting for the first time in January 1920 and running to eight weeks, it typically convened four times a year.[59] In time, it expanded to twelve weeks of instruction. The introduction of the Tactical School came partly at the expense of the SOTC, as the need to accommodate appointments to coincide with the offerings of the War Course and to stay within a fixed operating budget proved necessary. Of the pace of instruction, it proved more measured than grueling, as classes only began at 1400 hours on Monday ending by noon on Friday.[60] Indeed, so sedate was the pace

53   *King's Regulations, Volume II*, pp. 76-78.
54   ADM 239/142, *C.B. 03016/39*, p. 79.
55   Wells, 'Staff Training', p. 101.
56   Journal entry of 24 September 1925, Midshipman John Waldron Papers, Imperial War Museum, London, United Kingdom and ADM 182/74, Admiralty Fleet Order '574.— Models of British and Foreign Warships for Submarine Attack Teachers—Additional Models' of 18 March 1937.
57   *King's Regulations*, Article 326, p. 111.
58   ADM 196/125 (Casement).
59   ADM 196/90 (Collard).
60   ADM 1/9999.

of learning that Captain John Harper managed to serve as a representative of the Anglo-American Arbitration Board concurrent to his course in 1922.[61]

Though *King's Regulations* stipulated that the SOTC was for officers on the active list only, officers holding commissions in the Royal Naval Reserve frequently attended. In its initial guise, the course was limited to officers of Captain's rank only, with subordinate officers attending a separate venue. Given the ongoing retrenchment in naval funding and the emasculation of senior officer numbers, in July 1922 the decision was taken to abolish the separate Commanders' Course and to have these officers attend the SOTC as vacancies permitted.[62] The typical class was composed of about 15 officers with flag officers often present, though the primary aim remained that of preparing those recently promoted to Captain for their first command.[63] As was the case when attending other schools, Dominion naval officers regularly attended the SOTC with the cost of their tuition paid by the home establishment on a reimbursable basis. Numbers, in this case, however, were few and given the expense and time away from the parent force, a common practice was to have officers proceed immediately to the War Course following completion of their Technical Course. Captain George Moore,[64] RAN, is representative of this tendency. Having completed a prior sitting of the SOTC in early 1934, he returned to Portsmouth and completed it a second time in 1938 before proceeding to Greenwich and the War Course.[65]

In 1922, the core subjects covered included gunnery, torpedo, navigation, signaling, anti-submarine warfare, submarine warfare, and air work. The material was presented by officers from the staffs of the local technical schools or type commands with *Excellent* responsible for the gunnery portion of instruction (15 days), the *Vernon* for torpedo, mining, anti-mining and ASW, though the last named became the responsibility of the *Osprey* when that establishment was created (15 days). Officers from the *Dryad* addressed navigation (5 days) whilst instructors from the Signal School borne in the HMS *Victory* dealt with communications over seven days. Fort Blockhouse covered submarine operations during two days of the course with air topics limited to only a single day of instruction.[66] The last topic may be viewed as particularly shortchanged and probably was, but it must be remembered that the course was intended as a review of topics of relevance to the prospective commanding officer and aviation was

61   ADM 196/89 (Harper).
62   'Naval Notes, Commanders Technical Courses', *Journal of the Royal United Services Institution*, Vol. LXVII, No. 468, November 1922, p. 745.
63   *Navy Estimates 1926*, p. 78b.
64   Later Acting Rear Admiral George Dunbar Moore (1893-1979). Tactical Course, 1930; Training and Staff Duties Division, 1931-32; Senior Officers' Technical Course, 1934 and 1938; Captain, 1935; Senior Officers' War Course, 1938-39; Acting Rear Admiral, 1944; Acting First Naval Member, 1945 and retired, 1953.
65   National Archives of Australia, hereafter, NAA, A6769, Moore, G. D.
66   'Courses for Senior Officers', *The Navy List* (London: His Majesty's Stationery Office, October 1922), p. 2402.

in an acute doldrums at the moment with the withdrawal of embarked aircraft from most heavy ships. Additionally, officers attending the SOTC benefited from having outside lecturers appear and cover topics not unlike their counterparts meeting in Greenwich. Still, these lectures never approached in scale what was the norm at the Royal Naval College and the primary means of delivering instruction remained using serving officers from the technical establishments.[67]

As the topics were matters of some fluidity, it was only natural that some officers attended the course multiple times during their career though the suspicion was raised that sending officers to the SOTC was also means of keeping them employed between assignments.[68] In attending the course on three separate occasions, Andrew Cunningham was atypical, but with command of flotilla forces or the Battle Cruiser Squadron immediately following these occurrences, the claim that unemployment was the alternative in his case is not sustainable. Cunningham as a salt horse lacked a strong technical grounding, though his seamanship and leadership skills honed in command of destroyers doubtlessly prepared him in other ways. This was the strong suit of the salt horse where practical experience in the ranks of Lieutenant or Lieutenant Commander captaining an escort vessel was the norm.[69] Not having specialized, though, meant that Cunningham had not acquired an in-depth knowledge in gunnery or a like subject and his periodic attendance at the SOTC was probably compensation for his lack of prior technical grounding before flying his flag afloat. In-depth is the operative word here, as even the salt horse had to pass fleet examinations in gunnery and torpedo work before taking command of his ship. As a statement of general practice, it can be said that some officers in the initial postwar period did take the SOTC and SOWC multiple times and this for two primary reasons. First, it kept officers employed absent a ship to send them to, and, secondly, disruptions of courses occurred when aid to the civil power was required. In time, the practice was more orderly and an officer after five years from his last attendance could be assigned to the same course.

The SOTC was a very practical effort on the part of the Service to ensure that officers had a rudimentary grounding in the latest methods employed by all the combat arms and this knowledge would prove useful whether they were destined for command or staff duties. Though the trained military staff officer received instruction in the separate combat arms of his Service, he did not receive periodic refresher training afterwards. A consequence was that their knowledge was limited outside their own field of specialization and diminished as time advanced. Walter Guinness from his time serving with the British Army during the World War observed that 'One disadvantage of our comparatively amateur staff officers is that they cannot stand

67   *Navy Estimates 1935*, p. 82.
68   Anon., 'Need for Economy', *Naval Review*, p. 665.
69   Comma, 'The Training of Officers', *The Naval Review*, Vol. XXII, No. 1, February 1935, p. 41.

up to the superior technical knowledge of the gunner. When the latter say a thing is impossible the British Staff Officer is too ignorant to be able to do anything but collapse.'[70] Though the naval officer leaving the SOTC could not compete in knowledge with the trained specialist he at least appreciated the limits of his ignorance.

Geoffrey Blake, as a gunnery specialist, was not an officer lacking technical proficiency; yet, he too passed through the Tactical Course twice. In his case, Blake first sat the course in 1935 returning the following year to complete it again before assuming command of the Battle Cruiser Squadron. Professor Michael Simpson has noted that Chatfield eyed Blake to be First Sea Lord around 1943.[71] If so, this may offer the strongest evidence that multiple attendances of the course occurred if for no other reason than to keep a very promising officer employed between appointments. Whether Blake would have become First Sea Lord if ill-health had not forced his early retirement is questionable, as serving as Director of the Staff College was not an assignment leading to the office during the period between the wars and would only be achieved by another Admiral in the 1960s.[72] Still, command of the Battle Cruiser Squadron was not the typical assignment for a previous Staff College Director either so it may be that Blake's star was especially ascendant. Certainly, his previous time serving with Chatfield in the *Queen Elizabeth* during the war meant that his talents were known in the highest quarters and successfully fulfilling successive assignments made his future prospects promising.

An additional consideration why officers attended the SOTC more than once was to maintain currency with the latest technical practices. Until the AFO issued in 1930 mandating officers attendance at the SOTC, SOWC and Tactical Course, officers had no specific continuing education requirements after their promotion to Lieutenant. This now changed and though not established in regulation, it appears officers were expected to complete the SOTC at least once every five years. This certainly was the case for officers holding RNR commissions which given the vagaries of their civilian careers is understandable. For the RNR officer, initial completion of the SOTC and Tactical Course was required within six years following promotion to Commander with re-qualification necessary at intervals of five years.[73]

Just as some argued that attendance at the Staff College should be required before promotion to Commander or Captain, others held that all executive officers should be a specialist of some branch and deprecated the presence of the salt horse.[74] Their

70   Brian Bond and Simon Robbins, eds., *Staff Officer: The Diaries of Walter Guinness (First Lord Moyne) 1914-1918* (London: Leo Cooper, 1987), p. 117.

71   Michael Simpson, ed., *The Cunningham Papers: Selections from the private and official correspondence of Admiral of the Fleet Viscount Cunningham of Hyndhope, O.M., K.T., G.C.B., D.S.O. and two bars, Volume II, The Triumph of Allied Seapower, 1942-1946* (Bodmin: Ashgate, 2006), p. 210.

72   Admiral Sir John Luce in 1963.

73   Rainbow, 'Red, White and Blue', *Naval Review*, May 1934, p. 246.

74   First Principles, 'Specialists', *Naval Review*, February 1939, p. 17.

argument was informed by the increasing contribution that science and industry were making in marine engineering and the appliances of war. The thousand and one inventions flowing from this inventiveness and now found in the modern naval vessel were nothing but ballast, if they could not be maintained and used properly. This debate lies outside the scope of this work and compelling arguments from advocates of both sides were advanced. Still, Cunningham's multiple attendances at the SOTC without ever proceeding to the War Course indicates that those in favour of technical specialization had the future if not the weight of history on their side.

As the SOTC only covered its specialist topics in a cursory manner given the time available, the officer attending was at best securing only a general appreciation for the weapons, techniques and procedures discussed. Still, having last received instruction in most of these areas when preparing for his Lieutenant's examinations—probably some 15 – 20 years previous—a refresher of sorts was not without merit. However, the SOTC was no mere rehash of previous instruction, as the opportunity now arose to address other topics more appropriate to the senior officer. Foremost, was the vital importance of secure communications and this subject was covered by a representative of the Naval Intelligence Division in a lecture to successive classes of the SOTC from 1932.[75] Such an emphasis was natural in light of the Navy's own success in attacking the codes of the German Navy in the late war and the only question arising was whether the subject should be broached by the Admiralty's NID, the Signal Department or from a member of the Government Code and Cypher School.

Similar instruction was provided to the qualifiers of the Staff Course and the proper handling of codes and ciphers was a central element of what was taught in the Accountant Officers' Technical Course. As a commanding officer would only have occasion to decode signal traffic rarely, its prominence in the SOTC was more as a reminder of the importance of ensuring the safety of British codes, the perils if they were compromised, and securing an understanding of the responsibilities of his subordinate officers before assuming afloat command. Such was also true for the rationale of instruction for much of the material covered at the SOTC and the senior officer upon leaving Portsmouth and the *Vernon* had a clear appreciation of the roles and responsibilities of the officers comprising a ship's wardroom, if not the actual knowledge to perform those tasks himself.

In 1938, an officer attending the SOTC received four lectures on engineering covering the principles of damage control and the technical features of the warships then under construction with instruction provided by a member of the Naval Equipment Department. A member from the *Excellent* addressed the recent gunnery practices conducted within the fleet whilst his counterpart from the *Vernon* reviewed minesweeping, torpedo control both by submarine and destroyer attack along with the latest capabilities of the Whitehead torpedo. The Naval Air Division and the Air Personnel Department of the Admiralty combined to cover the role of the Fleet

75   ADM 1/9999.

Air Arm. Here, the separate functions of the several types of ship-borne aircraft employed received consideration, as did the nature of pilot training and the training afforded to Naval Observers and Telegraphist Air Gunners.[76] These lectures were augmented by a visit to RAF Gosport, the nearby air station, where officers had the opportunity to witness air operations of the types to be found in the fleet and to see examples of the aircraft assigned to Coastal Command and supporting naval cooperation. A representative of the accountant branch addressed the training provided to officers attending the Accountant Officers' Technical Course whose qualifiers would soon graduate to coordinate the affairs of a Captain's office where overseeing the correspondence flowing between a ship, the fleet, and the Admiralty was foremost. Meanwhile, a member of the Signal School covered the increasing use of sound and photography occurring within the fleet as an aid to morale, but, more importantly, also in capturing the results of at-sea exercises. The Navy retained a significant library of training films available for use within the fleet which the commanding officer was free to request. Provided through the Service's Cinema Scheme and soon augmented by commercial films from British studios, topics as diverse as personal hygiene and common communicable diseases were featured next to professional subjects such as the Fleet Air Arm, the exercises of the fleet, and the workings of specific weapons.[77] A final set of lectures and demonstrations covered defending a ship against the various forms of chemical attack through both collective and individual protective means and how to decontaminate a vessel once exposed.

By 1938, a subject receiving renewed emphasis at the SOTC was the adoption, organization, and controlling of convoys. Here, the Trade Section of the Admiralty Plans Division assisted matters. The slow response of the Admiralty in adopting convoy as a means of negating the submarine peril in the late war and the dire consequences that followed were lessons well appreciated. Now, though, the problem facing Britain was not limited to the submarine and officers attending the SOTC learnt:

> Convoy has only one advantage, but that is one of paramount importance, namely, that it is the surest means of protecting shipping in the face of intensive and persistent attack, whether by surface vessels, submarines or aircraft. History has shown the value of convoy as a protection against surface vessels; the late war showed it to be the best antidote to ruthless submarine attack; against unrestricted air attack in narrow waters the Air Staff agree with the Naval Staff that there is no alternative but to putting shipping into convoy, and to provide escorts armed to protect convoy against air attack.[78]

76   *Ibid.*
77   *Ibid.*
78   *Ibid*, 'Lecture on Convoys for Senior Officers' Technical Course', no date but *c.* 1938. Original emphasis.

Though officers assigned for duties under the Naval Control of Shipping (NCS) programme were the primary agents for arranging convoys, others including those serving as Staff Officers Intelligence needed to be cognizant of this work for the NCS was not maintained on a permanent footing. Intelligence Officers during the crisis period or upon the immediate outbreak of war might have to perform these duties in the first instance until the NCS scheme was properly established.[79] If this were to prove so, the Intelligence Officer need not fear that he was trying to herd cats, as many of the officers and Masters of British shipping were receiving much the same instruction on convoying as he was getting in the SOTC through a separate training initiative.[80]

Thus, the Royal Navy did not ignore the lessons of the past, at least, with respect to value of convoy in the face of submarine warfare, but that was only a part of the equation. What it failed to do was appreciate that its offensive anti-submarine methods needed to be better coordinated with its defensive practices. Operational research in the next war would prove this conclusively, of which the efforts of P. M. S. Blackett were at the fore in this regard. The late Vice Admiral Sir Peter Gretton[81] argued that contributing to Britain's poor performance initially against the U-boat in the Second World War was the doctrinal emphasis on fleet action, as espoused in the *Naval War Manual,* and the divided writing of the World War's official histories covering air and maritime operations. This division in writing the war's history precluded consideration of the anti-submarine warfare (ASW) problem in its entirety. Contributing to the failure was the lack of studying the entire ASW question at the Staff and War Colleges between the wars with the consequence that Britain found herself facing a fresh submarine crisis in a new war.[82]

These are grave charges made from a serious and most capable practitioner of the naval art. In the main, this writer accepts Gretton's case—with reservations—and will address the issue shortly. Certainly, the Navy's separation of tactical instruction and investigation from the Greenwich Staff and War Courses and placing it at Portsmouth made a holistic approach to investigating any problem such as ASW and anti-air warfare more difficult. And, while the writing of separate Official Histories may have made the public record of the anti-submarine campaign more opaque, the Navy did prepare internal Technical Histories covering the air and submarine campaign, its system of convoy, and the role of the Anti-Submarine Division of the

---

79   *Ibid.*
80   *Ibid.*
81   Vice Admiral Sir Peter William Gretton (1912-1992). Captain, 1948; Joint Services Staff College, 1948-49; Naval Assistant to First Sea Lord, 1950-52; Commanding Officer, HMS *Gambia,* 1952-53; Real Admiral, 1958; Imperial Defence College, 1958-60; Flag Officer Sea Training, 1960-61; Vice Admiral, 1961; DCNS and Fifth Sea Lord, 1962-63 and retired, 1963.
82   Gretton, 'Why Don't We Learn from History?' *The Naval Review,* January 1958, pp. 18-24.

Naval Staff soon after the war.[83] At the operational level, the Navy accepted the need to convoy shipping and understood that the safe passage of goods and not the destruction of submarines was the fundamental necessity.[84] Yet, understanding the problem if an important step was insufficient by itself.

As with all the courses surveyed in this study, formal instruction only served as a means of serving up the question at hand. For it was in the period following the set lecture, when questions, discussions, and the arguments arising from the topic were raised by those attending that a true understanding of any issue and its associated complexities became known. Here, the officers with their varying technical backgrounds and operational experiences amplified and discounted what they just heard or witnessed and given the seniority and experiences of the officers attending, their perspective made the course invaluable.[85]

Providing perspective, especially with regard to the operative nature of the *Battle Instructions* was an important aspect of the Tactical School and not surprisingly, Jutland was a prominent topic at the school. Officers attending the eight-week course examined the battle in depth and a demonstration of its lines of action was the culminating event to mark the conclusion of each sitting of the course.[86] Analysis of the battle by the school was broad, deep, and ongoing with its major findings summarized in booklet form, *New Light on Jutland*, and issued to the fleet in 1934.[87] It was also in 1934 that Admirals of the Fleet Earl Jellicoe and Sir Charles Madden, his Chief of Staff during the battle and, moreover, his brother-in-law, visited the school to offer their perspective of the engagement. For Jellicoe appearing at the school was something of a regular sojourn that year; it was a sojourn that other senior officers of the engagement willingly made, too, including Admiral Sir William Goodenough and Admiral of the Fleet Sir Osmond Brock.[88]

---

83   Amongst the imprints of the Technical History Series covering convoy were *C.B. 1515(4), Aircraft and the Submarine Campaign*; *C.B. 1515(7), The Anti–Submarine Division*; *C.B. 1515(8), The Convoy System*; *C.B. 1515(14), The Atlantic Convoy System, C.B. 1515(15)*; *Convoy Statistics, C.B. 1515(16)*; *Convoy Defence*; *C.B. 1515(17),Convoy Supplement and C.B. 1515(18), Convoy Supplement*. In addition, three separate technical histories were written on mercantile control, a further technical history addressed convoy in general whilst *C.B. 1515(40),Anti–Submarine Development* reviewed the progress in counter-measures and counter-tactics. Portions of the Admiralty Technical History Series may be found under the ADM 189 and ADM 275 series. The specific titles referenced above have not been located but have been identified through their reference in the 'Monthly Intelligence Report' series published during the period and found under the ADM 223 series.

84   ADM 186/72, *Battle Instructions*, p. 71 and ADM 186/106, *Battle Instructions*, p. 61.

85   ADM 1/9999, Vice Admiral Wilfrid French letter regarding 'Suggestions for Improvement of the Senior Officers' Technical Course' of 28 October 1938.

86   ADM 116/6364, Tactical School, 'Annual Report, 1937'.

87   Earl Mountbatten of Burma comments of 1967, NMM/BLE/12, Vice Admiral Sir Geoffrey Blake Papers, National Maritime Museum, Greenwich.

88   Admiral of the Fleet Sir Osmond de Beauvoir Brock (1869-1947). Five firsts in examinations for Lieutenant; Captain, 1904; War Course, 1904-05; Assistant Director, Naval

A common theme of Jellicoe's comments to the reenactment of the battle given was for officers to remember that what perforce they were witnessing of the engagement was not necessarily what the Commander-in-Chief and his principals had seen. Jellicoe recounted the numerous errors made during the battle when subordinates failed to appreciate that what was unfolding before them might be of interest to their superiors. The necessity for providing scouting reports was essential, a point also emphasized contemporaneously at the Staff College. Though the battle suggested many lessons, the need for individual initiative may be thought the foremost one to be derived from the several courses across the Service where Jutland was studied.[89]

Madden, for his part, told the officers attending the Tactical Course that Jellicoe's deployment was a singular achievement. It had never been attempted in the manner executed by the Grand Fleet until that moment. Oral testimony many years after the fact is notoriously suspect. Still, his observation that if deployment on the starboard wing had been made, as some were arguing was the more preferable option, it would have come at the fleet's peril and indicates that the issue remained one of the central what-ifs of the encounter.[90] Thus, William Tait's recording of the American naval view of the question recounted earlier would have been read with interest by his superiors coming as it did from an independent source and addressing one of the more intractable unknowns of the battle.

For visiting dignitaries, the school frequently played an abbreviated version of the battle with the Director conducting the action.[91] In 1937, Samuel Hoare, the First Lord, inspected the school and viewed a reenactment of the battle; an opportunity that 65 other officers not attending the school as qualifiers also received at some point during the year.[92] This is unsurprising for the battle, if known by many was understood properly by few. Such demonstrations beyond indicating the many changes in tactics undertaken since the war also served a corollary purpose. They became the naval equivalent to the Hendon air displays of the Royal Air Force or the annual Royal Tournament held at Olympia. This is not to trivialize the demonstration's importance. Yet, they became a practical means of telling the Service's story, highlighting the ongoing progress occurring in the fleet, whilst making the Royal Navy appear relevant to political masters. Hence, when Hoare spoke in parliament in support of the *Estimates* in 1937, he buttressed his argument by recounting his recent visit:

Intelligence, 1907-09; Rear Admiral, 1915; Commander, First Battle Cruiser Squadron, 1915-16; Chief of Staff to Commander-in-Chief, Grand Fleet, 1916-19; Vice Admiral, 1919; DCNS, 1919-21; Commander-in-Chief, Mediterranean, 1922-25; Admiral, 1924; Commander-in-Chief, Portsmouth, 1926-29; Admiral of the Fleet, 1929 and retired, 1934. Recalled for war service.

89   Dreyer, *Sea Heritage*, p. 168.
90   *Ibid*, p. 169.
91   *Ibid*, p. 168.
92   Mountbatten comments in Blake Papers, NMM/BLE/2.

> I was very much struck by this fact the other day when I went to Portsmouth and saw in the Tactical School the Battle of Jutland worked out in great detail, with model ships, each of them exactly in its place during the course of a long-drawn-out engagement, with electric lights showing the visibility and with a senior naval officer, who was himself present at the engagement, explaining at the end of each phase of the battle exactly what did happen and contrasting what did happen with what might happen to-day under modern conditions and the many changes which have taken place since the Battle of Jutland.[93]

The Tactical School's full work-up of Jutland was enacted in three separate phases corresponding to the battle cruiser action, the battlefleet engagement, and those operations occurring during the night of 31 May and 1 June as the High Sea Fleet attempted to disengage from the Grand Fleet and return to its home ports.[94] Drawing on the Official History, German sources, and Jellicoe's memoirs, the Tactical School's review of the battle was not afraid to indicate where errors were present even in the presence of those who commanded. The initial demonstrations of the battle were rather crude affairs judging by the later standards Hoare and others observed. Initially, lighting to indicate the likely arcs of visibility that each fleet enjoyed was lacking, and, instead, a circular pattern overlaid to a diagram of the battle depicted the probable fields of view available from that point.[95] In time, a beam of light focusing on Jellicoe's flagship, the *Iron Duke*, was introduced and allowed viewers to gauge the limited span of battle that the Admiral presumably possessed.

Spectators watched as the action unfolded before them with informed commentary noting where changes in current method were now different from 1916 practices. Thus, those observing were told that the previous manner of firing single salvoes and waiting to observe the fall of shot before correcting for the next round had been replaced with the firing of double salvoes in rapid succession. By firing one salvo long and a second one short to the estimated range of the target and then proceeding to correct for the observed error, the true range could be determined more rapidly.[96] During the phase of the game demonstrating the battlefleet action, observers were reminded that Jellicoe could only deploy the Grand Fleet on a port or starboard wing column. On the day of battle, he elected to deploy to port with the signal Equal Speed Charlie London announcing his decision to the rest of the Grand Fleet.[97] Following

93   *Hansard*, House of Commons debate of 11 March 1937, v. 321, cc1367-508.
94   Jutland Demonstration Tactical Course of 1935, 'Jutland I', Admiral of the Fleet Lord Chatfield Papers, National Maritime Museum, Greenwich, NMM/CHT/8/2.
95   Chalmers, *David Beatty*, p. 221.
96   Jutland Demonstration Tactical Course of 1935, 'Jutland I', Chatfield Papers, NMM/CHT/8/2.
97   Geoffrey Bennett, *The Battle of Jutland* (London: B. T. Batsford, Ltd., 1964), p. 105.

the battle, the fleet perfected the means to deploy on the centre column of the force and audiences were able to see how the action flowed when using the evolution.[98]

When commenting on the action at night, the tendency for British forces to fire their torpedoes singly rather than in salvoes, and to treat unknown ships as friendly and not hostile, were highlighted.[99] The latter point was also a key reason why the Staff College expended so much effort in training qualifiers in plotting. If a ship's Captain understood where British forces were located in any encounter, then he would not necessarily have to make a challenge when approaching an unknown ship, but could proceed to engage immediately knowing that any unidentified ship had perforce to be hostile.[100] Accordingly, the initiative in action would be maintained by the British fleet and not surrendered. As for firing torpedoes at night, British doctrine emphasized the need to employ all weapons and to develop the maximum fire available to ensure the enemy's destruction. Firing torpedoes singly wasted opportunities, as the chances of achieving hits were greater when fire was maximized and concentrated. Finally, in common with what was also being stressed at the Staff College, the Tactical School called attention to the need for the fleet's scouting and reconnaissance forces to report the presence of enemy forces. If damage precluded the senior officer of a flotilla from making the required report, then it remained for any subordinates in company to accomplish the task. Touch with the enemy once gained must not be surrendered without a compelling reason, for there was no guarantee that once lost it could be reestablished.

Following the demonstration, the student was left with a strong impression where British practice had been found wanting and how matters now stood with respect to the current *Battle Instructions*. Yet, changes in doctrine could only be part of the answer. It remained for officers to prepare themselves, too, for the next encounter by securing a strong appreciation of the relative silhouettes of British and foreign men-of-war, ensuring lookouts were posted at the appropriate points of advantage, and remembering that the need for initiative was no less a requirement of a ship's company as it was for its commanding officer. Indeed, in 1935 the Tactical School was warning qualifiers that the *Battle Instructions*, notwithstanding their postwar revisions, still had omissions, of which, the need for a commander to ask for information in the absence of its automatic forwarding by subordinate forces was but one.[101]

Given the prominence of Jutland in the course's syllabus, by 1935, some officers were questioning whether more pressing contemporary problems ought not to be considered

98   'Jutland Demonstration Tactical Course of 1935, Jutland II', Chatfield Papers, NMM/CHT/8/2.
99   *Ibid*, 'Jutland III'.
100 *Ibid*.
101 'Night Operations at Jutland', pp. 33-36, Tactical School demonstration of 6 September 1935, DS/MISC/4, Admiral Sir Harold Martin Burrough Papers, Imperial War Museum, London.

in its place. Commander Frank Walton,[102] a Naval Observer, believing the days of the battleline were largely past, argued for fleet problems to be investigated where destroyers screening aircraft carriers was now the norm. Another school of thought, equally questioning the prominence of the battlefleet but not nearly so committed to Walton's view, pushed for investigating actions where formations of differing types of warships were involved and sought to explore the possibilities of battle at night. Other problems they wished examined included the best means of shadowing an enemy force in addition to the carrier operations advocated by Walton.[103] Lionel Wells, the Director, was unwilling to take this step and opted to stress the importance of the battlefleet. While hindsight might view Wells' reticence as mistaken or even hidebound, a degree of sympathy is appropriate. The primary task of the school was to train officers in fleet tactics based on the *Battle Instructions* with a subsidiary mission to investigate current problems posed by the fleet or the Admiralty. Until the creation of the Tactical Investigation Section in 1939, a tension remained over the relative weight afforded between these two tasks and Directors faced a bit of a poisoned chalice in meeting the two requirements.

Such issues as battle by night were topics of white-hot argument within the fleet and it was difficult for the Tactical School to take a lead where consensus did not yet exist. Drax, during his time in the Mediterranean Fleet, was probably the greatest exponent of night-action, yet his was an uphill battle and his greatest opponent was no less than the Second-in-Command, Vice Admiral Howard Kelly.[104] A suitable exercise was arranged to test the merits of each officer's respective views and Drax definitely had the measure of Kelly in the ensuing demonstration. This did not settle matters and in the ensuring hot-wash Kelly belittled Drax's achievement in biting, sarcastic terms.[105] This was 1929 and the *Battle Instructions* were subsequently amended. Thus, it is difficult not to conclude that Rear Admiral Wells was orthodox in the direction he provided at the school when compared to both his predecessor and his successor. Captain James Troup, preceding Wells as Director had during his

---

102 Later Captain Frank Michael Walton (1900–1982). Observer in *Eagle*, 1924-25 and 1928-29, *Furious*, 1926 and 1929, *Argus*, 1927-28; Meteorological Course, 1929; Observer in *Courageous*, 1930; School of Naval Co-operation, 1931-34; Tactical Course, 1932 and 1935; Observer in *Eagle*, 1934; Tactical School, 1935-37; Commanding Officer, HMS *Dainty*, 1938-40; Captain, 1941; Commanding Officer, HMS *Striker*, 1942-44 and retired, 1951.

103 Mark Williams, *Captain Gilbert Roberts R. N. and the Anti-U-boat School* (London: Cassell, 1979), pp. 71-72.

104 Later Admiral Sir William Archibald Howard Kelly (1873-1952). Captain, 1911; Naval Attaché Paris, 1911-14; Naval Intelligence Division, 1916; Liaison Officer, Paris, 1916-17; Commander, Third Light Cruiser Squadron, 1917-19; British Naval Mission to Greece, 1919-21; Rear Admiral, 1922; Commander, First Battle Squadron, 1923-24; Commander, Second Cruiser Squadron, 1925-27; Vice Admiral, 1927; League of Nations, 1927-29; Commander, First Battle Squadron and Second in Command, Mediterranean Fleet, 1929-30; Admiral, 1931; Commander-in-Chief, China, 1931-33 and retired, 1936. Recalled for war service.

105 Unpublished manuscript, pp. 49-50, Waymouth Papers, IWM/87/16/1.

regime made a practice of stressing the importance of being prepared to fight at night. Lecturing to this theme, he ended his talks quoting with effect from Psalm 91, 'Thou shall not be afraid for the terror by night.'[106] Meanwhile, under Captain Ion Tower,[107] Wells successor, the balance shifted away from Jutland and the battleline concept to the consideration of other problems. This move was made easier by the rise in naval problems closer to home where navies, not so heavily centered on the capital ship as the long-viewed Japanese threat, operated and the broader view that Tower brought to the school. A Greenwich *psc* and holder of the *idc*, Tower had also taken the War Course, the other required preparatory courses required of Captains, and the short Aviation Course offered to selected senior officers of Commander's rank. This he did during his attachment with the Air Ministry in its Directorate of Armaments. An officer of proven merit, Tower qualified in carrier deck landings notwithstanding a visual impairment. A veteran of Jutland, Tower had clearly moved beyond the limits of that engagement.[108]

If Jutland continued to receive full treatment at the Tactical School, then it was hardly the only topic covered. Nor was its demonstration the only offering provided to visiting officials. In 1937, the same year that Hoare was seeing the battle reenacted, 135 military officers and 67 members of the RAF witnessed a demonstration on the table or attended special lectures provided by the school. Those attending these events were qualifiers from the IDC, Camberley, or others under instruction at the RAF Torpedo Training Course.[109] Still, the core of its instruction was providing naval officers a solid understanding of current naval tactics based on the doctrine of the several manuals used within the Service. Preliminary work in the first two weeks of the course covered the *Manoeuvring Manual* and those publications dealing with signals, followed by the specific orders used within the Atlantic Fleet, plotting, and making contact reports.[110] Next, officers learned how to play the game in both its strategical and tactical versions, though the tactical game was the preferred instrument of instruction. This is unremarkable, as the strategical game was the version most

---

106 James Troup to Stephen Roskill letter, no date, but from June 1964, Roskill Papers, ROSK/7/163. *King Jane's Version*, Psalm 91, v. 5, 'Thou shalt not be afraid for the terror by night; *nor* for the arrow *that* flieth by day.'

107 Rear Admiral Ion Beauchamp Butler Tower (1889-1940). War Staff Course, 1923-24; Air Ministry, 1924-26; Captain, 1929; Tactical Course, 1930 and 1937; Captain (D), Fourth Destroyer Flotilla, 1930-31; Senior Officers' Technical Course, 1931-32; Senior Officers' War Course, 1932; Imperial Defence Course, 1933; Flag Captain in *Kent*, 1934-35; Assistant Director and Director, Tactical School, 1937-38; Commanding Officer, *Malaya*, 1938-39; Rear Admiral and Liaison Officer, Commander-in-Chief, Home Forces Command, 1940 and killed in bombing raid, 1940.

108 ADM 196/52 (Tower) and ADM 196/127 (Tower).

109 ADM 116/364, 'Tactical School Annual Report, 1937'

110 Tactical School, 'Course (X)' syllabus, Admiral Sir Frederic Dreyer Papers, Churchill Archives Centre, Cambridge, DRYR/7/2.

frequently played in both the War and Staff Courses where the writing of orders and appreciations to support its play was central to their studies.[111]

Playing the tactical game offered officers the opportunity to test their mettle against the navies of the United States, Japan, France and Italy, while later editions allowed them to conduct play with the post-Versailles German Navy in mind.[112] Having mastered the game, officers learnt the tactics of cruisers, destroyers, submarines, and naval aircraft, the employment of smoke and gas, and the *Battle Instructions* of the Atlantic Fleet. By the seventh week, officers were finally studying the tactics of the World War, of which the Battle of Jutland was central, along with the most recent exercises conducted at sea. However, the progress made since that time was the essential lesson offered and demonstrating Jutland simply facilitated this understanding and was the culmination of all previous efforts pursued.[113]

The Tactical School made a practice of recording the results of each game played and benefited from having multiple syndicates test the concepts or problems posed which doubtless informed the conclusions that they drew and the recommendations that they proposed to the Admiralty.[114] Towards the end of the course, officers were examining any problems that the Admiralty or fleet had forwarded for investigation along with the tactics of cruiser searches, patrols, how to shadow an enemy force, reconnaissance, and divided tactics. These subjects were more speculative or tentative in nature and suffered if the length of the course was curtailed for any reason. Included amongst William Tennant's papers are his notes on the several types of searches investigated at the Tactical School and their relative merits for differing types of operations. These demonstrate that more than just teaching the mechanics of any operation, the course aimed to instill in officers the underlying reason why certain courses of action were favoured over others.[115]

Truly, investigating specific tactical problems posed by the Admiralty or the fleet was a major advantage of having a school and with multiple classes operating in most years, the problems investigated benefited from the perspective of many senior officers. This was a vital contribution of the school, as there were few officers actually attached to the Tactical Division of the Naval Staff to engage in the systematic evaluation of any issue. It was problematic for the Tactical Division just keeping pace with the routine business of the Admiralty and the resources of the school were a boon to

111 ADM 186/76, *C.B. 3011. War Game Rules, 1929*, Admiralty, Naval Staff, Tactical Division, April 1929.
112 ADM 1/8668/177, Admiralty Fleet Order '1541.—Models of British and Foreign War Vessels for General Instructional purposes, Small Torpedo Attack Tables, and Gunnery Spotting Tables' of 8 June 1923.
113 Tactical School, 'Course (X)' syllabus Dreyer Papers, DRYR/7/2.
114 Williams, *Gilbert Roberts*, p. 72.
115 'T.S. 66, Tactical School Example', Tennant Papers, NMM/TEN/8.

its efforts.[116] Of necessity, the relationship of the Tactical Division to the school was close, but the flow was not of a single direction with Division members lecturing at the school on current problems and efforts to resolve.[117]

If the results obtained were not always repeated in the actual exercises of the fleet, this did not detract from the value of the school's endeavours. Variations in results were bound to exist in a field more often viewed as an art than a science. The real usefulness of the Tactical School defining and solving current problems was that it could examine them with a detached eye and in a reflective manner not possible within the cauldron of the fleet or the docket passing ways of the Admiralty. Meanwhile, the school also supported joint command post exercises with the Army and the Air Force. The yearly coast defence drills running to several days are representative of the latter case that saw the Services coming together to run an exercise where the forestalling of a raid, feint, or invasion against the British coast was a common theme. The school also had a corollary duty of maintaining version control of the war game used throughout the Service, as recommending possible changes in doctrine was only possible in an environment where a degree of commonality actually reigned. They oversaw improvements in the game's physical layout and assets, suggested changes to its rules and kept establishments operating at the correct revision level.[118] Interestingly, during Keyes command of the Mediterranean Fleet responsibility for overseeing the ashore war game at Malta and president of that fleet's Tactical Committee was Rear Admiral Bernard Collard,[119] he of the *Royal Oak* tempest.[120] Given the very pronounced opinions held by Dewar on current tactical questions was this, indeed, the underlying source of the two officers' differences?

Though Sir John Fisher might hold that the best scale for an experiment was twelve inches to the foot, the war game was valued because it could quickly demonstrate the soundness or otherwise of new ideas without resorting to a full scale exercise and the expense such evolutions entailed. Just as the Staff College might offer suggestions based on reviewing the reports of the exercises actually conducted in the main fleets, the Tactical School performed a similar role. A case in point is provided by its recommendation covering the possible use of flotilla forces during a fleet action. Fleet exercises had shown that one side was frequently able to achieve a position where a torpedo

116 In 1924, the Tactical Section consisted of two officers rising to five officers in a Tactical Division by 1936.

117 ADM 196/92 (Scott).

118 ADM 116/6364, Tactical School, 'Annual Report, 1937'.

119 Later Vice Admiral Bernard St George Collard (1876–1962). War Course, 1909-10; designated War Staff Officer, 1912; Naval Intelligence Division, 1912-15; Captain, 1915; Director, Operations Division (Home), 1918-20; Senior Officers' Technical Course, 1920 and 1927; Senior Officers' War Course, 1920, 1921 and 1927; Director, Gunnery Division, 1922-24; Rear Admiral, 1926; Tactical Course, 1926; Commander, First Battle Squadron, 1927-28; retired, 1928 and Vice Admiral, 1931.

120 Halpern, ed., *Keyes Papers, Vol. II*, pp. 245-46.

attack by destroyers was possible and that the opposing gunfire of the targeted fleet was at times minimal. Playing the exercise again at the table highlighted that dividing the flotillas into sub-divisions or even proceeding with attacks by single ships allowed the attacking force to approach much closer. Attacks launched from 1,500 yards were thought possible under such conditions and the unstated conclusion reached was that a corresponding increase in hits achieved would now result.[121] What the Tactical School report did avow was that the present *Destroyer Attack Instructions* issued as *C.B. 01721* were insufficient. To that extent and in common with the Staff College, the Tactical School followed the practice of raising areas where present doctrinal guidance was lacking or obscure.

The Tactical School acted in a similar capacity when it came to the local supplements that the fleet commanders issued. It reviewed these to ensure that their meaning and intent were clear and that ambiguity was minimal. Thus, in 1932 'Home Fleet Amendment No. 1' was criticized for its vagueness about the function of British battle cruisers, if the battlefleet of the enemy was weaker to the British one. Its recommendation was that:

> From experience on the tactical table it is suggested that it should be made rather more clear that this Amendment does not apply when we possess a superiority over the enemy in capital ships, and that in those circumstances there is every advantage in not limiting the freedom of action of the battle cruisers in any way.[122]

Similarly, in 1934 James Troup, following a review of the current *Battle Instructions* and using the casualty rules found in *C.B. 3011: The War Game Rules*,[123] reasoned that British heavy ships must adopt a better scale of anti-aircraft defense in the face of the current air threat. Arguing before the advent of radar and considering whether battle occurred either by day or at night, he wrote:

> The fleet, during the approach, has to be prepared for the attack of enemy striking forces. It is believed that air interception is not reliable. If this view is accepted, it follows that our ships, and particularly battleships and battle cruisers, must be rendered as invulnerable as possible. To this end, provisions of the heaviest possible anti-aircraft armament, combined with the fitting of armoured decks, appears to be the appropriate measure.[124]

121 ADM 1/9007/79, Director of Tactical School to Secretary of the Admiralty letter No. 862/25 of 8 May 1934.

122 *C.B. 3016/31*, p. 42, British Sources, Box 12, Naval Historical Center, Washington, DC.

123 See ADM 186/78, *C.B. 3011. War Game Rules, 1929*. Admiralty, Naval Staff, Tactical Division of April 1929.

124 ADM 1/9007/79, Director of Tactical School to Secretary of the Admiralty letter No. 862/25 of 8 May 1934.

By playing games against standardized rules, basing its results on findings demonstrated within the fleet, and using the Navy's own formal doctrine as its guide, the school served the corollary purpose of acting as an independent verification authority for the adequacy of current fleet dogma and practice.

Though lectures were a vital component of instruction at the Tactical School, the real lessons were distilled at the table where the war game was played. Testing problems and theories on the tactical table was invaluable as the results likely to arise were rapid and economical. In the time taken to conduct a single exercise at sea, the tactical table allowed six games testing the same concepts to be played. What was discovered at the table and at the Tactical School was reinforced elsewhere as the ships and shore establishments used the very same war game as the Tactical School.[125] More than any publication or fleet exercise, the war game and the tactical table transferred a common doctrine throughout the Navy. That may have not been its original intent, but it certainly was its greatest achievement. For the game made naval doctrine demonstrably alive in a manner that manuals, instructions, lectures, and even *The Naval Review* could not. This value was recognized with one writer claiming:

> [W]hile the written word is incapable of imparting any but the barest principles to be followed, the conduct of games on the tactical table is, I would suggest, of inestimable value to young officers in developing his appreciation of a given tactical situation; his knowledge of the limitations of the gun or torpedo fire and their effect on the handling of a ship or squadron; and that sixth sense which, later on, may transform his course of action in battle from a conventional move into a stroke of genius.[126]

The object of the tactical game, then, was to instill a proper sense of space and time when facing a problem at sea and to acquaint officers with the British plan for naval battle. The game allowed decentralization of command to be practiced without risking ships or sailors and the officers playing were expected to demonstrate the full power of initiative 'in accordance with common tactical doctrine'.[127] As the British fleet would perforce be engaged in battle against a foreign adversary, the tactical game exposed officers to the presumed capabilities of those forces. In this regard, naval officers enjoyed an advantage largely missing in the training of their military brethren, as the British Army did not attempt to instill an appreciation of the capabilities of foreign forces in their training until 1936.[128]

Just as the War Course and the Staff Course examined future problems for the Naval Staff to inform future material acquisition, it should not be surprising to learn that the Tactical School performed a similar role. In 1939, it was investigating tactical

125 *Ibid.*
126 Kay, 'Tactics and the Two-Stripper', *Naval Review*, May 1939, p. 242.
127 ADM 186/78, *C.B. 3011*, p. 10.
128 French, *Churchill's Army*, p. 46.

problems of the presumed British and Italian fleets as constituted in 1942. How much this was to support material acquisition as opposed to strategic force distribution decisions cannot be determined from the records in extant, but what is clear is that independent analysis was being employed by the Admiralty to inform its recommendations to Government.[129]

The instructors set the parameters of the game providing common information such as the weather conditions prevailing, especially wind and visibility, the time of day, the geographical setting of the war, and the object of the game to the teams playing Red and Blue. Unique cloths highlighting the geographical features of the game's setting and of variable scale were also commonly employed. This innovation allowed students to visualize the specific limits governing play and allowed the game to alternate between strategic and tactical actions. Meanwhile, a card was placed on the table and was used to indicate relative orientation of the compass. Still other aspects of the game were made known only to one side with each team receiving a unique summation known to them and the umpires overseeing play. This specified the objectives, the forces available, their tactical designations, the general idea governing operations, and any specific intelligence.[130] Of the scenarios practiced, a demonstration of the tactics during the approach phase of battle was always a central lesson.

Meanwhile, those assigned as instructors came from the principal specialist branches of the Navy so that it was common to have a representative from the gunnery, torpedo, signals, navigation, submarine and salt horse communities on the staff. The emphasis placed on plotting and following proper communications procedure ensured the presence of Signals and Navigating Officers though having members from the Fleet Air Arm or Coastal Command on the staff was a more difficult proposition. In this, the problems the Navy faced in general when it came to integrating the air into its operations were largely mirrored. Commander Vivian Voelcker,[131] an Observer, completed his Tactical Course in March 1933 and then stayed on as member of the instructional staff.[132] Ion Tower, the Director from 1937-38 and a trained gunnery specialist, was also a qualified pilot and a member of the Royal Aeronautical Society. In this he was a singular exception and the practice of having air officers on the school's staff proved infrequent and confirms that, in this area as in so many other areas of aviation support, divided control meant loss of control. This was not all the fault of the Air

---

129 ADM 239/142, *C.B. 03016/39*, pp. 79-80.
130 ADM 186/78, *C.B. 3011*, p. 23 and Burrough Papers, IWM/DS/MISC/4.
131 Later Captain Vivian John Voelcker (1897–1962). Short Course, Cambridge University, 1919; Observer in *Eagle*, 1925-28 and *Furious*, 1928-29; Staff Course, 1929; Staff Officer Operations in *Osprey*, 1930-31; Observer in *Courageous*, 1931-32; Senior Officers' Technical Course, 1932 and 1935; Tactical Course, 1933; Tactical School, 1933-35; Commanding Officer, *Thruster*, 1935, HMS *Walpole*, 1935-36; and HMS *Crescent*, 1936; Staff Officer Operations in *Pembroke*, 1938-41 and retired, 1941. Recalled for war service. Acting Captain, 1943 and Captain, 1946.
132 ADM 196/120 (Voelcker).

Ministry for the Navy failed to assign an officer to the directing staff at Andover as the former had sought.[133] Both the Army and the Air Force from an early date made a practice of having officers on the directing staffs of their respective Staff Colleges with the Navy seemingly the odd man out. A root cause was that officers attached to the RAF in time reverted to general service duties within the Royal Navy. Until it established a career path for those in the naval aviation community that recognized their duties as the specialist and *permanent* work that it was, progress was bound to be slow and uneven. Yet, as true as that all remains, the Navy did attempt to ensure that from 1932 onwards officers with air experience did take the Tactical Course even if a corresponding representative was missing from the instructional staff.

In common with the other shore establishments, the Tactical School issued a yearly report highlighting the findings of its work. Though only one example is in extant—that for 1937—the partial contents of others are available. They appear as secondary references in other reports that still exist. Accordingly, in 1932 Rear Admiral George Chetwode,[134] commanding the First Cruiser Squadron, noted that a significant finding from exercise 'R.E.', a Mediterranean Fleet evolution conducted 2-3 September 1932, was the continuous threat that Blue Fleet faced in protecting his aircraft carrier from the surface forces of Red Fleet. The risks encountered were so severe that Blue Fleet was unable to carry out its strategic plan. More significantly, the Tactical School in one report drew attention that over the previous two years, no surface force had carried out an attack on an aircraft carrier in any game played on the table.[135] Clearly, results of the Tactical School were not definitive and proved that testing at sea required confirmation of any findings found.

Again, in 1932, the school examined how a Red Fleet desiring to give battle should operate against an enemy force markedly superior in gun-range. In this instance, the problem was:

If *Blue* opens fire on Red at (say) 30,000 yards, when *Red's* guns are outranged and cannot reply, it is obvious that *Red* must get through the danger zone as quickly as possible. If the tactical table is a correct guide, *Red's* manoeuvre to close to effective range must be a bold one, involving possibly the loss of "A" arcs and certainly

<hr>

133 Air Marshal P. B. Joubert de la Ferté speech to the Annual Reunion of the Air Force Staff College, 19 February 1937 cited in *The Hawk: The Journal of the Royal Air Force Staff College*, No. 10, December 1937, p. 85.

134 Later Admiral Sir George Knightley Chetwode (1877-1957). Captain, 1917; British Naval Mission to Greece, 1919-22; Deputy Director, Naval Intelligence Division, 1923-25; Flag Captain in *Queen Elizabeth*, 1926 and *Warspite*, 1927; Rear Admiral, 1928; Senior Officers' Technical Course, 1928; Senior Officers' War Course, 1928-29; Tactical Course, 1929 and 1932; Naval Secretary to First Lord, 1929-32; Commander, First Cruiser Squadron, 1932-33; Vice Admiral, 1933; Commander, Reserve Fleet, 1933-36; Admiral and retired, 1936.

135 ADM 186/152, *C.B. 1769/32(2): Exercises & Operations, 1932, Volume II.* Admiralty, Naval Staff, Tactical Division, July 1933, pp. 39-40.

considerable loss of bearing. The amount of bearing lost may well place *Red* in a position of disadvantage, and it appears that the correct tactical move by *Red* is to deploy on the opposite course and adopt form of battle "B". By such a manoeuvre he will force *Blue* either to accept the heavy concentration on his rear ships, or to turn inwards to support them, and so accede to *Red's* object of closing the range. *Blue* can only maintain the range by a similarly drastic turn away, which will lose him his "A" arcs also and so reduce *Blue's* advantage.[136]

With the above scenario and the Tactical School's recommendations in hand the Mediterranean Fleet investigated the possibilities at sea. Exercise 'R.R' held on 26 April 1933 pitted a British fleet having an advantage in speed and with one more heavy ship but limited to a gun-range of 23,800 yards against a Blue force enjoying gun-ranges of 32,000 yards. The at-sea demonstration saw the Red fleet lose one of its battleships (*Revenge*) whilst attempting to close the range and concluded that unless Red enjoyed a marked superiority in speed, or visibility was limited, smoke had to be used to forestall the serious losses awaiting the Red fleet.[137]

The following year saw the combined Mediterranean and Home Fleets test the concept of closing the range and attempting Form of Battle 'B', a pattern of deploying the battlefleet on a parallel and opposite course to the enemy fleet, and first raised by the Tactical School.[138] This expanded exercise, 'Z.J.' conducted on 15 March 1934 confirmed that Red's battlefleet would face hard going in any engagement where Blue enjoyed such a pronounced advantage in gun-range. She lost two heavy ships outright—*Revenge* and *Royal Sovereign*—whilst Blue lost only the battleship *Rodney*.[139]

The problem of an out-ranged British battlefleet was particularly vexing in view of the Washington Naval Treaty and its follow-on provisions as codified in the London Naval Agreement where one-upmanship and opting for increasingly larger capital ships were supposedly suspended. In truth, these agreements merely directed navies to seek qualitative advantage where previously they had sought both qualitative and quantitative advantages. One key advantage sought was to out-range an opposing navy by modifying the maximum elevations achievable in naval rifles. The Americans were pursuing such changes and arguing that such was not precluded by the terms of

---

136 British Sources, Box 12, Naval Historical Center, Washington, DC, USA, *C.B. 3016/32: Progress in Tactics*, 1932, pp. 43-44. 'A' arcs were the relative bearings that allowed a ship's main armament to deliver the greatest weight of fire without obstruction. The Royal Navy defined several variations of deployment that could be executed rapidly with minimal communications. Form of Battle 'B' was an encounter fought as a circling action when the British fleet was deemed to be the superior force.

137 ADM 186/154, *C.B.1769/33(2), Exercises & Operations, 1933, Volume II*. Admiralty, Naval Staff, Tactical Division, April 1934, p. 9.

138 ADM 186/106, *C.B. 01821, Battle Instructions*, p. 2.

139 ADM 186/155, *C.B. 1769/34(1), Exercises & Operations, 1934, Volume I*. Admiralty, Naval Staff, Tactical Division, November 1934, pp. 19-23.

the original agreement. Further justification in their eyes for the decision to modify their battleships was the cruiser programme that the British were soon pursuing after the agreement. Though cruiser totals resided outside the scope of the Washington Naval Agreement, Arthur Balfour had said in 1922 that the Washington Naval Agreement had established a new status quo between the contracting parties in all types of warships.[140] If his claim was true (and as the senior Empire negotiator during the talks his musings could not be dismissed as irrelevant), then the British were, at a minimum, demonstrating bad faith in implementing the recently concluded agreement. The British thought otherwise, but not to the point of abrogating the treaty and took steps to respond in kind. These, however, were not simple measures to implement. In the interim, recourse to any tactical palliatives available in an attempt to negate their new vulnerability was one expedient pursued. Here, the Tactical School provided an ace card, and whilst adjusting the 15-inch guns of British heavy ships from 20° to 30° maximum elevation and adopting shells with a 6-calibre radius head (crh) rather the previous 4-crh would allow ranges of over 32,000 yards to be achieved, all this took much time and money.[141]

Whether the American charges of bad faith were true or not, the claim had been registered and a degree of caution now entered into any British calculation about the nature and scale of any new changes proposed potentially touching upon her treaty obligations. When in 1923 the Air Ministry suggested modifying merchant ships to operate aircraft in wartime to support future convoy operations, Pound as the Director of Plans, noted the previous American claims of bad faith. Writing against the proposal, he advised it would make it that much easier for Congress to authorize funds for the pending battleship modifications in gun elevations to proceed.[142] For both navies, then, claims of bad faith proved something of a running sore. Having established a strategic balance, tactical proficiency and securing a nominal advantage became increasingly important. For just as the competition in naval building soon migrated to those classes not covered by the tonnage ratios of the Washington and London Naval Agreements, so, too, did competition at the tactical level assume a greater remit. Consequently, the Tactical School, where the war game was central, became the linchpin of the Royal Navy's efforts to maintain its competitive edge. Its founding and the concurrent upgrade of the Tactical Section to a Division within the Naval Staff were two measures adopted in an effort to retain British sea supremacy through enhanced tactical prowess though formally conceding parity in numbers.

The war game played on the table was a model; it was not reality and only representative of what should occur. Perhaps its weakest feature was that it was limited by

140 W. N. Medlicott *et al*, eds., *Documents on British Foreign Policy 1919–1939: European and Security Questions 1927–8* (London: Her Majesty's Stationery Office, 1971), pp. 463–64.
141 V. E. Tarrant, *Battleship Warspite* (Annapolis: Naval Institute Press, 1990), p. 63.
142 ADM 1/8687/178, Director of Plans minute of 28 June 1923 to Air Ministry letter 428933/23/S.9.A of 20 May 1923.

its rules, assumptions, and tabulated results and was designed to provide a reasoned result. It could not account for the inspired genius of a leader such as Admiral Nelson who could fathom a possibility where others could not.[143] The same applies, too, to the exercises of the fleet which were scripted to operate over a limited time and area, had restrictions imposed either to foster safety or economy, and were run without recourse to the logistical constraints, such as in ammunition expenditure, that a fleet would face in war. Still, exercises at sea beyond testing tactical precepts fostered unit proficiency and highlighted factors such as fatigue, stress, and equipment failures that the war game played on the table was less able to demonstrate.[144] Both, then, were complimentary to each other, had their strengths and weaknesses and were useful within their limits. Moreover, the danger was ever present that games played repeatedly at the tactical table and reaching much the same results offered a distorted picture of what actually might occur if the assumptions and rules underlying the exercises were false.

That the assumptions and rules could paint a false picture is clearly shown by the events of 1931 when the battleship *Marlborough*, now surplus and available for disposal, was used in a test to measure the effects of aerial bombing. The rules as specified in *C.B. 3011* indicated that her deck armour should withstand penetration by the 500-lbs semi-armour piercing bombs used by the RAF. Yet, of the seven direct hits received, two penetrated her armour with one passing right through the ship and a second ending up in her engineering spaces.[145] The rules were equally optimistic on the presumed ability of British heavy ships to withstand torpedo attack or enemy shellfire. Six torpedo hits were thought necessary to sink or totally disable a battleship,[146] whilst a battle cruiser would need to receive fifteen hits of 15-inch shellfire before considered lost.[147] Hindsight being perfect, the reader may appreciate that the test of war demonstrated otherwise.[148] Still, the assumptions made and the tactics that followed were

143 Grenfell, *Art of the Admiral*, pp. 24-25.
144 Less able but not unable. By 1932, the Tactical School was including morale as a factor affecting the effectiveness of a fleet's gunnery performance by having each side cut for high or low morale before the beginning of play. Though neither opponent knew the results of the draw, as they remained with the game's umpire, gunnery results were assessed based on the morale factor drawn. See *C.B. 3016/31: Progress in Tactics, 1931*, Admiralty, Naval Staff, Tactical Division, August, 1932, p. 40, British Sources, Box 12, Naval Historical Center.
145 Geoffrey Till, 'The Impact of Airpower on the Royal Navy in the 1920s', Unpublished Ph. D. Dissertation, University of London, March 1976, p. 106.
146 ADM 1/9007/79, Director of Tactical School to Secretary of the Admiralty letter No. 862/25 of 8 May 1934.
147 ADM 186/78, *C.B. 3011*, Table 3, p. 38.
148 *Barham* was sunk by three torpedoes on 25 November 1941 whilst *Royal Oak* was lost after being hit by four torpedoes on 14 October 1939. *Hood* suffered a cataclysmic explosion on 24 May 1941 during her engagement with *Bismarck* and *Prinz Eugen*. Though the exact cause of her sinking is unknown, it was not due to receiving fifteen hits from *Bismarck*'s main armament.

based on the best estimates of what was to be expected. If the means adopted proved insufficient to the task, then the tactics recommended by the Tactical School may be faulted. Yet, the greater fault would have been not to contemplate the problem in the first instance.

In this effort, *Naval Tactical Notes, Part II*, was critical in assisting the commander in formulating the appropriate response when facing an enemy battlefleet.[149] This confidential addendum to the unclassified publication *O.U. 6183* specified the anticipated vulnerability for British heavy ships against a defined foreign counterpart based on estimates of inclination, range, and the probable zones of immunity enjoyed. Corresponding summaries described the supposed armour efficiency of Japanese and American heavy ships against differing British shellfire. In short, *C.B. 01847A* allowed the British commander to tailor his tactics during the approach phase of a naval action to maximize his relative strengths against the presumed vulnerabilities of the enemy fleet. Unfortunately, given the differing fighting characteristics of British heavy ships no solution was ever ideal.

For the Admiralty, the primary role of the Tactical School was to have officers assimilate the precepts of the *Battle Instructions* and from 1939, the *Fighting Instructions*.[150] The Tactical School proposed the suggested change in style in 1937 and evoking the spirit of those *Instructions* employed in Nelson's day and making a break with the *Battle Orders* of the World War was the likely objective.[151] Captain Terence Back,[152] an innovative torpedo officer, was especially commended for his efforts in preparing the new edition as was Lionel Wells, the onetime Director, highlighting that whilst the *Instructions* remained the keystone of fleet doctrine, their evolution now owed much to the Tactical School.[153]

However styled, the *Instructions* were primarily written to cover naval battle on the largest scale.[154] Training officers at the Tactical School to that end ensured that they were prepared to deal with any worse-case scenario that they might have to face. This was not necessarily the most likely event and shifts in the perception of the most likely foe influenced the instruction received. By the late 1930s, the proximity

149 *C.B. 01847A. Naval Tactical Notes, Part II.* This publication is not on file at the National Archives. Its existence is confirmed though as extracts of its contents are present in ADM 1/9387.
150 Admiral Sir Geoffrey Miles letter to Captain Stephen Roskill of 30 August 1964, Roskill Papers, ROSK/ 7/163.
151 ADM 116/6364, Annual Report of the Tactical School, 1937.
152 Later Captain Terence Hugh Back (1895-1968). Staff Course, 1931; Chemical Defence Course, 1932; Tactical Division, 1932-34; Commanding Officer, *Lupin*, 1934-35; Tactical School, 1936-38; Captain, 1938; Special Staff Officer in *Warspite*, 1938; Tactical Division, 1939; Senior Officers' Technical Course, 1939; Flag Captain and Chief Staff Officer in *Capetown*, 1939-40; Commanding Officer, *Bermuda*, 1942-44 and retired, 1946.
153 ADM 196/117 (Back) and ADM 196/49 (Wells).
154 ADM 186/106, *Battle Instructions*, Mediterranean Fleet covering memorandum of 15 June 1930.

of the Japanese threat receded in the presence of a rearming and increasingly belligerent Germany close to Britain proper. She did not yet possess a battlefeet and if the Anglo-German Naval Agreement, negotiated in June 1935, held would not ever have one of a scale to threaten Britain.[155] A cruiser war against British trade was thought the more likely risk and officers attending their Tactical Course exercised against this threat. One manner of dealing with the new pocket-battleships entering the German Navy in an uneven encounter was for British cruisers to close and turn in succession allowing them to approach within range but still offering only the narrowest of targets themselves. The combination of fire and movement allowed the British cruisers to open their 'A' arcs to fire and then resume the chase between salvoes against an adversary that would enjoy a marked superiority in gun-power.[156] Whether Terence Back and Lieutenant Commander Gilbert Roberts[157] developed this corkscrew tactic as Mark Williams claims is unknown, but the above scenario and the engagement of a German battle cruiser by a flotilla of British destroyers were being gamed at the table by 1936.[158] Moreover, in early 1939 the Tactical School was examining how two light cruisers and one heavy cruiser could best operate against a German Group centred on a pocket battleship with no aircraft present on either side. The conclusions of the school were that a *Deutschland*-class ship could be defeated for the probable loss of one British cruiser.[159]

Similarly, the Tactical School developed a game by the late 1930s with Italy as the possible foe where only light forces engaged each other and the battlefleets remained off-stage. The purpose of the game was to analyze which tactics offered the best scope for achieving success in minor actions where only cruisers and destroyers were present.[160] The Tactical School, then, while instilling the fundamentals of the *Battle Instructions* and *Fighting Instructions* came to modify its teaching to reflect the threats most likely to be faced and did not limit its work merely to the letter of Admiralty tasking. That said, by focusing the curriculum of the Tactical School around the *Battle Instructions*, the Admiralty emphasized fleet action to the neglect of examining many of the subsidiary tasks that now featured in naval operations. With hindsight, it is easy to find fault in this emphasis, but in an introductory course lasting usually eight weeks, this approach is understandable. Speaking of a different problem in the postwar stra-

---

155 The agreement, itself, did not come into force until 4 November 1937 upon its ratification by both countries. See ADM 182/76, Admiralty Fleet Order '2580.—Anglo German Naval Agreement (1937)' of 2 December 1937.
156 Miles letter to Roskill of 30 August 1964, Roskill Papers, ROSK/7/163.
157 Later Captain Gilbert Howland Roberts (1900-1986). Short Course, Cambridge University, 1920; to RAN, 1928-30; Senior Officers' Technical Course, 1935; Tactical School, 1936-37; Commanding Officer, HMS *Fearless*, 1937-38 and retired, 1938. Recalled for war service; Acting Captain, 1942; Captain, 1945 and retired, 1956.
158 Williams, *Gilbert Roberts*, p. 72.
159 ADM 239/142, *C.B. 03016/39*, pp. 45-46.
160 ADM 116/6364, Annual Report of the Tactical School, 1937.

tegic environment, Marshal of the Royal Air Force John Slessor once said, 'The dog that takes care of the cat can take care of the kittens too.' Perhaps a corresponding view informed the utility of the Tactical School with the Navy believing that if it could prevail in a fleet action, then it could prevail in any other naval operation. This was, after all, the view held by some at the Tactical School including Lionel Wells.[161]

When initially formed, the course lasted two months with approximately ten officers attending.[162] By 1937, the course had become one of eleven weeks.[163] In common with the Staff College, the Tactical School forwarded recommendations and findings to the Admiralty with emphasis on the doctrinal tactical publications used within the Service. In 1934, Captain Troup noted that the fleet now operating the greater number of aircraft enjoyed an advantage over its adversary. Such an advantage should allow the first blow to be delivered against the battlefleet when not in gun-range and was in keeping with the primary objective of the *Battle Instructions*.[164] That the aim of the Fleet Air Arm was to strike at the enemy's battlefleet and not his carrier force, if present, indicates the power of doctrine and how difficult it is to move beyond its precepts. For all their testing and analysis and given their understanding of the respective fighting ranges available to heavy ships and carriers, the idea that destroying the enemy battlefleet remained the primary object in naval battle if the first encounter occurred over-the-horizon and outside of gun-ranges, is evidence that progress in tactics was extremely hard even when aided by a Tactical School. If sometimes the school reached an unsound conclusion, then criticism must be tempered in the knowledge that they were groping to find the correct method that would apply for most situations. The scientist in his laboratory is thought a genius when after much trial and error he arrives at a new discovery. For the naval officer facing battle, his decisions are final and the results achieved do not admit of recasting.

Amongst the first group of officers to attend the course was Rear Admiral Vernon Haggard, who had been in the same War Course as Frederic Dreyer, and Commander Henry Blagrove.[165] Blagrove would return twice more to take the Tactical Course. An officer of ability and highly regarded by his superiors, he was lost in the battleship *Royal Oak* in 1939 when she was torpedoed whilst at anchor in Scapa Flow. Meanwhile,

161 Williams, *Gilbert Roberts*, p. 71.
162 ADM 182/83, Confidential Admiralty Fleet Order '3168a.—Tactical Courses at Portsmouth for Senior Officers' issued 28 November 1924.
163 ADM 116/6364, Tactical School, Annual Report, 1937.
164 ADM 1/9007/79, Director of Tactical School to Secretary of the Admiralty letter No. 862/25 of 8 May 1934.
165 Later Rear Admiral Henry Evelyn Charles Blagrove (1887-1939). Commanding Officer, *Ceres*, 1924-25; Tactical Course, 1925, 1928 and 1932; Captain, 1927; Senior Officers' Technical Course, 1928 and 1932; Senior Officers' War Course, 1928; Flag Captain and Chief of Staff in *Norfolk*, 1932-34; Naval Assistant, Second Sea Lord, 1934-37; Commanding Officer, HMS *Sussex*, 1937-39; Rear Admiral, 1939; Rear Admiral, Second Battle Squadron in *Royal Oak* and lost with ship, 1939.

preceding Harold Collins to the Tactical School and one of the first Air Force officers attending was Group Captain Albert Daly, also from Coastal Command in 1935-36.

Each of the officers selected to oversee the Tactical School between 1925-39 eventually reached flag rank, perhaps the ablest refutation of the Shavian adage that, 'Those who can – do; those who can't – teach.' If he had not been killed in a bombing raid in 1940 whilst serving as a liaison officer to General Sir Alan Brooke, General Officer Commanding Home Forces, Rear Admiral Ion Tower would likely have seen another promotion. As it was, he remained the only Director not to achieve the rank of at least Vice Admiral. Of the ten officers to head the school, seven specialized in gunnery (e.g., Usborne, Colvin, Bailey, Binney, Patterson,[166] Wells, and Tower), two were Navigating Officers (i.e., Troup and Miles[167]) with Captain Francis Barry[168] proving the lone torpedo specialist. Though each was decorated with the War Medal for service in the Great War, Binney, Miles, Troup, Wells and Barry failed to see action at Jutland, emphasizing that for all its supposed centrality to naval education, during the period between the wars more important was it to fill a posting with the best officer presently available.

The appointment of Usborne as the Tactical School's first Director represented the marriage of opportunity and ability in his selection and was probably due to Dreyer's influence.[169] Highly capable and possessed of an inventive mind, if the device ultimately becoming the paravane (an underwater appliance streamed by vessels to cut

166 Later Rear Admiral Julian Francis Chichester Patterson (1884-1972). Five firsts in examinations for Lieutenant; War Staff Course, 1913-14; Naval Ordnance Department, 1917-18; Squadron Gunnery Officer in *Barham*, 1918-20; Captain, 1921; Assistant Director, Naval Ordnance Department, 1922-24; Senior Officers' Technical Course, 1924 and 1928; Senior Officers' War Course, 1924-25 and 1927-28; to RAN, 1925-27; Tactical Course, 1927; Director, Naval Ordnance Department, 1928-31; Flag Captain, *Hood*, 1931-32; Director, Tactical School, 1932-33; retired and Rear Admiral, 1933.

167 Admiral Sir Geoffrey John Audley Miles (1890-1986). Staff Course, 1919-20; War Staff Officer in *Coventry*, 1920-22; Squadron Navigating Officer in *Coventry*, 1922-24; Tactical Course, 1925 and 1933; Operations Duties in *Iron Duke*, 1925-26, in *Barham*, 1926-27; Navigating Officer and Staff Officer Operations in *Hood*, 1927-29; Royal Naval Staff College, 1929; Plans Division, 1929-31; Captain, 1931; Commanding Officer, HMS *Pangbourne*, 1931-32; Deputy Director, Royal Naval Staff College, 1933-34 and 1934-35; Director, Royal Naval Staff College, 1934; Captain (D), Third Flotilla, in HMS *Codrington*, 1935-37; Deputy Director, Tactical School, 1937-38; Director, Tactical School, 1938-39; Flag Captain in *Nelson*, 1939; Rear Admiral, 1941; Vice Admiral, 1944; Commander-in-Chief, Royal Indian Navy, 1946; Admiral and retired, 1948.

168 Later Rear Admiral Francis Reginald Barry (1887- ?). Torpedo and Mining Department, 1921-23; Squadron Torpedo Officer in *Hood*, 1923-25; Captain, 1928; Commanding Officer, HMS *Valentine*, 1929-30; Short Anti-Submarine Course, 1929; Senior Officers' War Course, 1930; Tactical Course, 1931 and 1935; Commanding Officer, *Carlisle*, 1931-32 and as Flag Captain and Chief Staff Officer, 1933; Assistant Director, Tactical School, 1935-37; Director, Tactical School, 1937; Commanding Officer, HMS *Dorsetshire*, 1937-39; Rear Admiral and retired, 1940. Recalled for war service.

169 Dreyer, *Sea Heritage*, p. 279.

tethered mines from their moorings) was not of his design, he could still lay claim to a large measure of the credit for the direction pursued in finding the solution adopted.[170] Before then and his time in the battleship *Colossus*, Usborne had worked in the Naval Ordnance Department where he was partially responsible for overseeing the development of the new range of fire control instruments finding their way into the Service before the World War. In this capacity, he had worked with Dreyer teaming with him to prepare a portion of an Admiralty report reviewing the history of Arthur Pollen's fire control efforts and the merits of that inventor's approach to solving this pressing problem.[171] Usborne's efforts in fire control were, however, not simply limited to analyzing the work of others, as he invented a useful fall-of-shot indicator adopted by the Navy. This appliance proved itself superior to Commander Patrick Macnamara's time-of-flight watches and, in time, came to equip many vessels in the fleet.[172]

Usborne would return to the Naval Ordnance Department after the Great War to serve as its Deputy Director and then moved to the Gunnery Division of the Admiralty before completing his War Course in 1923. Following his attendance of the SOWC, where it will be recalled he presented an essay titled appropriately, 'Naval Tactics, 1924', Usborne became the Navy's senior representative to the Chemical Warfare Committee, a joint body with members from the Army and Air Force charged with overseeing all British technical efforts within that field.[173] Before the adoption of the Geneva Protocol of 1925 committing Britain to a policy of no first use, chemical warfare assumed a greater prominence in her military strategy.[174] Usborne's service on the committee was not the professional backwater that it tended to become later but represented the cusp of current thinking and development. Thus, by the time he arrived at Portsmouth in February 1925 to head the Tactical School, Usborne had an established record of working and solving some of the most intractable technical and tactical problems facing the Service over the previous fifteen years whether coming from the air, the surface, or below.[175] Moreover, experience of chemical warfare matters became a foundational element in the officers serving as Director as they frequently attended the Army School at Porton Down immediately before talking up their appointment.

One of the first officers to join Usborne at the Tactical School and serving as his deputy was Troup who came from the Admiralty where he had headed the small and nascent Tactical Section.[176] Troup would return to the school in 1933 to serve as its Director balancing his responsibilities then with writing a practical guide to seaman-

170 Admiral Viscount Jellicoe, *The Grand Fleet, 1914-16: Its Creation, Development and Work* (London: Cassell and Company, Ltd., 1919), pp. 60-61.
171 Sumida, *Defence of Naval Supremacy*, p. 238.
172 Dreyer, *Sea Heritage*, p. 279.
173 *Navy List*, April 1924, p. 260
174 ADM 186/117, *C.B. 3042, Manual of Combined Operations, 1938*, p. xii.
175 *Navy List*, July 1925, p. 282A.
176 *Navy List*, April 1924, p. 260.

ship titled *On the Bridge*. This work was well received, and though written before the advent of radar and echo sounding, was, nevertheless, reissued in subsequent editions containing as it did much sound advice from an excellent Navigating Officer and was even quoted at length in the Admiralty's own *Manual of Navigation*.[177]

It was during Troup's watch and that of Wells when the Battle of Jutland achieved its greatest prominence in the course's curriculum. This was partly because of their innate interest in the battle and the steady refinements made in perfecting the demonstrations occurring year on year. Indeed, to Troup was the credit largely given for the place and style of instruction at the table in the school's early days.[178] Yet, perhaps just as importantly, it owed a great deal to timing. Principals such as Jellicoe, Beatty, and Madden were now passing from the stage, and though others continued to visit the school and share their thoughts on the battle, an era had passed. However, if Jutland had its greatest resonance during the regimes of Troup and Wells, then it is also the case that Troup continued the practice of attending the yearly combined exercises of the Home and Mediterranean Fleets doing so in both 1934 and 1935.[179] Binney appears to have been the first Director to adopt the practice and like him, such visits allowed Troup to see events at first hand whilst also appraising seagoing officers of what the school was seeing in its games.[180]

Rounding out the school's initial staff were Commanders Frederic Schurr,[181] a much-experienced destroyer officer who had served in the embryonic Tactical Section of the Training and Staff Duties Division, and Herbert Spreckley,[182] a qualified Navigating Officer who would soon revert to general service duties.[183] Meanwhile, Commander William Wake-Walker,[184] a Torpedo Officer, and Lieutenant Commander Angus Cunninghame-Graham, a former Fleet Signals Officer in the *Iron Duke*, joined the

---

177 Eeyore Smith, "'On the Bridge'", *The Naval Review*, Vol. XXXIX, No. 3, August 1951, p. 329.
178 ADM 196/91 (Troup).
179 ADM 196/47 (Troup).
180 ADM 196/48 (Binney).
181 Later Captain Frederic Graham Schurr (1886–1964). Five firsts in examinations for Lieutenant; Tactical Section, Training and Staff Duties Division, 1920-21; Tactical Section, 1921-22 and 1924-25; Tactical School, 1925-26; Torpedo and Mining Department, 1926-32; Captain and retired, 1932. Recalled for war service.
182 Captain Herbert Malcolm Spreckley (1887-1974). Navigating Officer in *Malaya*, 1920-21; Tactical Section, 1922-25; Tactical School, 1925-26; retired, 1930 and Captain, 1932. Recalled for war service.
183 ADM 196/127 (Spreckley).
184 Later Vice Admiral William Frederic Wake-Walker (1888–1945). Six firsts in examinations for Lieutenant; Squadron Torpedo Officer in *Coventry*, 1920-21; Staff Course, 1921-22; Operations Division, 1922-24; Tactical Section, 1924-25; Tactical School, 1925; Captain, 1927; Senior Officers' War Course, 1928 and 1939; Commanding Officer, *Castor*, 1929-30; Training and Staff Duties Division, 1930-32; Commanding Officer, *Dragon*, 1932-35; Director, Torpedo and Mining Department, 1935-38; Commanding Officer, *Revenge*, 1938; Rear Admiral, 1939 and Vice Admiral, 1942.

staff from the Tactical Section. Both were recognized as brilliant officers within their spheres of specialization. Indeed, Wake-Walker, in association with Thursfield, had developed a virtual plotting indicator in 1910 to be followed by a Torpedo Attack Trainer in 1915. It was Wake-Walker who had perfected the performance of the 'W' torpedo—a torpedo that ran in a weaving pattern making avoidance by surface ships extremely difficult. This act of inventiveness earned him a prize of 150 guineas from the Admiralty in 1922. All in all, he was seen as one of the cleverest torpedo officers in the Service and approached his arm's equivalent to Frederic Dreyer.[185] Yet, in a move that Richmond found annoying at the War College, Wake-Walker had soon left the school being replaced by another strong player, Commander Algernon Willis.[186]

As can be seen, from the beginning the school was composed of a balance of officers with differing technical skills and backgrounds and this feature was only enhanced as the staff grew in size over the rest of the period. Still, creation of the school, if an investment in the future, came at some immediate expense to the Naval Staff, as Troup, Spreckley, Wake-Walker, and Cunninghame-Graham, the only four officers of the Tactical Section, now found themselves in Portsmouth and racing with the others to put a course together.[187] Time was not a friend as the first sitting began on 7 March 1925 and ending 1 May. Though Schurr had previously served in the Training and Staff Duties Division, only Wake-Walker of the first officers assigned to the school as instructors was a qualified or designated staff officer. Hence, most had not received the instruction in educational theory provided to qualifiers of the Staff Course before arriving at Portsmouth and a fair conclusion is that professional acumen was the essential requirement with any thought of their inclination to teaching a distant second.[188]

The alacrity with which Beatty acted on Dreyer's suggestion for such a school meant that those appointed from the Tactical Section would, of consequence, be brief as a return to sea was necessary. This proved to be the case and after only eighteen months, most had departed. Troup bound to command the cruiser *Cairo* on the America and

185 ADM 196/144 (Wake-Walker) and ADM 196/127 (Wake-Walker).
186 Later Admiral of the Fleet Sir Algernon Usborne Willis (1889–1976). Six firsts in examinations for Lieutenant; War Staff Course, 1922-23; Squadron Torpedo Officer in *Coventry*, 1923-25; Tactical School, 1925-27; Commanding Officer, *Warwick*, 1927-29; Captain, 1929; Senior Officers' War Course, 1929-30; Royal Naval War College, 1930-32; Senior Officers' Technical Course, 1932; Tactical Course, 1932; Flag Captain, *Kent*, 1932-34 and *Nelson*, 1934-35; Commander, *Vernon*, 1935-38; Flag Captain and Chief Staff Officer in *Barham*, 1938-39; Chief of Staff to Commander-in- Chief, Mediterranean, 1939-41; Rear Admiral, 1940; Commander-in-Chief, Africa, 1941; Commander-in-Chief, South Atlantic, 1941-42; Vice Admiral, 1943; Second Sea Lord, 1944-45; Admiral, 1945; Commander-in-Chief, Mediterranean, 1946-48; Commander-in-Chief, Portsmouth, 1948-50; Admiral of the Fleet, 1949 and retired, 1950.
187 *Navy List*, April 1924, p. 260 and *Navy List*, July 1925, p. 282A.
188 Such courses were taught by Commander Arthur Dowding in 1920-21 and Captain Vernon Haggard in 1923, both of the Training and Staff Duties Division.

West Indies Station, whilst Spreckley went to the *Frobisher* on the China Station and Cunninghame-Graham to the *Royal Sovereign* on the Mediterranean Station.[189] Usborne was more fortunate in this regard, directing the school for two years before leaving for the Mediterranean Station to command *Malaya* of the First Battle Squadron.[190]

Succeeding Usborne as Director was Captain Ragnar Colvin in February 1927. Tall, of a quick wit, sure mind and proving himself a gifted administrator and leader, Colvin saw action at Jutland and following the war commanded the cruiser *Caradoc* in the Mediterranean Fleet. This was a period of near constant turbulence as Britain assisted anti-Soviet forces in the Black Sea area whilst concurrently attempting to hold Turkey to the terms of armistice and guaranteeing the free passage of the straits. Consequently, Colvin had considerable experience of not only fleet action, but also the lesser actions that British naval forces often faced.

Chatfield once described Captain Sidney Bailey as one of the ablest Gunnery Officers in the fleet and given that Chatfield was an officer of very pronounced views on the role of the staff this was praise indeed.[191] Before the war and as a Lieutenant, Bailey had taken a lead in working out the best means of defending the fleet against torpedo attack by flotilla forces at night.[192] Bailey came to the school from command of the *Renown* where he had secured high praise from Dreyer, then commanding the Battle Cruiser Squadron, for his skill as a lecturer and the care taken in training subordinate officers.[193] Previously, when commanding destroyers in the Atlantic Fleet Bailey had been at the fore in investigating destroyer tactics where his gunnery knowledge was especially prized.[194] Still, it did not hurt that Bailey was also an officer keen on his sports and sociable. A staff serving afloat was a breed apart—more *of* the ship than *in* the ship. Doubtless, these qualities in Bailey were valued by others, as much as his known gunnery skills, but command of the Tactical School and later of Greenwich College testify that he was also a serious naval thinker.[195]

Following Bailey at the school was Captain Hugh Binney, but not before immediately completing his own sitting of the Tactical Course for a second time. In this, he left his War Course before completing it though his performance had in all respects been held superior until that point. That Binney should leave his War Course to attend a course which he had already sat previously offers the strongest evidence that such training formed part of the routine turnover conducted for not only the Tactical School but the other venues of training.[196] Though not a graduate of the Staff College,

189 *Navy List* (London: His Majesty's Stationery Office, October 1927), pp. 220, 240, and 265a.
190 *Ibid*, p. 253.
191 Chatfield, *Navy and Defence*, p. 137.
192 ADM 196/142 (Bailey).
193 ADM 196/47 (Bailey).
194 ADM 196/91 (Bailey).
195 Chatfield, *Navy and Defence*, p. 242.
196 ADM 196/91 (Binney).

Binney, whilst serving as the Naval Assistant to the ACNS, visited Greenwich and lectured to its qualifiers on the question of employing 'Covering Fire in Combined Operations'.[197] Having served in *Queen Elizabeth* during her support to allied forces in the Dardanelles campaign, it was a subject rich in experience for him. Binney also has the distinction of being the only Director of the school to return subsequently as one of its students. Returning in 1936 as a flag officer, Binney completed the course for a third time in preparation for assuming command of the First Battle Squadron. Indeed, Binney's career was marked by success following upon success and his time as Director of the Tactical School only added to his laurels.[198]

Relieving Binney was Julian Patterson who had an abbreviated tenure owing to his perceived failure to take a strong hand during the mutiny at Invergordon.[199] Patterson's appointment only ran from August 1932 until May 1933 when relieved by James Troup. Patterson, too, continued the example of attending the combined fleet exercises visiting the fleets in Gibraltar in March 1933. Patterson was thought by one senior officer to be the best Gunnery Officer in the Service and was respected by seniors and juniors alike, accordingly.[200] This had been in 1920 and the years ahead would bring challenges where technical proficiency was not the only skill required. Now, the handwriting was on the wall and, doubtlessly, knowing he would be relieved for cause because of Invergordon, he requested to be placed on half pay before retiring in October 1933.[201]

Troup, in turn, was relieved by Rear Admiral Lionel Wells in 1935. Wells, already a product of the Tactical School having sat the course in 1930, preceded his assumption of command by taking the course yet a second time. William Fisher in assessing Wells period in charge of the Tactical School believed his appointment had been a wise and sagacious one. A prior instructor at the War Course, he possessed firmness, yet tact, and was able to keep control of the lectures where debates could often become heated and emotional.[202] This was no small feat as Wells, himself, could display flashes of a hot temper. Now, though, he was warmly praised by many senior officers passing through the course for the excellent lines upon which the course was being run.[203]

Relieving Wells in March 1937 was Captain Francis Barry who had been serving as Assistant Director since April 1935. Barry was the sole Torpedo Officer to serve as Director and was to hold the briefest of the interwar appointments as Director.[204] A good lecturer with a sound grasp of subject, Barry spoke with an unpleasant accent that put strangers off. The source of this defect was thought a degree of nervousness

197 ADM 203/100.
198 ADM 196/91 (Binney).
199 ADM 196/91 (Patterson).
200 ADM 196/126 (Patterson).
201 ADM 196/91 (Patterson).
202 ADM 196/91 (Wells).
203 *Ibid.*
204 ADM 196/51 (Barry).

when addressing classes.[205] As Barry had been in the school for two years already, his tenure was but a stopgap until Ion Tower assumed responsibility in July 1937.[206]

Geoffrey Miles was the second navigation specialist to be appointed to oversee the school becoming Director in 1938. Miles had served for a time as both Navigating Officer and Staff Officer Operations in the *Hood* when Dreyer commanded the Battle Cruiser Squadron in the late 1920s where Dreyer thought him an excellent officer in both capacities.[207] During his time in the *Hood*, Miles had developed many of the tactical and strategical problems being investigated at the time. He had a quick brain, was good on paper and was excellent at plotting. A qualifier in the first postwar Staff Course, Miles also sat the first session of the Tactical Course in March 1925 securing perhaps a naval version of a double-first.[208] He returned to the Staff College in 1921 whilst serving as a War Staff Officer in the cruiser *Coventry*, then flying the flag of Rear Admiral Destroyer Flotillas, to lecture on 'Destroyer Tactics in the Atlantic Fleet'.[209] Finally, mentioned should made of Gerald Muirhead-Gould, a Signals Officer with a strong intelligence background who was slated to relieve Miles as Director on 1 September 1939 before events dictated otherwise.[210] Muirhead-Gould's selection as Miles' replacement probably owed something to the latter as he served in his flotilla in the Mediterranean in 1932 and was believed a very good Divisional Leader in the flotilla.[211]

Officers were introduced to the war game and its nuances in all its varieties as part of their instruction. Games of varying complexity from the very basic to the strategic set-piece were played on a large, squared-table to the typical scale of 1-inch to a cable, or 202 yards. Ships were depicted by small, scale models and the performance of weapons was as confirmed during fleet exercises.[212] Over time, modifications were made to the physical layout of the game. Using special optics and lighting, players secured an appreciation of the fleet commander's likely view during the approach phase of the encounter while factors of increasing complexity were added to the rules to ensure as accurate play as possible.[213] That said, chance was ever an element in the game's play and the range of results possible guaranteed that what followed was not a prediction of a specific outcome, but, rather, an estimate of what was most apt to occur.

205 AND 196/92 (Barry).
206 ADM 196/92 (Towers).
207 Dreyer, *Sea Heritage*, p.286.
208 ADM 196/52 (Miles).
209 ADM 203/100.
210 ADM 196/52 (Muirhead-Gould).
211 ADM 196/92 (Muirhead-Gould).
212 Anon., 'Some Notes on the Early Days of the Royal Naval War College', *The Naval Review*, Vol. XIX, No. 2, May 1931, p. 240.
213 Narrative, Volume VI, p. 21, Haggard Papers, IWM/85/21/2.

In 1930, attendance at the War Course and the Senior Officers' Technical Course ceased to be voluntary and hereafter officers were 'appointed to take the course as the exigencies of the Service permit.'[214] Though the order is silent on the matter, presumably this applied to the Tactical Course as well. That said, even by 1939—nine years after mandating officer attendance—not all executive officers holding their captaincy had attended even one of the courses. The question remains if the courses were required for officers of their rank, what explanation exists for why officers failed to attend? Appendix V offers evidence as to the specific officers, as it were, 'missing in action' but a word on the matter is appropriate at this point. Here, two answers suggest themselves. First, the drafting margin was now so narrow for senior officers that after filling operational assignments, few officers of the necessary type actually remained to attend the courses. Hence, officers were completing two or even three classes of one or both courses to make up the numbers. The more likely explanation, though, is that while regulation might specify that attendance was mandatory, the Navy did not consider attendance an absolute bar before assuming command afloat. Class numbers were deliberately kept small with the ratio of staff to students nearly at times one to one, but, more frequently, one to two.[215] The low ratio of students to instructors for the Tactical Course can be contrasted with the ratios existing for both the Staff and War Courses. Doubtless, this low ratio facilitated instruction and the quality of games enacted, but it came at the expense of denying others the opportunity to attend. From its inception through 1937, 44 separate classes of the Tactical Course met so it was not the case that the number of sessions was being slighted.[216] Still, many officers having attended soon found themselves surplus to the needs of the Service and retired. Given the small size of its classes and the numbers requiring its training, for the Tactical Course it was ever a struggle just to keep pace with the promotion cycle.

With regard to the SOTC, the cancellation of a separate Commanders' Technical Course in 1922 and assigning those officers now to the SOTC discussed previously probably had the unintended consequence of shifting the priority from educating junior Captains to educating senior Commanders. In August 1939, not counting officers of the Royal Household, there were over 250 officers holding their captaincy on the active list. When the most senior officers holding Commanders' rank are considered, as well as Commonwealth officers of equivalent rank, those RNR officers

---

214 ADM 182/58, Admiral Fleet Order '132—Captains, R.N.—Technical and War Courses' of 17 June 1930.

215 As the *Navy Lists* only specify officer assignments, the number of ratings employed supporting the Tactical Course cannot be determined. That ratings were present and providing key support is confirmed in Williams' biography of Captain Gilbert Roberts who called upon Chief Petty Officer Bernard Rayner, a veteran hand of the Tactical School, to assist him at Western Approaches Command during the Second World War in establishing the Western Approaches Tactical Unit. In any event, the ratio of staff to students cited would not have been greater. See Williams, *Gilbert Roberts*, p. 87.

216 ADM 116/6364. Annual Report of the Tactical School, 1937.

required to attend, and the occasional flag officer taking the two courses, it becomes apparent that the Royal Navy possessed insufficient capacity to train its officers to the standard required. By late 1937, the Navy was operating three Tactical Courses a year where before 1935 four courses had been the norm. This reduction was entirely due to the press of operational commitments when the Service responded to the Abyssinian Crisis and the Spanish Civil War. Such pressures would only worsen as the Navy mobilized to respond to the Munich Crisis of 1938 when only two courses convened and officers joined the fleet.[217]

If the Tactical Course and the Senior Officers' Technical Course failed to address certain problems, such as anti-submarine warfare, the air threat, and amphibious operations with the depth and rigor required of another war, it should be remembered that the strategic environment when it changed, changed very quickly. Against a first-rate power, it was difficult to envision Britain undertaking an opposed landing and specialized training was thought unnecessary, whilst if required to operate against a lesser power, such as the Chinese, the training and equipment required were no more than that which they already possessed. A measure of hubris, perhaps, but it was also an appreciation that the funds to meet every contingency were simply not available. By 1939, when the prospect that Britain might actually need to conduct an opposed entry was recognized, the ISTDC started lecturing on the possibilities of amphibious assault at both the SOTC and the Army's Senior Officers' School.[218]

The naval technical and tactical courses taught were largely introductory in nature. Fully half of the Tactical Course curriculum was dedicated to topics that officers of the rank attending should already have mastered covering as it did the Service's published doctrine.[219] This grounding, though necessary, came at the expense of the time that was available to pursue more in depth issues. The possible solutions to this state of affairs such as having officers attend a longer course or establishing an advanced course dedicated to the examination of theoretical problems were eventually pursued. However, the formation of the Tactical Investigation Section only came very late in the day whilst increasing the length of the Tactical Course from eight to eleven weeks came at the expense of the SOTC and channeling more officers through the training regime. The expedient of simply having officers attend a course lasting more than the 4½ hours of study required each day was not pursued, and it is difficult to escape the conclusion that the peacetime Navy was reticent to break with its standing routine absence the pressure of a crisis. While such a regime was sustainable at the Staff and War Courses where outside reading, research, and writing were expected, the same did not apply at the Tactical Course whose source material and deliberations were of a confidential nature.

---

217 *Ibid.*
218 Maund, *Assault from the Sea*, p. 19.
219 'Course (X)', Dreyer Papers, DRYR/7/2.

Still, if the Tactical School failed to address some of the key problems of the day this is not to say they were not examined elsewhere. Frequently, anti-submarine exercises were carried out within the fleet and so, too, anti-aircraft shoots. The object of these evolutions, however, was to provide practical experience to crews and foster weapon use proficiency; they largely failed to investigate the range of tactics to be employed in war against a given threat scenario. Typical of such exercises was the case of the submarine HMS *Oberon* streaming fisherman's buffs, inflatable canvass markers, which allowed the destroyers HMS *Glowworm* and HMS *Greyhound* to observe the path *Oberon* was making whilst operating at depths of 60-75 feet against the results that they were seeing from their Asdic displays.[220] In truth, it was difficult to do more as the manning policy of the Navy on the eve of the Second World War was to relieve one-third of the ship's company during the three primary leave periods of the year making sustained tactical investigations difficult.[221]

Roskill's claim that the Navy failed to conduct a single exercise where the protection of a mercantile convoy featured is correct only in so far as merchant ships were not employed.[222] A regular exercise conducted was to have the fleet screen and protect a slow moving convoy which was represented by a single warship or a fleet auxiliary from an enemy fleet or submarine attack.[223] Additionally, by the mid-1930s exercises to test whether it was safer to evasively route merchant shipping rather than depend solely on convoys appeared with exercises 'R.T.' of the Mediterranean Fleet and 'D.A.' of the Home Fleet featuring the theme.[224] Though conducted with the thought of facing enemy surface raiders, amongst the conclusions reached were the presence of submarines made the use of heavy ships dangerous in screening such routes whilst aircraft carriers were ever at risk from surface forces.[225]

At-sea exercises tested the fleet against attack by both air and submarines, but danger was ever present. In 1920, Admiral Madden reported to the Admiralty that he had 'exceeded the safety limit in allowing the ships to zigzag slightly when screened by

220 Midshipman journal entry for 10 March 1936, Rear Admiral Ottokar Harold Mojmir St John Steiner Papers, Imperial War Museum, London, United Kingdom. The 'buffs' were usually red in colour and streamed from the submarine's conning tower.

221 Captain M. W. S. L. Searle, 'The Organisation of the Practices of the Fleet in Peace & War', Royal Naval Staff College paper of 1 July 1948, Roskill Papers, CCC/ROSK/7/163.

222 Roskill, *Naval Policy, Vol. I*, p. 536.

223 Contrary to received wisdom, many exercises featuring convoy and trade route protection were conducted between the wars whether in the face of surface, air, or submarine attack. Of those which featured submarine defence, the more notable were the Mediterranean Fleet 'L.A.' held in January 1928, 'L.C.' conducted in March 1928 and 'N.P.' in June 1928. Atlantic Fleet serials included 'F.A.' held in January 1923; 'L.D.' conducted in May 1928 and 'A.F.', a serial run in September 1930.

224 ADM 186/153, *C.B. 1769/33(1). Exercises & Operations, 1933, Volume I*. Admiralty, Naval Staff, Tactical Division, November 1933, p. i and ADM 186/154, *C.B. 1769/33(2). Exercises & Operations, 1933, Volume II*. Admiralty, Naval Staff, Tactical Division, April 1934, p. ii.

225 ADM 186/154, *C.B. 1769/33(2)*, p. 26.

destroyers, and on two occasions submarines were run into by capital ship.'[226] After the losses of submarine *K 5* in January 1921 and *H 42* in March 1922 combined with the near of loss of submarine *H 24* in February 1922, safety restrictions followed severely affecting the ability of anti-submarine evolutions from approximating wartime conditions. This was not a failure to heed a vital lesson from a past war, but a necessary compromise on the risks acceptable in the fleet's exercise programme to prevent the undue loss of life. Robert Elkins, a midshipman in the *Hood*, recorded the risks facing submarines in exercises following the loss of *H 42*:

> Personally, I do not think it is quite fair on either destroyers or submarines in this type of exercise since the destroyer cannot avoid ramming an unseen submarine while the submarine itself is "blind". I should imagine the Press will have something to say about this, since it is the fourth submarine to be sunk or damaged during exercises (*K5, M3, H24, H42*) during the last year.[227]

Other navies, of which the Imperial Japanese Navy comes to mind, pushed the envelope in the risks that they were willing to pursue with a consequent loss of life in their peace exercises.[228] This, however, was not the British way.

Gretton's charge that the utility of convoy had been forgotten after the war is mistaken and misleading.[229] Operationally throughout the period between the wars, the Royal Navy recognized that convoy was the preferred manner of protecting British merchant shipping and this view prevailed even in the absence of the Trade Division which the Admiralty abolished all too readily after the war with its responsibilities transferred to the Plans Division. Historians often endorse the tenor of Gretton's view and refer to the public pronouncements of Ministers made in parliament as late as 1935 that convoying need not be implemented in the opening stages of a war or crisis and depended on whether an adversary followed the laws of commerce warfare.[230] Whilst officials did make such claims, what is lacking in citing such evidence is the context of their statements. Typical of the evidence cited is the pronouncement by Lord Stanley, the Parliamentary and Financial Secretary:

> You will not know in the first place whether the ships are going to be in any great danger. It may be that it will be safer for them to sail by themselves. They will be

---

226 ADM 1/8597/9, Commander-in-Chief, Atlantic Fleet to Secretary of the Admiralty letter No. 36/A.H.0013 of 10 January 1921.
227 Journal entry of 24 March 1922, Elkins Papers, NMM/ELK/1.
228 David C. Evans and Mark R. Peattie, *Kaigun: Strategy, Tactics, and Technology in the Imperial Japanese Navy, 1877–1941* (Annapolis: Naval Institute Press, 1997), pp. 210-11.
229 Gretton, 'Why Don't We Learn from History?', *Naval Review*, January 1958, p. 13.
230 See Herwig, 'Innovation Ignored' in Millett and Williamson, eds., *Military Innovation*, p. 248 and Gretton, 'Why Don't We Learn from History?', *Naval Review*, January 1958, p. 13.

a smaller target. The enemy ships would not know where they were to be found. If raiders were about we should have to institute the convoy system at once. It is simply a matter of expediency. We should be ready to put the scheme into operation but we should wait until we thought that the proper moment had arrived. Having got to the point when it is considered that the ships ought not to sail by themselves but should be protected by an escort, we have to decide what is the best form of protection for the convoys, and I think it is agreed by everybody that what is known as the general convoy is the best system. That is the convoy which has an escort ready to protect its ships from surface attack, from submarines and possibly from the air.[231]

The first comment to note about the above statement is that when read in its entirety, it is not so nearly a definitive exposition against the implementation of convoys as later writers make it to be, for in the case of surface raiders, it is conceded that convoying will be adopted immediately.

Secondly, convoy was a means to secure an end—the protection and safe passage of British commerce and military forces. Whether it was adopted, or adopted immediately, were very much contingent on the theatre of operations in any crisis or war. If Britain found herself fighting in Asia against Japan or facing Italy in the Mediterranean, would it be necessary to convoy her shipping in theatres removed from the crisis point? Only the moment and an assessment of the actual maritime threat would decide the issue, as she simply did not have resources to guard all her merchant shipping and support all the other tasks that a navy was expected to meet. Additionally, the implementation of convoy was executed as part of the Naval Control of Shipping (NCS) regime. Its full implementation required the participation of Commonwealth navies who remained responsible for those portions of the NCS occurring in their waters. Instituting convoy, then, was more than just an operational decision to counter a threat; it was also a political decision for governments that made them partners in any belligerency. Following the Statute of Westminster, the Dominions reserved the right to decide when and how to support Britain as the previous discussion of the Chanak Crisis makes abundantly clear. Events would show that Canada and South Africa were wont to accept their role in the execution of convoy under the NCS without a prior request made through political, rather than naval, channels. The most that they would concede was to allow the evasive routing of shipping to proceed, as this did not require any belligerent actions on their part nor infringe on their rights if deciding to remain neutral in any situation.[232]

Moreover, and just as importantly, the statement came in support of the debates on the *Navy Estimates*. If the government conceded that convoy was always to be implemented, it would have been savaged by its critics for not providing the funds required to procure the aircraft and escort ships necessary to meet its declared policy. This it was not willing to do. Expense was one reason, but a second cause was it

231 *Hansard,* House of Commons debates, 14 March 1935, v. 299, cc. 609-729.
232 Tracy, ed., *Collective Naval Defence,* pp. 597-601.

still entertained hopes that further naval arms reductions agreements might yet be negotiated with its rivals. This underlying political context of the convoy question is either ignored or forgotten by many observers. The Navy preferred to implement convoy, but it could not guarantee that its operation would be palatable to all involved. Knowing that the implementation of convoy was not always going to be possible, the Royal Navy began testing the evasive routing of shipping in fleet exercises and these evolutions, such as 'D.A.' and 'R.T.' mentioned earlier, confirmed that convoy was the preferred manner of protecting her shipping.

Others argued that the conditions favouring convoy in the World War were now absent. This was due to the range and lethality of modern airpower and amongst the critics voicing such fears was Wing Commander John Slessor of the Royal Air Force. Slessor saw dispersion as necessary in the future, as concentration of shipping under convoy only offered the airman a target easier to locate and attack. Slessor's views on convoy were formed as part of his greater consideration of the protection of logistics in a future war in the face of the air threat. Armies would need to disperse stores and reliance on a few lines of communications in the face of an enemy's air offensive would no longer suffice. Seeing the few focal points available to receive British shipping and noting the extreme vulnerability of the port of London, Slessor easily believed that the same held true in the maritime dimension. The Navy thought otherwise demonstrating that the convoy question was no longer a simple matter between the Board of Trade and the Admiralty. As officers attending the SOTC on the eve of war discovered, the Royal Air Force came to accept the Royal Navy's position. For Slessor, there is a bit of irony in all this, as he would direct Coastal Command during a critical phase of the Battle of the Atlantic in 1943.[233]

Finally, that the pronouncements of Ministers cannot be taken at their face value one need look no further than the contemporaneous actions of the Navy and the Air Force during the Abyssinian Crisis. Admiral Sir William Fisher, Commander-in-Chief, Mediterranean anticipated convoying his own ships and merchant shipping in late 1935 when war with Italy appeared imminent. Plans were concerted between Brooke-Popham, the Air Officer Commanding, Royal Air Force, Middle East, and Fisher to this end to support the required air reconnaissance of these convoys.[234] Previously, the Joint Planning Sub-Committee had advised the Chiefs of Staff that:

> The maintenance of our shipping through the Mediterranean would present very serious difficulties in the face of attack by Italian forces especially should Italy decide to use her submarines and aircraft in an unrestricted manner. The provision of adequate anti-submarine vessels and aircraft would take considerable time.

233 J. C. Slessor, *Air Power and Armies* (Tuscaloosa: University of Alabama Press, 2009), pp. 117-18.
234 Brooke-Popham to Sir Edward Ellington letter of 16 November 1935, Brooke-Popham Papers, 4/3/11, LHCMA.

In the event of the situation reaching actual hostilities we consider that until the
necessary counter measures could be improvised and the situation is clearer, British
trade in the Mediterranean should be suspended and it might be necessary to subsequently to divert it round the Cape and accept the resultant dislocation in shipping.[235]

Meanwhile, Italy's use of chemical agents, including mustard gas, in contravention of
her prior unrestricted commitment to forego such use made her likelihood of resorting
to unrestricted commerce warfare more probable in Admiralty eyes. Whilst the final
decision to convoy would have been one subject to political review, it does confirm
that even in 1935 the value and necessity of adopting convoy was understood at the
operational level.

Turning to the technical means of defeating the submarine, it is doubtlessly true
that some were prone to over-sale the promises of Asdic or what is commonly called
sonar today. This was especially so at the Chief of Staff level though others such as
Grenfell, a veteran of the Tactical Course and more recently from teaching at the
Staff College, also hinted that the threat was not what it had been previously. Writing
in 1937, he was arguing that surface escorts had largely mastered the submarine
threat.[236] Others, though, were not as sanguine. In 1925, the Commander-in-Chief,
Mediterranean cautioned the Admiralty just on this point:

> The impression conveyed by the Admiralty letter is that Asdics have passed the
> experimental stage, and while this is most welcome news, it does not entirely agree
> with experience gained by the Fleet. Exercises so far conducted have not led to the
> belief that much reliance can be placed on asdics for the detection of submarines.[237]

If the Admiralty was prone to egging on the merits of Asdic, then a practical seaman
such as Admiral Keyes, the Commander-in-Chief, Mediterranean, ensured a coun-
tering view was registered. As with most inventions showing promise, supporters
tended to sing praises for their faith just a little too quickly. Time demonstrated that
Asdic's performance varied for a host of physical and environmental reasons. This was
understood within the fleet. Here, operating squadrons on a global basis provided a
means to compare results.[238] Admiral Pound, Commander-in-Chief, Mediterranean,
noted the varying effects of Asdic during his fleet's operations at the time of the
Spanish Civil War:

235 CAB 53/25, Joint Planning Sub-Committee report of 9 August 1935.
236 Grenfell, *Art of the Admiral*, p. 57.
237 ADM 1/8685/151, Commander-in-Chief, Mediterranean to Secretary of the Admiralty
     letter No. 1184/1240 of 5 June 1925.
238 *C.B. 3016/30, Progress in Tactics, 1930.* Admiralty, Naval Staff, Tactical Division, July 1931,
     British Sources, Box 12, NHC.

> Up to 6 weeks ago we were getting <u>very</u> good results but the summer water condi-
> tions are now on us and results are not nearly so good. If we have to cross swords with
> our friends across the way I must say I should prefer to do it in winter both from the
> submarine and air point of view.[239]

Notwithstanding the several improved models of Asdic developed between the wars to deal with the submarine peril, George Franklin's conclusion that it was not seen as a panacea by the greater Navy has the ring of the truth.[240] The suspicion remains, however, that the results being achieved in exercises when personnel were alert and only having to face a threat of a limited duration were accepted rather too readily as indicative of what wartime conditions would bring when crews faced long, tedious hours at watch with boredom as much a foe as the enemy.

The poor response to a new submarine peril that Gretton lamented was fundamentally caused by four problems. Chiefly, the treaty regime the Royal Navy operated under for most of the interwar period limited the overall carrier tonnage she could operate. Thus, carriers to support trade protection were not built and did not exist when the war occurred in 1939. Secondly, the responsibility for Coastal Command rested with a Service committed to having a general-purpose force able to meet most operational taskings, but wary of developing specialist units detracting from this greater need. Thirdly, financial stringency compounded with the dual-control of naval aviation made the solution of any maritime problem inordinately difficult. Funds to develop reserves in personnel and material were lacking and most of the flying assets supporting the fleet at war's opening were antiquated. The Royal Navy was fortunate that the European powers she faced in the Second World War had done even less than her in developing a naval air arm. Meanwhile, she failed to develop key components such as the hedgehog and shipboard direction finding systems which could rightly have been anticipated during peacetime exercises. Finally, having insufficient land-based assets dedicated to the trade protection task, the Navy ill-employed the few carriers that she possessed on offensive sweeps needlessly risking her one operational advantage. This was a failing in doctrine, but when life hands one lemons one is apt to make lemonade. Of the four fundamental problems governing the defence of trade problem, doctrine was the least of these.

The Royal Navy's approach to tactical investigation was systematic and thorough. Elements of the Naval Staff set issues to be resolved and the Tactical School and the major fleets worked in tandem to discover the answers. The school and the fleet provided feedback to the Admiralty with the Tactical Division in its yearly report,

239 Pound to Chatfield letter of 7 August 1937, Chatfield Papers, NMM/CHT/4/10. Original
     emphasis.
240 George Franklin, *Britain's Anti-Submarine Capability 1919–1939* (London: Frank Cass,
     2003), p. 71.

C.B. 03016, *Progress in Tactics*, publishing the summaries of the conclusions reached.[241] Reviewing the summaries one must conclude that far from ignoring the lessons of the past, the Service was committed to cataloguing and testing these issues and adopting new measures based on the empirical data reached when warranted. Battlefleet tactics were at the heart of these investigations, but so, too, were convoy, anti-submarine prosecution, air defence, and carrier operations.

Gretton's charge that the pre-Second World War Navy was too focused on waging fleet action at the expense of other tasks is accepted, but the more fundamental problem was the fleet's total emphasis on offensive action. This orientation was natural, even largely necessary, where terrain provided little protection and the probable effects of the N-square law in the naval encounter were understood. The value of offensive action was enshrined as a principle in the *Naval War Manual* whilst the *Battle Instructions* placed a similar emphasis on officer initiative. This emphasis was not mistaken— only incomplete. Situations might arise where these traits needed to yield to other principles—security or maintenance of the object. The Tactical School's emphasis on fleet action, particularly, surface action, came at a price: it reduced all other naval tasks to the second eleven. This need not be a fatal error if all officers were destined to Greenwich and the Staff and War Courses with their broader view of the naval tapestry. Immediate tactical considerations, then, could be weighed against strategic and operational requirements and the dictates of policy. Yet, not all would complete the full regime of higher officer education and this, then, was the greatest shortcoming of the Royal Navy's efforts: it produced too few, well and fully trained officers.

The committee chaired by Admiral William James examining the future style of officer higher education focused mainly on who should receive such education and when to deliver the training specified; it failed to consider the actual content of naval education.[242] The committee's terms of reference shaped the report delivered, so Admiral James and his associates are largely blameless for this omission. Not so the Admiralty; its primary concern for the Portsmouth portion of officer education was for the committee to report on whether the Tactical Course should be obligatory for both commanding officers and Staff Officers Operations. Though both Captain Macnamara and Commander Norman had experience serving as afloat staff officers and were naval *psc*'s, neither had attended the Tactical Course—an omission, thankfully, which did not apply to Admiral James. Still, the report's conclusions focused mainly on evolving the Staff and War Courses where the committee members' interests and experiences primarily resided and largely ignored recommending any changes to the Tactical Course and the SOTC.

241 The National Archives, Kew retain only one example of *C.B. 03016* and that for the year 1939 found as ADM 239/142. Other editions are held at the Naval Historical Center, Washington, DC.
242 ADM 1/9041, Committee to Investigate the Future Training of Naval Officers for War report of 21 May 1935.

No more than the earlier creation of the War and Staff Courses, the institution of the Tactical Course was never a silver bullet. Just as it was difficult to sustain the quality of education delivered at Greenwich after the period of Richmond and Drax, so it proved equally challenging maintaining the Tactical School's edge following the departure of Usborne. The officers that followed were eminently capable practitioners of their craft, but not deep, reflective thinkers of it. Troup wrote a practical book on navigation and pilotage, which pays homage to his technical training, but says nothing of his tactical ideas. Of the other officers to direct the school, forgiving the limited output of Colvin who wrote a single article in 1919 and two brief letters in reply to articles penned by others in 1927, from its founding until 1930, when the trail of authorship runs cold, the pages of *The Naval Review* are void of their thoughts.[243]

The two advanced courses taught at Portsmouth for executive officers were excellent grounding for an officer about to assume command of his ship or fly his flag afloat for the first time. They forced him to think about employing naval forces in battle and taught him to do so following the precepts of British naval doctrine. Focused as they were on the technical and tactical aspects of his craft, they went a long way to correcting the vacuum which had previously existed in naval education which either made an officer an excellent tradesman in his chosen field or assumed that he would acquire his tactical acumen through constant practice at sea. The courses were not perfect and a large measure of duplication existed in the higher aspects of naval education. Though some might argue for eschewing the SOTC altogether and merging the Tactical Course into the War Course proper, keeping them separate made drafting officers to individual courses an easier affair.[244]

Gaps remained, though, for the instruction provided focused on solving largely naval problems. The Navy recognized this fact and from an early stage, even predating the World War, began sending some of its officers to military schools as Drax's experience demonstrates. Creation of a new arm, the Royal Air Force, made the matter more complicated still. The Imperial Defence College promised to address some aspects of how to coordinate and operate in a tri-Service environment, but its limited scale and intended audience meant that not all naval and marine officers needing the skills to operate in the joint arena would find their way there. The solution was to have a cadre of officers attend the Staff Colleges of the other arms, send others to Army courses intended for their senior officers, and attach officers to the respective headquarters of the other arms. Thus, it is to the joint environment that this story now turns to investigate how the Royal Navy used the training opportunities available from the British Army and the Royal Air Force to fashion a better officer.

243 Ragnar Colvin, 'Propaganda on Board Ship in Action', *The Naval Review*, Vol. VII, No. 2, pp. 225-27, 'Naval Education' *The Naval Review*, Vol. XV, No. 2, pp. 428-29 and 'Jutland', *The Naval Review*, Vol. XV, No.4, pp. 861-63.
244 H. H. Smith, 'Education of the Naval Office', *Naval Review*, November 1928, p. 723.

5

Amongst the Owls and the Hawks: The Joint Environment and the Limits of Inter-Service Education

His knowledge and experience of the Air Force combined with his qualities of a naval officer have proved most valuable.[1]

Air Vice Marshal Ivo Vesey, 1925

I should think he is rather inclined to run his Squadron from his desk.[2]

Admiral Sir Dudley Pound, 1938

The Royal Navy may have been viewed as the sure shield of Empire, but a shield without a corresponding javelin was a hoplite at best only half-prepared for battle. True in Achilles day, it was even more the case once the air became a factor in war. Until the advent of the Imperial Defence College, the primary means for naval or Royal Marine officers to secure an appreciation for the capabilities, organization, and roles of the other forces of the Crown was through temporary attachment to another arm, attending courses, but, more especially, through an exercise where the Army or Air Force played a part in a joint scheme with the Navy. If the officer was lucky to serve an attachment, then his experience might prove only the briefest of periods, say ten days, to observe a tactical unit on exercise. A more substantial period was possible for a few as part of the Navy's efforts to better liaise with another Service, but it was also entirely possible that an officer might never have a significant encounter with those serving outside the naval service.

Officers serving in Britain in a shore assignment were encouraged, however, to seek attachment, and though this allowed an officer to witness a unit for only an abbreviated period, the desire that officers should know something of the other Services was a strong lesson of the recent war. This was perhaps only natural as the operational experiences of Africa, the Near East, Russia, and Antwerp were still fresh in the

1   ADM 196/127 (Boddam-Whetham).
2   ADM 196/91 (Kennedy-Purvis).

minds of leaders. Of course, the benefits of a temporary attachment were necessarily limited and prone to extending no further than to the actual officer in question. An article might follow in *The Naval Review* just as surely as a report to the Admiralty noting the officer's impressions. If not quite a lark, then this was training of a very informal and relaxed nature where the real benefit may have secured nothing more than an appreciation that the ways of the Navy were far from the only ways possible.[3] Such attachments, particularly, to Army elements, were a very frequent occurrence for junior naval officers between the wars and the impression remains that the Admiralty pursued them with the idea of broadening one's horizons much as a midshipman spent time in destroyers and aircraft carriers.

Attachments of a more formal nature were those where an officer spent a rotational tour at the War Office or Air Ministry. The opportunities for these were more limited as the positions were few. Additionally, specialist skills allowing the naval member to serve as a productive partner might be required precluding most officers from consideration. Other opportunities to form an impression and gain an appreciation of Army and Air Force personnel, if not actually of the other Services, could arise from serving on a joint committee. Increasingly, supporting such committees became a fact of naval life. Whether the body was one formed to address a specific technical issue and was temporary, or was established on a more permanent footing, such as the Committee of Imperial Defence, naval and marine officers now faced new challenges outside pure Service lines, where the skills required were not so much martial as mental.

It has been seen already that officers of varying ranks such as Richmond, Usborne, and Moore were part of this growing trend towards inter-Service cooperation and even dependency. If the World War was not the cause of such, it certainly hastened the process. Yet, others factors were at play such as technology and, of course, the evolving political climate which made the consideration of any defence problem a question of growing complexity. Witness the tensions of Empire and the League of Nations played out during the Chanak Crisis and the Locarno agreement, whilst aviation and the Royal Air Force made almost every defence consideration a sore spot with the older Services. The Navy's fight for control of the Fleet Air Arm was the most obvious example of this tension in a maritime setting, yet the Army faced its own share of issues. In this instance, the problem was the concept of substitution—replacing conventional military forces with air and mobile ground units in imperial policing which offered to reduce the costs of military support and the corresponding influence of the British and Indian Armies.

Such tensions existed and are recognized by this observer providing the context of much of what follows. Yet, what follows is not an analysis of these issues, *per se*, but

---

3   Sirius, 'Ten Days with the Army', *The Naval Review*, Vol. XXI, No. 2, May 1933, pp. 291-97 and R.P.T., 'Nine Days with the Army', *The Naval Review*, Vol. XXI, No. 4, November 1933, pp. 737-40.

rather an investigation into how the Royal Navy employed the training and educational regimes of the other Services to maximize its own effectiveness in this new environment. The record is an uneven one where if it was not quite one step backward for every two steps forward, then at times it must have felt so. This is not surprising as a master plan did not exist. Writing in *The Naval Review* one commentator observed:

> Soldiers say before the late war they experienced great difficulty in making naval officers realise that troops landed on a hostile beach were of little military value without artillery, ammunition and transport. Now, happily, all this is changed. Yet one often hears the question asked "What on earth is the good of naval officers going to Army senior officers' schools and vice versa? How easily the lessons of war can be forgotten![4]

Fortunately, the Royal Navy discounted such criticisms and continued sending its officers to the schools of the British Army and the Royal Air Force. This is not to say that more was not desired of the Service, as events will demonstrate. Still, its efforts, if appearing paltry by today's standards, were not inconsiderable for the age and this when economy might have easily prompted it to do otherwise. Much the same can be said of the other Services, facing as they did fiscal perils every bit as challenging as the Royal Navy with their record, too, proving uneven. That said, the Services recognized that they needed to work together in their educational efforts, though the larger policy issues dividing them often obscures this from immediate view. Noting the recent reestablishment of the Royal Naval Staff College, the First Lord called attention to its joint composition of '16 Naval, two Military, and two Air Force Officers. So far as can be judged at the present stage it promises to prove of great benefit, not only to the Navy directly, but also in furtherance of the co-operation between the Services.'[5] This view was, indeed, mirrored by at least one of the other Services for Air Vice Marshal Brooke-Popham in his introductory address to the officers attending the Air Force Staff College in 1925 encouraged the qualifiers to 'pay particular attention to co-operation between the three services. A great deal can be effected, in fact is being effected in this direction by the close connection that exists between the three Staff Colleges.'[6]

Amongst the plethora of issues the James Committee investigated when reviewing the higher education of naval officers was the question whether the current regime offered sufficient scope for an officer to acquire knowledge of the other Services. Recognizing that few of its officers would attend the IDC, the James Committee considered whether an additional course emphasizing combined operations for naval

---

4   John Knox, 'Some Comments on the Relations between the Services', *The Naval Review*, Vol. XVI, No. 3, August 1928, p. 539.
5   ADM 1/8744/140, Statement of the First Lord of the Admiralty of 1 December 1919.
6   Brooke-Popham 'Fourth Course' lecture of 5 May 1925, Brooke-Popham Papers, 1/5/10, LHCMA.

psc holders and lasting six months should be pursued. The benefits of having officers attend the planned course at Camberley were obvious, as a cadre of officers would in time develop knowing the intricacies of amphibious assault and secure a better appreciation of the military and air contributions. The more fundamental concern for the Navy was, if adopted, the effect it would have on the practice of attaching officers to the other forces.[7] In the near term, little was done to advance the issue and though creation of the Inter-Service Training & Development Centre can be seen as signpost along the way, it would take the pressure of another war to make real inroads into furthering the practice of joint training.

For the Navy, then, the need to develop a cadre of officers familiar with military problems was an issue of longstanding and became even more pressing once the Royal Air Force was established, as many of its officers with experience of the air transferred to the new arm. Over 5,000 officers had been part of the Royal Naval Air Service (RNAS) supported by nearly 50,000 ratings responsible for supporting the approximately 3,000 aircraft of all types in 1918. The appliances included not only regular aeroplanes and specialized types such as float-planes, but also kites and balloons employed in observation duties.[8] This had now been lost and whilst it is true that the totals when measured against the Navy of the time were but a fraction, it was a fraction with very unique skills and possessing very unique capabilities. It was also a fraction of the greater Royal Air Force which in November 1918 mustered over 31,000 officers and at least 22,00 aircraft of all types.[9]

Of course, most of the officers belonging to the RNAS were not holding permanent commissions and peace would have seen retrenchment in their numbers in any event. In this regard, officers who were previously ratings would have been especially vulnerable including those eventually achieving high office in the Royal Air Force. Therefore, one must not leap to the conclusion that maintaining the status quo was a golden path either.[10] Yet, not all officers transferring to RAF were temporary, and whether Rear Admiral Mark Kerr's subsequent career would have been as short if he had not found his himself under Trenchard's command remains a curious unknown.[11] Certainly, Captain Murray Sueter,[12] a former head of the Naval Air Department,

<hr>

7   ADM 1/9041, 'Committee to Investigate the Future Training of Naval Officers for War Terms of Reference'.

8   Peter Lewis, *Squadron Histories: R.F.C., R.N.A.S. and R.A.F. 1912-59* (London: Putnam, 1959), p. 183.

9   J. M. Spaight, *The Beginnings of Organised Air Power* (London: Longmans, Green and Co. Ltd, 1927), pp. 272-75.

10   For example, Air Commodore Oswald Gayford, formerly a rating employed with the Seventh Destroyer Flotilla before commissioned into the RNVR in 1916.

11   Later Vice Admiral Mark Edward Frederic Kerr (1864–1944). Captain, 1903; Rear Admiral, 1913; Commander-in-Chief, Adriatic Squadron, 1916-17; Major General, 1918; Deputy Chief of the Air Staff, 1918; Vice Admiral, 1918; General Officer Commanding, South-Western Air Force Area, 1918 and retired, 1918.

12   Later Rear Admiral Sir Murray Frazer Sueter (1872-1960). Ordnance Department, 1904-06

found in remaining with the Navy an equally unforgiving environment, but then writing self-promotional letters to one's Sovereign was not a sin easily forgiven by the Admiralty.[13] Now the problem was not only to work with the new Service, but also to reconstitute a body of officers technically competent and understanding the potential uses of airpower, its limitations in the maritime environment, and all the while possessing knowledge of their own Service's requirements and capabilities.

Responsibility, for a time, for the air defence of Great Britain and operations of the RNAS in France had fostered a unique closeness in military and naval cooperation. Creation of the RAF now threatened that which had been gained at so much cost. This is not to avow that the founding of a unified air force was a mistake—only that it came with another sort of a cost. That cost, moreover, was apt to require more immediate payment while the benefits of the new Service accrued more slowly as it established itself and developed along the lines desired by its political masters.

In the near term, the Royal Air Force was dependent on the two older Services for the provision of key support, as it lacked much of its own organic infrastructure. The armament works of Woolwich and Enfield operated by the War Office were emblematic of this dependency and to these establishments might be added her need to send officers to the older Staff Colleges.[14] Though the Command Paper outlining the peace organization of the RAF specified the early creation of a separate Staff College for the new arm,[15] it was not an immediate priority. Thus, not until 1922 did it become a reality when the first course met in April under Brooke-Popham's tutelage.[16] In the interim, officers such as Wilfrid Freeman attended the Royal Naval Staff College whilst Charles Samson sat the Senior Officers' War Course. When at last a course began in Andover, Hampshire it was not until its second convening in May 1923 that naval and military officers first appeared. This respite allowed the Air Force to establish the school along the lines it required without being under the critical eye of either the Navy or the Army. Attending the second course were the first of two naval officers, Commander Edye Boddam-Whetham,[17] a trained Observer who

and 1908-10; Commanding Officer, *Barham*, 1906-08; Captain, 1909; Commanding Officer, HMS *Hermione* and Inspector of Airships, 1910-12; Director, Admiralty Air Department, 1912-17; Rear Admiral and retired, 1920.

13   P. K. Kemp, 'Sir Murray Frazer Sueter', E. T. Williams and Helen M. Palmer, eds., *Dictionary of National Biography 1951-1960* (Oxford: Oxford University Press, 1971), pp. 940-41.

14   Lord Tedder, *With Prejudice: The War Memoirs of Marshal of the Royal Air Force Lord Tedder G.C.B.* (London: Cassell, 1966), p. 4.

15   Air Historical Branch, *Command Paper 467, An Outline of the Scheme for the Permanent Organisation of the Royal Air Force*, 25 November 1919.

16   *Ibid.*

17   Later Rear Admiral Edye Kington Boddam-Whetham (1887–1944). Commanding Officer, HMS *Swallow*, 1918-21; Senior Officers' Technical Course, 1921 and 1929; Observer in *Queen Elizabeth*, 1922; Staff Course Andover, 1923-24; Observer in *Argus*, 1924; Air Ministry, 1924-25; Commanding Officer, HMS *Witherington*, 1926-27; Captain, 1928; Senior Officers' War Course, 1929; Senior Naval Officer, West River, 1930-31;

nevertheless spent most of his career in destroyers, and Lieutenant Alfred Sprott.[18] Significantly, neither officer had previously attended the Greenwich Staff Course, a feature requiring further consideration anon.[19]

The typical Andover course ran for twelve months but a change to a May matriculation date began with the second sitting. This pattern held until 1926 when the fifth course ran from May of that year and ended in July 1927. The next sitting, the sixth session, then started in September and completed in December 1928 with the seventh course immediately following, beginning in January 1929 and running to following December. Henceforward, the course would run from January through December, and, thus, a pattern congruent with the Royal Naval Staff College, the Army Staff College, and the IDC emerged once the last-named institution opened.[20] Though the successful naval and marine qualifier did not receive the *psa* distinction upon leaving Andover, the report provided to the Admiralty following an officer's studies indicated whether a *psa* would have been awarded if the officer had been a member of the RAF. Accordingly, the period's *Navy Lists* reflected successful completion of the course.

This pattern prevailed for the Navy until 1932 when it became customary to send only qualifiers who had previously secured the naval *psc*. There was a sound reason for this change. Naval and Royal Marine officers attending the Air Force Staff College were more than just qualifiers, they were seen as quasi-members of the directing staff and were expected to speak to the Navy's role in the schemes and appreciations considered.[21] A move all the more desirable as the Admiralty never posted an officer to serve as a teacher on the Andover directing staff. The Air Force desired this experience and stressed to the Admiralty and the War Office the benefits of having their qualifiers attend their own Service courses first before arriving at Andover. This requirement the Admiralty generally followed as 26 of the officers attending had previously taken the equivalent naval course. After 1930, the custom was to send its officers to Andover immediately following completion of the Greenwich course and testimony to the general soundness of this approach is the experience of Lieutenant Commander Eric Davis.[22] Sitting the 1926-27 course, Davis had not passed through the naval course

---

Anti-Submarine Course, 1932; Captain (D), Second Destroyer Flotilla, 1932-33; Captain, Dockyard Bermuda in HMS *Malabar*, 1935-36; Captain of Dockyard Chatham, 1937-39; Rear Admiral and retired, 1939. Recalled for war service.

18   Later Commander Alfred William Sprott (1893-1942). Staff Course Andover, 1923-24; Commanding Officer, *Dorking*, 1924; *Argus*, 1924-25; *Furious*, 1928-30; Commander and retired, 1936. Recalled for war service and killed in action whilst commanding HMS *Dragonfly*, 1942.

19   *Hawk*, 1936, p. 134.

20   *The Hawk: The Journal of the Royal Air Force Staff College*, No. 11, December 1938, pp. 110-17.

21   Royal Air Force Staff College to Air Ministry letter SC/60/Air.1 of 30 March 1926, Brooke-Popham Papers, 1/5/13. LHCMA and CAB 53/5, Chiefs of Staff Sub-Committee minutes of 1 November 1934.

22   Later Commander Eric Alfred Davis (1892-1960). Navigating Officer in HMS *Temeraire*, 1920 and *Argus*, 1924-26; Staff Course Andover, 1926-27; Navigating Officer in

and struggled to meet the standard required at Andover. Indeed, he struggled in his examinations to Lieutenant failing engineering twice and then was given accelerated promotion to forestall being passed over by more junior officers. Though he was a first-rate Navigating Officer and a good teacher of pilotage, his appointment to the Staff Course could have been bettered given the role expected of qualifiers attending another Service's course.[23] Davis, though, had sought the course and it may be that his appointment owed something that others were not so inclined to apply.

That said, having one of its Staff Course graduates attend the Greenwich course was not a practice the Air Force always followed itself. Of those attending Greenwich after the formation of the Royal Air Force Staff College, both André Walser and Squadron Leader Alan Gillmore[24] had preceded their attendance at Greenwich by securing the *psa*. Yet, just as significantly Wing Commanders Henry Busteed,[25] who came to the Air Force via the RNAS, and Alec Coryton, coming via the Royal Flying Corps, had not.

Of the three Services, the Air Force Staff College typically accepted the fewest qualifiers each year with Camberley easily exceeding Greenwich. Hard and fast rules for Andover do not apply, but each class usually saw officers attending from both the British and Indian Armies alongside two naval officers. Additionally, four officers from the Dominions typically rounded out the course.[26] Godfrey Wildman-Lushington was the sole Royal Marine officer and the only qualified pilot to attend, completing his course in 1930 before proceeding to Greenwich and becoming an instructor at the Royal Naval Staff College.[27] Tellingly, no officers from the fledgling imperial naval forces went to Andover between the two wars, but then those forces were hard-pressed just sending eligible officers to Greenwich.[28]

Returning to a comparison of officers trained for staff duties, the military officer attending Camberley or Quetta sat a two-year course. Though this inflated the total number of officers qualifying, even when allowances for this are duly made, the numbers enrolled each year exceeded the pace of naval and air instruction and averaged sixty qualifiers in each division at Camberley and thirty for each class at Quetta.[29] By the end of the late 1930s, the trend in officer staff training changed with the Air Force surpassing the Navy in the numbers under instruction. Thus, in 1938

Marlborough, 1927-28 and *Cyclops*, 1929-31; retired, 1931 and Commander, 1932. Recalled for war service.

23   ADM 196/54 (Davis).

24   Later Air Vice Marshal Alan David Gillmore (1905-1996). Staff Course Greenwich, 1939.

25   Later Air Commodore Henry Richard Busteed (1887-1965). Staff Course Greenwich, 1926-27 and Senior Officers' Technical Course, 1929.

26   Ian M. Philpott, *The Royal Air Force: An Encyclopedia of the Inter-War Years, Volume One, The Trenchard Years, 1918-1929* (Barnsley: Pen and Sword, 2005), p. 97.

27   ADM 196/64 (Wildman-Lushington) and *Hawk*, 1938, pp. 110-17.

28   *Hawk*, 1938, pp. 110-17.

29   Slessor, *Central Blue*, p. 85.

where the Navy had 44 officers from all Services qualifying at Greenwich, the Air Force was teaching 45 officers with 65 planned for 1939.[30]

Where Greenwich could run sections with at least one member from another Service, the syndicates at Camberley were overwhelmingly composed of military members only, as beyond the limited number of naval and marine officers appearing, the RAF contribution was usually two or three officers in attendance. In this regard, Andover with its lower number of air officers under instruction approximated Greenwich in the prominence that officers from the other Services contributed to the course's work and style. Of course, both Camberley and Quetta benefited by having an Air Force officer assigned as an integral member of their directing staffs during portions of the interwar period. Consequently, for the naval officer attending Camberley, his impressions of the Army were likely to be far more robust than his military counterpart's views of the Navy given the diluted naval presence found.

Judged by their later careers, the quality of officers assigned by the Air Ministry to serve on the Camberley directing staff was extremely high. Wing Commanders Leslie Gossage,[31] serving at Camberley from 1922-25, and Owen Boyd,[32] late Royal Flying Corps and Indian Army and serving from 1928-30, both reached air rank whilst Slessor, the air member from 1931-34, won a RUSI Gold Medal in 1936 and ultimately rose to become Chief of the Air Staff.[33] Nor would it appear that the Army failed to return the favour, as Lieutenant Colonel John Eldridge[34] served in a similar capacity at Andover from 1937.[35] Army officers normally relished a posting to the Camberley directing staff as promotion often rapidly followed.[36] However, he still needed to be mindful not to overstay his welcome and balance the time spent at the Staff College with other assignments. Dill is an excellent case in point. When told by Liddell Hart that he was being tipped to return to Camberley as Commandant, he answered:

> I see you say – in much too flattering terms – that I am likely to go to the Staff College – I wonder! I also wonder if perhaps I have not already done quite enough

---

30 *Hawk*, 1938, p. 10.

31 Later Air Marshal Sir Ernest Leslie Gossage (1991-1949). School of Army Co-operation, 1921-23 and Staff College Camberley, 1922-25.

32 Later Air Marshal Owen Tudor Boyd (1889-1944). Commandant, School of Army Co-operation, 1923-26; Staff Course Camberley, 1926-27 and Staff College Camberley, 1928-30.

33 *London Gazette*, No. 33042, 28 April 1925, p. 2843 and Slessor, *Air Power and Armies*, p. i.

34 Later Lieutenant General Sir William John Eldridge (1898-1985). Staff College Andover, 1937-38; Commandant, Military College of Science, 1948-51 and founding member Institute of Strategic Studies later restyled as the International Institute of Strategic Studies.

35 *Hawk*, 1937, p. 10.

36 David French, *Military Identities: The Regimental System, the British Army, and the British People c. 1870-2000* (Oxford: Oxford University Press, 2005), p. 162.

of that sort of thing. S.C. 1919 – Easter 1922 & I.D.C. 1927-28. However, I shall be highly honoured if I do get 'the firm offer'.[37]

In practice, the Air Force was closer to the Army in how it esteemed its Staff College, though the ethos of the individual airman frequently approached that of naval officers. Freeman for one was a reluctant instructor at Andover though viewed by Slessor as the best Air Force officer between the wars.[38] He sought a prompt return to flying duties; a view heartily endorsed by Arthur Harris,[39] the future wartime leader of Bomber Command, who evidenced little desire to remain at Camberley as the air member of its directing staff.[40] Shaping these views was the concern that their careers would suffer if too long absent from the *real* Air Force. Officially, the mission of Andover mirrored its naval counterpart: to train officers in staff duties whilst also providing a higher education in war.[41] In practice, it followed Camberley's objective of also developing the command skills of higher officers. This owed much to Brooke-Popham and his grounding previously at the Army Staff College probably was the primary reason why this was so.

A major reason why the Services trained such varying numbers of staff officers was that their requirements were so different. Fundamentally, it was a function of their operational environments. The Army and the Air Force, operating on and from land, had fixed lines of communications drawing heavily on road and rail networks to maintain their forces in the field. Staff officers were required to plan and oversee associated transportation arrangements and these problems were a regular feature of Staff College exercises. In contrast, naval forces enjoyed a degree of mobility and sustainment measured in days and weeks, and not in hours, and far-removed from the practices of the Army and Air Force. Consequently, its administrative and logistical infrastructures were of a different order. Additionally, the Army required staff officers to maintain command and control of its forces in battle in an age when wireless communication was not yet pervasive at the lowest tactical levels. Moreover, these differences influenced the command ethos that evolved over time in the respective Services and became a contributing factor to how the staff and the staff officer were viewed. If the Naval Staff element of the Admiralty organization was small, its administrative

37 Dill to Liddell Hart letter of 22 May 1930, Liddell Hart Papers, LH 1/238/6, LHCMA. Dill was also having a bit of fun in his reply, as the Army Staff College was known colloquially at times as 'The Firm'.

38 Marshal of the Royal Air Force Sir John Cotesworth Slessor Sound Recording 3176 of 8 March 1978, Reel 1, Imperial War Museum, London.

39 Later Marshal of the Royal Air Force Sir Arthur Travers Harris (1892-1984). Senior Officers' School Sheerness, 1924 and Staff Course Camberley, 1927-29.

40 Philpott, *Royal Air Force*, p. 321 and Henry Probert, *Bomber Harris: His Life and Times, The Biography of Marshal of the Royal Air Force Sir Arthur Harris, the Wartime Chief of Bomber Command* (London: Greenhill Books, 2001), p. 61.

41 English, 'RAF Staff College and the Evolution of British Strategic Bombing Policy', *Journal of Strategic Studies*, Vol. 16, No. 3, p. 410.

and functional arms were very developed, indeed. In 1928, the headquarters' element of the Navy measured over 3,100 officers, ratings, and civil servants compared to approximately 2,100 for the War Office and 1,800 for the Air Ministry.[42] Further, the Service enjoyed a host of dockyards and overseas ports where it could draw sustenance and support and though these were fixed, largely known establishments, the paths to them were inherently more flexible than the roads and rails networks supporting military and air operations.

Frequently removed from the threat of immediate attack, these bases allowed naval forces to focus on offensive operations—or in the vernacular of today, power projection—in a manner unknown to British military and air forces. Those arms had to transport much of their supporting tail with them on active service and then needed to ensure the security of their rear operating areas. This luxury of not having to look after their rear area was fast slipping away with the advent of the air. Even if a dockyard might not be captured, then its use might become untenable in the face of air attack. The implications of this threat to the strategic utility of naval power was openly acknowledged in appreciations covering any Far Eastern operation, but rarely mentioned when the threat considered arose closer to home. This threat from the air goes far to explain why amphibious operations were always going to be hazardous affairs and it also explains why military forces required such sizeable staffs. Protecting their lines of communications to the base area and ensuring that the beans, bullets, and other ancillary support elements so vital to the conduct of any operation were available were no small tasks for armies and air units. Consequently, a key question asked in any military or air appreciation developed was: Is the plan administratively sound? This may have been thought a particular bugbear for the Army given that it might have to fight in desert, mountain, or similar hostile terrain where essentials such as an adequate and safe water supply might be lacking. Nor could one assume away how the troops and horses were to be feed and sustained, yet, no Service thoroughly imbued with a 'staff culture' was immune to emphasizing the logistical requirements of any scheme.[43] For a corps sized operation, the supporting headquarters in the British Army was over 400 men and 40 vehicles, and, if the entire expeditionary force were to deploy, then the headquarters element amounted to over 800 men and 70 vehicles.[44]

As their counterparts found whilst attending that year's Greenwich Staff Course, Commanders Geoffrey Robson[45] and Lennox Boswell[46] had their 1938 Andover

---

42 CAB 24/193, Chancellor of the Exchequer note C.P. 92 (28) of 20 March 1928.
43 Slessor, *Central Blue*, pp. 94-95.
44 Slessor, *Air Power and Armies*, p. 97.
45 Later Vice Admiral Sir William Geoffrey Arthur Robson (1902–1989). Short Course, Cambridge University, 1922; Commanding Officer, HMS *Wishart*, 1936; Staff Course, 1937; Staff Course Andover, 1938; Tactical Investigation Section, 1939; Tactical Course, 1939; Captain, 1941; Rear Admiral, 1951; Vice Admiral, 1954 and retired, 1958.
46 Later Captain Lennox Albert Knox Boswell (1898-1975). Five firsts in examinations for Lieutenant; Gunnery Officer in *Furious*, 1926-28; Staff Course, 1935; Fleet Gunnery

course temporarily interrupted for ten days due to the partial mobilization conducted during the height of the Munich Crisis. This saw qualifiers and directing staff both reassigned to operational duties, and, if the hiatus proved to have little impact on their overall studies, then it did at least remind the wayward what the greater objective remained.[47] The same circumstances arose again the following year when war against Germany was declared by Britain on 3 September 1939. Thus, 1937 proved to be the last Air Force Staff College session not curtailed by real-world events.[48]

As was common at Greenwich, Andover students researched papers and presented their work as part of the coursework completed. The tendency for naval officers to write of which they knew most and not move the stakes in their learning may have been a strong one. Alternatively, it may be that the naval and military officer attending was in a better position to enlighten the air qualifier on aspects of their respective Services and so they were encouraged to write along such lines. Whatever the true reason, for Royal Navy officers, naval topics predominated in the papers presented. Commander Charles Daniel[49] is a case in point. Attending Andover's tenth course in 1932, Daniel delivered two papers on the Battle of Jutland.[50] Having only completed his Greenwich Staff Course the previous year and having witnessed Commander Tennant's fulsome treatment of the battle, one option facing Daniel would have been to capture the essential points for his more limited foray. Yet, Daniel's survey was hardly a recasting of Tennant's and showed him possessing original thoughts in his own right. Here, Daniel benefited from being a veteran of the encounter having served as a turret officer in the *Orion* and being specially recommended and receiving promotion as a consequence—something Tennant could never claim.[51] Lecturing to a joint audience, Daniel reminded his fellow officers why studying the battle was important, for it remained the only example of a modern fleet action. He stressed that the Navy had taken on board its lessons—lessons the other Services would be wise to grasp as well.[52]

Officer in *Nelson* and *Rodney*, 1936-38; Staff Course Andover, 1938; Commanding Officer, HMS *Pelican*, 1939; Captain, 1940 and retired, 1950.

47   *Hawk*, 1938, p. 10.

48   Unpublished manuscript, Air Commodore Colin Simpson Cadell Papers, Liddell Hart Centre for Military Archives, King's College, London.

49   Later Admiral Sir Charles Saumarez Daniel (1894–1981). Signal School, 1923-24; Fleet Wireless Officer in *Revenge*, 1926-27 and *Nelson*, 1927-28; Signal School, 1928-30; Staff Course, 1931; Staff Course Andover, 1932; *Glorious*, 1933-34; Tactical Course, 1934 and 1938; Captain, 1934; Imperial Defence Course, 1935; Senior Officers' Technical Course, 1936; Plans Division, 1936-38; Short Anti-Submarine Course, 1938; Captain (D) Sixth Flotilla, 1938-39; Director, Plans Division, 1940-41; Commanding Officer, *Renown*, 1941-43; Rear Admiral, 1943; Combined Operations, 1943-44; British Pacific Fleet, 1944-45; Third Sea Lord, 1945-49; Vice Admiral, 1946; Commandant, Imperial Defence College, 1949-52; Admiral, 1950 and retired, 1952.

50   'Jutland I', Charles Daniel lecture delivered in 1932, Tennant Papers, NMM/TEN/41/5.

51   ADM 196/56 (Daniel).

52   'Jutland I', Charles Daniel lecture delivered in 1932, Tennant Papers, NMM/TEN/41/5, p. 1.

Daniel was an extremely proficient Signals Officer who shined in his duties. Beyond sitting the Greenwich Staff Course and the one at Andover, he also would complete the Imperial Defence Course. What he did not complete were his examinations for Lieutenant taking only his seamanship assessment and passing with a first. Nor did he attend the abbreviated course at Cambridge University that other officers of his seniority had. Instead, from the fleet he took the Long Signals Course in 1918 returning to HMS *Minotaur* on completion of his course at war's end.[53]

In keeping with what he had learned at Greenwich, Daniel noted that initiative was the important lesson of the battle. Where it had been displayed, invariably the results had been amply rewarded. Still, Daniel reminded those surveying the battle after the fact to make allowance when initiative was not always present for Jutland highlighted something the limited peacetime exercises frequently masked: exhaustion was a factor in war 'that enters largely into the conduct of operations. This has to be remembered when we criticize the lack of alertness at night after a day of battle.'[54] Daniel also noted the need for a common doctrine to guide the actions of officers and, if such was the case when only a single Service was engaged, then the audience could draw their own conclusions how much more this was needed in the joint environment now facing them.

The RAF officer was also apt to be younger and a rank lower than the normal naval officer who trained as a staff officer. Moreover, in his preparatory background leading to his commission, his path was very much akin to his Army counterpart coming frequently as he did from the public schools of Britain and then attending a two-year course of professional study at the RAF College at Cranwell. The entrance examination he sat for Cranwell was the same as for the budding military officer entering Sandhurst or Woolwich and his test score in large measure determined whether he became an engineer or gunner, wore the light blue of the RAF, or served in the infantry or cavalry.[55] Officers of Wing Commander rank at Andover were the exception and, after 1931, completely absent save for the presence of officers from the medical branch, of which, more remains to be said. The initial presence of officers holding Wing Commander rank may have reflected their availability and presence in the Royal Air Force following the war and the fact that they had missed the opportunity to attend a staff college while the World War was raging. Typically, the fashion was to send its more senior mid-level officers requiring staff training to Camberley or Greenwich and not to Andover. Thus, Group Captain Arthur Longmore and Wing Commander Arthur Harris received their staff training with the Army and not at Andover, whilst Busteed, as previously discussed, went to Greenwich.[56]

53   ADM 196/145 (Daniel).
54   'Jutland II', Charles Daniel Lecture delivered in 1932, Tennant Papers, NMM/TEN/41/5.
55   Longmore, *From Sea to Sky*, pp. 138-39.
56   Probert, *Bomber Harris*, pp. 58-59.

Meanwhile, by 1931, only Royal Navy officers holding the rank of Commander went to Air Force Staff College, which may be contrasted to the practice of sending officers holding their captaincy to Camberley. Lieutenant Commander George Oldham,[57] RAN, an officer with much air experience sought to attend the Air Force Staff College in 1938 following his year at the Royal Naval Staff College.[58] If he had obtained an appointment, then he would have become the first naval officer of any Dominion to do so. In the event, he failed in his efforts and returned to Australia for duties with that Commonwealth's Navy Office. Notwithstanding his air background, Oldham's lack of seniority was probably the greatest hindrance to attending Andover rather than his RAN status. Only a recently promoted Lieutenant Commander (his rank dated from 15 July 1937), he lacked the perceived gravitas of Commanders Boswell and Robson who eventually represented the Navy at the 1938 course.[59]

Having officers of some seniority no doubt made the naval contribution to the course more prominent, but it came with a degree of discomfort to the directing staff who were mostly of Squadron Leader rank. Of course, this contrast in the typical age between the air and naval officer qualifying for staff duties had much to do with the physical demands and requirements of their Services. Flying (and most of the officers attending Andover were qualified air personnel) in the 1920s was a young man's pursuit where reflexes, eyesight, and a degree of bravado counted as much in an airman's survival as equipment maintenance, weather, and a strong dose of luck. In this, the airman was closer to the submariner. The airman's flying career typically ended by the time he reached the rank of Squadron Leader and the transition to staff duties with command assignments later following were what remained. In the Navy and the Royal Marines, if his specialist duties ended at much the same point as the air officer, then the call of the sea was ever present. Command of a lesser vessel was the norm for the non-specialist officer whilst serving in staff or divisional duties afloat remained for the specialist. Finally, formal staff training was apt to follow and *not* precede the naval officer's initial posting to a staff; a pattern similar to that adopted within the Army and frequently practiced in the RAF.

Though the air officer seeking to attend Andover soon had to pass a competitive examination, he was spared the pain of sitting a final examination to gauge his overall mastery of course material.[60] Taking the examination was not always a simple and

57   Later Rear Admiral George Carmichael Oldham (1906-1974). Observer's Course, 1930; *Glorious*, 1930-33; Meteorology Course, 1933; *Eagle*, 1933; Staff Course, 1937; Radio Warfare Course, 1938; Staff Officer Operations and Intelligence, 1940-41; Commanding Officer, HMAS *Swan*, 1942-43; Captain, 1948; Director of Naval Intelligence in Navy Office, 1948; Commanding Officer, *Australia*, 1949-51; Imperial Defence Course, 1957; Rear Admiral and retired, 1962.
58   NAA: A6769, Oldham, G. C.
59   *Navy List*, March 1939, p. 52.
60   Brooke-Popham, 'Fourth Course', lecture of 5 May 1925, Brooke-Popham Papers, 1/5/10, LHCMA.

pain free matter as the experience of Flight Lieutenant William Long[61] demonstrates. Serving in Aden, Long sought his commanding officer's permission to return to Britain to sit the examination. This was given but on the condition that he meet his own expenses of returning home, as it meant that his tour would, of necessity, prematurely end. Long sat his examination at Uxbridge Station in the winter of 1933 and was earmarked for the January 1935 course. In the meantime, he was posted to a squadron of the Auxiliary Air Force at RAF Hendon and bided his time teaching reserve officers to fly.[62]

Once arrived at Andover, the directing staff employed a process of continuous assessment rendering an immediate verdict to the many papers, schemes, and exercises submitted by the qualifier. Where the work fell short, critical comments returned in green ink was a certainty. Additionally, interim reports were issued to qualifiers indicating their overall progress and an officer was in a fair position to judge whether one's prospects were favorable. Consequently, passing was not automatic with some officers, including Eric Davis, failing in their efforts to reach the required standard.[63] The Air Force was not shy to fail an officer at Andover and academic performance was rarely the proximate cause. Rather, if an officer displayed a certain pigheadedness in his approach or was too argumentative with staff, then failure was a strong possibility. This had nothing to do with competence but everything to do with developing those traits so esteemed in a staff officer: tact, patience, and an ability to work as part of a team.[64]

Exactly why the naval member going to Andover was now a more senior officer than was the custom when the course initially formed is unknown, but a number of reasons suggest themselves. First, the Admiralty gradually deferred formal staff training until later in an officer's career and this tendency only grew as the age for consideration increased during the period. A corollary to this phenomenon is that the officer now under instruction likely held a more substantive rank as a result. Another possibility is that the officer not possessing a specialist skill was at a disadvantage to his counterpart who did when the operational tempo of the fleet receded in the face of fiscal stringency. After all, the specialist could mark time as an instructor in his parent establishment, such as the *Excellent* or the *Vernon*, which was a possibility largely missing for the salt horse with no identifiable professional home other than a ship. Sending the salt horse to Andover was one means of mitigating this problem, came with the benefit that it did not prejudice the Navy's attempts to increase the size of the FAA, and kept a good

---

61   Later Air Vice Marshal Francis William Long (1889-1983). Staff Course Andover, 1935.
62   Air Vice Marshal Francis William Long, Sound Recording 4799 of December 1980, Imperial War Museum, London.
63   Marshal of the Royal Air Force Sir William Forster Dickson, Reel 5, Sound Recording 3168 of February 1978, Imperial War Museum, London and RAF Staff College Andover to Air Ministry letter SC/60/Air.1 of 30 March 1926, Brooke-Popham Papers, 1/5/13, LHCMA.
64   Elmhirst, Sound Recording 998 of 20 December 1977, IWM.

officer employed. Unlike many of the officers who had completed the War Course, the Tactical Course, or the SOTC and soon found themselves without further employment, officers attending the Air Force Staff College were not so soon cast aside. This says something about those selected to attend even, if they did not always appear much higher in the *Navy Lists*. Finally, the Air Force requirement that the officer attending from another Service should be a graduate of their Staff College meant that the naval member would likely come as a more senior student. This posed another sort of problem *vis-à-vis* his Air Force counterpart as will be related.

The naval officer studying at Andover need not be an aviator—indeed, few actually present were—only five attending during the period.[65] As for other officers who had specialized and attended, the distribution was fairly even amongst Gunnery, Signals, Navigating Officers, and submariners. Only one anti-submarine specialist attended Andover and none from the torpedo and PRT branches. The dearth of aviators may be thought unusual and can be attributed to two factors. First, sending such officers to the Air Force Staff College was the naval equivalent of carrying coals to Newcastle, as these personnel already were in close contact with the RAF. Secondly, their form of specialization was of a unique type. They actually practiced the skills they had acquired on a regular basis and did not simply manage others. Where the Gunnery Officer oversaw his turret crew or the Signals Officer supervised his telegraphists, aviators had to actually fly their aircraft or observe for fall of shot. Committed as the Navy was to expansion of the FAA and desiring to meet its required quota of aviators, having these officers attend the course was always problematic. Moreover, as they were required to return to general naval duties on a periodic basis, any window for attending the course was necessarily a narrow one. Consequently, where the vast majority of naval officers attending Andover held the naval *psc*, no naval or marine aviator attending did, though Wildman-Lushington would eventually secure one by virtue of his later employment on the directing staff of the Staff College. In all events, what emerges clearly is that of the naval officers attending, it was that proverbial all-rounder, the salt horse, whose presence was most frequently felt at Andover.[66]

Unlike the Army and the Navy who could look to a long record of serious discourse on the place of armies and navies in warfare, the RAF faced the immediate challenge that its operational experience was much more limited. Perhaps more than any other reason, this circumscribed experience gave rise to the Air Force Staff College's motto, '*Visu et Nisu*', which was loosely translated by the Air Force as, 'By Vision and Effort'.[67] Consequently, foresight played a strong part in shaping the initial doctrine of

65   Namely, Commander Edye Boddam-Whetham, Lieutenant Commanders John Garland, James Hawkins, and Henry Johnstone and Captain Godfrey Wildman-Lushington, RM.

66   Of the 33 officers traced, five specialized in Navigation, four in Gunnery, four were qualified in Signals, one was an Anti-Submarine Warfare specialist, four were Naval Observers, one was a Pilot, four served in Submarines, and ten officers had no defined specialist skills at the time of their attendance.

67   Air 2/2376, Air Ministry to College of Arms letter of 21 February 1938.

the Royal Air Force and the teachings of the course. This too applied to its theoretical underpinnings, and, though this would change, of course, as the period progressed, in the immediate aftermath of the World War a doctrinal vacuum existed. Air Marshal Sir Victor Goddard[68] having experience of both the Naval and Air Staff Colleges held that Andover was the more forwarding looking in approach whilst Greenwich had its eye firmly set on the past.[69] This observation was not wrong, but it is misleading for the simple reason that the Navy had recourse to a longer historical record upon which to draw when framing its instruction.

The relative lack of operational experience for the Air Force was virgin territory and one means of alleviating this shortcoming was to have each qualifier recount their experiences from the war. Difficult it was to say what the true lessons of air power were and the correct means of applying it. Having qualifiers recount their own stories allowed the directing staff to mine these experiences for nuggets casting light on the proper role of the new arm.[70] This was all to the good, but, unfortunately, from a naval perspective the experiences captured all too often focused on the role of air forces in supporting military operations in France, Belgium, and Palestine, or underscored the missions of the Independent Air Force which had been formed to act as a strategic bombing element. Missing was the vital role aviation played in defeating the submarine threat, an omission that was equally present in the lectures provided by the directing staff. Again, these lectures emphasized the role of the Royal Air Force and Royal Flying Corps in France. Fleeting references only appear indicating the breadth of contribution made to the anti-submarine patrols supporting the Navy or of the Christmas 1914 air raid on Cuxhaven by the Royal Naval Air Service.[71]

In fairness, this was in the Staff College's early days and time would show that the maritime role of airpower would receive greater prominence in the course. If nothing else, naval officers attending, such as Commander Angus Cunninghame-Graham delivering a paper on the role of aircraft in the anti-submarine war of 1914-17 whilst sitting the course in 1934 would see to that. Yet, that even in its initial stages the Air Force Staff College neglected the air's contribution to the naval war is the more remarkable because contrasting the differences between the traditional military and naval strategies to that of the air was a cardinal feature of the Andover entrance examination. So, too, was how to employ air forces when working with a fleet with both topics featuring in the 1924 screening assessment.[72]

68   Air Marshal Sir Robert Victor Goddard (1897-1987). Staff Course Andover, 1929 and Staff Course Greenwich, 1934.
69   Air Marshal Sir Robert Victor Goddard, Reel 7, Sound Recording 3189 of 17 April 1978, Imperial War Museum, London.
70   Air 10/973. *Selected Lectures and Essays First Staff College Course*, Air Ministry.
71   Brooke-Popham, 1/8, LHCMA.
72   Air 10/123, *Air Publication 1118*, March 1925, 'Report on the Qualifying Examination for R.A.F. Staff College, Andover'.

Brooke-Popham registered the view that the continuing use of airships in support of fleet operations was a lesson in danger of being lost and said as much in a letter to the Air Ministry in December 1923. His concern arose from the results of two exercises held jointly between Andover and Greenwich remarking that no other means existed to afford the Navy the kind of continuous reconnaissance it desired other than by lighter-than-air airships. This may have been so, but it was not a point that Trenchard welcomed from the new Staff College. In fairness to Trenchard, his rejection was based primarily on the lack of funds to develop both fixed-wing and lighter-than air capabilities with the Air Ministry opting for the former.[73]

One hindrance that plagued the Air Force in its review of naval air co-operation was that the appropriate section of *C.D. 22, Operations Manual, Royal Air Force* dealing with the matter was classified Secret at the request of the Admiralty. This chapter drew heavily upon the *Battle Instructions* of the Atlantic Fleet and on Confidential Air Orders. Indeed, if the naval co-operation was not treated at all in *C.D. 22*, then it would have been possible to issue the *Manual* as an Official Use Only publication. This would have suited the RAF, as it was desirous of imparting its doctrine across the greater air force and faced the challenge of recalling the copies that had been issued directly to Staff College qualifiers, including those naval ones, when it had not collected them at the end of the course.[74]

Rendering an exact judgment how extensively naval air requirements and naval air co-operation were covered at Andover is extremely difficult. Exercises such as the one where a sub-committee formed by 'Pug' Ismay,[75] Indian Army, Slessor, and Flight Lieutenant Thomas Elmhirst[76] prepared a written appreciation on how best to pass a convoy through the Mediterranean in a presumed war with France and its anticipated air support were conducted, though they were peripheral to the course's broader studies. Ismay, already a graduate of the Quetta course and Slessor with previous experience in staff duties courtesy of time spent at the Air Ministry were joined by Elmhirst, a veteran of the RNAS who had specialized in airships after seeing action in a battle cruiser at the Dogger Bank. Their appreciation was thought superior to

73   Air 5/339, Commandant Royal Air Force Staff College to Secretary, Air Ministry letter of
      8 December 1923.
74   Air 5/299, Assistant Chief of the Naval Staff to Director of Operations and Intelligence
      letter of 16 March 1922. *Operations Manual, Royal Air Force* was issued in 1922 and was
      supplanted by the *RAF War Manual* in 1928. See Peter Gray, *The Leadership, Direction and
      Legitimacy of the RAF Bomber Offensive from Inception to 1945* (London: Continuum, 2012),
      p. 62.
75   Later General Hastings Lionel Ismay, Baron Ismay (1887-1965). Staff Course Quetta,
      1922-23; Staff Course Andover, 1924-25; Assistant Secretary, CID, 1926-30; Deputy
      Secretary, CID, 1936-38; Secretary, CID, 1938-40 and Deputy Secretary, War Cabinet,
      1940-45.
76   Later Air Marshal Sir Thomas Walker Elmhirst (1895-1982). Staff Course Andover,
      1924-25 and Senior Officers' War Course, 1934.

the school solution previously prepared by the directing staff and received very good marks as a consequence.[77]

If the content of the Air Force Staff College instruction was focused on the needs of the new Service and its potential value in war, then its style in many respects mirrored what was taught at Camberley and Greenwich. With Brooke-Popham, the first Commandant, and Squadron Leader Bertine Sutton,[78] a member of the directing staff, both graduates of Camberley, and with Freeman serving as the senior instructor a graduate from the Royal Naval Staff College, this was only natural. Doubtless, the uncertainty of formal Air Force doctrine made the use of more traditional teachings an imperative. Still, Brooke-Popham and Freeman were extremely gifted expositors and others would reap where they had sown to the benefit of future courses.[79] Air Commodore Edgar Ludlow-Hewitt[80] and John Slessor added to their output and Freeman would return to serve as a future Commandant.

The initial curriculum at Andover concentrated on air strategy, tactics, the organization of the RAF and its associated training. Imperial strategy was also addressed, as were elements of naval and military strategy and tactics. As with his Greenwich counterpart, the air officer attending learned something of the organization of the Navy, Army and the forces of the Dominions. Additionally, training in staff duties was common to the curricula of all the Staff Colleges. This included the familiar topics of drafting orders, writing appreciations, preparing correspondence and securing a strong foundation of the role and purpose of the staff and its work in general.[81] Still, at Andover strong emphasis was placed on the proper framing of verbal orders once one had grasped the essentials of how to draft properly written signals. This facilitated speed of tactical execution in air operations and was a form of instruction missing at Greenwich.[82] Moreover, the Andover officer was explicitly trained in the duties of both higher command and unit command and how to train other officers—something Greenwich expressly avoided.[83]

It might be thought that in an arm where all its senior officers began their careers in one of the older Services the temptation to stop sending some of its junior officers to Greenwich, Camberley and Quetta would have been a strong one. This was not the case—at least for officers destined to serve at Andover—and the custom for one of the

77   Air Marshal Sir Thomas Walker Elmhirst, Sound Recording 998 of 20 December 1977, Imperial War Museum, London.

78   Later Air Marshal Sir Bertine Entwisle Sutton (1886-1946). Staff Course Camberley, 1921; Staff College Andover, 1922-26; Imperial Defence College, 1929-32 and Commandant, Staff College Andover, 1941.

79   Slessor, *Central Blue*, pp. 40-41.

80   Later Air Chief Marshal Sir Edgar Rainey Ludlow-Hewitt (1886-1973). Senior Officers' War Course, 1925-26 and Commandant, Staff College Andover, 1926-30.

81   CAB 53/1, Committee of Imperial Defence memorandum of 12 October 1923.

82   Air Chief Marshal Sir Aubrey Beauclerk Ellwood, Sound Recording 3167 of 20 February 1978, Imperial War Museum, London.

83   Air 5/486, 'Staff College Courses of Instruction'.

members of the Andover directing staff to have taken the Greenwich Staff Course was a firmly established one. In addition to Freeman, whose presence at Greenwich has been noted already, Group Captain André Walser, the senior instructor between the years 1932-36, attended the Royal Naval Staff College in 1929 whilst Group Captain Arthur Tedder, his predecessor, had qualified at Greenwich during the 1923-24 course. Squadron Leader Frederick Hards,[84] an instructor from 1927-29, attended the naval course in 1921-22 presenting as his qualifying lecture, 'Long Distance Air Communications'.[85] A former pilot in the RNAS, Hards also attended the Greenwich-based SOWC before arriving at Andover. He subsequently served in the Mediterranean Fleet, Malta, and with RAF Coastal Command making him one of the most qualified and experienced air officers assigned to naval duties. Finally, Ludlow-Hewitt, Andover's second Commandant and bereft of a higher military education, prepared himself for his new duties by sitting the Navy's Senior Officers' War Course from October 1925 through February 1926.

Of course, the majority of officers on the directing staff at Andover had not attended either the naval Staff Course or the SOWC, and, by the late 1930s, finding even one instructor possessing such qualifications was proving problematic. Yet, that does not mean that the Air Ministry neglected to assign officers with naval experience; only that the experience was now of a more practical nature. Group Captain Grahame Donald,[86] the senior instructor from 1935-37, began his career as probationary surgeon in the Royal Navy transferring to the RNAS and more martial duties in 1915. Flying shipboard and amphibian aircraft with RAF Coastal Area emphasized the maritime orientation of his Air Force career and he eventually assumed command of the School of Naval Co-operation in 1929. Meanwhile, Group Captains Christopher Courtney,[87] the senior instructor between 1925-28 who did so much to shape air training in later years, and Ronald Graham,[88] a member of the 1937-38 directing staff, along with Wing Commander Robert Saundby,[89] an instructor from 1934-37, began their careers

---

84   Later Air Vice Marshal Frederick George Darby Hards (1889-1963). Staff Course Greenwich, 1921-22; Senior Officers' War Course, 1924; Staff College Andover, 1927-30; Imperial Defence Course, 1932; Fleet Aviation Officer, Mediterranean Fleet, 1933-35 and Costal Command, 1938-40.

85   ADM 203/100.

86   Later Air Marshal Sir David Grahame Donald (1891-1976). Staff Course Andover, 1923-24; Staff College Andover, 1924-27; Commander, School of Naval Co-operation, 1929-31; Imperial Defence Course, 1931, Staff College Andover, 1935-37 and Imperial Defence College, 1937-38.

87   Later Air Chief Marshal Christopher Courtney (1890-1976). Staff Course Quetta, 1924 and Staff College Andover, 1925-28.

88   Later Air Vice Marshal Ronald Graham (1896-1967). Staff College Quetta, 1922-23; Staff College Andover, 1926-30 and 1937; Imperial Defence Course, 1936 and Imperial Defence College, 1938-39

89   Later Air Marshal Sir Robert Henry Magnus Spence Saundby (1896-1971). Staff Course Andover, 1927-28; Imperial Defence Course, 1933 and Staff College Andover, 1934-37.

with the RNAS. Each having practical experience of naval aviation and operations, if not of Royal Navy's higher educational courses, brought a measure of perspective to their teaching largely absent from their Greenwich counterparts.

Exactly why directing staff members no longer were graduates of the Royal Naval Staff College or failed to attend the War Course is unknown, but the reasons probably owe much to the creation of Service's own Staff College and, subsequently, the formation of the Imperial Defence College. Thus, from 1936 the more senior members of the college—though not, interestingly, the Commandant, Air Vice Marshal Arthur Barratt—usually possessed the *idc* post-nominal whilst many instructors were now prior graduates of their own Staff College.[90] This highlights what was perhaps, then, an unintended consequence of the IDC's creation. For in furthering joint training, doctrine, and co-operation in a tri-Service environment, it exacted a price in the existing bilateral relationships of the Services. As with the earlier creation of the Royal Air Force, this does not mean that the benefits did not outweigh the costs. After all, the officer attending the IDC secured an appreciation for the other two Services, the overall needs of imperial defence, and the economic underpinnings of British policy. Yet, unlike the earlier creation of the Air Force, the price exacted went largely unnoticed. Certainly, this writer has found no contemporaneous commentary alluding to its effects.

If the Army might despair of the typical Air Force staff officer, then there was no mistaking that the RAF sent some of its best to serve as instructors at Andover. Still, the view that the Air Force deprecated proper staff procedure was not an opinion limited merely to the Army, as even RAF officers admitted that relying on the telephone rather than arranging their thoughts first and committing them to paper was a curse of their Service.[91] So, too, it should be noted was it a vice amongst the FAA which had the reputation of having too much of an oral tradition in how it handled staff matters.[92] In this, the Air Force was closer to its military counterpart in how it approached staff work and prepared for battle. The speed of air operations was the controlling factor for the RAF whilst for the Army it was the scale of operations once battle had been joined which necessitated a looseness in command. For the Navy, the scale of battle was smaller than for land forces and though its speed was now quicker than for military forces, it was sedate compared to the air. Consequently, the Navy relied more on set piece orders and standing doctrine to guide its actions and increasingly relied on the initiative of the subordinate officers to anticipate events and act in the knowledge of the Admiral's intent

Even following his training at Andover, the Air Force recognized that its budding staff officer was not necessarily the equal of his Camberley equivalent. Speaking for

90   *Hawk*, 1938, pp. 116-17.
91   Squadron Leader Cedric Bell to Richmond letter of 22 November 1939, Richmond Papers, NMM/RIC/7/4/B.
92   Till, *Air Power*, p. 125.

the final time to the officers of the first course, Brooke-Popham conceded as much reminding them:

> The other point of difference between you and the Camberley student is the fact that you start from a lower average level of general and professional education. Most officers at Camberley have spent at least a year's strenuous work preparing for the entrance exam., and this is an assistance to them. There is therefore all the more reason for you continuing to read and study and think after you leave here.[93]

As few RAF officers were university graduates, a healthy dose of writing, speaking, and argument featured. Gerald Gibbs,[94] a future Air Marshal and at twenty-nine the youngest Flight Lieutenant of his term, was struck by the amount of work required of the qualifiers when he arrived in 1926.[95] The initial part of the course introduced the basic precepts of air warfare followed by examining the duties of the Air Force staff officer and concluded with examining the higher direction of war.[96] Of warfare, officers were told that the bomb was the principle weapon and airpower was primarily exercised through a fleet of bombers.[97] This was a key tenet; otherwise, the Air Force would be but an ancillary arm of the Navy and Army.

Surveys of morale, psychology, and the nature of war would not have been out of place at Greenwich, though emphasizing the leadership skills of Marlborough and Napoleon rather than Nelson and St Vincent was perhaps where paths seemingly diverged. Brooke-Popham during his period as Commandant directed the sessions covering morale, though in common with Greenwich, the topic proved a fertile ground for contributions by qualifiers as many wrote essays on the question. Of the essays written by qualifiers and issued in a booklet following the first course, one-third touched upon some aspect of morale as a factor in war with the balance of the publication consisting of officer operational experiences from the late conflict.[98]

If the emphasis on morale as a factor in war was treated equally amongst the Services, its place was not. For the Army and the Navy, morale was a key tactical consideration and the temperament of an enemy nation was something that one should consider when developing an appreciation. All other factors being equal, a battle's outcome might be decided on the side possessing the greater *élan* and able to sustain its morale in the presence of serious losses. This was a major reason why the *Naval War Manual* called attention to developing moral qualities in its training and exercise regime. This

---

93   Brooke-Popham, 'Final Address', lecture of 29 March 1923, p. 2, Brooke-Popham Papers, 1/5/4, LHCMA.
94   Air Marshal Sir Gerald Ernest Gibbs (1896-1992). Staff Course Andover, 1926-27.
95   Gerald Gibbs, *Survivor's Story* (London: Hutchinson & Co. Ltd., 1956), pp. 48-49.
96   Dickson, Reel 3, Sound Recording of February 1978, Imperial War Museum, London.
97   'Air Warfare I', Air Vice Marshal A. S. Barratt lecture of July 1936, Air Commodore John Francis Titmas Papers, 76/198/1/10/3, Imperial War Museum, London.
98   Air 10/973.

the Royal Air Force conceded as well, but morale for it also retained a strategic import largely missing in the writings of the older Services. For the RAF held that the morale of a people was a legitimate target for air action. The upshot of this emphasis on morale was that targeting the military forces of an adversary, long viewed as the surest way of achieving one's policy objectives, was in time minimized by the RAF. Its view was that airpower was better employed in breaking an adversary's will and ability to resist, and the 'Air Force will contribute to this aim by attacks on objectives calculated to achieve this end in addition to direct co-operation with the Navy and Army in furtherance of the policy of His Majesty's Government at the time.'[99] Trenchard was emphatic on this point and one can be sure that qualifiers were left in no doubt on its centrality to air doctrine.

Slessor in his lectures and writings emphasized that attacks against an enemy's vital centres, rather against an opponent's air forces, had first call upon the Royal Air Force.[100] Exactly what constituted a vital centre was an open question, as it would in large measure reside in an understanding of what an actual enemy needed to safeguard. The important point was that it need not be his military forces. Slessor put it thusly:

> The whole art of air warfare is first the capacity to select the correct objective *at the time*, namely that on which attack is likely to be decisive, or to contribute most effectively to an ultimate decision; and then to concentrate against it the maximum possible force, leaving only the essential minimum elsewhere for security—and possibly to contain superior enemy detachments.[101]

Though there is much in the above formulation that a naval officer would find attractive and nod his head in agreement given its underpinnings of the offensive, concentration, maintenance of the object, security and economy of force, ultimately, the above was contrary to the Navy's understanding of how military force should be applied. It held a more traditional view believing that 'Destruction of the enemy is the object of battle. When it has been decided to seek decisive battle, every means must be concentrated on the destruction of the enemy's forces.'[102] Of course, some actions by a navy, such as blockade, were acts of war but not necessarily acts of battle, proving to be not so much acts of destruction as acts of denial. Further, if battle did follow from blockade, as was hoped, not all forces would participate, as some would continue to carry out their existing duties. Thus, the *Naval War Manual* cannot be

99   Air 5/169, Air Staff memorandum No. 43, S. 28279, 'The War Aim of the Royal Air Force', no date, but *c.* October 1928, p. 3.
100 Slessor, *Air Power and Armies*, pp. 25-27.
101 *Ibid*, pp. 82-83. Original emphasis.
102 ADM 186/86, *Naval War Manual*, p. 27.

read too literally as exceptions to every edict existed. Destruction was the stated aim, but capture, if possible, was at times the better course else why have a *Navy Prize Manual*.[103]

Of course, where engaging the military force of the enemy was central to air doctrine was when standing on the defensive in meeting any air attacks he might launch against the British homeland. Fighter interception, it was taught as late as 1939 at the Staff College, was better served by annihilating the few than intercepting the many.[104] The naval officer shared the view that annihilation was central to battle and the desirability of laying waste to a portion rather than engaging the whole of an enemy fleet was a principle at the heart of naval tactics. The actual application, though, owed much to the circumstances prevailing at the moment of battle, but the *Battle Instructions* of 1934 stressed:

> The fire of the British fleet will be concentrated on a few of the enemy's ships when at long range, in order to obtain a decisive result. As the range is reduced, the degree of concentration will be lessened and, when a range is reached at which the fire of individual ships becomes effective, each ship may be ordered to engage a separate enemy ship.[105]

The question remains, though, was a tactical precept understood and practiced by Nelson and central to contemporary naval doctrine appropriate for air forces when defending what today would be considered a counter-value target?

The same caveats were no doubt present in air power doctrine, but its underlying tenets were expressed in a vehemence and absolutist manner that was disquieting to the Navy. Wing Commander Alec Coryton found whilst studying at the Greenwich Staff Course in the late 1930s that recourse to assertion without supporting facts was deeply frowned upon by the naval instructors.[106] A situation that was more tolerated at Andover. Perhaps if there was a longer record of operational experience in employing airpower this would not been so, as the historical pattern of air warfare would have highlighted how the exceptions operated. This, however, was not the case and the fact remained that airmen might actually attempt that which they believed. The not too subtle message was that cooperation had its limits. Concerting naval and military plans had always been a difficult proposition as their needs at times conflicted. Study of the Dardanelles campaign and the yearly combined exercises held amongst the three Staff Colleges only emphasized the problem. Still, past difficulties in effecting cooperation were nothing compared to the difficulties now inherent in operations involving the Royal Air Force. In the first joint Staff College exercise conducted

103 *C.B. 3, Navy Prize Manual* and *B.R. 2074, Navy Prize Manual, 1923.*
104 Orange, *Park*, p. 72.
105 ADM 186/106, *Battle Instructions*, Section I, Paragraph 9, p. 3.
106 Coryton, Sound Recording 3190 of 18 April 1978, IWM.

after the World War in November 1919, Major General Hastings Anderson, the Commandant at Camberley, made this very point in his report to the War Office observing, 'It has become apparent that, difficult as this particular operation has been in the past, the complications of modern war in or on all three elements have made the task before us even more difficult than ever.'[107] To be fair, the RAF recognized that conforming its actions to the other Services was a *sine qua non* in amphibious operations, and so neutralizing an opposing air force was the one exception conceded to the norm of executing strikes against distant vital centres.[108] That was a start, but what of the myriad other tasks a navy might execute where air support was crucial? Time would show that the doctrine and procedures governing such were lacking.

As a unified air service, the RAF had responsibility for all air matters throughout most of the period under review and fielded both Army Co-operation and Naval Co-operation squadrons as specialized formations to work with the older arms. At each of the Staff Colleges, the respective roles of these squadrons featured in the curriculum. The mission of those squadrons earmarked for the British Army was primarily to provide tactical reconnaissance and support divisional fires operating under the direct control of the military element being supported. Direct support of a more active nature was beyond their remit and requests for air support were passed between the headquarters of the two deployed forces.[109] Speaking many years afterwards, Alec Coryton thought a fundamental flaw in the Staff College coverage of how the co-operation squadrons were to work was that it assumed the presence of a negligible air threat.[110] In theory, the Navy was in a slightly better position as beyond the squadrons committed to Coastal Area and, subsequently, styled Coastal Command, the embarked units of the FAA operating on its carriers were dedicated to maritime needs. This, however, was not so nearly true, as the Air Ministry reserved the right to limit the number of aircraft and crews embarked to but half a carrier's presumed complement and deploying the remainder only when operationally necessary.[111]

Notwithstanding the probable efficacy of air strategy, doubts even then were voiced about its moral implications. Once military forces are not the expressed aim in operations, the place of the non-combatant becomes problematic. If attacking vital centres does not explicitly target civilians, then it does ensure that strikes are likely on locations where their presence is greater. Refugees in the past streaming to the rear areas during battle may have not been immune to war's deprivations, but it was at least a practical attempt to sanitize the battlefield for armies. The official Air Ministry position was that such bombing and strafing attacks would conform to the existing

---

107 Air 1/516/16/4/17, Report by Commandant on Naval and Military Exercise conducted November 1919.
108 Slessor, *Air Power and Armies*, p. 38.
109 Shelford Bidwell and Dominick Graham, *Fire-Power: British Army Weapons and Theories of War 1904–1945* (Boston: George Allen & Unwin, 1985), p. 263.
110 Coryton, Sound Recording 3190 of 18 April 1978, IWM.
111 Till, *Air Power*, p. 194.

rules governing naval bombardment and would not be merely indiscriminate attacks on civilians.[112] Whether this was truly believed or mere eyewash cannot be said, but the practicalities of air operations given the limitations of the day's technology meant that it was not a sustainable tenet. Air Commodore Lionel Charlton serving as Chief Staff Officer in Iraq resigned his post in 1926 after observing the implications of the policy at firsthand.[113] To be sure, in Iraq, air operations were in aid of the civil power and were only to be employed after the lesser means available to the state had been exhausted or proved ineffective.[114] What would follow when air power was applied with vigour when the gloves were off remained to be seen, but contemporary commentators and writers, including Charlton himself, left little to the imagination and the analogy of Hiroshima and Nagasaki in 1945 to his presumed war of 1942 shares an eerie similarity.[115]

The naval officer attending Andover would have enjoyed another feature common to all the Staff Colleges: a healthy regime of tours to other establishments and practical demonstrations. In 1935, the members of the course visited the Ford Motor Company at Dagenham.[116] Given the importance that industrial capacity and production capabilities assumed in fielding modern aircraft and the responsibility the Air Ministry assumed for defining requirements in light of what was achievable, such visits had a practical side and should not be viewed as a mere dog and pony show. Sometimes, though, a tour or demonstration produced the unexpected and when a parachute practice went terribly wrong, Squadron Leader Sutton, witnessing the tragedy, asked for a nearby aircraft, strapped on a parachute rig, and finished the display.[117]

Needless to say, most activities held outside the lecture hall, including the common exercises held at Camberley and the yearly visit to Greenwich previously discussed, were not as eventful. Still, aviation pursuits were inherently risky and visits to the fleet where progress in naval flying were observed at firsthand, such as to the carrier *Courageous* in both July 1937 and July 1938, should not be discounted as mere sideshows. This is surely the primary reason why the Admiralty initially restricted naval qualifiers at Greenwich from taking such flights when witnessing similar proceedings and only relaxed the restriction at a later date. Losing an officer was wasteful, but another consideration was that during most of the period between the wars, the Admiralty did

112 Air 5/169, Air Staff memorandum No. 43, S. 28279, 'The War Aim of the Royal Air Force', no date, but *c.* October 1928 p. 5.

113 Powers, *Strategy Without Slide-Rule*, p. 172.

114 Vincent Orange, *Dowding of Fighter Command: Victor of the Battle of Britain* (London: Grubb Street, 2008), p. 43.

115 L. E. O. Charlton, *The Menace of the Clouds* (London: William Hodge and Company Limited, 1937), pp. 286-93.

116 Anthony Furse, *Wilfrid Freeman: The Genius Behind Allied Survival and Air Supremacy 1939 to 1945* (Staplehurst: Spellmount Limited, 2000), p. 103.

117 P. B. Joubert, 'Sir Bertine Entwisle Sutton', eds., L. G. Wickham Legg and E.T. Williams, *Dictionary of National Biography 1941-1950* (Oxford: Oxford University Press, 1967), p. 854.

not provide insurance benefits to officers who died whilst serving. Thus, one officer was allowed to fly in a RAF aircraft as long as he accepted all the risks associated with the flight.[118] Still, such visits were very much the stock in trade for Andover and the previous year's course had visited RAF Calshot whilst the Supermarine aircraft facility featured in June 1937.[119] Moreover, if a tour to a European battlefield was a common enough occurrence at Camberley, then the Andover qualifier during his summer recess was expected to visit a foreign counterpart to learn something of another air force. Visits to Polish, Turkish, German, French and American establishments all featured at one time or another as part of the Air Ministry's Foreign Travel Scheme and qualifiers submitted their observations in written reports. The actual intelligence secured was negligible but it did establish personal links with other countries' air elements, and broadened one's horizons to consider how others were fostering such forces.[120]

When not engaged in attending lectures, tours, or preparing his own work for submission, the officer attending Andover had ample opportunity for pursuing sports and, of course, flying. The last named not merely restricted to qualified pilots, as it was possible to catch a ride for those whose responsibilities were anchored more firmly on earth. Major Hastings Ismay, already a graduate of Quetta, attended Andover's third course during 1924-25, thus, becoming the first Indian Army officer to do so. Frequently after class, he proceeded to the nearby airfield in the hope of securing a flight from a friendly face. After a fellow student warned him that not all Staff College pilots were of equal ability, Ismay tempered his enthusiasm for flying by opting only to share his destiny with the better aviators his classmate had identified.[121]

Where Andover did broach new ground was in assigning a medical officer to attend the third, last term of each course.[122] This innovation dating from 1933 saw an officer of Wing Commander rank present whilst a representative from Porton Down, the primary chemical warfare research and testing facility operated by the War Office, visited. These lectures lasting for a week reviewed the trends, threats, capabilities, and defences available.[123] Though every bit a student, the medical officer served a purpose akin to having naval and military officers at Andover: he acted as a quasi-member of the directing staff enlightening qualifiers on the effects of chemical warfare during the course's culminating exercises. Conversely, he became acquainted with current opinion on the efficacy of chemical weapons and any of the other subjects studied

118 ADM 196/53 (Hamilton).
119 *Hawk*, 1938, p. 73 and *Hawk*, 1937, p. 24.
120 Goddard, Reel 4, Sound Recording 3189 of 17 April 1978, IWM.
121 Hastings Ismay, *The Memoirs of General the Lord Ismay, K.G., P.C., G.C.B., D.S.O.* (London: Heinemann, 1960), p. 41.
122 *Hawk*, 1938, pp. 115-117.
123 Philpott, *Royal Air Force*, p. 97 and G. B. Carter, *Porton Down: 75 years of Chemical and Biological Research* (London: Her Majesty's Stationery Office, 1992), p. 27.

during his four-month presence at the Staff College. To be sure, he did not leave as a qualified staff officer, but he no longer remained a simple flight surgeon.

Having little operational experience to draw upon, the Air Force qualifier was treated to the presumed benefits and probable role of air power in a manner that was largely foreign to those studying at Greenwich. As such, the support that RAF Iraq Command was providing in imperial policing was a central element of this instruction as, too, was the role of the RAF during the recent standoff with Italy in the Near East. Brooke-Popham, fresh from duty in Egypt, returned to Andover and lectured on both of these themes. Exercises conducted in association with lectures introducing the topics were a regular feature of Staff College training and schemes and those where officers demonstrated their leadership to critical assessment were especially prominent.[124] The ethos that evolved at Andover and within the Royal Air Force emphasized the primacy of the offensive, and this, too, a naval officer trained in the principles of war would have recognized as familiar turf.[125] This is not say that the role of defence was not acknowledged. It was. Yet, it was apt only to be a phase until the time arrived to assume the offensive. Whilst a student at Andover, Gerald Gibbs presented as his qualifying lecture a paper entitled 'Fighter Squadrons in Air Defence' which was later published as part of the contributions of *Air Publication 1308*. Hence, it was not the case that defensive side of operations were neglected, merely that the place of defence was limited.[126]

Yet, one cannot but be struck by the similarities in thinking shared between the RAF and the Navy on the primacy of the offensive occurring at this moment. The Air Force eschewed the role of tying its fighter escort too closely to its bomber force and the Navy, if accepting the need for convoy, sought to have its surface forces act in the most aggressive manner possible against the submarine through offensive patrolling. Slessor summed up the Air Force view in one of his lectures:

> In the British Service a basic principle of bomber training is that squadrons must be capable of looking after themselves by close and steady formation flying, and the resultant mutually supporting fire from the rear guns. We have never favoured a policy of close escort for bomber formations, for usually very good reasons.[127]

Just as the Navy understood that the safe passage of shipping was the ultimate aim and purpose of convoy, so the Air Force accepted that the bomber reaching its target and bombing *accurately* was what was required.[128] Yet, each stopped short of providing the means for achieving its ends, and though many reasons may be cited, a strict

---

124 'Exercise No, 25, Leadership and Morale', Air Marshal Sir Gerald Ernest Gibbs Papers, GB99, Liddell Hart Centre for Military Archives, King's College, London.
125 Slessor, *Central Blue*, p. 41.
126 Gibbs Papers, LHCMA.
127 Slessor, *Air Power and Armies*, p. 46.
128 *Ibid*. Original emphasis.

interpretation of the principles of war with its emphasis on offensive action remains a factor common to both.

This is not to claim that the Services interpreted the relative importance of the separate principles of war equally. This they did not. The weight afforded to the principle of concentration was one area where significant variance existed. The Navy accepted the principle and, indeed, concentration on the enemy battlefleet was emphasized in its doctrine and training. Richmond lecturing on strategy told the officers attending the War Course, 'Our doctrine is that the business of the main fleet is to destroy the main fleet of the enemy.'[129] Yet, for the Navy the concept's utility never approached the degree to which it featured in air doctrine where it served as the cornerstone of its strategy.[130]

In application, the Air Force was closer to the Army's enunciation of the principles of war specifying concentration as *primus inter pares*. Certainly, Dill subscribed to the pre-eminence of the principle of concentrations telling Liddell Hart:

> I am interested to see how you exalt Economy of Force as a principle of war. I look upon concentration as the dominant principle. Is it not by economy of force, mobility & surprise that one aims at concentrating moral & material force at the decisive time & place.[131]

Officially, though, the Army's *FSR* explicitly warned that the relative importance of the individual principles was not static reminding officers:

> They can be defined, but their relative importance and the method of their application are constantly varying. The outstanding cause of this variation in the value and in the application of the principles of war is that, in war, the human factor predominates, and the human factor is an ever varying quantity.[132]

Comparing the *FSR* of 1929 to the *Naval War Manual* of 1925 sheds light if not so much on the relative weight afforded to each of the principles, then to their slightly differing formulations:

British Army	Royal Navy
i. The principle of concentration	Maintenance of the object
ii. The principle of economy of force	Offensive action
iii. The principle of surprise	Surprise
iv. The principle of mobility	Concentration

---

129 Richmond, 'Strategy XI. America and Atlantic Strategy, Part III', p. 14, lecture of 31 March 1921, Richmond Papers, NMM/RIC/10/5a.
130 Slessor, *Air Power and Armies*, pp. 62-63.
131 Dill to Liddell Hart letter of 24 April 1926, LH 1/238/2, LHCMA.
132 *FSR, Vol. II*, p. 7.

v.	The principle of offensive action	Economy of force
vi.	The principle of co-operation	Security
vii.	The principle of security	Mobility
viii.		Co-operation.[133]

For its part, the Royal Air Force followed the same formulation as the Navy saved that it referred to 'maintenance of the objective' rather than 'maintenance of the object'.[134] The short-cut in mental thinking that the principles of war provided became a danger when others not so clever as Fuller, their putative author in the *FSR*, forgot the accompanying caveats, discarded that which they found troubling, and simply accepted them as specified.[135] Slessor was guilty of this in the Air Force, but the Army itself was, too, in rescinding object as a principle without explaining why.

Exactly why the British Army rescinded 'object' as a principle of war is unknown. If no longer featuring as a principle, it still appeared in the 1929 edition of the *FSR* operating as it did at two levels. Firstly, it represented the overarching political goal of the nation at war with object expressly given as 'the subjugation of the opponent's will'. Secondly, it prefaced the introduction to the *FSR*'s discussion of the other principles of war. In this instance, the 'object' acted as the foundation of operational planning influencing where military force was directed.[136] It may be that having it as an explicit principle conflicted with its higher formulation. Removing it allowed the remaining seven tenets to serve as the essential planning elements to be applied in any appreciation developed at the operational level.[137] If this was the reasoning for its removal, then it could only lead to confusion elsewhere, as it was a short stretch in logic to conclude that striking at the enemy's will rather than his military forces was now the more direct and *correct* manner of waging war. To Richmond and the broader Navy, if there was a dominating tenet then it was the principle of the object as 'the object dominates everything & can't be left undiscussed whether we treat of Grand Strategy, Major & Minor Strategy, or Tactics. There is the object of a battle and the object in a battle – different things which I think are too often confused.'[138]

In hindsight, it is easy to see that the drafters of the *FSR* would have been better served to restrict their usage of 'object' to one level of war employing alternate terminology such as objective, goal, or aim when addressing the other. The Army's removal of 'object' as a principle did not cause the Air Force to now espouse attacks on the will; Trenchard had already reached his views previously. Yet, it did now make it harder to resolve the latent conflicts in the interpretation and formulation of the principles

---

133 *Ibid*, p. 9 and ADM 186/66, *Naval War Manual, 1925*, p. 1.

134 Air 5/299, *Operations*.

135 On Fuller's authorship, see Bidwell and Graham, *Fire-Power*, p. 171.

136 *FSR, Vol. II*, pp. 2 and 7.

137 *Ibid*, p. 7.

138 Richmond to Liddell Hart letter of 7 September 1932, Liddell Hart Papers, LH 1/598/29, LHCMA. Original emphasis.

of war held amongst the Services. For most officers if they were even aware of these conflicts, it would have been thought a tempest in a teacup. Lord Gort,[139] for one, believing it was 'useless and in fact unreasonable to expect junior officers to read the F.S.R. The fact remains they won't and if they do they don't understand it.'[140] A view not too dissimilar from one naval officer's observation that those desiring to take the naval Staff Course should read 'the N.W.M., but carefully and in small doses. It is the sort of book which should be "chewed and digested."'[141] Where, though, such differences and nuances in interpretation of the principles were apt to result in problems was in the planning staffs of the Services.

Unlike the Navy which largely eschewed employing the concept of doctrine in its publications, the Air Ministry explicitly accepted doctrine with *RAF Operations* emphasizing 'The Royal Air Force will be trained in peace and led in war in accordance with the doctrine contained in this volume.'[142] With the first six chapters of *Operations* adapted from the 1920 edition of the *FSR* and with the *Naval War Manual* following its form if not is content, the *FSR* was a guiding hand in furthering common doctrine even if that doctrine was not truly joint. The Army, too, accepted doctrine explicitly with the *FSR* advising military officers that it:

> sets forth the principles which govern the employment of armed forces and inspires their action in war. It has as its primary object the elucidation of the doctrine of strategy and tactics by which military commanders will be guided; it also contains instructions on matters of organization and procedure which concern all arms of the service.[143]

This is not without a little irony as the one Service that formally embraced doctrine, did not have a unified arms tactical school and left much of its training to the individual commanding officers of their regiments. Meanwhile, the Service that avoided speaking openly of doctrine had established a tactical school based on the precepts of the standardized *Battle Instructions* and trained its senior officers to the *Naval War Manual*.

For the Royal Air Force, the embrace of doctrine was not without risk as its operational experience was so limited. Moreover, its guiding manual attempted to be prescriptive in a manner that the *NWM* did not which was supported by a host of subsidiary publications such as the *Battle Instructions*. The RAF explicitly recognized that the principles to be followed when conducting war against a first-class power were not the same as those to be employed when facing an 'uncivilised enemy'. Though it

139 Later Field Marshal John Standish Surtees Prendergast Vereker, Viscount Gort (1886-1946). Staff Course Camberley, 1919-20; Staff College Camberley, 1921-23; Senior Officers' School Sheerness, 1926 and Commandant, Staff College Camberley, 1936-37.
140 Lord Gort to Liddell Hart letter of 24 August 1924, Liddell Hart Papers, LH 1/322/15, LHCMA.
141 'Naval Staff College', *Naval Review*. February 1932, p. 31.
142 Air 5/299, *Operations*, p. 1.
143 *FSR, Vol II*, p. 1.

cannot be proven, the suspicion remains that its doctrine and operational experience in Iraq and elsewhere, where attacks on morale were central, heavily influenced its subsequent approach in how to fight wars against highly advanced, industrialized states.

Colonel Pope-Hennessy believed that the operational failings of the British Army during the South African War led it to embrace doctrine.[144] As failure is often the handmaiden to military innovation, this may have been so, but also of relevance was the rich body of literature and experience now existing. In this, military theory was ahead of naval thought and much beyond what existed for the air. This void was probably a strong reason why the Royal Air Force demurred from claiming its precepts were settled doctrine with outsiders even if internally it believed they were. Its organizational existence, precarious as it was, offered a strong secondary reason. Where air power was to be concentrated was a matter of both the moment and the object to be achieved in any war, but its prominence in air warfare theory was a recipe for future conflict with the other Services, as their tactical view of battle was always narrower and more immediate.[145] This points to another fundamental problem that naval officers now had to consider which Andover exposed to them: the very speed of air operations conflated tactical, operational, and strategic objectives in a manner that the traditional arms of war were slow to appreciate. The warship employed on blockade duty or the squadron steaming over the horizon mirrored this blurring, but they could not approach in speed and timeliness the effects that air power now brought to the theatre of operations.

The conflicting views offered over their interpretation to the principles of war were recognized early on by planners with attempts made to resolve the outstanding differences. The compromises that the Royal Navy accepted when drafting its *War Manual* have been discussed previously, but given the presence now of a third arm, further efforts at resolving the matter were thought imperative. In 1928, an attempt to resolve these discrepancies was made, but this effort proved nugatory. In truth, until a higher authority imposed a common doctrine, the attempt to resolve it from below was largely premature. Though the Services largely accepted a common body of principles, their view of their hierarchy and application was very much tied to the medium that they served. Time would demonstrate that this authority would be a Ministry of Defence, but a prime minister actually interested in the minutiae of defence problems might have cracked the nut too.

Given the efforts made in 1928 to resolve their varying interpretations to the principles of war, it is all the more striking that when the Army issued a revised edition of its *FSR* in 1929 object as an explicit principle of war ceased.[146] Much of the histo-

---

144 ADM 1/8710/138, Pope-Hennessy, 'The Services and a Common Doctrine of War', paper of 17 January 1927, pp. 3-4.
145 Air 5/169, Air Staff memorandum No. 43, S. 28279, 'The War Aim of the Royal Air Force', no date, but *c.* October 1928.
146 Brooke-Popham, 'Co-operation', lecture delivered 29 March 1932 to Royal Naval War College, Brook-Popham Papers, 9/4/5, LHCMA.

riography of the interwar period when discussing British inter-Service relations is set against the backdrop of the Air Force's attempts to establish itself in the face of the resistance voiced by the Army and the Navy. This is not wrong, but it is incomplete, and the Army's 1929 revision to the *FSR* highlights that the differences existing amongst the Services had antecedents beyond simply the formation of a new arm. From a naval perspective, any disagreements it had with the Army may have faded to insignificance when compared to those it faced with the Air Force, but the corollary was never the case. Combined operations required the wholehearted commitment of the Royal Navy in support of the British Army in a manner not too dissimilar to the Navy's need of the air. Thus, doctrinal variations existing between any two of the Services were always a source of future problems lying just beneath the surface.

Unlike Andover, which was mainly the preserve of naval officers attending, Royal Marine officers went to Camberley on an equal basis with their naval brethren. Nay, they outpaced the naval officer, as some attended the full course, received a scholarship to do so, and secured the military *psc*.[147] The marine officer was also required to successfully negotiate the Camberley entrance exam, an act naval officers avoided, and this requirement was in keeping with the overall objective of aligning military and Royal Marine professional development.[148] The naval officer most often attended for a single year sitting in the Senior Division, having previously attended the Greenwich Staff Course. Again, it is impossible to claim a firm rule in these matters for officers as junior as Lieutenant and even as senior as Captain appeared at Camberley whilst others preceded to the Army Staff College without sitting the Greenwich course first.[149] Nevertheless, from 1919-39 it was customary for two naval officers to attend each year's Senior Division joined by one or two marine officers. Lieutenant Arthur Allen[150] was the sole naval officer to attend the complete Camberley course between the wars. Though nominally a salt horse, a more accurate portrayal is that he was a specialist in staff duties having already attended the 1920-21 War Staff Course and filling many staff appointments in his subsequent career. A graduate of the second postwar War Staff Course, Allen sought to take the Andover course as well because of his duty in *Courageous*. This he failed to do and, though the reason cannot be claimed definitive, increasingly medical problems faced this officer.[151]

147 ADM 196/64 (Campbell).
148 Major General James Louis Moulton, Sound Recording 6818 of 18 May 1983, Imperial War Museum, London.
149 For example, Admiral the Hon. Sir Cyril Douglas-Pennant studied in the Junior Division of Camberley in 1923 whilst a Lieutenant before sitting the Greenwich Staff Course in 1924-25 as a Lieutenant Commander.
150 Later Commander Arthur Cyril Allen (1892-1938). Staff Course, 1920-21; Staff Course Camberley, 1922-23; Staff Officer Operations in *Queen Elizabeth*, 1925-26; Training and Staff Duties Division, 1931-33; Tactical Course, 1934; Staff Officer Operations in *Hood*, 1934-35 and retired, 1937.
151 ADM 196/127 (Allen).

Of the naval members attending Camberley, a strong preference existed towards sending gunnery specialists or salt horse officers. Thirteen of the former and nine of the latter appeared between the wars whilst the corresponding figures for officers of other technical backgrounds quickly slipped into single digits with five submariners, five signals, and four torpedo specialists appearing. Though one Pilot attended, no Observers, Navigating Officers or Physical and Recreational Training specialists went to Camberley. In contrast, four Naval Observers and one Pilot attended the Air Force Staff College between the wars representing, at least on the surface, a poor effort at ensuring that officers proceeding to Andover had some direct experience of the air. Loss of pay may have been a consideration as some found themselves attracted to the FAA with its promise of higher allowances. Lieutenant Henry Fancourt[152] was one such officer having no particular desire to become a pilot except that he was recently married and ill in Malta. As a salt horse, his commanding officer said that a brass hat and a baldhead might yet be his if he remained in destroyers, but flying and its allowances offered an important safety net as a newlywed.[153] Fancourt did not go to a Staff College though he had sought to go to Andover. Nor did he attend the other higher courses expected of a rising naval officer, though he did spend time in staff duties at the Admiralty.[154] This highlights the difficulty facing the Navy in balancing aviator assignments between attachments to the RAF, returning to general naval duties for a time, whilst also trying to expand the FAA at the same moment. Contrasted with the RAF practice of sending officers with prior naval experience to attend the Staff and War Courses, the four Observers attending the Air Force Staff College (e.g., Commander Boddam-Whetham and Lieutenant Commanders John Garland,[155] James Hawkins,[156] and Henry Johnstone[157]) were notable exceptions in naval practice.

---

152 Later Captain Henry Lockhart St John Fancourt (1900-2004). Short Course, Cambridge University, 1919-20; Pilot in *Eagle*, 1925-26; in *Argus*, 1927; in *Courageous*, 1929-31; Naval Air Division, 1931-32 and 1934-36; Officer-in-Charge, 822 Squadron in *Furious*, 1932-34; Naval Intelligence Division, 1937; Commanding Officer, HMS *Weston*, 1937-38; Air Personnel Department, 1938-39; Captain, 1940 and retired, 1950.

153 Captain Henry Lockhart St John Fancourt, Reel 3, Sound Recording 12274 of 30 September 1991, Imperial War Museum, London.

154 ADM 196147 (Fancourt) and ADM 196/123 (Fancourt).

155 Later Commander John Alfred Garland (1894-1981).Observer in *Pegasus*, 1922-23, in *Argus*, 1923-24; in *Hermes*, 1924-26; Staff Course Andover, 1926-27; *Furious*, 1930-31; Attached to RAF 1932-34; Commanding Officer, *Ceres*, 1938-39 and retired, 1939. Though rated as an Observer, Garland qualified as a pilot in 1924 and flew service types whilst attending Andover. Recalled for war service.

156 Later Acting Captain James Frederick William Cyril Hawkins (1897-1973). Short Course, Cambridge University, 1919; Observer in *Queen Elizabeth*, 1923, in *Argus*, 1923-24, and in *Eagle*, 1924-25; Staff Course Andover, 1925-26; Observer in *Furious*, 1926-27; in *Argus*, 1927-28, in *Eagle*, 1928-31; Tactical Course, 1931 and Observer in *Furious*, 1931-34; Acting Captain, 1943 and retired, 1947.

157 Later Captain Henry James Johnstone (1895–1947). Observer in *Argus*, 1922, in *Queen Elizabeth*, 1922-23 and in *Argus*, 1923-24; Staff Course Andover, 1924-25; Observer in

In keeping with the prominence that equitation enjoyed in the Army, naval and marine officers attending Camberley were expected to compete in the Drag. This may have put some officers off from seeking to attend, but as many officers during the period were also keen hunters or played polo, the requirement was not as limiting as might be thought to the contemporary eye. This, though, was not without its own slightly amusing difficulties, as the Army provided all officers with chargers to use in the Drag, but not necessarily stabling facilities at Camberley. Lieutenant Commander Robin Jeffreys,[158] accordingly, sought reimbursement from the Admiralty for the sum of 3/9d a week to meet the expense; a measure that was only approved after the usual round of docket passing between the Admiralty and the Treasury.[159] Jeffreys is also unusual in that his report from Camberley is more fulsome of his abilities than his earlier report from Greenwich. Drax thought that he would be a staff officer of only average ability—something later confidential reports substantiated—whilst Ironside reported his knowledge as equal to any soldier of similar seniority. Still, a naval officer who rode to hounds and was a keen swordsman was sure to be prized by the military, but then his father had been an officer in the Royal Horse Artillery and taught his son well.[160]

The prominence of the Drag at Camberley (and to a lesser extent at Andover) to the modern eye smacks of just that strain of conservatism that marks the military mind: steeped in tradition and more focused on the social aspects of the profession than on its official. This view is not entirely wrong, but it is not entirely correct either, neglecting as it does the importance leaders attached to developing the perspective of terrain and the balance of risks presented to the rider. Of course, these considerations owe more to the nuances of tactics than to strategy, and Ironside certainly believed that Dill, a strong supporter of the Drag,[161] was much too absorbed in the finer points of tactics at

Argus, 1925, in *Furious*, 1925-27; Naval Air Division, 1928-30; Tactical Course, 1930; Commanding Officer, HMS *Whitshed*, 1930-31 and HMS *Brazen*, 1931-32; Staff Officer Operations in *Courageous*, 1932-34; Senior Officers' Technical Course, 1935 and 1937; in Command, *Argus*, 1936; Naval Intelligence Division, 1937-38; Captain, 1938 and retired, 1940. Recalled for war service.

158   Later Captain Robin Edmund Jeffreys (1890–1963). Flag Lieutenant in HMS *Russell*, 1916; Torpedo and Mining Department, 1919-20; Flag Lieutenant Commander in *Delhi*, 1920-21; War Staff Course, 1921-22; Commanding Officer, HMS *Verdun*, 1922; Staff Course Camberley, 1923; Signal School, 1923-24; Squadron Wireless Telegraphy Officer in *Iron Duke*, 1924-25; Fleet Wireless Telegraphy Officer in *Queen Elizabeth* and *Warspite*, 1925-26; Naval Intelligence Division, 1927-28; New Zealand Division, 1928-31; Tactical Course, 1931; Staff Officer Operations and Intelligence to Commander-in-Chief, The Nore, 1932-34; Attached to Senior Officers' School Sheerness, 1933; Liaison to Naval Reserve and Merchant Navy, 1935-36; Sea Officers' Training Course, 1936; Captain and retired, 1937. Recalled for war service.

159   ADM 1/8705/184, Acting Director of Training and Staff Duties minute of 3 December 1923.

160   ADM 196/145 (Jeffreys) and ADM 196/53 (Jeffreys).

161   Slessor, *Central Blue*, p. 83.

the expense of the broader view.[162] Though not directly analogous, the debate on sail training was of a similar type of purpose. For both Services, as for the Air Force, the problem remained how best to prepare officers in peace for the pressures and uncertainties of war. By 1938, the Drag was no longer compulsory at Camberley and this was the last year when officers were assigned chargers for their personal use. If an era had passed, then it was but a reflection of what was occurring in the greater British Army, as cavalry regiments began their conversion to armoured cars.[163]

With only ten percent of military officers holding the *psc* designation and the necessity for acquiring such seen as a key stepping-stone to higher command, the race to attend either Camberley or Quetta was severe. Of the two Staff Colleges, competition to Camberley, and, consequently, the score required to pass the first hurdle was the keener.[164] Passing the entrance examination outright secured entrance for an officer, however, the more normal route was to qualify by exam and then secure entrance by receiving a nomination and negotiating the subsequent interview process. By 1926, only five percent of infantry applicants were passing the examination outright (down from a high of 75% in 1870), so reaching the Staff Course by one's own efforts was becoming a distinctly minority approach.[165] Still, too much can be read into the falling numbers passing. For one thing, more candidates were taking the examination in the twentieth century than previously confirming the importance the *psc* designation was assuming within the British Army's ethos.

Whether the examination continued to capture one's capacity for succeeding at Camberley and Quetta is difficult to conclude. It may have been an adequate assessment of overall mental acumen, but it did not account for the many other skills required of the successful staff officer or commander. To wit, tact, perseverance, good humour, physical fitness, and military comportment remained essential elements not amenable to assessment by examination. Wavell, a man of no mean intellect and a veteran of the World War, questioned aspects of the examinations given and reached a view endorsed by many naval officers when he concluded 'that character is worth more than brains in war.'[166] Still, securing a nomination, receiving laudatory reports from previous commanding officers, and how one handled oneself during the interview were every bit as vital as the examination itself and allowed authorities to balance these wider traits. The first measure, as ever, remained to do reasonably well on the qualifying examination. Well could Brooke-Popham, in assessing the entrance process for Camberley and Quetta, marvel at its scope, rigor, and scale and conclude that the Air Force entrant was not the equal of his military brethren.

<hr>

162 Ironside to Liddell Hart letter of 8 November 1931, Liddell Hart Papers, LH 1/401/72, LHCMA.
163 Moulton, Sound Recording, IWM.
164 Ronald Lewin, *Slim: The Standardbearer, A Biography of Field-Marshal The Viscount Slim, KG, GCB, GCMG, GCVO, GBE, DSO, MC* (London: Leo Cooper, 1990), p. 46.
165 Godwin-Austen, *Staff College*, p. 277.
166 Wavell to Liddell Hart letter of 30 October 1931, LH 1/733/37, LHCMA.

Given the prominence of the examination process and the efforts officers pursued to do reasonably well at it, suspicion existed by some in authority whether the balance was correct. Perhaps surprisingly given his own published record of historical discourse, Wavell, believed that military history was an inappropriate subject for the entrance exam though he did not deprecate the need for intellectual assessment itself.[167] Others, though, did raise the broader question of the examination's role and emphasis. During Meyric ap Rhys-Pryce's[168] first year at Quetta, one exercise saw qualifiers articulate an alternative manner of selection rather than relying on the exam to determine entry into the Staff Colleges. In the words of the assignment's preamble, 'The candidates have served in the Army for 12 years or more; we should be able to pick out the right man by a more simple method and with a better chance of ensuring that they really are the right men.'[169] To this, the Navy would concur.

Normally, the military officer taking the Staff Course was of Captain's rank on the active list, though an officer of higher and even lower rank was accepted. Initially, his age was not to exceed thirty-five at the time of entry, but by 1923 this was reduced to thirty-three years.[170] Prior service as an adjutant, on a staff, or serving as an instructor was esteemed over purely regimental duties and, from 1927, quotas for each of the principle arms or corps were abolished.[171] In the late 1930s, Leslie Hore-Belisha, the reforming Secretary of State for War, introduced a change allowing officers of the Territorial Army to attend the Staff Course on a part-time basis.[172] Failure to attend the Staff College did not mean ruin, but it did raise the bar that much higher to enjoying a highly successful career. Captain Orde Wingate failed the entrance examination on his first attempt and only qualified at his second effort. His marks though were below the standard that would have netted him an automatic appointment. At this moment, a quota system operated with each arm allocated a number of openings to the course and Wingate had to rely on gaining one of the eight vacancies reserved for gunners through nomination. In the event, he was unable to secure one and, thus, failed to attend. However, after being pigeonholed by this brash officer, the Chief of the Imperial General Staff agreed to support Wingate's appointment to staff duties at a future date.[173]

That Wingate and Fuller had both failed the qualifying exam at their first sitting and were permitted to retake the test was an option precluded to others by 1929.

167 Wavell to Liddell Hart letter of 16 September 1931, LH 1/733/30, LHCMA.
168 Brigadier Meyric ap Rhys Pryce (1902–1996). Staff Course Quetta, 1933-34 and Staff College Camberley, 1939-40.
169 Staff College Quetta, 1933, Junior Division, Commandant's Paper No. 9, Selection of Officers for the Staff College, Brigadier Meyric Home ap Rhys Pryce Papers, National Army Museum, London, 9204-164-18.
170 Godwin-Austen, *Staff College*, p. 284.
171 *Ibid*, pp. 279-84.
172 Liddell Hart, *The Defence of Britain* (New York: Random House, 1939), p. 331.
173 Trevor Royle, *Orde Wingate: Irregular Soldier* (London: Phoenix, 1995), pp. 89-93.

Failure to secure the minimum of marks necessary for qualifying meant failure to attend with no chance to re-sit the examination.[174] Wingate's failure to pass the examination at his first attempt is understandable given the depth of knowledge that was required. The examination held in 1921 for the 1922 course required one to have a thorough grasp of military history, the organization and administration of the Army from 1868, a grasp of how imperial forces were organized, and a solid grounding in military law. Beyond this, the officer needed to know the geography and make-up of the Empire and be able to write credibly on those foreign states influencing British strategy. Mastery of French, German, or the Russian language was a given, whilst the history of either Europe or the United States since 1848, expositing on a scientific subject, addressing political economy, or writing on the principles of business rounded out the examination.[175]

The Camberley qualifier attending the complete course secured a strong appreciation of the responsibilities a staff officer faced in both peace and war. The administrative foundations of the British Army received strong emphasis during the course's first year when qualifiers mastered the logistics' requirements and organizational composition of the various arms, the role and application of military law, the provision of medical services, and the replacement and reconstitution of units. Introductory lessons covered the object of war, strategy, preparing appreciations, and the proper drafting of signals and orders. A timed exercise with students given the rudiments of a situation and a fixed period to prepare the necessary orders and messages frequently assessed their mastery of the last named topic. Lectures covering each of the Army's principle arms also featured during the course's first year, as did expositions on specific campaigns of the 1914-18 war such as Palestine, Mesopotamia and the military operations in France and Belgium.[176]

Recalling that Palestine, Mesopotamia and the Western Front were also lectures prominent at Greenwich highlights that qualifiers, no matter their Service or school, shared much in common in the lessons being derived from the World War. Of course, fundamental differences remained—the Navy caring little of cavalry and the Army and Air Force eschewing a need to understand destroyer work. Yet, the greater point is that though largely training and educating their officers separately, the Services did not think about war in isolation. Thus, those who have argued that the Navy was fixated on Jutland in their staff training can only have reached such a conclusion by ignoring the many other strands of instruction underway. Whether seen from the cross-lecturing occurring at the Staff Colleges, the regular debates featuring at the RUSI, or, in its most tangible form, the yearly combined exercise, a commonality,

<hr>

174 F. W. Young, *The Story of the Staff College, 1858-1958* (Aldershot: Wellington Press, 1958), p. 25.

175 Godwin-Austen, *Staff College*, pp. 273-74.

176 Brigadier Leslie Skipp Lloyd Papers, National Army Museum, London, 9405-19 and 9405-19-9.

if not a common, training regime was practiced. Indeed, the best demonstration of this commonality is Fuller who won the RUSI Naval Prize in 1921 for his essay addressing 'The Communications across the oceans of the world being essential to the Empire, how best can they be safeguarded.' The embarrassment of a soldier wining the contest was palpable and, consequently, the actual Gold Medal eluded Fuller.[177] This, however, should not mask the important lesson that Staff College training was of a common enough standard for a soldier to write the winning essay.[178]

Common but not the same. Camberley qualifiers were spared the requirement of researching and lecturing on a set-piece topic—so common a practice for the qualifier attending Greenwich and Andover. With a strong body of teachers always present this was less of a requirement, but it also reflected that what Camberley taught was expressly approved by the General Staff. This did not mean that qualifiers did not at time present their work orally. Rather, such presentations were but a part of the learning exercise of the syndicate and not a means of imparting received wisdom across the course.

It was during his first year at Camberley that an officer studied the role of the division and the staff duties necessary to support it whilst the investigation of corps-level operations consumed one's second year.[179] Professor Bond notes that with so many officers arriving at Camberley fresh from the recent war possessing experience of fighting in large formations a valuable opportunity was missed in not capturing the lessons of handling formations the size of armies or the British Expeditionary Force itself.[180] This is no doubt true, but the purpose of the Staff College was primarily to prepare officers for acting in the capacity of a Brigade Major, essentially a brigade's operations officer by another named, with training clearly focused on the echelons of command residing below armies.[181] If a common criticism of the interwar Royal Navy was that it was consumed with the idea of fleet action, then the British Army stands innocent of a corresponding charge, as its horizons were more circumspect. This contrast in operational outlook reflected the standing the two forces of the Crown enjoyed in funding, prominence, and anticipated roles. The educational grounding naval and military officers received strongly mirrored and supported the missions envisioned for their arms, though the same cannot be said of air officers, whose Service always aimed to punch above its weight.

Tangentially, much was common in the subject matter of the several colleges, but a closer inspection reveals just how consumed the Army qualifier was concerned with the minutiae surrounding and supporting military operations. The provision of water, messing, sanitation arrangements, route marches, trains, medical organization, river

---

177 The essay appeared in February 1922 issue of *The Naval Review*.
178 Fuller to Liddell Hart letter of 31 January 1949, LH 1/302/370, LHCMA.
179 Godwin-Austen, *Staff College*, p. 288.
180 Bond, *Military Policy*, pp. 37-38.
181 Nigel Hamilton, *Monty: The Making of a General, 1887-1943* (New York: McGraw-Hill Book Company, 1981), p. 153.

crossings, and reconnaissance were all topics featuring in the course of instruction along with their associated exercises and schemes.[182] It is not that these matters, or their corresponding equivalents, were not important to naval officers, but they were largely absent from the Greenwich curriculum. The reason for this is not hard to find, as the Navy operated a separate Accountant Officers' Technical Course to address logistical issues leaving executive officers freer to focus on the broader picture.

As one progressed to the Senior Division at Camberley or Quetta, the course become broader in scope with the subject of strategy coming to the fore. Where the Junior Division focused on tactics and operations, the officer now lifted his eyes to examine the strategic setting of the British Army and its associated partners. Here, the strategy of the World War dominated and it was during this second year that officers studied Gallipoli in preparation for their participation in the yearly combined exercise. This same pattern followed at Quetta, though, of course, the scale was different and the place of the Naval and Air Force Staff Colleges was replaced by working with the East Indies Squadron and local elements of the Quetta air station.

As at Andover and Greenwich, working in small groups was a central component of the learning process, though with far more qualifiers present at Camberley, it tended to be a somewhat larger body with the syndicate at times reaching ten members.[183] Depending on the scheme being considered, qualifiers, acting in turn, served as the syndicate's chairman allowing successive officers to develop skills in critical analysis, discourse, and leadership. A member of the directing staff was ever present to advise on points of knowledge and to ensure that discussions did not diverge too astray and the objectives were maintained. The process was at times brutal with pungent criticisms issued by peers and instructors—made all the more striking as use of Christian names was prohibited during working hours—frequently following one's efforts.[184] In this, some naval officers thought that the balance was not always correct at Camberley. This may have owed something to the seniority of the average naval officer attending relative to the member of the directing staff. Receiving criticism is never easy and taken from an officer of lesser rank rarely appreciated. True as this is, this observer tends to accept that in the cases of Captain George Philip[185] and Commander William Friedberger[186] their comments on the instructional process were of more a substantive

<hr>

182 Lloyd Papers, NAM, 9405-19-1.

183 Dominick Graham, *The Price of Command: A Biography of General Guy Simonds* (Toronto: Stoddart Publishing Limited, 1993), p. 27.

184 'Notes on D.S. Conference held in the Henderson Library 22nd January, 1939', Rhys-Pryce Papers, NAM 9204-164-11-4.

185 Captain George Tothill Philip (1895-1966). Squadron Gunnery Officer in *Cardiff*, 1928-29; Senior Officers' Technical Course, 1930; Tactical Course, 1931; Squadron Gunnery Officer in *Coventry*, 1931-33; Staff Course, 1934; Captain, 1938; Commanding Officer, *Resource*, 1938; Senior Division, Staff Course Camberley, 1938; British Mission to Egypt, 1939-41; Commanding Officer, *Argus*, 1941-43 and *Furious*, 1943-44 and retired, 1947.

186 Later Captain William Howard Denis Friedberger (1896-1963). Commanding Officer, *H 34*, 1921-23 and *H 44*, 1925; Staff Officer Operations, First Submarine Flotilla, 1927; Staff

nature. They were both naval *psc's* and submitted detailed observations of their impressions of the 1938 Senior Division course. Writing to the Commandant, Friedberger observed:

> The work done in syndicate naturally made the most impression on one's mind. A particular point in this form of work was the relationship produced between the student and the member of the directing staff.
>
> The syndicate discussion was the occasion when the character and method of the "tutor" and the capacity and judgment of the student were tested. As one saw it, the discussion gave all students the opportunity of seeing for themselves how the solution could be or should have been evolved, and what alternatives existed. The skill of the tutor determined whether the discussion ended with a realisation of general principles by all students, or in an argument on the justice of the tutor's red ink marks.[187]

Philip's criticisms largely mirrored those of Friedberger and the extent to which both officers coordinated their comments is unknown, though the supposition must be that they did. Concluding his assessment of the course, Philip believed:

> In matters that are purely dogmatic or doctrinal the role of the D.S. [Directing Staff] should be that of a teacher. In the second year, to allow the student more latitude and to afford him more opportunity of exercising his own imagination and initiative, this role should be modified to that more of a leader than a teacher, and further the D.S. should be more closely associated with the student in certain major exercises. In less important exercises more opportunity should be given to students to act and "correct" their own schemes under D.S. direction.[188]

Implicit in Philip's advice which Bernard Paget took to heart, was that though Camberley offered an effective course of instruction, the objective of developing future commanders was hindered by the manner of instruction. Instructors controlled the workings of the syndicates far too much stifling the development of initiative in the qualifiers. As the Greenwich course stressed the 'vital importance of initiative' and

Course, 1927-28; Commanding Officer, HMS *Odin*, 1930-31 Senior Officers' Technical Course, 1932; Operations Division, 1932-34; Commanding Officer, *Delhi*, 1936; Staff Officer Operations in *Dolphin*, 1937; Tactical Course, 1937; Commanding Officer, *Resource*, 1938; Senior Division, Staff Course Camberley, 1938; Plans Division, 1939-40; Captain, 1939; Assistant Director, Plans Division, 1943; Deputy Director, Plans Division, 1944 and retired, 1949.

187 W.H.D. Friedberger comments to Senior Division Course, December 1938 contained in 'Notes on D.S. Conference held in the Henderson Library 22nd January, 1939', Rhys-Pryce Papers, NAM 9204-164-11-4.

188 *Ibid*, G.T. Philip, 'Remarks on the Method of Teaching Second Year Students', 20 December 1938.

the *Battle Instructions* codified the same at the tactical level, Philip was only reflecting the value that that tenet assumed within the Royal Navy. Thus, the conclusion that Camberley—and Quetta—proved more effective at preparing competent staff officers than developing future commanders is if, un-measurable, probably nevertheless, true.[189] Paget, himself, had recently changed the manner by which the directing staff member oversaw his qualifiers and the feedback provided demonstrated that kinks in the process remained. Yet, for all that, the syndicate was also a highly effective form of instruction emphasizing composure and rational analysis in solving any military problem.[190] Perhaps as a safety valve, Camberley enjoyed an 'eleven o'clock' rule where qualifiers and directing staff mixed together in the evening after the day's studies and riding to share a drink and reflect. Students were free to criticize and the instructors accepted any barbs directed in the positive spirit offered.[191]

Visiting lecturers was a common pattern to all the Staff Colleges and at Camberley, air and naval officers familiarized the qualifiers on the capabilities of their respective Services. Thus, Captain Barry Domvile,[192] the Director of Plans, spoke at length in April 1920 on 'Future Policy and its Relation to Military Forces' a topic closely aligned to what he was also presenting to the War Staff Course at the same moment on 'Future Naval Policy'.[193] Such expositions, though, were not limited to the military family and frequently an outsider was invited to address both divisions of the college following dinner. This allowed other views to be aired in an informal setting and Liddell Hart, the military correspondent of *The Times*, spoke to such effect in December 1937 to Camberley qualifiers on the present military problem facing Europe.[194] A gifted scribe who borrowed freely from others, Liddell Hart's position with the leading journals of the day allowed him to reach audiences beyond the original purveyor's grasp. If his views were not universally admired and accepted—an earlier Camberley class lampooned him as 'Captain Diddle Dart'—the invitation was recognition enough of his prominence in status, if not in the soundness of his

189 Major General Nigel William Duncan Sound Recording 829 of 24 September 1976, Imperial War Museum, London.
190 Dominick Graham and Shelford Bidwell, *Coalition, Politicians and Generals: Some Aspects of Command in Two World Wars* (London: Brassey's, 1993), pp. 15-16.
191 Lewin, *Slim*, p. 56.
192 Admiral Sir Barry Edward Domvile (1878-1971). Five firsts in examinations for Lieutenant; War Course, 1910; Appointed War Staff Officer, 1912; Assistant Secretary, CID, 1912-14; Captain, 1916; Commanding Officer, *Curacoa*, 1918-19; Deputy Director and Director, Plans Division, 1919-22; Chief of Staff Atlantic Fleet, 1922-24; Commanding Officer, *Royal Sovereign*, 1925-26; Rear Admiral, 1927; Senior Officers' Technical Course, 1927; Director, Naval Intelligence Division, 1927-30; Vice Admiral, 1930; Commander, Third Cruiser Squadron, 1930-32; Vice Admiral, 1931; Senior Officers' War Course, 1932; Admiral-President, Royal College Greenwich and Vice Admiral Commanding, Royal Naval War College, 1932-34; Admiral and retired, 1936.
193 ADM 196/44 (Domvile).
194 Liddell Hart, *Defence of Britain*, p. 51.

prescriptive views.[195] Such venues aired contrarian views and provoked debate, an important aspect of Staff College training where the acceptance of doctrine must always be done with a critical eye.

It was also the custom of Camberley to include one officer of the Indian Army on its directing staff. This kept the Indian Army perspective of any military problem at the fore, but it also ensured that officers serving in British regiments were familiar with conditions applying on the sub-continent if they had not served there already.[196] Still, when contrasted with what his peers were learning concurrently at Greenwich, the military officer's horizon was limited with studies focused almost exclusively on divisional and corps level operations and tactics. It is not that the Camberley qualifier failed to study the higher direction of war, for this he surely did. With industrial mobilization, inter-Service cooperation, and theoretical wars featuring in the course's final term the subject was tackled.[197] Yet, the weighting of instruction was skewed to the consideration of lesser actions. Of course, this merely reflected how the Army was employed between the wars and also reflected the immediate duties an officer was likely to assume upon leaving. That said, Brigadier Cyril Barclay[198] believed that Camberley's instruction was focused at too high a level when it would be many years before a graduate would have to face the higher problems of war.[199] As an interwar qualifier and noted author on the British Army his view carries much weight indicating that it would have been better served by dividing the course into two parts and returning the qualifier to consider the higher direction of war at the moment he was closer to having to face the problem. In fact, this was the thrust of the changes introduced to the Staff Course during 1938-39.

As to the higher direction of war, syndicates tackled the problem in strategic war games with officers posing as senior political leaders. Common to what was also being heard at the IDC, the natural advantage a totalitarian power enjoyed in directing war was a frequent observation. Major General Lord Gort, the Commandant in 1937, writing to Liddell Hart stressed the strength of Mussolini's hand *vis-à-vis* Britain noting a 'totalitarian state certainly has advantages when it comes to the direction of war.'[200] Meanwhile, the emphasis at Camberley was on mastering war from the bottom up whilst the Greenwich qualifier tackled the subject from first principles. Thus, from the beginning the naval qualifier was asked to consider all he learned in relation to the higher direction of war. Both, then, covered much the same ground, but the emphasis was subtly different. Lord Tedder, with firsthand knowledge of how

---

195 Danchev, *Alchemist of War*, p. 223.
196 Lewin, *Slim*, p. 55.
197 French, 'Officer Education in the British Regular Army, 1919-39 in Kennedy and Neilson, eds., *Military Education*, p. 118.
198 Brigadier Cyril Nelson Barclay. (1896–1979). Staff Course Camberley, 1930-31.
199 Young, *Staff College*, p. 26.
200 Lord Gort to Liddell Hart letter of 6 March 1937, Liddell Hart Papers, LH 1/322/38, LHCMA.

naval and military officers typically faced the same problem, appreciated their differences in approach. When the problem of defending Iraq arose whilst also engaged in operations in the North Africa, he observed that it was easier to secure the agreement of Admirals than of Generals when the question was one of mobility. In short, the former were comfortable considering movements of hundreds of miles whilst military officers thought little further than the range of their guns.[201]

Tedder's testimony is amplified by that of Russell Grenfell who believed that the training and experience of naval officers forced them to think much further ahead in time than their military counterparts commenting:

> The destroyer officer who is used to high speed night attacks when thought has to be projected a long way ahead & decisions have often to be taken in a fraction of a second would seem almost specially cut out for success in tank command.[202]

Thus, if the Navy and the Army approached the subject of staff training from different angles other influences of a more practical nature were also present shaping how they faced and treated any problem.

Historical study to enlighten the enduring lessons of successful military operations was a salient feature of Camberley instruction making due allowances for their current application as codified in the *Field Service Regulations*. Clausewitzian principles such as striking at the decisive point, the employment of mass, and the importance of morale were instilled at Camberley, even if the qualifier did not necessarily read *On War*. So, too, were qualifiers reminded that war was not fought for its own sake but served only as a means to an end—a formulation his Greenwich counterpart would have understood even if its origins remained more opaque.[203] Similarly, the campaigns of Napoleon and of Marshal Foch, the Commander-in-Chief of Allied Armies on the Western Front during the Great War, received strong treatment. The Army Staff College relied on the Official Histories of the World War in a manner that was mostly missing at Greenwich where they were employed to set the scene for any number of schemes run. One example of this in operation was an exercise conducted in 1925. Qualifiers of the Junior Division, including Captain Robin Campbell, RM,[204] were required to prepare an appreciation for the Commander II Corps, British Expeditionary Force on

201 Tedder, *With Prejudice*, pp. 89-90.
202 Grenfell to Liddell Hart letter of 5 July 1942, Liddell Hart Papers, LH 1/330/61, LHCMA.
203 French, *Churchill's Army*, p. 14.
204 Later Major General Robin Husluck Campbell (1894–1964). GSO3 and Assistant District Intelligence Officer in *Malabar*, 1919-21; Staff Course Camberley, 1925-27; Seconded to Army, 1927-31 and 1934-38; Squadron Marine Officer in *Curacoa*, 1932; Lieutenant Colonel, 1938; Fleet Marine Officer in *Warspite*, 1938-39; Deputy Assistant Adjutant General, 1939-40; Temporary Brigadier, 1940; Acting Major General, 1942; General Officer Commanding, Royal Marine Division, 1942; Major General, 1944; Chief of Staff, Royal Marines, 1944 and retired, 1946.

the situation arising on the evening of 25 August 1914 to serve as an aide memoir for a meeting with General Allenby. The narrative outlining the task advised that 'For the purpose of this appreciation it may be assumed that you were in possession of all the information given in Chap. VI of Vol. I of the "Official History of the War" and in the maps and sketches referred to in that chapter.'[205] Moreover, officers attending were required to be conversant with its battle summaries and this sanctioned use of the Official History indicates that for the moment, any criticism, if acknowledged, was not openly expressed and can be contrasted to the naval experience.[206]

The Camberley qualifier, and more so his Quetta peer, however, spent a large portion of their time weighing the military's role in imperial policing. This was practical rather than theoretical education, as pacifying the indigenous peoples found in the Sudan, Iraq, and the Northwest Frontier of India was more likely to be the next call of an officer's time rather than facing a new European threat if public opinion and governing politicians had any say in the matter. The emphasis on internal security was deplored by some officers including Percy Hobart[207] who subscribed to the view that 'the Empire can police itself sufficiently against pilferers: against vital attack is what we must consider.'[208] To a Service committed to the maintenance of British India this was an unrealistic view and officers of the East Indies Squadron when visiting Quetta in 1934 to assist in that year's combined exercise with the Indian Staff College were provided a demonstration in how communal riots were to be resolved with qualifiers playing both military and civil roles.[209] If the topic of internal security was the more pronounced at Quetta, its presence was, nevertheless, felt at Camberley too. This was only natural as a substantial portion of the military forces serving in India came from the Home establishment. Tellingly, Major General Charles Gwynn,[210] the Camberley Commandant from 1926-31, treated the subject at length in his 1934 work *Imperial Policing* following his retirement.

During his time at Camberley, Major William Slim,[211] the Indian Army member of the Camberley directing staff from 1934-36, was in his element covering internal

205 'Appreciation of a Situation', Lloyd Papers, NAM, 9405-19-9.
206 For example, before leaving for France to tour the Aisne battlefield officers were directed to read '*Official History, Military Operations, France and Belgium, 1914, Vol. I*'. Alanbrooke Papers, 3/16/1, LHCMA.
207 Later Major General Sir Percy Cleghorn Stanley Hobart (1885-1957). Staff Course Camberley, 1919; Staff College Quetta, 1923-27; Senior Commanders' Course, 1933 and Director of Military Training, 1937-38.
208 Hobart to Liddell Hart letter of 7 October 1934, Liddell Hart Papers, LH 1/376/9, LHCMA.
209 C.K.S.A [Charles Aylwin]., 'A Naval Visit to Quetta, November, 1934', *The Naval Review*, Vol. XXIII, No. 1, February 1935, p. 73.
210 Later Major General Sir Charles William Gwynn (1870-1962). Commandant, Staff College Camberley, 1926-30.
211 Later Field Marshal William Joseph Slim, Viscount Slim (1891-1970). Quetta Staff Course, 1926-28; Staff College Camberley, 1934-36; Imperial Defence Course, 1937;

security and mountain warfare. Taking the qualifiers on a tactical exercise without troops to the Welsh hills to instill an understanding of how to move forces through terrain meant to represent the mountains of Waziristan, qualifiers learnt to picket the controlling tops and pass a vulnerable column through the recesses below. This was a made to order practical for Slim, as it was a staple of his earlier Quetta Staff Course.[212] Not all were buying it though at this late date with one officer questioning why armoured cars supported by top cover provided by the RAF did not achieve the same result at less cost and more quickly.[213]

Outdoor exercises, such as the time in the Welsh hills, demonstrate a practical difference between the Navy and Army approach to staff training. The Greenwich approach was more 'academic' in the sense that it was entirely theoretical supported by historical method and based on the Service's doctrine and publications. At Camberley and Quetta, the qualifier had a mix of theoretical and practical work with a fair portion of his time spent investigating problems first indoors at lectures, conferences, or at the sand-table followed by considering the same problem outdoors in a natural setting. Of course, the weather might not cooperate when the latter evolutions were attempted, as qualifiers found to their discomfort in 1937, but then weather was ever a factor to be considered in any military problem.[214]

A week in the Welsh hills was a common event for members of the Junior Division, but so also was visiting the primary military training establishments whether it be the Small Arms School at Netheravon, Wiltshire or the Tank School at Wool, Dorset. Venues successive Greenwich qualifiers came to know well during their tours at Camberley. How much an officer actually took away from such cursory visits is difficult to gauge, but exposure to the rudiments of other arms not of his own background before considering the larger issues at play should not be discounted out of hand. The visits also served a dual-purpose, as they were conducted as a military operation with qualifiers working in syndicates preparing the necessary tasking orders to officers detailed to oversee a portion of the work. Overkill? Perhaps, but it served a practical purpose.[215] Some questioned the amount of written work required at Camberley in an age when radio was now available to direct forces, but the emphasis was not mistaken, only misunderstood.[216] Small tactical operations might be conducted verbally, but the student was also being prepared for something greater. Formal appreciations, orders, and directives had a place at the higher levels of war for which he was being trained

Senior Officers' School Belgaum, 1938; Commandant, Senior Officers' School Belgaum, 1939 and Commandant, Imperial Defence College, 1946-48.

212 Lewin, *Slim*, pp. 56-57 and General Lord Hastings Lionel Ismay Papers, Liddell Hart Centre for Military Archives, King's College, London, United Kingdom, 3/2/14.

213 Lewin, *Slim*, p. 57.

214 Lord Gort to Liddell Hart letter of 36 May 1937, Liddell Hart Papers, LH 1/322/43, LHCMA.

215 Godwin-Austen, *Staff College*, pp. 288-89.

216 Graham, *Simonds*, pp. 28-29.

and recourse to verbal operation's orders and off-the-cuff preparation was a recipe for disaster during set-piece operations. Those criticizing the emphasis on the written word forgot that the Staff College was not interested in making an officer a better company commander but a better officer at the brigade-level and echelons above. Flexibility and battle-drill had its place, but it was no substitute for the prior preparation required for the largest of endeavours.

Unlike the Royal Naval Staff College qualifier who was prohibited for a time from flying in military aircraft whilst studying, the military officer was required at some point to fly with the RAF during his time at Camberley.[217] The origins of the naval prohibition probably owed much to the frequency of early air accidents though a review of the period's *Navy Lists* specifies drowning as a more common reason for an officer's death. Whatever its exact origins, the prohibition would have done little to encourage the naval officer to view flying in a more positive light. This conservative strain in naval practice can be contrasted with the Army's approach and indicates that polo and the horse-set ever present at the Staff College were not a bar in themselves to more progressive thinking. This is not to imply that the colleges wholeheartedly embraced the place of armour in its instruction, but it did not neglect it either. When the Experimental Force was established—incomplete as it was—Junior Division qualifiers watched it in action in July 1927 as an advance guard of the main force swept aside cavalry opposition to cross the River Avon.[218] Moreover, instructors such as Fuller at Camberley and Hobart and Martel at Quetta ensured that such would never be the case even if the latter institution only reluctantly accepted its role. Yet, the prominence it should reflect in British military thinking and its actual capabilities at that moment were issues upon which learned officers disagreed. Brooke,[219] the gunner and practitioner, and Fuller, the infantryman *cum* tanker *cum* proselytizer represented an arch-example of this discord. Contemporaries serving on the Camberley directing staff, both officers wrote to effect on the lessons of the recent war and the respective merits of artillery and armour.[220] If their conclusions in the style of Scottish Law were 'not proven', then the qualifier listening to their separate discourses must have relished the intellectual atmosphere then present at Camberley. Notwithstanding their differing views, Fuller privately rated Brooke a very good gunner and an officer prepared to argue and not accept as cant any proposition was an officer that Fuller could fully respect even if largely disagreeing with his views.[221]

217 *Ibid*, p. 291.
218 Wavell to Liddell Hart letter of 27 June 1927, LH 1/733/7, LHCMA.
219 Alan Francis Brook, later Field Marshal Viscount Alanbrooke of Brookeborough (1883-1963). Staff Course Camberley, 1919-20; Staff College Camberley, 1923-26; Imperial Defence Course, 1927 and Imperial Defence College, 1932-34.
220 David Fraser, *Alanbrooke* (London: Harper Collins Publishers, 1997), pp. 62-64.
221 Fuller to Liddell Hart letter of 8 January 1926, Liddell Hart Papers, LH 1/302/85, LHCMA.

Arthur Harris after sitting through a week's coursework dedicated to the place of armour in modern war, concluded that the British Army would never really accept the tank until it could do all the things required of it including making the noises of a horse and eating hay.[222] Harris' criticism must be treated with caution, as he was a person of very pronounced views who was nearly asked to quit Camberley. An officer of more dispassionate perspective was Lieutenant Commander King-Hall who even before his arrival at Camberley in 1924 had written on the future possibilities of armoured warfare predicting that tank battles would share many of the same characteristics of naval battle.[223] King-Hall was by no means the first officer to suggest armoured engagements might approach naval actions and to conclude that his foray into future war was a deciding factor in his selection to attend Camberley would be speculative. Yet, it does indicate that the naval officer attending Camberley had much to offer within his own sphere that the Army might find useful. Ironside, the intelligent and imposing Commandant committed to preparing for a future war, and Fuller, the prophet of armour, would both have looked on King-Hall's writing with a measure of favour.

A similar cleavage at Greenwich amongst naval officers is difficult to cite. Richmond's separate contretemps with Drax and Usborne previously discussed were never so much about first principles as historical usage and care. The absence of evidence is not the absence of occurrence, though, and a definitive judgment remains problematic. Notwithstanding this evidentiary void, the Royal Navy had its Tactical School and, in time, its Tactical Investigation Section, where such controversies could be aired, leaving the Royal Naval Staff College to focus on educating officers in staff procedure and administration. Moreover, the Navy operated the Senior Officers' War Course and used it to examine the strategic implications of force structure. The British Army did not have a Tactical School and only late in the day sought to shape officer higher education by having the Senior Wing of the Staff College, as it was now styled, approach in substance what the SOWC accomplished for the Navy when the basic course was reduced to a single year in 1939. Now, the age of the typical Army qualifier became thirty whilst a higher course for more experienced officers convened at nearby Minley Manor.[224] Thus, the Army ultimately accepted the arguments in military educational reform previously advanced by Nye and Slessor.

If the war game played on the strategic or tactical table was a salient feature of naval instruction, then the use of the sand-table played a correspondingly familiar place in the lessons of Camberley and Quetta. In essence, the sand-table was a three-dimensional map over which military forces were represented. By portraying the terrain and tactical units together in a lifelike representation, qualifiers secured an appreciation of the actual conditions governing an operation in a manner not conveyed

222 Arthur Harris, *Bomber Offensive* (London: Greenhill Books, 1990), p. 24.
223 Trythall, *Fuller* p. 101.
224 Liddell Hart, *Defence of Britain*, p. 330.

through conventional topographical maps where engaged units were merely indicated by coloured markers. The sand-table facilitated planning, execution, and after action assessment and Bernard Montgomery,[225] the future Field Marshal, was a recognized advocate and teacher of this form of instruction using it to good effect to show tactical precepts of platoon-sized operations emphasizing fire and movement.[226] Montgomery's forte was teaching tactics and battlefield leadership at Camberley with his quality of instruction so esteemed that one officer has ranked it, preparing as it did many of the corps and divisional commanders of the next war, the equal of his contributions on the field of battle in securing victory.[227]

The staff ride was another form of instruction employed to good effect at the two military Staff Colleges being largely absent at Greenwich. Visits to continental battle-fields of the Great War were a staple of the period and restricting one's vision merely to Belgium and France was far from the case. Cyril Barclay and his syndicate traveled to East Prussia surveying the scene of the 1914 Russo-German campaign.[228] The benefit of such tours was that the previously acquainted qualifier actually viewed the ground where an action had occurred. This developed an officer's appreciation for factors such as terrain, weather, and time and their influence in battle in a manner mere lecturing and reading could not.

When studying past battles one purpose may have been to highlight a particular precept, but a second reason was to lay the groundwork for considering the tactical circumstances the contemporary commander faced in light of the weapons now available. Brigadier Brooke, surveying the first actions of the British Expeditionary Force in the war, returned to Camberley in 1933 to lecture on the Battle of Le Cateau. He asked qualifiers to consider in their answers not only the general lines of the action arising, but also how a modern commander would approach the battle.[229] This approach in analyzing a previous battle to inform the use of contemporary means was a common practice and the syndicates considering the Aisne during their 1934 tour of that battlefield were similarly tasked to weigh first its original lines and then to address its features applying 'up to date armament and organization.'[230] This is highly remi-niscent of the use of Jutland between the qualifiers of the air and naval Staff Colleges

225 Bernard Law Montgomery, Field Marshal Viscount Montgomery of Alamein (1887-1976). Staff Course Camberley, 1920-21; Staff College Camberley, 1926-28 and Staff College Quetta, 1934-37.

226 Hamilton, pp. 194-95 and '*49th (West Riding) Division. Precis of Lectures delivered throughout the Division in January, February & March, 1924*', enclosure to Montgomery to Liddell Hart letter of 16 July 1924, Liddell Hart Papers, LH 1/519/3, LHCMA.

227 Field Marshal Lord Harding, Reel 10, Sound Recording 8736 of 1984, Imperial War Museum, London.

228 Young, *Staff College*, p. 27.

229 Brooke, 'Le Cateau Exercise,-1933', Alanbrooke Papers, 3/15/4, LHCMA.

230 War Office memorandum 43/Training/1643 (M.T.) subject War Office Tour of the Aisne Battlefield 28th June to 3rd July, 1934', Alanbrooke Papers, 3/16/1, LHCMA.

meeting in common at Greenwich and demonstrates that the primary purpose of historical instruction in the military environment is to anticipate the future.

Employing a recently fought battle as the context for considering how newer methods and means might effectively be used in battle was an innovation that appears to have been introduced at Camberley only sometime after the end of the World War. A common criticism sounded by a few progressively inclined officers attending in the immediate aftermath of the World War was that the course owed too much to perfecting the shortcomings of previous campaigns rather than considering how to win the next one.[231] To this there is probably much truth, but given the abbreviated courses conducted for the first three sessions of the postwar Staff College attempting more was asking too much.[232] The criticisms leveled were not wrong, but the conundrum was that the first officers attending were also some of the Army's leading younger lights including Brooke, Hobart, Gort, and 'Jumbo' Wilson.[233] The abbreviated courses of necessity had to forego some topics and those presented touching on the 1914-18 war no doubt improved over-time as refinements were made to them by successive instructors and as the relevant volumes of the Official Histories appeared. The first course meeting in April 1919 ran to eight months whilst the second course ran to a year.[234] The instruction that the first officers attending received was disappointing on two fronts: it was a curtailed, rushed affair and it failed to meet the high expectations of those very capable officers attending who had secured such strong practical knowledge during the war.

Lieutenant Colonel Russell Luckock[235] lectured to the initial qualifiers on the Great War's Western Front covering the ebb and flow of the action in that theatre in nine lectures. The emphasis of his surveys was certainly on the war's early stages as the actions occurring in 1914 consumed his first five talks with the balance of the war covered in only four lectures.[236] From a historical perspective, one may question such an emphasis on 1914 when the mobilization of armies and states was only beginning. Yet, if the lectures were to have any but tactical value such a weighting is understandable, as this was the moment when Britain still retained the initiative on when, where, and how to commit herself to the war. Today, it is difficult to assess the degree that Luckock entertained such considerations in his talks, as his surviving lectures do not note any supplementary commentary offered by him and the comments of the qualifiers are missing. Still, circumstantial evidence is strong that he did resort to the

231 Bond, *Military Policy*, pp. 36-37.

232 Kenneth Macksey, *Armoured Crusader: The Biography of Major-General Sir Percy 'Hobo' Hobart* (London: Grubb Street, 2004), pp. 71-72.

233 Henry Maitland Wilson, later Field Marshal Lord Wilson (1881-1964). Staff Course Camberley, 1919-20 and Staff College Camberley, 1930-33.

234 Godwin-Austen, *Staff College*, pp. 270-72.

235 Later Major General Russell Mortimer Luckock (1877–1950). Staff Course Camberley, 1913–14 and Staff College Camberley, 1920–21.

236 Dill Papers, 2/1, LHCMA.

counter-factual telling his students when discussing Haig's actions during the March 1918 German offensive:

> Criticisms have been leveled at the Commander-in-Chief as regards the position in which he located his slender reserves. Criticism after the event is easy and is of little value, but consideration of the situation as it presented itself at the time to the responsible commander is of the greatest value to the military student and if conducted in the right spirit will always lead to good results.[237]

Common to the practices of Greenwich where examination of Jutland was employed in order to highlight deficiencies in British staff procedure, such a path also featured as a reason to examine the battles of the World War by the British Army. Brooke's use of the Aisne shows this trend. Emphasizing aspects of initiative, timeliness in execution, and resolution in the face of adversity that an officer listening to William Tennant's assessment of Commander Tovey's actions at Jutland would have found familiar, Brooke noted that:

> The Official Historian, after commenting on the failure of the Higher Command in the conduct of the operation on September 12th and 13th to appreciate the value of time and to make a resolute effort to forestall the enemy's reinforcements and to exploit the existing gap in the line, goes on to say-

> "By the evening of the 13th September the situation had completely changed. German reinforcements were known to have arrived, and serious resistance was to be expected on the 14th; yet G.H.Q. orders merely repeated the formula "the Army will continue pursuit........and act vigorously against the retreating enemy"; they gave no more tactical direction than to allot roads. There was no plan, no objective, no arrangements for co-operation, and the divisions blundered into battle."[238]

Attention to detail and the perils of poor staff procedure were common themes hammered at Camberley, Andover, and Greenwich in the schemes, exercises, and written assignments, though one might have a measure of pity for the unnamed officer who claimed nil marks out of 500 from Montgomery when his paper was assessed. Asked to explain such a poor score, Montgomery retorted that the directions clearly stated not to write in the margins of the paper and the officer had written in the margins.[239] On another occasion, Montgomery listened as a future Field Marshal and

---

237 'The Final Campaign, 1918', R.M. Luckock lecture No. 9, Dill Papers, 2/1. LHCMA.
238 Brooke, 'Aisne Exercise 1934', lecture, no date, but *c*. June 1934, Alanbrooke Papers, 3/16/1, LHCMA.
239 Hamilton, *Monty*, p. 195.

CIGS explained his syndicate's solution to a set problem. Offering faint praise as the solution was expounded, Monty concluded:

> Well Harding, I know that you and your syndicate have taken a great deal of trouble and have worked very hard to produce this solution, but I can tell you there would only be one consequence that would happen if your solution would be put into operation—it would end in a scene of intense military confusion.[240]

Nor was this simply the style of a unique and self-opinionated officer. Freeman whilst serving as Commandant at Andover after reviewing the work of qualifiers remarked that he found their work both interesting and astonishing; yet, it was interesting and astonishing for all the wrong reasons. He could only conclude that their thinking came from their backsides and their prose from their feet and, henceforth, he would disabuse them of such practices.[241]

On the surface, one might view the above examples as pettiness or rudeness of an inordinate degree and, doubtless, some officers receiving blinders of a similar nature thought so too. Still, the reasons behind such actions were clear enough when at another moment, lives could be at stake and an errant command or an ill-considered action might spell disaster. The qualifier was a cut above and knew it; the directing staff aimed to instill the value of working as part of a team and not to take himself more seriously than his commanding officer and a measure of cutting one down to size was practiced.[242] Moreover, it must be remarked that both Montgomery and Freeman went out their ways to encourage the more promising officers by suggesting alternate readings to those officially sanctioned or writing testimonials of their potential to higher authority. That said, Paget deprecated the practice of needless belittling and during his time as Commandant issued specific guidance that sarcasm was never an appropriate tool for remarking on a qualifier's work.[243]

Following the World War, formal examinations at Camberley of the material studied were abolished with assessment, nevertheless, made of the candidate through observation and review of the work submitted. Failure was an option and the qualifier not proving himself equal to the task might end his studies after the first year if warranted, or denied his *psc* at the end of the second, if deemed to be insufficiently prepared for staff duties. Unlike Andover, which did not rank its qualifiers on leaving, Camberley did and one's next assignment was an excellent indication of how well an officer had done. The qualifier was assessed whether he was likely to prove a good commander and if his talents were more suited to the Administrative rather than the General Staff.[244]

<hr>

240 Harding, Reel 10, Sound Recording 8736 of 1984, Imperial War Museum, London.
241 Furse, *Freeman*, p. 56.
242 Warner, *Auchinleck*, p. 33.
243 'Correction of Individual Work', Rhys-Pryce Papers, NAM 9204-164-11-7.
244 Major General Douglas Wimberley Papers, PP/MCR/182, p. 227, Imperial War Museum, London.

Brooke, the gunner, had been a member of Camberley's directing staff from 1923-27; a role he was also to enjoy at the IDC from 1932-34. Forceful and direct in style and destined to serve as CIGS during most of the Second World War, a naval counterpart to him, or equally Field Marshal Sir John Dill, is hard to suggest. Admiral of the Fleet Sir Henry Jackson is perhaps one, but then he features largely before the period covered by this review. This does not necessarily mean that the Navy deprecated mental ability, only that it did not look necessarily to those who had been instructors serving at its War and Staff Colleges when considering its choice of First Sea Lords. As a new arm of defence, those officers commanding the Air Force Staff College were too junior to be considered for promotion to Chief of the Air Staff between the wars. Later, two former Commandants would eventually reach the highest professional office in their Service. In practice, then, the RAF more closely followed the Navy when filling its highest offices between the wars.

Why Camberley enjoyed such prominence at the cusp of a new war is unknown, but with the first three professional heads of the British Army (i.e., Gort, 'Tiny' Ironside, and Dill) during the Second World War all former Commandants, its role as an incubator for the CIGS position is unmistakable. Graham and Bidwell avow that it was at this moment that Camberley reached the apogee of its influence and their conclusion is difficult to refute.[245] The talent of those studying and teaching there no doubt made this possible, but so, too, did the existing organizational structure of defence. It was essentially weak at the centre and slow in execution in the absence of war. Though Brooke severed the trend of having a Commandant of the Staff College as a CIGS, he retained strong ties to the college and in the normal scheme of things would have been considered a strong candidate for leading it, if the Second World War had not intervened. It may be that in the absence of robust operational commands and given the limited size of the British Army, Camberley's role and that of its Commandant enjoyed a special status due to its emphasis on training future commanders and its considerations regarding the application of force. Still, Ironside once conceded that in his four-year tenure he was never once consulted by the War Office on the preparation of the Army for war, so the Commandant's influence, if substantial, remained largely of an unofficial kind.[246] The same could also be said for Commandants of the Imperial Defence College with the distinction that their lines of authority extended from the Chiefs of Staff Sub-Committee and not direct from the War Office.

The combined exercise in time became the culminating experience for all who attended a Staff College and whether this exercise was held at Camberley or at Quetta, it made little difference in its scope, if rather more so only in its scale. Serving as the vehicle where qualifiers demonstrated all the elements of planning, administration,

---

245 Graham and Bidwell, *Coalition, Politicians and Generals*, p. 15.

246 Ironside to Rory Macleod letter, no date but *c.* September 1939, Colonel Rory Macleod Papers, 2/2, Liddell Hart Center for Military Archives, King's College, London, United Kingdom.

and command previously learned, it emphasized the necessity for cooperation amongst the forces and fostered relationships across Service lines. This was no trivial matter for the 'history of combined operations between the Royal Navy and Army is not on the whole complimentary to either Service. More often than not the true reason for the failure of this or that expedition has been a lack of sympathetic understanding between the commanders.'[247]

The 1921 Camberley exercise with Greenwich saw a Japanese assault on Hong Kong staged via hypothetical landings at Mirs Bay situated to the immediate northeast of the colony. The results proved inconclusive as the syndicate playing Blue claimed to have subjugated Hong Kong after a three-month siege conducted by Japanese naval, military, and air units, though the British syndicate thought otherwise.[248] As with the strategical war game played at Greenwich, the combined exercises served the additional purpose of informing Admiralty planning. The exercises influenced the force structure requests made of the Government by the Admiralty and shaped the scope of the strategic direction provided to its operational commanders. Thus, Keyes when serving as DCNS wrote to Richmond whilst the later commanded the East Indies Squadron that:

> The importance of Singapore, whether further developed or not, is obvious and we must do everything possible to safeguard it. Since you left England the Combined Staff Colleges have investigated the question of direct attack on the Island of Singapore, but the report has not reached us though I understand it was considered to be impracticable under the conditions on which the exercise was based, which assumed that the defences had been augmented to the degree contemplated for the Singapore base.
>
> We shall shortly be revising the War Memorandum and the points you mention will be carefully gone into.[249]

Still, it not just in London where the import of the annual Camberley combined exercise was studied. Thomas Blamey,[250] an Australian attached to the War Office, providing a summary of the 19-24 March 1923 exercise to General Sir Cyril White,[251] the Chief of the Australian General Staff, remarked how the recent Staff College exercise demonstrated the possibility of the Japanese investing Singapore with a force of 100,000 troops in less than two months and all before the main fleet could arrive

---

247 C.K.S.A., 'Visit to Quetta,' *Naval Review*, February 1935, p. 71.

248 Longmore, *From Sea to Sky*, p. 99.

249 Keyes to Richmond letter of 19 May 1924 cited in Halpern, ed., *Keyes Papers, Vol. II*, pp. 96-97.

250 Field Marshal Sir Thomas Albert Blamey (1884-1951). Staff Course, Quetta, 1912-13.

251 General Sir Cyril Brudenell Bingham White (1876-1940), Staff Course Camberley, 1906-07.

from Mediterranean and Home Waters. The implications for the Commonwealth were serious and deserved the considered attention of Australian authorities.[252]

Before holding the annual exercise, it was customary for qualifiers to receive preparatory lectures at their home establishments on how the Services viewed and approached combined operations. In 1923, Air Commodore Robert Clark-Hall,[253] a former naval officer who forewent a career as a gunnery specialist and then as a naval aviator and now serving as the senior Andover instructor, visited Greenwich and spoke about the role of air forces in supporting such operations while a military representative covered the topic from the Army's point of view. In turn, Captain Charles Forbes treated the responsibilities of the Navy during amphibious landings.[254] Additional subjects covered before the exercise's commencement included those on coastal batteries, armoured warfare, or as was styled at the time, 'mechanical warfare' given by Fuller, and the shore observation of naval gunfire covered by Brevet Major Gordon Seath, RM,[255] himself a qualifier of the current War Staff Course. The last topic proving one where firsthand experience applied, as Seath secured his DSO for observing the fleet's gunfire during the Dardanelles campaign.[256] Commander Frederic Bennett[257] of the Greenwich directing staff surveyed the Navy's Mobile Naval Base Defence Organization, while two external lecturers, Sir Frederick Black, previously Admiralty Director of Contracts, and Alfred Faulkner, the Director of Sea Transport at the Board of Trade, reviewed the possible shipping issues arising in any Pacific war.[258]

Following the exercise, the Admiralty proposed increasing the scale of naval forces available in the Far East. Hitherto, capital ships and aircraft carriers were not stationed in the Pacific on a regular basis, doubtlessly, influencing the perceived success of the Blue force in the recent game. Now, the Naval Staff recommended sending three battle cruisers, one aircraft carrier, eighteen destroyers, and to double

252 David Horner, *Blamey: The Commander-in-Chief* (St Leonards: Allen & Unwin, 1998), p. 68.
253 Later Air Marshal Sir Robert Hamilton Clark-Hall (1883-1964). Commanding Officer, *Ark Royal*, 1914-15; Staff College Andover, 1922-24; Chief Staff Officer, Middle East, 1924-25; and Air Officer Commanding Coastal Area, 1931.
254 ADM 203/100.
255 Later Major General Gordon Hamilton Seath (1886–1952). War Staff Course, 1922-23; District Intelligence Officer/Staff Officer Intelligence Gibraltar, 1923-26; Naval Intelligence Division, 1926-30; Fleet Royal Marine Officer, America and West Indies Station, 1930-32 Senior Officers' War Course, 1933; Senior Officers' School Sheerness, 1933-34; Combined Operations Course, 1934; Lieutenant Colonel, 1934; Portsmouth Division, 1934-39; Higher Commanders' Course Netheravon, 1936; Colonel 2nd Commandant, 1937; Colonel Commandant and Temporary Brigadier, 1939; Major General, 1941 and retired, 1942. Recalled for war service.
256 ADM 196/63 (Seath).
257 Captain Frederic Walter Bennett (1887–1969). War Staff Course, 1921-22; Royal Naval Staff College, 1922-23; Gunnery Division, 1923-26; retired, 1927 and Captain, 1932. Recalled for war service.
258 ADM 203/100.

the scale of submarines available. The conclusion of the Air Staff was that if the forces now proposed by the Admiralty had been present, the loss of Singapore would have been averted for the Japanese would not risk a landing in the absence of local sea control and while the RAF continued to offer a credible air threat. Moreover, if these proposals stood, then the Air Ministry need not augment its own forces as it had originally intended do in light of the exercise.[259] The story of the Singapore Naval Base and its subsequent history lies outside the scope of this survey. Yet, its relevance as a vehicle for officer training at the Staff Colleges, both individually and collectively, and the use the Services, in turn, made of these results are clear.

Finally, having gotten a force ashore, it remained to consider how to extricate it if the operation was only a raid or abandon the enterprise if it fell afoul. Here, Lieutenant Colonel William Godfrey, RM, lectured Greenwich qualifiers at the conclusion of the exercise during the years 1921-23 on how to evacuate a military force.[260] A veteran of the Dardanelles campaign, the one redeeming feature of that misadventure had been the re-embarkation of the British and Commonwealth forces. A cynic might suggest that such analysis was merely planning for defeat; yet, as every raid or amphibious operation must at some point end, the focus on evacuation was entirely sensible. Following in Godfrey's path, Lieutenant Colonel Richard Foster, RM,[261] the lead instructor for the Intelligence Course, continued to lecture on the question of evacuations just as Edward Altham had lectured on the evacuation of Archangel during the first postwar course. Moreover, in 1926, Foster whilst still assigned to the Staff College, attended the Mediterranean Fleet landing exercises held during July and August to witness the practical aspects of his subject.

The Royal Naval Staff College did not participate in the combined exercise held in 1927 and this omission owed everything to the fact that beginning in 1929 the course would now convene in January. Consequently, the 1927-28 Staff Course was the longest of the period running to fourteen months. In due course, the naval members prepared themselves for the exercise scheduled for 12–17 November 1928. Lieutenant Commander Charles Drage, an old Ludgrovian, spent the two weeks preceding the exercise at Camberley with Commander Thomas Fellowes.[262] For Drage, it was a chance to prepare himself for the role of Chief of Staff to Fellowes

<hr>

259 ADM 116/2394, Air Council letter S. 22781 of 4 September 1923.
260 ADM 203/100.
261 Later General Sir Richard Foster Carter Foster (1879–1965). Staff Course Camberley, 1912-13; War College, 1913-14; Attached to British Army, 1918-23; Senior Officers' School Woking, 1923; War Staff Course, 1923-24; Royal Naval Staff College, 1924-27; Temporary Brigadier, 1930; Colonel in Command Chatham Division, 1930-33; Higher Commanders' Course Nethervaon, 1931; Adjutant General Royal Marines, 1933-36; Lieutenant General, 1934 and retired, 1936.
262 Later Captain Thomas Balfour Fellowes (1891–1974). Commanding Officer, *Torch*, 1923-24 and *Voyager*, 1924-26; Staff Course, 1927-28; Commanding Officer, HMS *Daring*, 1932-34; Captain, 1934; Senior Officers' Technical Course, 1935; Senior Officers' War Course, 1935; Chief Staff Officer in *Cormorant*, 1936-38 and retired, 1938. Recalled for war service.

in the upcoming exercise, but also a chance to reestablish ties with Oliver Leese,[263] another old boy from Ludgrove Preparatory School.[264] Again, a representative from the Board of Trade supported the exercise highlighting that these evolutions, beyond preparing the officers under study, also served a corollary purpose for civil departments. This time, the member was Julian Foley, a former Director of Sea Transport in the Admiralty and a past lecturer at the Royal Naval Staff College. Foley concluded the week's festivities on a lighter note offering the sage advice to the officers attending the champagne dinner marking the close of the week's events, 'As her virginity is to a lassie, so is administrative purity to a government department.'[265]

The November 1935 exercise amongst the Staff Colleges conducted was of a pattern similar to the ones held previously. A significant departure, though, was that the scale of supplies for the air units earmarked for overseas deployment was increased by an additional month. The additional fuel, lubricants, and ammunition contemplated could not be accommodated in the shipping nominally reserved by the Board of Trade, whilst a subsidiary issue arose over its distribution once landed. This indicated that a handy container or drum and a special type of tanker to facilitate provisioning were desirable. However, beyond the myriad details of technical issues, the exercise highlighted once more the problems of command.[266] The existing reference point for planning such operations, the *Combined Operations Manual*, was deemed deficient in a number of areas and Major General Clement Armitage,[267] the Commandant of the Army Staff College, believed a fundamental rewrite was appropriate and not just the mere revision then contemplated. Armitage also endorsed Captain Bertram Watson's proposal for a joint body to investigate the further refinement of combined operations.[268]

In 1937, the Staff Colleges considered again a Japanese assault on Hong Kong though the venue for the final portion of the exercise was now Greenwich. Captain Guy Simonds,[269] a gunner of the Canadian Army, led the syndicate controlling the Japanese Army and introduced the novel concept of directing his actions from a command ship something, Dominick Graham records failed to win the universal approval of

---

263 Lieutenant General Sir Oliver William Hargreaves Leese (1894-1978). Staff Course Camberley, 1927-28 and Staff College Quetta, 1938-40.
264 Diary entries of 1 and 12 November 1928, Drage Papers, Reel 3, IWM, PP/MCR/99.
265 *Ibid* entry of 16 November 1928.
266 CAB 53/28, Commandant, Staff College to Secretary, War Office letter of 31 December 1935.
267 Later General Sir Charles Clement Armitage (1881-1973). Staff Course Camberley, 1914; Staff School Cambridge, 1919; Commandant, School of Artillery, 1927-29 and Commandant, Staff College Camberley, 1934-36.
268 CAB 53/28, Commandant, Staff College to Secretary, War Office letter of 31 December 1935.
269 Later General Guy Granville Simonds (1903-1974). Staff Course Camberley, 1936-37; Royal Military College Canada, 1938-39; Imperial Defence Course, 1946; Imperial Defence College, 1947-49 and Commandant, Royal Military College Canada, 1949-51.

all naval participants.[270] That may be so but as early as the 1931 course, Commander Charles Daniel, a trained Signals Officer, was advising that the practical problems of communications during the yearly combined evolution needed to be addressed. This recommendation the Admiralty endorsed and by 1934 the Portsmouth-based Signal School was sending a dedicated signals team to Camberley to assist in the running and evaluation of the yearly exercise.[271] Both instances indicate that the thinking ongoing during the yearly convocation of the Staff Colleges was not necessarily limited to stereotyped lessons of value only to those under instruction.

Simonds and his syndicate enhanced their presentation of the assault plan developed by including photographs of contemporary Japanese landing craft then in use in Chinese waters. On the exercise's final day, the First Lord of the Admiralty and the Secretary of State for War attended and listened as qualifiers summarized their findings.[272] Notwithstanding whatever Duff Cooper and Hore-Belisha may have taken away from the proceedings, there can be little doubt that the qualifiers had absorbed a great deal regarding the execution of amphibious operations whilst appreciating something of the presumed military and naval capabilities of the Japanese.

The continued prominence of the yearly joint exercise was not without critics. The Air Ministry, for one, questioned whether its emphasis was correct given the problems now arising closer to home. No doubt, participating in the Drag, competition in the many other sporting matches, and the concluding dinner fostered camaraderie, amongst the Services, but was it still relevant? One can appreciate their concern and nor was it entirely misplaced. The combined exercise normally saw the RAF relegated to a supporting arm given its usual formulation of a Pacific conflict. An exercise focused on a German threat would have found the Air Force more to the fore and even one considering war against Italy would have shed light on problems that were not so skewed to one Service. Contributing to this view was that few air officers had the direct experience of Gallipoli that many of those serving in the Navy and the Army had. William Dickson appreciated the value of the Dardanelles campaign as amplifying the problems associated with administration, strategy, and the higher direction of war though he could not testify directly to the problems exposed.[273] If the yearly exercise held by the colleges in common was the only work of a joint nature pursued, then the criticism would have more merit. As it was, the Staff Colleges met bilaterally during the year to consider problems unique to two of the Services. Accordingly, the problem was not so much the yearly Camberley exercise, as the yearly Camberley exercise conducted.

If the joint Camberley exercise was informing the strategic direction of defence, the Quetta exercise did much the same for the Government of India, whilst also

270 Graham, *Simonds*, p. 31.
271 Unpublished manuscript, p. 51, Waymouth Papers, IWM/87/16/1.
272 Graham, *Simonds*, p. 31.
273 Dickson Sound Recording 3168 of February 1978, Reel, 5, Imperial War Museum, London.

influencing the operational plans of the Navy. During Richmond's time commanding the East Indies Squadron, a yearly evolution with the Quetta Staff College featured with several naval officers of the station playing key roles. In late 1924, the exercise lasted ten days taking place at Salsette Island, Bombay. Officers of the squadron served as part of the exercise's directing staff and participated in two of the syndicates considering how to recapture the Malay Peninsula and relieve Singapore from an investing Japanese force.[274] Noel Laurence,[275] Richmond's Flag Captain in HMS *Chatham* and a former instructor at the War College, and Loben Maund, the station's War Staff Officer, served as members of the directing staff with additional officers serving in the syndicates considering the problems set by the controllers.

As beneficial as the exercise was, watching the results achieved prompted Richmond to propose that a naval officer be formally attached to the Indian Army Staff College to better acquaint its qualifiers on the functions of the Navy while also lecturing on trade, economics, and strategy.[276] For no matter how effective such exercises were as a learning tool in their own right, they could not accomplish all that was desired. Less the Admiralty turn a Nelsonic eye to his recommendation, Richmond was moved to forward an even more forceful letter on the same day in which he made his original proposal. Employing language that even his ardent supporters would find hard to defend when addressing one's superiors, the Admiral pleaded:

> I look on this as deplorable. It is not possible in a short week of co-operation to disabuse men of these ideas, and to bring home to them the Naval factor. I think the presence of an able Staff Officer, who should impress the Naval side of the larger national strategy in war, would tend to correct the narrow Military view. Such an officer would be available not only to work at Quetta but also at Belgaum – the Senior Officers' School, – where the need for instruction is no less pronounced, and for introducing sound notions into the forthcoming Indian Maritime Service.[277]

274 ADM 203/84, Commander-in-Chief, East Indies Station to Secretary of the Admiralty letter No. 49/3202 of 26 January 1925.

275 Later Admiral Noel Frank Laurence (1882-1970). Five firsts in examinations for Lieutenant; in Command, *D 1*, 1910, *E 1*, 1914-15, *J 1*, 1916 and *K 2*, 1916-18; Captain, 1919; Senior Officers' Technical Course, 1921, 1926 and 1935; Senior Officers' War Course, 1921; Royal Naval War College, 1921-23; Flag Captain and Chief Staff Officer in *Southampton* and *Chatham*, 1923-25; Assistant Director and then Director, Training and Staff Duties Division, 1926-28; Tactical Course, 1928 and 1935; Commanding Officer, HMS *Eagle*, 1928-29; Imperial Defence Course, 1932; Rear Admiral, 1932; Rear Admiral, Submarines, 1932-34; Army Air Co-operation Course, 1935; Commander, Aircraft Carriers, 1935-37; Vice Admiral, 1936; Commander, Reserve Fleet, 1938-40; Admiral, 1940 and retired, 1942. Recalled for war service.

276 ADM 203/84.

277 ADM 1/8714/169, Commander-in-Chief, East Indies to Secretary of the Admiralty letter No. 51/271 of 26 January 1925.

It may be the case that Richmond knowing of Lieutenant Commander Danckwerts' attachment to Camberley in 1923 for six months to support that course thought the question was only the more pressing in India where support from naval authorities was not so near at hand. Yet, the Danckwerts initiative owed everything to the fact that the Admiralty would not spare that officer to spend a year studying at Camberley and so appointed him for a more limited engagement.[278]

Ultimately, as previously shown, the Admiralty decided not to assign an officer to Quetta, and with an officer of Richmond's caliber already on station to look after its interests and coordinate affairs, the temptation was probably strong to conclude that this was a fix to a problem that did not exist. After all, the weight of the Quetta course was very much towards meeting the twin problems of internal security and defending the Northwest Frontier. Important issues, no doubt, but not of immediate concern to the Admiralty. Finally, the Navy believed it had a strong grasp of the military issues at play already, as some Royal Marines already attended Camberley and subsequently served on the Royal Naval Staff College directing staff. This represented a very real cadre of naval expertise well acquainted with military problems, procedures, tactics, doctrine, and strategic thinking.

Richmond's successor as Commander-in-Chief, East Indies Squadron was Rear Admiral Walter Ellerton. A former DTSD, as Richmond had been, whether this background and a desire to enhance the Navy's cooperation with Quetta influenced Ellerton's appointment cannot be said for it remains that no other DTSD became a future commander of the station. Still, those following in Ellerton's wake possessed other skills honed at the Senior Officers' War Course, the Tactical Course, and the SOTC, so cooperation was hardly a dead letter. It just was not as continuous and intimate as Richmond deemed necessary. Some indication of this is offered by the 1925 exercise held by the Quetta Staff College and supported by Ellerton's squadron. Though the record is far from complete, the report of proceedings forwarded to the Admiralty shows that the naval contribution this time included a landing on Kasid Beach near Bombay during the night of 7-8 December supported by *Cairo*, *Columbo*, and five ships of the Royal Indian Marine.[279] This landing witnessed over 1,000 men come ashore and though the results achieved were poor, this did not detract from the serials conducted. Indeed, this was the very purpose of training and Major General Sir Gerald Boyd,[280] the Commandant of Quetta, concluded that it had been an excellent exercise for the number and variety of lessons it had indicated and furthered their understanding of the underlying problems associated with combined operations.[281]

278 ADM 196/127 (Danckwerts).
279 ADM 203/74, Commander-in-Chief, East Indies Report on Combined Military Exercise.
280 Major General Sir Gerald Farrell Boyd (1877-1930). Commandant, Staff College Quetta, 1923-27.
281 Maund, *Assault from the Sea*, p. 254.

Now serving as Staff Officer Operations in the East Indies Squadron, Charles Drage inherited a major portion of the responsibilities in supporting the now annual evolution between Quetta and the squadron. Arriving on station in early February 1929, he began revising the naval parts of the planned scheme a full seven months before the exercise's start date for that year. Even at this, he was hard-pressed to complete the five separate updates governing the exercise before its October commencement. Embarked in a command ship supporting the full range of naval responsibilities by showing the flag to distant locales such as Ceylon, East Africa, and the Trucial States—all the while inspecting subordinate elements—made corresponding with Colonel Pat Mackesy of the Quetta Staff College slow and ponderous. In the event, Drage only completed his last amendments two days before the exercise began after arriving at the Staff College.[282]

Fostering ties with the Indian Army Staff College by relying on the ships and personnel of the East Indies Squadron may have had practical benefits for the Service and acquainted qualifiers with the personnel and capabilities of the Navy on the station, but it also demonstrated practical limitations. The ships, after all, were operational units and liable to have their programme of training preempted to meet more pressing taskings. Such occurred during the November 1926 combined exercise with Quetta when the light cruiser HMS *Enterprise* had to depart early and sail for China as Britain strengthened its forces in view of the ongoing turmoil in that country.[283] To *Enterprise*, the change of programme was nothing more than a demonstration of the flexibility of naval power and in many ways part of the normal routine of a ship operating in foreign waters. For the Staff College, it highlighted the limitations of imparting a maritime dimension to the course, and though officers from the East Indies Squadron cooperated in the combined exercise of 1930, no landings actually occurred that year, as had previously been the custom.[284] Thus, Richmond's quest to secure an officer to support Quetta was fundamentally sound.

The limitations of these exercises where few troops were actually employed, the landings largely gamed, and the bottlenecks associated with taking up commercial shipping mostly avoided were at once recognized, but for all that their usefulness remained as the preparatory work required was not unlike that executed for the real operation imagined.[285] Representative of the exercises held at Quetta between the wars was the 1934 one which lasted three weeks and saw seven officers of the Royal Navy and three from

282 Diary entries, 9 February, 16 February, 2 August, and 3 October 1929, Drage Papers, Reel 3, IWM/PP/MCR/99.

283 Harold E. Stevens, ed., *H.M.S. "Enterprise": Story of the First Commission, April, 7th 1926 to December 19th 1928* (Aldershot: Gale & Polden, Ltd., 1929), p. 7.

284 *C.B. 3016/30, Progress in Tactics, 1930'*, Admiralty, Naval Staff, Tactical Division, July 1931, p. 42, British Sources, Box 12, Naval Historical Center, Washington, DC.

285 Anon., 'Notes Bearing on the Staff Requirements of a Commander-in-Chief', *The Naval Review*, Vol. XXVII, No. 1, February 1929, p. 9. Though not credited, Richmond is clearly the author of the article based on the events depicted.

the Royal Indian Marine joining seven air officers and the thirty members of Quetta's Senior Division.[286] An initial series of lectures held in the morning and presented by each Service's ranking officer indicated the requirements and shortcomings that any combined operation posed to their arm. Following these surveys, subordinate officers delved into the technical problems faced in any joint operation. For the Navy, these briefings covered ship-to-shore gunfire support, the role of the Fleet Air Arm, the use of submarines, and the employment of the Mobile Naval Base Defence Organization: the fleet's readily deployable assets allowing it to operate from an advanced location which was undeveloped in itself but previously surveyed.[287] Supplementing these lectures were practical demonstrations with tanks negotiating wire entanglements and bombers attacking transports and cutters in the face of the presumed air supremacy afforded by the single-engine Wapiti bombers of the RAF. During the afternoons, the three syndicates operating met to consider the parameters of the specific scenario posed. Following completion of the work at the Staff College, the scene soon shifted to Karachi for the final week of the combined operation. Now members of the exercise witnessed company-sized landings with the assistance of the ships of the squadron supplemented with demonstrations of depth charge and torpedo work. Capping it all, RAF aircraft allowed the squadron's anti-aircraft defences to be demonstrated as fire was registered at a towed aerial target.[288] If the scale of the proceedings did not match the effort involved, it served the purpose, nevertheless, of acquainting officers of the likely hurdles to be faced and the pitfalls to be avoided in any real evolution.

Following on the 1934 exercise, the Quetta Staff College forwarded its recommendations to the Commander-in-Chief India and amongst its findings was that the present *Manual of Combined Operations* was deficient in a number of areas and required amendment. It also recommended against the appointment of any joint Commander-in-Chief suggesting, instead, that the land component commander be given the authority to decide any issues where a difference of views amongst commanders existed. This conclusion was slightly different from the recommendation voiced by the Commandant at Camberley at the same moment who conceded that a Supreme Commander might be required during some operations. The RAF Staff College held firmly to the belief that a Supreme Commander was a necessity, though it reversed itself on this point in 1935. Bertram Watson, the Director of the Royal Naval Staff College thought present command arrangements were adequate though whether this view was endorsed by the Naval Staff the record in extant is unclear.[289]

The scenario posited in the combined exercise was invariably one centred on the Far East with the defence of Hong Kong, Singapore, and Borneo all featuring at one time

<hr>

286 C.K.S.A., 'Naval Visit', *Naval Review*, February 1935, pp. 71-72.
287 *Ibid.*
288 *Ibid*, p. 74.
289 CAB 54/3, Lieutenant Colonel J. K. McNair to Deputy Chiefs of Staff note of 23 November 1936.

or another. It will be noticed that these exercises were essentially ones where Britain stood on the strategic defensive and while Blue syndicates would gain an appreciation of the offensive, those lessons were not always applicable to Red. This was a short-coming in the training of the period which probably had unintended consequences during the next war when Britain decided to conduct offensive amphibious operations in places like Norway and Dakar. It was not the case that consideration of opposed amphibious landings were neglected, but the lesson most often drawn was that diffi-cult as these type of operations were when naval mines and submarines were the likely dangers facing the British, the presence of enemy air forces now made them almost impossible. Using open boats was tantamount to suicide demanding that specialist craft be employed. Unfortunately, the funding to develop and procure such instru-ments was never a priority. This owed something to Britain's strategic posture, which sought to defend what it possessed whilst minimizing any future territorial acquisi-tions in light of the obligations they entailed. This, however, ignored a key strand of previous naval experience where Britain frequently aimed to isolate an enemy from his overseas trade and deny him the strategic mobility that his naval bases conferred.

In a sense, though, the greater purpose of the combined exercise was not to prove whether Singapore was at risk from the Japanese or whether amphibious operations remained possible. Rather, they were conducted primarily to develop the common staff procedures allowing the Services to operate with the least amount of friction. Thus, the actual scenario was tangential to the greater goal. A common manual supporting combined operations from before the war; revised in 1925 and again in 1938, its formal issuance waited until late 1939.[290] If an amphibious undertaking was the deepest and closest of operations requiring inter-Service cooperation, then it was far from the only one that might arise in war. Thus, it is for this reason that the schemes conducted were secondary to the greater objective of fostering mutual understanding and sympathy.

That said, the results of these exercises were captured and the lessons learned (or learned anew) were filed for future reference with perhaps the premier point being to plan properly, to stow tactically and to resist making changes once the force had sailed.[291] Thus, by the end of the period the colleges working collectively had an acute sense of what was achievable and what remained to be accomplished.[292] These lessons informed opinion on the practicality of combined operations and the wherewithal required to conduct and never the more so than with the establishment of the Inter-Service Training and Development Centre. They, for one, cited the results of the 1938 exercise of the Staff Colleges where the attacking Japanese force was able to capture and use the airfields of North Borneo to support their subsequent opera-tions. Having lost these airfields, the British force was hampered in trying to fore-stall further Japanese operations, as the airfields that she now relied upon were over

---

290 DEFE 2/709, *C.B. 3042, Manual of Combined Operations, 1938.*
291 Maund, *Assault From the Sea*, pp. 26-27.
292 *Ibid*, pp. 3-4.

500 miles from the enemy lodgment. The exercise highlighted that the defending British forces either required the use of carriers to aid in the land campaign or British bombers needed to be equipped with floats to allow them to operate from protected anchorages, as time would not permit temporary airfields to be constructed to replace those already captured.[293]

Recognizing that the strategic environment had changed, the Joint Planning Committee proposed to the Deputy Chiefs of Staff Sub-Committee in August that the 1939 combined exercise to be held at Minley Manor focus on an Anglo-French assault on the Italian Dodecanese Islands. This results of the exercise would be examined by the Service Ministries and by the British Middle East commanders to inform their discussions with the Turkish General Staff. In the event, this late initiative proved just that—late and the onset of a new war derailed the game.[294]

The combined exercises held at Camberley between the wars allowed problems to be considered in their widest sense as the numbers participating were greater than what could be achieved working separately. Syndicates covering all aspects of Red and Blue forces analyzed aspects of the problem first at their home establishments before converging on Camberley for the week set aside for the scenario to be considered *in toto*.[295] For the first postwar exercise, twenty-six naval and marine officers joined three air and approximately 140 military officers. The limited air participation owing much to the fact that it did not yet have its own Staff College and its officers were those then studying at Camberley and Greenwich.

From the above, it will be seen that the Quetta combined exercise differed from those held at Camberley in two vital respects. Firstly, a substantial portion of the participants were not qualifiers at all but naval and air officers serving in India as part of a routine overseas posting. Secondly, the evolution was more than just a command post exercise where troop movements were simulated, but went a step further and actually employed military, air, and naval forces. Late in the day, Camberley would employ the ITSDC much as Quetta worked with the East Indies Squadron with qualifiers on the home establishment embarking in ships to witness a limited landing near the Isle of Wight.[296] Thus, in one key area Quetta stole a march on Camberley and was not the poor step-child so often depicted.

Upon leaving Camberley or Andover, the naval and marine officer would be well acquainted with the doctrine and ethos of the British Army or Royal Air Force. Though attaching officers to the other forces might have secured the same results achieved, attending their Staff Colleges came with the benefit that the press of responsible

293 CAB 54/13, Inter-Services Training and Development Centre memorandum of 12 December 1938.

294 CAB 54/11, Joint Planning Sub-Committee to Deputy Chiefs of Staff Sub-Committee note of 17 August 1939.

295 Air 1/516/16/417, Commandant Army Staff College Report on Naval and Military Exercises of November 1919.

296 Maund, *Assault From the Sea*, p. 16.

work did not intrude on this nurturing process. Weighing the overall effectiveness of Camberley and Andover training for the period is problematic, as both had the twin missions of developing staff officers and imparting the essence of command, whilst Greenwich only attempted to prepare one for the duties of the former. Hobart an officer of no mean intellect rated Camberley very highly during the decade from 1927 believing it even exceeded a university education in many respects.[297] This is not surprising as many of those attending were in fact university graduates. Slessor writing long afterwards concluded that Army Staff College was excellent at teaching officers *how* to do things; they were less successful in teaching qualifiers *what* to do.[298] Slessor was thought to be an excellent instructor at Camberley and his analysis of the course's weakness is probably fair.[299] Much the same could be said of what was attempted at Andover and even for the Royal Naval Staff College. Where the Navy parted company though, was in instilling the *what* by sending promising officers to the SOWC and to the Tactical Course.

Beyond Camberley, though, a select cadre of eligible officers attended courses organized by the principle arms of the British Army, (i.e., infantry and artillery) found at Hythe and Netheravon or the Senior Officers' School. The last named course ran three times a year for three months with 40 officers attending.[300] Originally formed in France in 1916, it soon moved to Aldershot and aimed to prepare officers before assuming command duties.[301] Though the exact origins of the course remain obscure, the need for such training was not, as expansion and losses within the expeditionary force exacted their toll on its command experience and performance. Fuller was one of its earliest champions with his earlier *Training Soldiers for War* shaping the style and content of the curriculum.[302] Even though some questioned the need for the school following the end of hostilities, the War Office retained it establishing it at Inkerman Barracks Woking in the first instance before occupying the former barracks of the Royal Garrison Artillery at Sheerness in April 1924.[303] For the Navy, the officer attending was typically a recently promoted Captain who had not previously held command whilst the marine studying was a Major awaiting promotion to Lieutenant Colonel.[304] As with the naval officers attending Camberley and Andover, the gunnery specialist and the salt horse were the backgrounds of the overwhelming

297 Hobart to Liddell Hart letter of 21 March 1937, Liddell Hart Papers, LH 1/376/44, LHCMA.

298 Slessor, *Central Blue*, p. 98. Original emphasis.

299 Major General George Warren Richards, Sound Recording 866 of 1976, Imperial War Museum, London.

300 *Navy List*, October 1922, p. 2402.

301 'Courses at Army Schools', *Journal of the Royal United Services Institution*, Vol. LXXI, No. 482, May 1926, p. 387.

302 Fuller, *Memoirs*, pp. 55-56 and 60-62.

303 *Hansard*, House of Commons Debates of 26 February 1924, v. 170, cc231-32 and *The Army List* (London: His Majesty's Stationery Office, May 1924), p. 925.

304 *King's Regulations*, Article 327, p. 111.

number present and aviators were conspicuous by their total absence. Training officers on the active list for regimental command, Territorial Army officers also attended, as did equivalent officers from the Dominions along with members from the Royal Air Force. From 1921, the Indian Army operated a course on similar lines at Belgaum, though still sending officers to Sheerness.[305]

Attendance at the course was by appointment with the objective of instilling in a senior Major how to command a battalion and the operational and administrative duties associated with such formations. Exercises without troops and lectures shaped the instruction and, again, the syndicate was employed when conducting TEWTs. Here, the problem might be to assess the best means of negotiating a river in the face of enemy opposition, landing a raiding party on a coast, or how to defend or attack a prominent point. Typically, the syndicate, composed of 3-4 members, surveyed the land, assessed the forces available and the likely scale of opposition, and then presented their plan after deliberating for about two hours. Instructors would then evaluate the syndicates' proposals against a school solution previously developed.[306] What the marine officer attending took away from the course is obvious, but the more speculative matter is what did a naval officer seek to gain by attending? Closer touch with the Army, certainly, but the proper means and tactical issues surrounding the employment of a naval landing party were also likely benefits. What the Army gained from the two naval officers present in each class were subject matter experts in naval procedure, and, as with what was emphasized at the Staff Courses, combined operations was a strong element featuring in the curriculum.[307] In common with all the other courses discussed so far, students attending the Senior Officers' School offered their own lectures on special topics. Thus, if naval officers did not appear as members of the directing staff, its qualifiers attending still were able to impart the naval view to a broader audience. Accordingly, Captain Claude Hermon-Hodge, attending the course in 1933 and with much experience in minesweeping operations, lectured to his class on the subject.[308] Captain Richard Benson,[309] sitting the course immediately following Hermon-Hodge's, lectured on 'Naval Tactics' and was praised for his support of the Combined Operations Exercise.[310] It was also in 1933 that Commander

---

305 H. R. Sandilands, 'The Case for the Senior Officers' School', *The Journal of the Royal United Services Institution*, Vol. LXXIII, No. 491, May 1928, p. 236.
306 Major General George Warren Richards, Sound Recording 866, 1976, Imperial War Museum, London.
307 ADM 196/92 (Hermon-Hodge).
308 *Ibid.*
309 Captain Richard Stoddart Benson (1892-1940). Five firsts in examinations for Lieutenant; Naval Ordnance Department, 1928-30; Staff Course, 1930; Tactical Course, 1931; Captain, 1932; Senior Officers' School Sheerness, 1933; Senior Officers' Technical Course, 1933 and 1938; Imperial Defence Course, 1934; Captain (D), Eighth Destroyer Flotilla, 1935-37; Senior Officers' War Course, 1938; Royal Naval War College, 1938; Captain (D), Twelfth Destroyer Flotilla, 1939-40.
310 ADM 196/92 (Benson).

Robin Jeffreys was attached to the school to serve as a naval instructor, whilst holding the concurrent position of Staff Officer Operations and Intelligence in the Nore Command.[311] This was unusual, but not out of character as liaison between military and naval authorities was constant and close during the period,

With responsibility for unit proficiency and officer development residing with the regimental commanding officer, quality and conformity to a common standard across the British Army was difficult to ensure. Had a tactical school existed, the problem would not have been nearly as acute. Continuing the Senior Officers' School after the World War was a measure to remedy the worse aspects of this unevenness and can be seen as a compromise between those who resented the intrusion into internal regimental affairs and those desiring to establish a College of Tactics. It also highlights a major early lesson learned by the British and Indian Armies of the 1914-18 conflict: the need to prepare officers for battalion command and not to trust all to seniority. Thus, an important role of the schools as specified in *King's Regulations* was 'to report on officers as regards their ability to conduct the training and administration of a battalion or equivalent unit.'[312]

In common with the other venues of officer education, lectures, exercises, student papers, and debates on topical issues featured. At times, these served a broader purpose and a broader audience. When in 1920, Churchill, the Secretary of State for War, took the chair and moderated an exchange between Lieutenant General Sir Philip Chetwode and Fuller titled 'Tanks as a New Arm' the latter relished the opportunity presented. Treating the question as one of 'Tanks *versus* Cavalry', Chetwode, a noted cavalry officer, stood his ground with Fuller and gave as good as he got.[313] To Captain Frank Austin,[314] direct from Greenwich and the SOWC, the proceedings offered testimony that each Service had its challenges in defining a force structure for the future.

Finally, mention must be made of two other venues of senior military instruction provided to officers of Brigadier rank and above. The first was the Higher Commanders' Course run by the British Army whilst the second was its French Army counterpart held in Versailles. Both aimed to develop the higher skills required of senior commanders, though the French Army course was, of necessity, limited to those

311 ADM 196/53 (Jeffreys).
312 E. W. Brighten, 'The Senior Officers' School: The Case for a College of Tactics', *The Royal United Services Institution*, Vol. LXXIII, No. 490, February 1928, pp. 24-25.
313 Fuller, *Memoirs*, p. 395.
314 Later Vice Admiral Sir Francis Murray Austin (1881–1953). Commanding Officer, HMS *Astraea*, 1918-19; Captain, 1919; Senior Officers' War Course, 1921 and 1932; Senior Officers' School, Woking, 1920; Royal Naval Staff College, 1921; Senior Officers' Technical Course, 1923 and 1932; Commanding Officer, *Danae*, 1923-24; Director, Physical Training and Sports, 1925-26; Deputy Director, Operations Division, 1926-27; Flag Captain and Chief Staff Officer in *Iron Duke*, 1927; Rear Admiral, 1931; Tactical Course, 1932; Admiral-Superintendent, Gibraltar, 1933-35; Vice Admiral and retired, 1936. Recalled for war service.

officers able to deal with instruction in a foreign language. Wavell was one officer to attend the French Army class (in his case during late 1934) affording him the opportunity to learn something of that nation's strategic and tactical views while Brigadier Edward Tollemache,[315] late Coldstream Guards, sat the course in 1937.[316] Though no naval or marine officers is thought to have attended the French school, marine and naval officers including, D'Oyly-Hughes, Henry Inman and Lieutenant Colonel Charles Lucas did sit the one-week class held variously at Old Sarum, Netheravon and Aldershot. In Lucas' case, he attended the April 1936 session having previously passed through the Sheerness Senior Officers' School. Both Inman and Lucas had also taken previous specialist courses in artillery and combined operations testifying that the linkages between the Army and the Navy if not necessarily intimate were, nevertheless, substantial.[317]

Until the opening of the Imperial Defence College in 1927, higher level joint training was primarily a case of sending a limited number of officers to another Service's Staff College, to the Air Force's Schools of Army and Naval Co-operation, to the Army Senior Officers' School, or the Tactical and Senior Officers' War Courses of the Royal Navy. The RAF having no senior command and staff training establishment of its own was a prime advocate for the creation of the IDC.[318] Trenchard was also a resolute advocate for creating the Chiefs of Staff Sub-Committee and an overarching Ministry of Defence. For the moment, the last named was a thought too far. Individual Service attempts at joint training were constrained for a number of reasons, of which, their own requirements for such officers was a primary limiting factor and so too cost.

It has already been noted that the Air Force desired to have a naval officer on its directing staff at Andover; a measure the Indian Army Staff College concurred in when it sought much the same support. In 1923, the India Office proposed that the Royal Navy send one or two officers each year to Quetta for a one year course as was the custom at Camberley and that, in turn, two of its officers following the completion of their course at Quetta proceed to Greenwich to complete the naval Staff Course. This suggestion initially engendered a positive response, but the need to charge Indian Officers £75 per term for attending Greenwich proved insurmountable. In the eyes of Indian authorities, such a charge need not apply in view of the *quid pro quo* implied in the exchange of officers. Yet, the Navy saw only grief from the Treasury and resentment from the Dominions if the cost of the course was not collected as their officers were subject to the charge.[319]

---

315 Later Major General Edward Devereux Hamilton Tollemache (1885-1947). Staff Course Camberley, 1920 and Higher Commanders' Course Versailles, 1937.
316 Wavell to Liddell Hart letter of 3 February 1950, Liddell Hart Papers, LH 1/733/386, LHCMA.
317 ADM 196/64 (Lucas), ADM 106/63 (Inman) and ADM 196/53 (D'Oyly-Hughes).
318 Powers, *Strategy Without Slide-Rule*, p. 187
319 ADM 1/8640/112, India Officer to Secretary of the Admiralty letter M.5043/23 of 8 August 1923.

As for the proposal to attach a naval officer to Quetta, an idea highly recommended by Richmond after witnessing a number of joint exercises, Beatty, the First Sea Lord, endorsed the idea wholeheartedly. However, cost as ever proved insurmountable. Though the Indian Government was willing to meet the transportation expenses of an officer to Quetta from England and his subsequent return, they would not cover the additional allowances required once on station. Nor would the Treasury, for that matter, meet these costs.[320] Here, the India Office must have received the Navy's request with a measure of incredulity, as its own officers when serving in Britain— whether as qualifiers or as members of the Camberley directing staff—saw sizeable losses in their pay and allowances.[321]

That more was not achieved along these lines was largely the Navy's fault. As can be seen, expense was the nominal reason why this did not occur, but the broader conclusion that sending its officers to India was also thought a poor return on investment is hard to escape. The academic programme of Quetta was heavily oriented to the defence of the Northwest Frontier where the interests and role of the Navy was always remote. Some graduates of the college readily conceded its academic limitations when compared to Camberley, though noting that some of its instructors were very good nevertheless. Included amongst these were Hobart and Mackesy.[322] Meanwhile, General Ironside, already a former Commandant of Camberley, lamented the quality of Indian Army officer serving him whilst posted to the sub-continent in 1929 telling Liddell Hart:

> Worthy people with no inspiration and to my disgust I find every one of them was educated at Quetta. Ludicrous place. They spend their time training officers for a limited warfare in the Mountains and at the back of their heads is a pinching economy which they carry from peace to war.[323]

Yet, the more fundamental reason why more was not achieved is that Royal Navy always favoured attachment over formalized joint training as the primary means of fostering a closeness between the Services. This view owed something to ethos. Philosophically, the Navy deprecated specialization of any type and sought to produce capable, well-rounded officers able to meet any situation. Specialization was viewed as a necessary evil and made managing officer assignments challenging, as ensuring the right man with the appropriate skills was available became that much more difficult. As the Navy did not maintain operational staffs of comparable size to the Army

320 ADM 1/8714/169, First Sea Lord minute to Commander-in-Chief, East Indies to Secretary of the Admiralty letter No. 51/271 of 26 January 1925.
321 Lewin, *Slim*, p. 55.
322 Major General Horace Leslie Birks, Sound Recording 870, Reel 6, of 15 December 1976, Imperial War Museum, London.
323 Ironside to Liddell Hart letter of 26 January 1929, Liddell Hart Papers, LH 1/401/8, LHCMA.

and the Air Force, its training requirements for staff officers were always of a lesser scale. Finding qualified officers to instruct at Greenwich was proving difficult enough. Having to find others to support Quetta and Andover only added to this burden. If the Navy was wrong in its response, then it was certainly no more wrong than the Air Staff which long eschewed posting its officers to the Admiralty.[324]

Whether one attended the Royal Naval Staff College or one operated by the Army or Air Force, certain strands were common to all. Each sought to broaden the horizons of the qualifier and introduce him to the higher aspects and needs of his Service. For the Army and Air Force, the aim was to develop capable staff officers whilst also laying the groundwork for eventual command.[325] For the Navy, the objective was more circumscribed and merely sought to make an officer an effective cog in the greater machinery of naval planning, operations, and administration.[326] Writing, speaking, and developing powers of critical thought combined with employing syndicates to weigh problems and develop solutions featured prominently. Just as the Greenwich qualifier was told to accept nothing at face value, so, too, was his counterpart at Camberley advised much the same. Fuller, as Chief Instructor, stressed to the officers attending in 1923, including Lieutenant Commander the Honourable Cyril Douglas-Pennant[327] and Robin Jeffreys, that:

> There is a word in our language which I believe to be the most potent word in the dictionary. So that you do not forget it, I have written it on the blackboard—it is the word WHY. Whatever I say to you, whatever your instructors say to you, whatever you read, whatever you think, ask yourselves why. If you do not do so, however much you may strive to learn, you will be mentally standing at ease. Remember this: your brain is not a museum of the past, or a lumber room for the present; it is a laboratory for the future, even if the future is only five minutes ahead of you; a creative centre in which new discoveries are made and progress is fashioned.[328]

In the critical environment that the colleges sought to foster, the syndicate was largely an adoption of the tutorial method of learning grafted to the military environment with the important caveat that teamwork and not individual work was the

---

324 Till, *Air Power*, p. 131.
325 CAB 53/5, Chiefs of Staff Sub-Committee minutes of 1 November 1934.
326 *Ibid*, Chiefs of Staff Sub-Committee minutes of 9 October 1934.
327 Later Admiral the Honourable Sir Cyril Eustace Douglas-Pennant (1894-1961). In Command, *P 37*, 1918-19; Junior Division, Staff Course Camberley, 1923; Staff Course, 1924-25; Staff Officer Operations in *Barham* and *Resolution*, 1925-26; Training and Staff Duties Division, 1929-31; Naval Assistant, First Sea Lord, 1931; Tactical Course, 1931; Royal Naval Staff College, 1931-33; Captain, 1935; Commanding Officer, HMS *Deptford*, 1935; Flag Captain and Chief Staff Officer in *Exeter*, 1935-36 and *Iron Duke*, 1936-37; Royal Naval War College, 1937-39; Rear Admiral, 1944; Commandant, Joint Services Staff College, 1947-48; Vice Admiral, 1948; Admiral, 1952 and retired, 1953.
328 Fuller, *Memoirs*, p. 419. Original emphasis.

goal. Lectures provided by both a Service's own specialist officers and outside experts featured along with tours of the military and civilian institutions playing a part in the overall defence establishment.

What is striking in surveying the different approaches adopted in the education of officers is not how deeply rooted in the past it was, but how much it anticipated the requirements of the present and the near future. It is easy to claim that the Navy was fixated on Jutland and the Army on the spectre of the Western Front, but the truth is that both sought to anticipate how better to meet what was expected of them in the present and foreseeable future. It is also the case that as the interwar period was drawing to a close, the Army came increasingly to shape its Staff College training more along naval lines. At both Camberley and Quetta the courses were reformed in 1939. Instead of a two-year course, as had been the practise since 1921, a new one-year course was adopted with Minley Manor opening in January 1939 to accept more senior officers returning to complete a shorter, advanced course. Quetta, long following the Camberley model, if not the exact curriculum, did not establish a Senior Wing along similar lines as the one now operating at Minley Manor. Thus, the common format of both diverged. Meanwhile, a new course for those unsuccessful in negotiating the entrance exam to the Staff College was considered with the aim of preparing more officers for the war that appeared so imminent.[329]

Moreover, some military officers thought that it was a mistake for the Staff Colleges to have the twin responsibilities of preparing officers for staff duties and command and advocated limiting their purpose to that of staff officer development only. Instead, officers having secured their *psc* should be appointed to a revised Senior Officers' School jointly operated by the War Office and Air Ministry where the emphasis would be on developing qualities of command and eschewing tactical training. The course's new focus would be on the higher aspects of the profession drawing heavily on international and not national problems for consideration. Save for the joint nature of its operation, this was one officer's broad endorsement of the naval approach to staff training and education even if he did not acknowledge it as such.[330]

As the previous discussion has shown, the Services had done much already to improve the level of their cooperation on their own accord. By the mid-1930s, as Captain Watson was reminding the Chiefs of Staff, the summer period of the naval Staff Course witnessed naval and air staff college qualifiers working a common set of problems together for roughly three weeks. This was followed by each of the Staff Colleges during their final term supporting the combined exercise held each year.[331]

329 Percy Hobart to Liddell Hart letter of 31 May 1938, Liddell Hart Papers, LH 1/376/81, LHCMA.
330 Brevet Major Norman MacLeod, Royal Scots Fusilier essay of 1936, Frederick Pile to Liddell Hart Correspondence, LH 1/575/144, LHCMA.
331 CAB 53/5, Chiefs of Staff Sub-Committee minutes of 1 November 1934.

This degree of joint training demonstrates that Jutland was not a spectre but merely a spectrum of staff training.

Gaps remained, though, and advocates surveying the defence scene in the early 1920s called for additional measures and continued to press the matter. For some it was a matter of economic imperative. In a challenging fiscal climate where retrenchment in military spending was deemed necessary to meet more pressing needs, was it really necessary to have apparently duplicative branches seeing to the intelligence, logistics, medical, and educational needs of the Services and might not rationalization in these areas and still others provide the support required at lower operating costs? Certainly, this was the view of Geddes and his Committee on National Expenditure.[332]

For others desiring to see greater integration, the genesis of their views was not so much economic factors as the potential operational pitfalls now present. Difficult as it was previously for the Navy and the Army to harmonize their approaches to war, the presence of the Royal Air Force made the problem that much more intractable. If stronger machinery existed at the centre of Government, the risk that the Services would proceed along divergent paths in peace or fail to concert their actions in war would have not been so worrying. Yet, this was not the case. The command structure that eventually arose in the World War and controlling British and Imperial operations had evolved to meet the challenges of a global, total war. What arose was based very much on the art of the possible, being neither perfect nor unified. The success arising owed more to the strengths of personality of those occupying the critical positions of power and their innate abilities rather than the merits of the actual organizational structure adopted. These limitations were recognized by some of those advocating for closer defence integration, and though they too understood the economic imperatives, operational efficiency was to them always the more salient question. In time, a Chiefs of Staff Sub-Committee with a host of subordinate bodies operating under the CID appeared to better coordinate the problems of defence. Of those subordinate bodies, the Imperial Defence College was one. It sought to train a select body of senior personnel in the very highest aspects of defence planning, and including as it did members from the forces of the home establishment, the civil departments, and the greater Empire, its gamut was to be broad, indeed.

In these efforts, the Navy was a generally staunch and reluctant participant. It was staunch in its advocacy for creating those structures of learning and investigation, such as the IDC and the ISTDC, but remained reluctant to see changes in the command and control structures of defence. This owed something to the fear that creating higher structures only promised to introduce more points of friction into an environment that was difficult enough already. It also had the potential to reduce the influence and status of the Royal Navy within Government circles, though this fear is probably one easier for today's observer to see in hindsight than a peril recognized by contemporaries at the point of decision. Fundamentally, though, the Royal

332 Gray, *Imperial Defence College*, p. 1.

Navy believed it was both unnecessary and unsound. Not necessary because its officers were already being trained to consider defence matters across the entire spectrum of warfare. Granted not all its senior officers had received such training, but time would probably make this less of an issue as more officers benefited from the efforts already underway within the Service. Unsound because the personnel trained at any higher school of defence studies must not be allowed to consider problems and tender advice for matters where they did not hold executive authority and responsibility. This was a danger and merely promised to add a fifth wheel to the coach. These fears were not baseless, as the discussion will soon show, but the decision to adopt a joint defence college was always a balancing act of attempting to correct known deficiencies against the possibility of introducing further problems. In light of these reservations, how the Service adapted to the IDC is now appropriate to consider.

6

Living with the Beast: The Imperial Defence College and the Education of Naval Officers

It is hard to over-estimate the value of the close association with so many up and coming officers of the three Services and Civil Servants, specially selected to take the Course.[1]

Field Marshal Sir Claude Auchinleck, 1974

An offr. with sound knowledge & balanced views which he expresses clearly & concisely either on paper or verbally. He will be a good staff officer but I can say more. During the year no offr. has impressed me more by his personality & strength of character and though I have never seen him when charged with executive duties, I should pick him out as a natural leader of high order.[2]

Major General William Bartholomew, 1929

Creation of the Imperial Defence College must stand as one of the most farsighted and far-reaching decisions made by Britain during the interwar period. Taken by a Conservative Government, it was a decision that even a Labour administration committed to arms reductions and spending on the lowest scale possible with regard to defence did not seek to overturn when it came to power in 1929. This consensus across Party lines owed much to the arguments supporting its establishment—the twin promises of economy and enhancing operational effectiveness—which no Government could in principle deny as being worthy objectives. Surprisingly, once the decision was reached to institute a higher defence academy, it is remarkable how little controversy was engendered from the Services. Placing it operationally under the Chiefs of Staff Sub-Committee, no doubt, helped to minimize potential conflicts, but given the ongoing friction over budget levels and the status of military aviation,

---

1   Auchinleck cited in T. I. G. Gray, *Imperial Defence College and the Royal College of Defence Studies, 1927-1977* (Edinburgh: Her Majesty's Stationery Office, 1977), p. 32.

2   ADM 196/91 (A. B. Cunningham).

doctrinal reasons for continuing to oppose it after its creation remained just below the surface.

That these did not arise owed much to the fact that the Navy assumed responsibility for the administration of the new college. As the Service most pressing for the IDC, it accepted the lion's share of responsibility for setting it up and making a go of it. A naval officer was appointed as the first Commandant, and though ostensibly an independent office, Richmond had to be mindful that his future career was highly dependent on not offending Their Lordships during his tenure. Thus, pursuing a path at the IDC outside his own Service's views on any issue was not a move apt to enhance his prospects for future employment, and, assuredly, this was ever present in his thinking. Additionally, the risk that their own programmes of officer education might suffer at the expense of the IDC was forestalled in the near term by the Services. Finally, they were able to reach broad agreement on its overall lines of instruction and the doctrine imparted to officers attending. Had this not happened, then, doubtlessly, the IDC would have become another venue of disagreement. That the minefield of objections concerning the establishment of the IDC was negotiated successfully owed much to Hankey and here his intimate knowledge of the Services' views on the question was invaluable. Hankey, no doubt, kept the issue of the college alive because his office would be a major beneficiary of receiving officers so trained, yet adoption of the college remained a strong desire of Government.

The founding of the IDC was but one part of the ongoing debate over the organization of defence. Its gestation was a long one and the fear that a 'super-staff' would be imposed on the existing defence arrangements from its graduates was the crux of the matter. Thus, an early decision resolved to call it a defence college and not a joint staff college lest its purpose be misconstrued.[3] Though creation of a Ministry of Defence was eschewed for the moment, henceforth, the estimates of the Services would be considered collectively by parliament and rationalization in the provision of services common to each would be pursued.[4] The Navy's initial reservations over forming a defence college were matched even more so by the Earl of Cavan, the CIGS, of whom one officer once commented 'that as C.I.G.S. in the War Office he was as much out of place as a nun in a night club.'[5] Not surprisingly, Fuller was the officer voicing this low opinion of Cavan. An early and strong advocate of a joint college, his views often resided outside the bounds of current military orthodoxy and to Fuller belongs not a little credit for the name applied to the new college when he championed an 'Imperial Defence College' in early 1923.[6] Whether Fuller's advocacy for the college would have been better received if he had eschewed making his case in such a theoretical paper

3   CAB 53/1, Chiefs of Staff Sub-Committee minutes of 11 March 1926.
4   CAB 23/53, Cabinet minutes of 16 June 1926.
5   Fuller, *Memoirs*, p. 422.
6   J. F. C. Fuller, 'The Fiends of the Air: The Influence of Aircraft on Imperial Defence', *The Naval Review*, Vol. XI, No. 1, February 1923, p. 113.

that tilted at certain cherished windmills remains an interesting question and the fact remains that the issue of such a school was alive already. Not surprisingly, it was Churchill who set matters alight in 1920 when he said that future wars would demand officers trained across the full spectrum of war.[7]

As for Cavan, within the Army he was unique for his lack of staff experience, as most of his experience consisted in holding executive commands.[8] In this, his background was more akin to senior naval officers than perhaps any other CIGS of the interwar period. This background strongly shaped his views on the college writing, 'I do not believe that any school but that of experience will guide officers of middle rank of the three Services to the formation of correct and sound conclusions on "war in its widest aspects".'[9] Where some naval officers deprecated staff training because it was seen as Prussian in origin and not suited to British ways, Cavan deprecated the idea of an IDC because the Prussians had not found one necessary. Thus, Cavan's opposition to the college precluded any positive consideration of the matter by the Chiefs and it was only upon his replacement as CIGS by Sir George Milne,[10] the first gunner to hold the appointment, that Hankey broached the matter anew in 1926.[11]

Meanwhile, Trenchard though supportive of the college *per se* sought to fashion a more limited role and scope for it—certainly, in its initial period. Here, he desired to see it weigh the proper role of airpower as the basis of air strategy, resting as it did on such a limited historical basis. Additionally, he spoke against appointing a commandant, deprecated having civil servants attend in any numbers, and employing the college to weigh concrete proposals.[12] Ultimately, Beatty prevailed amongst his colleagues that the college should have a commandant and that the officer would hold flag, general officer, or air rank and not be a civilian, whilst Hankey sagely suggested securing Sir Warren Fisher's support in having civilians attend. This was a deft move as having the support of the Permanent Secretary of the Treasury would ease any financial issues whilst a nod to the needs of the Civil Service would not be remiss to its nominal head.

Thus, in 1926 the moving forces behind the IDC were the Navy and Hankey, though no progress would have been evidenced without the tacit support of Trenchard and Milne. Part of the problem was the COS Sub-Committee, dating from only 1923, was itself a new body and now it was being asked to accept a fresh institution the ultimate import of which it could only guess.[13] This reservation was met by placing the college

7   R. E. Stoehr, 'Combined Staff or Combined Staff College, *Journal of the Royal United Services Institution*, Vol. LXVI, No. 463, May 1921, p. 477.
8   Cyril Falls, 'Frederick Rudolph Lambert', *Dictionary of National Biography*, p. 469.
9   CAB 53/12, Chief of the Imperial General Staff note of 25 January 1924.
10 Field Marshal Sir George Francis Milne, later Baron Milne of Salonika and Rubislaw (1866-1948). Staff Course Camberley, 1898 and 1899.
11 CAB 53/1, Chiefs of Staff Sub-Committee minutes of 11 March 1926
12 *Ibid*, Chiefs of Staff Sub-Committee minutes of 22 April 1926.
13 CAB 53/12, Chief of the Imperial General Staff note of 25 January 1924.

under the Chiefs of Staff rather than having it report directly to the CID as originally envisioned. The Government was also pressed to provide fresh funds to establish the college and meet its yearly operating costs removing yet another objection of the Services that the college would come at the expense of their own regime of higher education. This was acceded to up to a point, but the Navy responsible for securing the leasehold and providing the Commandant in the first instance, still was required to met some of these costs from its own funds as discussed in an earlier chapter.

As it was, the primary sticking points arising after the decision was reached to move forward focused on the pay and emoluments of the officers assigned. The Navy and the Air Force paid their officers based on rank whilst the Army relied on the position an officer held. Compounding the issue was that naval officers did not in 1926-27 receive a marriage allowance—a bugbear for the Service for much of the period.[14] Other issues where differences remained, in retrospect, appear almost trivial. This included whether those passing through the defence college should be entitled to a post-nominal designation. Eventually, those graduating were styled as *i.d.c.*, but initially naval and air officers were only listed as having attended the college. This was more than even the Army allowed and not until Major General Robert Haining[15] raised the discrepancies in practice and the problems possibly arising during a future crisis did the War Office agree to note which of their officers had attended the college.[16]

The Imperial Defence College was one means of bridging the divide existing between the Services and the greater government though its limited physical facilities and restricted audience hindered its impact. The actual location of the college owed much to Rear Admiral Dreyer who recommended it to Major General Archibald Cameron, the Director Staff Duties, and then to Richmond, whilst serving as ACNS and chairing the committee charged with overseeing its establishment and opening.[17] Meanwhile, Richmond briefed the COS on the planned course of instruction in December 1926. In structure and method, it closely mirrored the lines of existing staff training conducted. Where it diverged was that officers did not reside in the college—a concession to their seniority and the limited facilities available and in its emphasis on the higher aspects of war.[18] With regard to the latter, Richmond suggested that, henceforward, the Staff Colleges should limit their examinations to tactical issues

---

14   ADM 116/2273, 'Instructions for the Commandant of the Imperial Defence College'.
15   Later General Sir Robert Hadden Haining (1882-1959). Staff College Camberley, 1922-23; Imperial Defence Course, 1927; Deputy Director and Director, Military Operations and Intelligence, 1933-34; Commandant, Imperial Defence College, 1935-37; Director, Military Operations and Intelligence, 1937-38; Vice Chief of the Imperial General Staff, 1940-41 and retired, 1942.
16   CAB 53/25, Commandant, Imperial Defence College to Secretary, Chiefs of Staff Sub-Committee letter of 23 July 1935.
17   Dreyer, *Sea Heritage*, p. 280 and ADM 116//2273.
18   'Biography of General Sir Eric de Burgh, Second son of Lt. Colonel T. J. de Burgh, J.P.D.L. of Oldtown, Naas, and Emily Anne, daughter of Baron de Robeck', 09/49/1, 22, Imperial War Museum, London.

whilst the IDC focused on strategy. In the main, Richmond's proposals were well received, but a sticking point was his request to have access to documents held by the CID. The COS advised that he should forward his requests for such documents to Hankey who would coordinate, as necessary, with the Chiefs.[19] When Richmond's subsequent request was received, the Chiefs were taken aback as he was seeking in excess of 120 papers. This they would not consider and their approval was limited to six specific papers with the stipulation that Hankey could release others to the Commandant at a future date at his discretion.[20]

For the three members of the directing staff, Richmond apportioned instructional responsibilities that closely aligned with their respective Services. Dickens had nominal responsibility for covering naval forces, trade, and sources of supply. Meanwhile, Dill covered military forces, manpower, and transportation, whilst Group Captain Philip Joubert de la Ferté[21] oversaw air forces and commercial aviation. Other topics to be addressed by visiting civil servants included finance, meteorology, and political geography. As for the course, itself, initially, no more than thirty candidates could attend each course, and, of these, only eighteen, apportioned evenly, were reserved for the forces of the home establishment. The balance of those attending came from the Commonwealth and senior civil servants in those departments, such as the Foreign Office and Treasury, concerned with strategic issues.[22] The latter group was broadly interpreted, as amongst those studying at the IDC between the wars were representatives from the Office of Works and the Government Post Office. Even this intake was proving a stretch and the first class met with only twenty-five members, the shortfall a result of the civil departments and Dominions not nominating their full quota of candidates.[23]

This was to prove something of a running sore and doubtlessly the period's economic woes were a contributing factor forestalling attendance by Commonwealth representatives. By 1929, the course had only twenty qualifiers and the balance between senior civil servants, British military members, and representation from the Empire was skewed heavily towards the military officers of the home establishment.[24] This was to be the smallest course of the period and had things continued along these lines it is difficult to see how the hoped-for promises of the college would ever be met. Still, other considerations were at play, of which, the very seniority of the course worked against some of the Dominions lacking as they did personnel with the requisite grounding to begin such studies. Newfoundland is certainly one such instance as

19   CAB 53/1, Chiefs of Staff Sub-Committee minutes of 2 December 1926.
20   *Ibid*, Chiefs of Staff Sub-Committee minutes of 25 January 1927.
21   Later Air Chief Marshal Sir Philip Bennet Joubert de la Ferté (1887-1965). Staff Course Camberley, 1920; Staff College Andover, 1922-23; Imperial Defence College, 1926-29; Commandant, Staff College Andover, 1930-34.
22   ADM 116/2273.
23   Gray, *Imperial Defence College*, p. 31.
24   *Ibid*, p. 8.

it failed to send even a single person to study during the period, but it must have been equally difficult for others, such as South Africa and New Zealand, to find two candidates each year meeting the general guidelines of entry. Through 1930, the Union had only sent a single candidate to the IDC whilst New Zealand had managed but three.[25] This was not a question of rank so much as experience and it is to the Chiefs' credit that they believed it would pay dividends in the future by allowing officers of lesser rank and corresponding backgrounds to attend and did not seek to restructure the course. As it was, not until 1938 did the course operate to its full quota of 32 qualifiers at which point discussions began about expanding the course.

Thus, attending the 1930 course, Lieutenant Commander Harold Farncomb,[26] RAN, was the only Dominion naval officer of his year joining five RN five officers.[27] A gifted young officer who had received five firsts in his examinations for Lieutenant, a qualified interpreter in both French and German, and a trained intelligence and staff officer, Farncomb subsequent career proved every bit as successful as his 22 years since commissioning.[28] Of course, that was unknowable to the Admiralty in 1930, but the error was not having the occasional younger officer of promise sit the course more regularly. Danckwerts who had demonstrated that such an officer, if carefully screened, could do well with the promise that more use could then be made of his training.

Of Commonwealth naval officers, Admiral Sir Frederick Field speaking in 1932 believed only Australia and Canada where in a position to send officers of the requisite standing to the IDC and the present contraction of the Royal Canadian Navy was not a measure likely to abet matters in the foreseeable future. For them, attendance at the Royal Naval Staff College was believed the more appropriate venue in the near term. This would allow the Dominions to further develop their naval officers, teach them of staff duties, and better prepare the officers to attend the IDC when they achieved the necessary standing.[29]

This last point was crucial as the custom was to send its officers to the IDC as a post Captain averaging about 43 years of age. This was somewhat older than his military counterpart attending and most definitely more senior than the airman who averaged but 38 and held Wing Commander or even Squadron Leader rank.[30] This practice was frowned upon by the other Services believing the Navy was leaving matters too late.

25   *Hansard*, House of Commons Debates, v. 240, cc966-7.
26   Later Rear Admiral Harold Bruce Farncomb (1899-1971). Intelligence Course, 1920; Staff Course, 1923-24; Training and Staff Duties Division, 1924; Staff Duties, Atlantic Fleet in *Barham*, 1924-25; Staff Officer Operations in *Sydney*, 1925-26; Staff Officer Operations and Intelligence in HMAS *Melbourne*,1927-28; Imperial Defence Course, 1930; Naval Intelligence Division, 1935-37; Captain, 1937; Senior Officers' Technical Course, 1939; Intelligence Division, 1939; Tactical Course, 1939; Commanding Officer, HMAS *Perth*, 1939-40, *Canberra*, 1940-41 and *Australia*, 1941-44; Rear Admiral, 1947 and retired, 1951.
27   *Navy List*, October 1930, p. 307.
28   NAA: A679, Farncomb, H. B.
29   CAB 53/4, Chiefs of Staff Sub-Committee minutes of 4 February 1932.
30   CAB 53/5, Chiefs of Staff Sub-Committee minutes of 1 November 1934.

One such consequence was that it prohibited naval graduates from returning to the Staff College to serve as instructors in order to foster the promulgation of a common doctrine. The Navy, though, always stressed the practical side of their profession and sending officers of Commander's rank was difficult, as they had only seven years from the time of their promotion to reach their captaincy or face mandatory retirement. Compounding the problem was that the Service typically sent its officers to its own Staff Course at Greenwich when they were a Commander. Attendance at the IDC for a year's course followed by a staff appointment ashore for two years made the window for securing the required sea time all too brief. A Captain, however, was allowed twelve years before reaching flag rank or facing compulsory retirement and so having these officers attend the IDC was thought the best approach for the Service. If its officers were too senior to return to the Royal Naval Staff College as instructors, the Navy increasingly ensured that the Director or Assistant Director, who were of Captain's rank, had graduated from the course.[31]

Recognizing that some desired to send more candidates to the IDC whilst others were not apt to meet their quota, the Chiefs of Staff Sub-Committee in 1938 agreed to raise to 36 the potential number of qualifiers. In part, this arose because the Government of India sought to send a civilian member to the IDC without infringing on the two military candidates normally nominated. Here, they were anxious to have senior members from the Indian Civil Service, Indian Political Service, or its police forces secure the benefits of an IDC education. As accommodation still only remained for supporting 32 students at any one time, a risk existed that curtailment in the number of officers appearing from the Royal Navy, British Army, and Royal Air Force might arise.[32]

Previously in 1935, the Services proposed to allow their own civil servants to study at the college though not at the expense of their military officers.[33] As the Services and not the civil departments met the operating costs of the IDC, vacancies, particularly from the Dominions that attended on a reimbursable basis, posed a fiscal challenge to the Admiralty. They, it will be recalled, were responsible for its administrative care. As the overall costs of the establishment were relatively fixed, ensuring that the IDC operated at capacity was a major concern and broadening its pool of candidates was one means of guaranteeing its efficient running. Beyond educating additional civil servants, the Chiefs considered opening the college to quasi-autonomous Government bodies such as the British Broadcasting Corporation and even commercial banking.[34] Financially, this must have been an attractive proposition, but the charter of the college, as discussed anon, made such a change difficult given the classified nature of much of its proceedings. Though civilian students from industry could be vetted

31   *Ibid.*
32   CAB 53/9, Chiefs of Staff Sub-Committee minutes of 29 March 1938.
33   CAB 53/5, Chiefs of Staff Sub-Committee minutes of 29 October 1935.
34   CAB 53/8, Chiefs of Staff Sub-Committee minutes of 11 October 1937.

and nominally attached to the Civil Service for their year's study, their presence at the college might raise unnecessary questions about the actual independence of their parent organizations. General Haining, the Commandant, spoke against the prospect and this was probably sufficient to ensure the proposal advanced no further, though a surprising supporter given the pejorative views attributed to him by others on staffs and the staff system was Chatfield confirming that the financial soundness of the IDC was ever a concern for the Admiralty.[35] Still, the actual basis of Chatfield's reasoning is unknown and, as will be seen, he may have backed the proposal on its merits having regard for the role such independent bodies as the BBC played in current strategic planning.

The actions of the Government of India offer strong evidence that the IDC, despite any limitations in physical property, was increasingly seen as a valuable tool and, never the more so, than in the role it was playing in preparing senior civilians having defence responsibilities. Moreover, the Army and the Air Force by the end of the period were moving to expand their quota of officers sent, a move that, of necessity, would have brought benefit to the Navy as the numbers enrolled needed to be balanced amongst the Services so that syndicates were not skewed in any one direction.[36] Its actual views towards expansion remain opaque, as Sir Roger Backhouse did not comment on the merits of expansion directly except to say that it might require the Navy to start sending Commanders to meet their portion of the quota. As the Army and Air Force already sent officers a step or two below in rank, such a move would have been received cordially. What did exercise Backhouse was that no action be taken until 1940 when the existing lease on the building at Buckingham Gate lapsed indicating that for the Navy any administrative issues surrounding the college were always a pressing concern. In the event, war intervened and the matter fell by the wayside for the moment.

From the previous, it can be seen that the Services treated the college seriously sending their best and brightest. This was no less true for the Navy though the yardstick that it employed was very much a practical one. Of the six initial qualifiers in 1927, all eventually reached flag or general rank on the active list though for Ralph Leatham,[37] a much respected torpedo specialist, the Imperial Defence College was his first foray into a higher educational course.[38] This understandably was not always

---

35 *Ibid.*

36 CAB 53/10, Chiefs of Staff Sub-Committee minutes of 21 December 1938.

37 Admiral Sir Ralph Leatham (1886–1954). Torpedo and Mining Department, 1918-21; Commanding Officer, *Winchester*, 1921-22; Fleet Torpedo Officer in *Iron Duke*, 1922-23; Captain, 1924; Naval Air Section, 1925-27; Imperial Defence Course, 1927; Senior Officers' Technical Course, 1930; Tactical Course, 1931 and 1938; Senior Officers' War Course, 1931; Royal Naval War College, 1931-33; Commanding Officer, *Ramillies*, 1934-35; Flag Captain in *Valiant*, 1935-36; Rear Admiral, 1936; Vice Admiral, 1939; Commander-in-Chief, East Indies, 1939-41; Admiral, 1943 and retired, 1945.

38 ADM 196/91 (Leatham). The first British naval entrants were Captains J. C. Tovey, R. Leatham, H. R. Moore, E. L. S. King, and A. F. Pridham. Commanders C. J. Pope, RAN, and C. T. Beard, RCN, were the two Commonwealth naval officers sitting the first course

the case and some in parliament questioned whether the Service was employing the officers trained or assigned there to useful ends. In late 1937, the First Lord, Duff Cooper, defended the Service's record though promising that the matter would receive his attention.[39] Ultimately, he accepted that where 17% of those who had been to the IDC were now either unemployed or retired—a much higher figure than for the Army and the Air Force—the sum had to be seen in the light those qualities which bore on one's fitness for promotion. Here, the test was suitability for command at sea.[40] This emphasis was not wrong for the present, but it did place her in an inferior position *vis-à-vis* the other Services, if operations other than purely naval ones arose.[41]

That more naval officers had separated owed much to the point in their careers when they attended the course. Coming to the IDC as a post-Captain, it was but natural that not all would secure flag rank and further employment. More is to be said of this below, but here one must also conclude that the Navy did a poor case of articulating its differences in officer accession, education, employment, promotion, and retention to parliament and to the public at large. Because naval officers were engaged at a younger age and trained to different standards, aligning their careers into a common joint education regime was only going to be achieved through a compromise of competing goals. The Admiralty primarily selected its officers for attending the IDC based on the recommendation of a reporting officer commenting on the fitness of a subordinate when considering his suitability for employment at the highest levels. Thus, William Boyle, proposed Lancelot Holland, then serving as Deputy Director of the Staff College, in 1929 which was noted by the Commissioned and Warrants branch of the Admiralty. The process of attending was not immediate and in Holland's case nearly four years elapsed before he took up his studies spending the intervening time as Chief Staff Officer in the *Hawkins* and then as a naval adviser to the Hellenic Navy.[42]

During his time as Admiral-President, Boyle made an effort to espy officers suitable for the IDC though not all, as in the case of Holland, eventually found themselves fortunate to attend. Captain Francis Goolden received a similar testimonial in 1931, and though held to be a very capable officer by his previous superiors, time in the First Cruiser Squadron under Rear Admiral Max Horton[43] ended with a poor

and Brevet Lieutenant Colonel R. D. Ormsby, RM, was the first marine officer attending. Only Beard failed to achieve flag or general officer rank retiring in 1941 as a Commander.

39   *Hansard*, House of Commons Debate of 15 December 1937, v. 330, cc1144-5.

40   *Ibid*, House of Commons Debate of 2 February 1938, v. 331, c216.

41   Actually, the numbers reported to parliament were worse than given, as Duff Cooper only accounted for those qualifying at the college. When Directing Staff and former Commandants are considered, the number was 20%. The corresponding figures were 8.4% for the British Army (excluding Indian Army) and 10% for the Royal Air Force. See CAB 53/43, Report by the Commandant of the 1938 Course of the Imperial Defence College, 22 December 1938

42   ADM 196/92 (Holland).

43   Later Admiral Sir Max Kennedy Horton (1883–1951). In Command, *E 9*, 1914-15, *J 6*, 1916-17 and *K 18*, 1917-18; Commander, Submarine Flotilla, 1918-20 and 1922-24;

evaluation killing any prospect of sitting the course. Horton thought Goolden too reserved in nature and, consequently, the fine morale of *London* owed everything to the ship's executive officer.[44] Nor was Goolden seen as a good handler of his cruiser. Whether this was a fair view of Goolden by Horton is difficult to weigh and Pound, the Commander-in-Chief, believed the problem may have been no more than the basic natures of two officers who were so different to each other.

Another officer highly touted for the IDC but failing to attend was Geoffrey Miles. Rear Admiral Charles Little, Director of the Staff College in 1929, was the nominating officer in this instance making his recommendation on the strength of Miles' brief interlude at the college before joining the Plans Division. Why Miles failed to attend is easy to see for between mandatory service afloat and appointments to the Staff College and Tactical School time was simply not available.[45] Just as Admiral Boyle had recommended him for the IDC, Kennedy-Purvis continued the tradition of proposing officers completing their War Course for the IDC, though with a new war arising in 1939 such recommendations had little chance of registering a favourable result.[46] Lord Louis Mountbatten is a case in point receiving a strong endorsement from Kennedy-Purvis in 1939 on the strength of his very successful War Course. Given his worldly views, deep professional knowledge, and strong social skills Mountbatten seemed a natural choice for the IDC, particularly, as he had not previously attended the Staff College, though war put paid to his attending.[47] As for Goolden, it may be that he was too senior a Captain by this point in his career to attend the IDC and this problem ever faced selectors when considering officers for attendance. Compounding the issue was that if an officer reached flag rank whilst attending, he continued to receive only the pay and allowances of a Captain.

Captain, 1920; Senior Officers' Technical Course, 1925 and 1933; Senior Officers' War Course, 1925-26 and 1933; Assistant Director, Mobilisation Department, 1926-28; Tactical Course, 1930; Flag Captain and Chief Staff Officer in *Resolution*, 1930-32; Rear Admiral, 1932; Commander, Second Battle Squadron, 1933-35; Commander, First Cruiser Squadron, 1935-36; Vice Admiral, 1936; Commander, Reserve Fleet, 1937-39; Flag Officer Submarines, 1940-42; Admiral, 1941; Commander-in-Chief, Western Approaches, 1942-45 and retired, 1945.

44   ADM 196/91 (Goolden).
45   ADM 196/127 (Miles).
46   ADM 196/91 (Kennedy-Purvis).
47   ADM 196/93 (Mountbatten).

Typical of the naval officers attending the IDC, but subsequently retired were Hugh England,[48] a graduate of the second course, and Leonard Potter,[49] a torpedo specialist and a member of the 1932 course. Both officers had commanded cruisers before attending their courses and both would command cruisers again upon leaving the IDC. While attendance at the course did not abet their career prospects, it cannot be said it hindered them either. England's lineal peers were highly capable including John Tovey, Robert Raikes, Bertram Ramsay, and Lachlan MacKinnon.[50] All reached flag rank and only MacKinnon failed to have a connection to the college. Richmond was satisfied with England's performance whilst attending the IDC and William Fisher thought England a very good ship handler and an ideal officer for wartime. However, his lack of technical knowledge and his temperament were seen as handicaps in peacetime and, here, the lingering effects of a war wound were thought part of the problem.[51] He was seen as a sound officer but, as always, an officer's future progress depended as much on the capacities of his peers as much as on the abilities of himself.

With the exception of William Whitworth,[52] Potter's peers failed to achieve the success of the 1923 cadre of Captains and most retired in 1936 having completed

48   Later Rear Admiral Hugh Turnour England (1884-1978). Appointed for War Staff, 1919; Operations Division, 1918-20; Commanding Officer, HMS *Valerian,* 1920-21; Captain, 1923; Commanding Officer, HMS *Weymouth,* 1923-24; Senior Officers' Technical Course, 1924; Senior Officers' War Course, 1924-25 and 1934-35; Commanding Officer, *Columbo,* 1925-27; Imperial Defence Course, 1928; Command of Mechanical Training Establishments in HMS *Fisgard,* 1929-31; Tactical Course, 1931; Flag Captain and Chief Staff Officer in *Sussex,* 1931-34; Rear Admiral and retired, 1935. Recalled for war service.
49   Later Rear Admiral Leonard Fosbrooke Potter (1885–1942). Plans Division, 1919-21; Staff Officer, Rear Admiral Submarines, 1921-23; Fleet Torpedo Officer in Mediterranean Fleet, 1923-25; Captain, 1925; Senior Officers' Technical Course, 1926; Tactical Course, 1926; Deputy Director, Torpedo and Mining Department, 1926-29; Commanding Officer, *Cumberland,* 1929-31; Imperial Defence Course, 1932; Commanding Officer, *Devonshire,* 1933, *Royal Oak,* 1934 and HMS *Defiance,* 1934-36; Rear Admiral and retired, 1936. Recalled for war service.
50   Later Vice Admiral Lachlan Donald Ian MacKinnon (1882-1948). Captain, 1923; Senior Officers' Technical Course, 1924 and 1928; Senior Officers' War Course, 1924, 1928 and 1937; Commanding Officer, *Assistance,* 1924-26; Tactical Course, 1927 and 1936; Commanding Officer, *Danae,* 1927-28; Deputy Director, Training and Staff Duties Division, 1928-30; Captain of the Fleet, Mediterranean 1930-32; Flag Captain and Chief Staff Officer in *Warspite,* 1932-33; Rear Admiral, 1935; Commander, Second Battle Squadron, 1937; Vice Admiral, 1938 and retired, 1939. Recalled for war service.
51   ADM 196/91 (England).
52   Later Admiral Sir William Jock Whitworth (1884-1973). Commanding Officer, *Woolwich,* 1918-19; Naval Intelligence Division, 1919-21; Captain, 1925; Senior Officers' Technical Course, 1926 and 1931; Tactical Course, 1926 and 1939; Captain (D), Second Destroyer Flotilla, 1928-31; Senior Officers' War Course, 1931 and 1936-37; Deputy Director, Personal Services Department, 1932-33; Captain of the Fleet, Mediterranean, 1933-35; Flag Captain in *Rodney,* 1936; Rear Admiral, 1936; Naval Secretary to First Lord, 1937-39; Commander, Battle Cruiser Squadron, 1939-41; Vice Admiral, 1940; Second Sea Lord, 1941-44; Admiral, 1943 and retired, 1946.

their allowed time in rank. That Whitworth achieved flag rank is a reflection that Roger Backhouse rated him as the leading Captain of his lineal peers in 1936.[53] Those failing to find promotion included Claude Dobson,[54] a holder of the DSO and subsequently of the Victoria Cross for his actions during the undeclared war against Russia in 1919 who retired early due to failing eyesight, yet for that, had not distinguished himself in his most recent appointments.[55] As for Potter, in 1925, Captain Domvile assessed him as the best staff officer he had ever met; a view Vice Admiral Brock generally accepted.[56] It remained that having trained a cadre of officers, the Navy still needed vacancies to match their talents and, if not found, retirement loomed. Compounding Potter's case was the recent death of his wife and sister whilst serving in *Devonshire* on the China Station. *Devonshire* was already viewed as a less than satisfactory quantity when Potter took command and his personal circumstances did not abet matters. Though he eventually remarried in 1936 shortly before his retirement, his performance during the intervening period was not up to his contemporaries.[57]

Gerald Warner,[58] a former Signals Officer before reverting to general duties in 1926, was an officer where parliamentary criticism was perhaps more nearer the mark. A very capable officer within his sphere of specialization, he also had a pronounced stammer that made its presence more evident when facing stressful conditions.[59] As it is difficult to imagine a more stressful condition than active service, Warner's likelihood of advancing further was always questionable. As it was, medical issues also arose in time and this, too, points to a problem with leaving training to officers of a certain seniority when chronic medical problems are apt to become more evident.

The parliamentary intervention concerning the employment of *idc* holders prompted Liddell Hart of *The Times* to solicit Richmond for his views on the question. The

53   ADM 196/91 (Whitworth).
54   Later Rear Admiral Claude Congreve Dobson (1885-1940). Five firsts in examinations for Lieutenant; Commanding Officer, *K 10*, 1917-18; Anti-Submarine Division, 1918-19; to RAN, 1922-24; Captain, 1925; Senior Officers' Technical Course, 1926; Senior Officers' War Course, 1926 and 1930; Tactical Course, 1927; Commanding Officer, HMS *Canterbury*, 1930-31; Senior Officers' School Sheerness, 1931; Commanding Officer, HMS *Cambrian*, 1931-32; retired, 1935 and Rear Admiral, 1936.
55   Undated *Daily Express* article held within ADM 196/50 (Dobson) and ADM 196/91 (Dobson).
56   ADM 196/126 (Potter).
57   ADM 196/49 (Potter) and ADM 196/91 (Potter).
58   Captain Gerald Harman Warner (1893-1979). Flag Lieutenant in *Revenge*, 1918-19 and *Queen Elizabeth*, 1919-20, *Coventry*, 1920-21 and *Iron Duke*, 1922-24; Flag Lieutenant Commander and Fleet Signals Officer, 1924; Royal Naval College Dartmouth, 1925-27; Staff Course, 1931; Tactical Course, 1932; Captain, 1936; Commanding Officer, *St Vincent*, 1935-37; Senior Officers' Technical Course, 1937; Imperial Defence Course, 1938; Commanding Officer, HMS *Tartar*, 1939-40 and retired, 1946.
59   ADM 196/93 (Warner).

Admiral's response echoes with pain as he contrasted the Navy's apparent disinterest in the college with that displayed by the Army's 'Uncle George' Milne complaining:

> In my time, only once did a First Sea Lord show even a vestige of interest in it, by one isolated visit. Its institution was condemned – it was described as unnecessary. The officers sent were, generally speaking, those for whom there happened to be no employment at the moment. Whereas George Milne took great personal care in picking out officers upon whom he and his Staff had their eyes for future high promotion, the naval staff were uninterested.[60]

Richmond at this point was a deeply disappointed man firmly believing that his efforts, including his time at the IDC, had not been understood, appreciated, or sufficiently rewarded. Accordingly, his views must be treated with caution. The documentary evidence does not confirm Richmond's view on the college, though whether that surviving record accurately reflects Admiralty views is an interesting question. Its practice in officer selection was not directly comparable to the other Services, as the officer attending was largely at a different stage in his career. As to the care that Milne demonstrated in selection, this, too, must be treated guardedly. Doing much to restore the status of the trained staff officer at the War Office during his tenure as CIGS, he failed to appoint Fuller to the college. This notwithstanding that Fuller was held by many to be the most talented officer in the Army and had served for a time as Milne's Military Assistant.

Moreover, the six officers who served as the naval member of the IDC directing staff during the period were of a high standard with all eventually reaching flag rank. Though their backgrounds were not as one, each brought a unique perspective of value to the college. Dickens, the first naval instructor, was not an obvious choice, as he had attended neither Greenwich course at this stage of his career. Highly talented though he was, his selection must have owed something to the fact that Captain Drax, probably the officer most qualified to serve, needed to return to sea duty if he was to reach flag rank.[61] Astley-Rushton, in command of *Malaya* and nominally available having done the minimum time required for promotion to flag rank, was now beyond the point of consideration. This highlights that managing appointments to the IDC or to any other staff position, for that matter, was always a balance against competing requirements in the naval service.

60   Richmond to Liddell Hart letter of 17 December 1937, Liddell Hart Papers, LH 1/598/115, LHCMA.

61   Following time with the Naval Inter-Allied Control Commission, Drax had been unemployed throughout 1925 until appointed to *Marlborough* as her captain on 9 April 1926. He relinquished command of that ship on 12 April 1927 and returned to unemployed status, thus, serving the minimum time required for consideration of promotion to flag rank. See ADM 1/8776/146.

Dickens, as the naval member of staff, operated in the shadow of Richmond when time came to consider those schemes of a primarily maritime orientation such as war with Japan. This may have been inevitable given the latter's experience and knowledge. Consequently, a great deal of his efforts was consumed in developing the statistical data required to consider the economic implications of the four major exercises posited each year during Richmond's tenure.[62]

William Tennant was the only one to have been a prior graduate of the course, but this reflected only the college's recent creation more than any other factor. His place as a naval educator has already been demonstrated—he of the Jutland lecture series at the Royal Naval Staff College. Of the remaining officers who served on the directing staff, Bertram Ramsay had already qualified as a staff officer and taken the SOTC and SOWC classes before arriving at the IDC. Group Captain Sholto Douglas,[63] his air counterpart on the directing staff, rated him as brilliant, though of an over-quick temper matched to a sharp and sarcastic tongue. A bit of a non-conformist, Ramsay was all business in his approach to work and a firm believer in the staff system.[64] If there was a weakness in Ramsay, then it was that he was not at his best in a formal setting being more at ease talking about an issue when sitting with qualifiers round a table.[65]

Somerville, an early graduate of the Tactical Course, had not been to either of the Greenwich courses, though he had a full career already in staff assignments due principally to his specialization in signals. When the Admiralty suggested Somerville as a possible replacement to Dickens, Richmond sought to forestall the appointment. Nor was he happy with an alternative candidate suggested: Captain Thomas Calvert. Richmond, for his part, was soon touting Captain Brian Egerton, an officer then serving at the Royal Naval War College.[66] It is easy to see what attracted Richmond to Egerton as a possible replacement to Dickens. Possessing a first class intellect, he secured the highest marks amongst his peers in the examinations sat for Lieutenant netting the Ronald Megaw Prize in the process before finding himself, alternately, Assistant Director and then Deputy Director of the Royal Naval Staff College.[67] As Somerville began his career as a torpedo specialist and Egerton was a torpedo specialist, Richmond arguably was in an excellent position to weigh the relative merits

62   CAB 53/16, Commandant, Imperial Defence College to Secretary, Committee of Imperial Defence letter of 6 November 1928.
63   Later Marshal of the Royal Air Force William Sholto Douglas, Lord Douglas of Kirtleside (1893-1969). Staff Course Andover, 1922-23; Imperial Defence Course, 1927; Imperial Defence College, 1932-36.
64   Sholto Douglas, *Years of Command: The Second Volume of the Autobiography of Sholto Douglas, Marshal of the Royal Air Force Lord Douglas of Kirtleside, G.C.B., M.C., D.F.C.* (London: Collins, 1966), pp. 28 and 256 and *Dictionary of National Biography*, pp. 706-07.
65   ADM 196/91 (Ramsay).
66   Richmond to Madden letter of 11 July 1928, Richmond Papers NMM/RIC//7/2/1.
67   ADM 196/50 (Egerton).

of each officer. Up to that point, there was little to indicate in his career assignments that Somerville was anything but a good, very capable seagoing officer who had yet to weigh the higher aspects of war.

As for Calvert, Hunt in his biography of Richmond relates a further consideration raised by Kenneth Dewar. Calvert, when head of the Admiralty's Air Section, had sought to subject *The Naval Review* to censorship following coverage of naval aviation in that closed, but independent journal.[68] The specific issue had been an article highly critical of the RAF, which Calvert believed would only exacerbate the Admiralty's relations with the Air Ministry. Richmond, as the officer most responsible for the creation of *The Naval Review*, clearly had a strong motive to forestall the appointment of Calvert to the IDC in light of this history, but the actual degree this influenced matters is difficult to weigh. That Calvert was not, it seems, a member of the Naval Society, the parent establishment of *The Naval Review*, could only have told against him in the eyes of Dewar and Richmond.[69]

Richmond had emphasized the need in his letter to Rear Admiral Eric Fullerton, the Naval Assistant to the First Lord, that the Captain destined to replace Dickens needed to be of the requisite seniority (i.e., promotion dating from 1921 or 1922) and to possess the naval *psc*. Calvert was of the appropriate seniority but lacked the *psc*, as was true also with Somerville. Richmond, meanwhile, also took soundings from the current President of Greenwich, Richard Webb, who sang of Somerville's capabilities whilst warning against the appointment of Calvert who 'is a good fellow in many ways, but too much wedded to material.'[70] One must respect Webb's view of Calvert given his direct knowledge, but an officer trained as a submariner, controlling the Director Control Tower in *Iron Duke* at Jutland, and with experience in the Naval Air Section was hardly a narrow specialist either.[71] It may be that association with Jellicoe and Dreyer—and he served with Jellicoe yet again during that officer's Empire tour of 1919-20—was his greatest sin. Still, the criticism was sure to weigh with Richmond, given his life's battle against the materialist approach to naval warfare and to Webb's forebodings about Calvert, he suggested an alternative candidate: Brian Egerton. Richmond seized this offering with alacrity and why not; he appeared to possess all the qualities required of a naval staff officer for the IDC but one—the requisite seniority. Richmond had earlier deemed this an essential facet of any officer aspiring to replace Dickens to ensure the Navy did not suffer unduly when the RAF nominated their candidate as Commandant. With Egerton's rank only dating from December 1923, a climb down of sorts was now in order. In all of this a measure of sympathy is due Madden, the Chief of Naval Staff, who honestly tried to provide

68   Hunt, *Sailor-Scholar*, pp. 163-64.
69   'List of Members', *The Naval Review*, Vol. XXII, No. 2, May 1934, pp. i-xi.
70   Richard Webb to Richmond letter of 9 July 1928, Richmond Papers, NMM/RIC/72/1.
71   ADM 196/143 (Calvert).

candidates of the required stature and brains, if deficient in formal training. This Richmond would not concede.

That Madden eschewed appointing Calvert who had the strong backing of Pound, the ACNS and the officer directly responsible to the First Sea Lord for the principles of naval training and staff duties, indicates that he was willing to defer to the judgment of the Commandant on the strong case made against Calvert.[72] Yet, he would not relent in the case of James Somerville, a proven officer of merit, absent a compelling justification from Richmond. As it was, Madden had other plans for Egerton earmarking him to serve as the Director of Torpedo and Mining with the charter to resolve the latent problems of British torpedoes which the fleet continued to experience. This was news not to Richmond's liking, as it only confirmed the emphasis of all things material within the Service. Yet, it was essentially a correct choice by the First Sea Lord for, notwithstanding Egerton's obvious academic achievements, he was seen as lacking in personality and powers of command and it was those very attributes Richmond deemed necessary if the officer replacing Dickens was to hold the Service's corner at the IDC.[73] Eventually, Somerville got the nod. Grenfell knowing something of Somerville's abilities having served as a staff officer beneath him rated his mental acumen only fair. Considered every bit a strong and effective officer, he held him to be a naval equivalent of Field Marshal Sir Henry Wilson.[74]

Beyond their merits as a possible successor to Gerald Dickens, Calvert, Somerville, and Egerton were linked by another connection unspoken of by Dewar in his correspondence to Richmond—the *Royal Oak* affair. Calvert was the prosecutor in the court-martial of Commander Henry Daniel, Dewar's erstwhile executive officer, while Somerville and Egerton both served as members of the jury in the proceedings.[75] Whether Richmond was aware of these connections is unknown, but Dewar certainly did as the events were less than six months past. As Daniel was the first of the pair to undergo trial, his guilt or innocence would establish the tenor in Dewar's own case. How much this weighed on Dewar, if at all, cannot be established but, at a minimum, he was guilty of not declaring an interest to Richmond in the advice tendered unless Richmond already knew of the linkage. Calvert, for his part, was more sinned against than sinner as his role as prosecutor was but a duty. As for his intended censorship of *The Naval Review*, what he really was arguing for was that the journal's editor exercise more discretion in what was published to ensure that the interests of the Service were protected. From the above, it is readily apparent that officer appointments cannot always judged at face value. Conflicting interests and perspectives influenced who was

<hr>

72 Eric Fullerton to Richmond letter of 4 July 1928, Richmond Papers, NMM/RIC/72/1.
73 ADM 196/91 (Egerton).
74 Grenfell to Liddell Hart letter of 1 September 1942, Liddell Hart Papers, LH 1/330/67, LHCMA.
75 Robert Glenton, *The Royal Oak Affair: The Saga of Admiral Collard Bandmaster Barnacle* (London: Leo Cooper, 1991), pp. 82-83.

deemed acceptable and weighed very much in the balance an officer's suitability for the IDC no less than any other assignment. More remains to be said on this subject.

Arthur Power,[76] following upon Ramsay as the naval member of staff, had garnered five first class certificates in his examinations for Lieutenant before receiving another first class pass in the Long Gunnery Course.[77] A trained staff officer, he served as Flag Captain and Chief Staff Officer to Astley-Rushton at the time of the Invergordon Mutiny, though escaping any ill-effects of that episode. Power's ship, *Dorsetshire*, had not been without disaffection, but he had taken a lead in communicating to his men the measures the Navy were pursuing to rectify matters and in the main, discipline held. In the words of Rushton, his flag officer, Power "Handled his men with good judgment, great tact, & determination. As a result the 'Dorsetshires' were all on duty by 1000 on 15/9, where they remained. Period of defection – 4 hours. No man could have done better."[78] This was Power at his best—decisive and not afraid to assume responsibility at the moment of crisis. Credited with a first-class mind and possessing high organizational skills, Power had been recommended to attend the IDC upon completing his tour as an instructor at the Staff College. His subsequent career is testimony that luck and timing are every bit as important, as innate talent and opportunity, in advancing one's prospects. Power proved to be an excellent member of the college where his outstanding knowledge and clarity of expression were much appreciated by Haining who believed that the success of the 1935 course owed everything to him.[79]

Captain Henry Horan,[80] but commonly known as Pat, came to the IDC from serving as Flag Captain in *Coventry*, taking over from Power during the third term

---

76   Later Admiral of the Fleet Sir Arthur John Power (1889-1960). Five firsts in examinations for Lieutenant; Staff Course, 1924-25; Royal Naval Staff College, 1927-29; Captain, 1929; Flag Captain and Chief Staff Officer in HMS *Dorsetshire*, 1931-33; Senior Officers' Technical Course, 1933; Imperial Defence College, 1933-35; Commanding Officer, *Excellent*, 1935-37; Tactical Course, 1937; Commanding Officer, *Ark Royal*, 1938-40; ACNS (Home), 1940-42; Rear Admiral, 1940; Commander, Fifteenth Cruiser Squadron, 1942-43; Vice Admiral, 1943; Commander-in-Chief, East Indies, 1944-46; Admiral, 1946; Second Sea Lord, 1946-48; Commander-in-Chief, Mediterranean, 1948-50; Commander-in-Chief, Portsmouth, 1950-52 and Admiral of the Fleet, 1952.
77   ADM 196/144 (Power).
78   ADM 196/92 (Power).
79   *Ibid.*
80   Later Rear Admiral Henry Edward Horan (1890–1961). Six firsts in examinations to Lieutenant; Commanding Officer, *TB 29*, 1914-15, HMS *Archer*, 1916, HMS *Nemus*, 1917 and HMS *Urchin*, 1917-18; Staff Course, 1920-21; Assistant War Staff Officer in *Barham*, 1921; Flag Lieutenant in Barham, 1921-22; Commanding Officer, HMS *Wolsey*, 1922-24 and *Whitehall*, 1924; Tactical Section, 1924-25; Royal Naval Staff College, 1925-27; Captain, 1930; Senior Officers' War Course, 1930-31; Senior Officers' Technical Course, 1931; Tactical Course, 1931; Flag Captain in *Coventry*, 1931-34; Imperial Defence College, 1935-37; Flag Captain and Chief Staff Officer in *Barham*, 1937-38; First Naval Member New Zealand Naval Board, 1938-39; Commanding Officer, HMS *Leander*, 1940; retired and Rear Admiral, 1941. Recalled for war service.

of the 1935 session. He, too, had shone in his examinations for Lieutenant taking six firsts, though unlike Power he elected not to specialize.[81] A graduate of both Greenwich courses and also a former member of the Staff College directing staff, Horan's qualifying lecture whilst taking the War Staff Course has not been located though he did subsequently lecture at Camberley in 1923 on 'The Function of the Navy'.[82] Horan also served for a brief period in the Tactical Section of the Naval Staff under Troup before assuming his appointment at Greenwich and Chatfield at a later date testified as to his *bona fides* claiming Horan was one the finest executive officers he had met.[83] His time in the Tactical Section and his completion of the Tactical Course stood him in good stead as Haining was struck by his grasp of tactical issues. Whilst this may not be the most ringing endorsement for a course dedicated to the higher aspects of war, the Commandant expressed himself fortunate to have been served by two officers of such proven ability: Power and Horan.[84]

Of these six, four had prior experience teaching at the Staff or War Course and a fifth, Somerville, had taught at the Signal School following the World War. The Navy, then, did not subscribe to the view current in some circles that before an officer came to the IDC he should have graduated from his own Staff College. Such had been a recommendation of the Wood Committee of 1923 formed to define the nature of a new college to train officers in joint planning.[85] To the Navy, this made sense if the purpose of the IDC was simply to train a higher-grade of staff officer, but this was not their view. Rather, they eyed the new course as a supplement to its own War Course with the further aim of preparing senior officers to consider the broader considerations of strategy in a joint environment. Thus, officers frequently recommended for attending the IDC were those studying at the SOWC and not the Staff Course and the committee's view showed the difficulties of overlaying a new course applicable to all the Services when they did not operate to a common level of instruction for mid-ranking officers.

Though more often mentioned as a perquisite for the potential qualifier, it was most assuredly assumed that it applied even the more so to any instructor appointed. Turning to any technical bias evidenced it can truly be said the Service did not discriminate. Two officers (Somerville and Ramsay) were of the signals branch, two others were salt horse officers (Dickens and Horan), and the remainder gunnery (Power) and navigation (Tennant) specialists. Somerville, it should be noted, was also a trained torpedo officer and only became a signals specialist when that discipline was separated from the torpedo branch and the *Vernon* in 1917.[86]

<hr>

81   ADM 196/144 (Horan).
82   ADM 203/100 does not specify Horan as one of the lecturers for the 1920-21 War Staff Course while Horan's lecture at Camberley is noted in the Alanbrooke Papers.
83   ADM 196/127 (Horan).
84   ADM 196/92 (Horan).
85   ADM 1/8644/169, E. F. L. Wood Sub-Committee of the Committee of Imperial Defence *Report*, 19 December 1923.
86   Poland, *Torpedomen*, p. 69.

Where the Royal Navy did diverge from the other Services was in the seniority of its officers attending whether as an instructor or as a qualifier. The naval member of the directing staff was apt to be junior in rank to his counterparts, whilst if a qualifier, he was frequently more senior. Both the Army and Air Force nominated officers to the directing staff of Brigadier or Air rank something the Navy never did. This owed something to the rank structure and style of titles employed in the naval service, as the nearest equivalent, Commodore, typically featured only in appointments afloat.[87] This created the anomaly that both its instructor and naval officers studying at the IDC held the same rank. Whether to appoint an officer holding Commodore's rank was considered, but Richmond, in truth, was a bit of an agnostic on the matter and the proposal fell flat.

The desire that those assigned to the IDC should already have passed through the Staff College was a logical condition of entry, as the course was very much focused on the higher aspects of war and eschewed training in general staff duties. Attendance was by appointment for all Services and offices of government. Thus, it was they who decided, and not the college, the composition of the course's membership. While no naval or marine officer sent to IDC is known to have failed, Captain Reginald Darke,[88] a very experienced submariner, withdrew early owing to illness during the 1935 course and William Houston-Boswall, a Foreign Office qualifier of the same year, left after the second term due to the ongoing crisis with Italy.[89] In the case of the former, Captain Henry Bovell,[90] newly promoted and from the cruiser *Norfolk*, left the ship he was set to command and began his studies from April.[91]

The actual circumstances surrounding Darke's departure are unknown, though the surviving evidence testifies that a breakdown was the proximate cause. Haining observed that Darke felt some trepidation about pursuing the course of study as he

87   *King's Regulations*, Art. 198, p. 66.
88   Later Rear Admiral Reginald Burnard Darke (1885-1962). Commanding Officer, HMS *Onyx*, 1910-12; *C 1*, 1915; *G 2*, 1916; *L 1*, 1917-18; *K 2*, 1918-19; HMS *Ross*, 1921-23; *Maidstone*, 1924-26; Senior Officers' Technical Course, 1926 and 1932; Captain 1926; Commanding Officer, *Lucia* and Commodore 2nd Submarine Flotilla, 1926; Commanding Officer, *Conquest* and Commodore, 1st Submarine Flotilla, 1926-28; Tactical Course, 1931; Senior Officers' War Course, 1932; Commanding Officer, *Enterprise*, 1932-34; Imperial Defence Course, 1935 and Rear Admiral and retired, 1937. Recalled for war service.
89   CAB 53/27, Commandant, Imperial Defence College to Secretary, Chiefs of Staff Sub-Committee letter of January 1936.
90   Later Vice Admiral Henry Cecil Bovell (1893-1963). War Staff Course, 1923-24; Squadron Gunnery Officer and Staff Officer Operations in *Douglas*, 1924-25; Gunnery Division, 1928-31; Staff Course Camberley, 1931; Captain, 1934; Imperial Defence Course, 1935; Senior Officers' Technical Course, 1936; Deputy Director, Training and Staff Duties Division, 1936-38; Tactical Course, 1938; Senior Officers' School Sheerness, 1938; Commanding Officer, *Vindictive*, 1938-39 and *Argus*, 1940; Imperial Defence College, 1940; Commanding Officer, HMS *Victorious*, 1941; Rear Admiral, 1944; retired, 1947 and Vice Admiral, 1948.
91   ADM 196/92 (Bovell).

was not a staff college graduate. Still, if he did not talk freely during sessions, his views were sound based on his store of practical knowledge and Darke's performance had compared favourably thus far to the other students.[92] His military performance to that point had been exceptional, but narrow focused as it was on the submarine service with the occasional time spent in heavy ships required of all officers. Being a bachelor and older than the typical qualifier, Darke lacked the support structure of a wife to share his burdens with nor the congenial surroundings of a wardroom as officers did not reside in college. Following his recovery, he performed his new duties well but with a degree of uncertainty and retirement soon followed. Having been rated midshipman without examination, perhaps success had come too early and his career had been too circumspect before entry.[93]

In late 1937, the War Office announced its desire to send an officer of the Territorial Army each year to the IDC, though whether they actually needed the concurrence of the other Services before adopting the measure is unclear. The previous investigations leading to the creation of the IDC were silent on the matter and practice had made the question of officer appointments a matter of the parent establishment only. Chatfield, nevertheless, endorsed the move noting that the Navy should like to raise the question of appointing an officer of the RNR or RNVR at a future date.[94] Chatfield's wish to see the IDC officer pool expanded may have only been prompted by the prospect of seeing the Navy falling behind in the training afforded to its officers. Yet, the Service had a strong record of RNR officers attending the SOWC and one officer of the RNVR did attend the naval Staff Course in 1934, so the idea was not as far-fetched as it may appear. Moreover, it was a practical means of ensuring that the numbers qualifying were maintained at or near the capacity of the college.

Whatever the status of the officer, the IDC qualifier was already expected to know how to draft orders, the general organization and workings of the Service Ministries, have an understanding of the moral element in war, and appreciate the importance of leadership. These strands of education were common to each of the Staff Colleges and were not initially covered at the IDC. Time would demonstrate that how each of the Services approached these subjects varied and Brooke-Popham was soon lecturing on the unique requirements of preparing a combined appreciation and how it differed from one prepared for use by a single component. The topic proved something of a forte for him as he returned to the IDC following his time as Commandant to lecture on the subject time and again.

The combined appreciation was both simpler and more complex than its Service counterpart. Simpler, in that the intended audience was not necessarily a defence expert and so the language employed had to convey the desired meaning absent the

92   ADM 196/92 (Darke).
93   ADM 196/50 (Darke).
94   CAB 53/34, First Sea Lord to Chief of the Imperial General Staff letter of 9 November
      1937.

jargon typically found in the papers read by the professional officer. Yet, the combined appreciation was more complex in the breadth of its survey and needed the input of experts beyond those of a single Service. Its conclusion was the recommendation of the appropriate policy to be adopted, and, unlike the Service appreciation, was not written to specify the actual operation required. This would follow in due course.[95] In teaching how to prepare a properly reasoned combined appreciation the IDC was breaking new ground and the:

> Military Advisers consequently have the heavy responsibility of providing their advice in such a form as to provide a clear picture of the probable course of events, the dangers incurred, the results to be expected, the effort entailed, and the multitude of other considerations which affect the problem.[96]

Part of that multitude was to consider the options available if an opposing power were to employ means proscribed by international agreement. Such actions might include the use of chemical agents, the bombing of non-military targets, or the waging of unrestricted submarine warfare. The planner preparing the appreciation needed to consider the advantages and disadvantages of such use, the likely response of neutral opinion, and the policy Britain should adopt in the event. One response the British might choose to adopt could be to initiate a propaganda campaign and not take the seemingly obvious and legal step of resorting to reprisal. Britain's experience in the World War offered guidance on the fallout of each of these specific threats, but the planner needed to weigh whether the responses taken then remained valid for the present.[97]

Of course, preparing an appreciation was very much tied to another subject that was taught at the Service colleges: The Higher Direction of War. Yet, fundamentally the combined appreciation differed from a Service appreciation because its intended audience differed. In style, scope, and content it took a sweeping view. This is discussed shortly when the overall curriculum of the IDC is assessed, but the point to emphasize here is that officers and their civilian counterparts were now asked to think and consider problems along the broadest lines. Naval officers used to dealing in terms of oceans rather than fronts and across command boundaries were not so nearly disadvantaged by failing to attend their Staff College as might be supposed when it came to considering such problems.

Sometimes in a given year, a civil servant was apt to attend the IDC from a Dominion rather than a military officer and this was recognition that their development as full-fledged nations owed just as much to civil considerations as any military.

95   'The Preparation of Combined Appreciations for a Major War', p.2, IDC lecture, 1933, Alanbrooke Papers, 3/12, LHCMA.
96   *Ibid*, p. 3.
97   *Ibid*, p. 13.

In 1928, Frederick Shedden of the Commonwealth's Department of Defence Finance Branch, became the first Australian civil servant to attend the course. Becoming Defence Secretary in 1937, eventually Shedden added the additional portfolio of Cabinet Secretary to his growing resume of responsibilities and only vacated office in 1956. Shedden's attendance, thus, proved a sound investment for both Australia and Britain.[98] It has been said that he modeled his career on Maurice Hankey, the ubiquitous power behind the throne of the British Cabinet, but it can also be said that Australia, in turn, modeled its fledgling defence structure along British lines which the Imperial Defence College did little to discourage.[99] Beyond Shedden another civilian of note attending the IDC was Francis Dwyer of New Zealand. Sitting the 1938 course, Dwyer returned home to serve first as Assistant Army Secretary in 1939 before becoming Army Secretary in 1941.[100]

In light of the complaints of subsequent Commandants concerning the college's facilities, the rush to establish the new school as quickly as possible had lasting, if not grievous, repercussions. From the moment the decision was confirmed to proceed with its establishment until its first meeting in January 1927 little time remained available to set matters right. Richmond, the designated Commandant was only confirmed in his appointment in September 1926.[101] Though he understood that there was a strong chance he would be nominated if the project advanced, he lacked official status to commit the institution and concentrated on activities incurring no financial obligations. Perforce, there was much to be done and time was not a friend. A suitable location had to be secured, staff acquired, and a curriculum prepared to the acceptance of the Chiefs of Staff Committee, the body responsible for the college's governance.[102] Little wonder that the first year's efforts when measured by the pace of later courses appears relaxed with only 82 lectures occurring to the 140 of 1938.[103] All was confusion simply trying to coordinate the many strands of the initial course and though Colonel John Dill, the senior member of the directing staff, soon took a firm hand in matters, even his appointment only dated from 10 November 1926.[104]

The site ultimately secured, 9 Buckingham Gate, was minimally acceptable offering, as it did, little scope for future growth. It served as the college's home until the outbreak of war in 1939 when courses were suspended for the duration of hostilities.[105] Though not an immediate limitation in 1926 and 1927, the unsuitability of the

98   Horner, David, 'Shedden, Sir Frederick Geoffrey (1893–1971)', *Australian Dictionary of Biography*, National Centre of Biography, Australian National University, http://adb.anu.edu.au/biography/shedden-sir-frederick-geoffrey-11670/text20853, accessed 7 October 2011.
99   Grey, *Military History of Australia*, p. 219.
100   New Zealand Archives, Francis Bernard Dwyer Papers, Dwyer 5.
101   *Navy List*, October 1927, p. 73.
102   ADM 116/2273, 'Instructions for the Commandant of the Imperial Defence College'.
103   Gray, *Imperial Defence College*, p. 8.
104   ADM 116/2273, Account General minute of 10 November 1926.
105   Gray, *Imperial Defence College*, p. 4.

premises become pronounced as the college developed momentum and others not studying at the college desired to attend its lectures on an *ad hoc* basis. This could not be accommodated for the lecture theatre simply did not support attendance beyond those taking the course much to the chagrin of Haining, Commandant from 1935-36.[106]

Of course, Richmond, Dill, and the other two members of the directing staff—Captain Gerald Dickens, RN, and Group Captain Philip Joubert de la Ferté, RAF, were not exactly working from a fresh slate, as the Chiefs of Staff Committee had specified the overall parameters of instruction. Though Dickens was unique amongst current directing staff members in not having previously served at the Staff College or for that matter even taking a Staff Course, he had extensive experience considering defence problems. With previous appointments in both the Intelligence and Plans Divisions of the Admiralty, he was also a qualified interpreter in both French and German. If not a Staff College graduate, Dickens, nevertheless held the *qs* designation marking him 'Qualified for War Staff Duties' based on experience. Dickens was just the sort of all-round officer that the Service traditionally favoured: broad in outlook, varied in experience, and not lacking in intellectual talent having mastered a couple of languages along the way.

Where it was hoped the college would break new ground was in having a senior member from the civil service seconded to serve as an equal member of the directing staff to address the political and economic issues of war planning.[107] This, alas, did not happen and initially lectures from the Civil Departments were provided from a pool of six civil servants.[108] This was a serious limitation partially made good by Dickens and particularly vexing given that the economic element in any defence problem was a unique aspect of British war planning. In one of his last official acts as Commandant, Richmond addressed this shortcoming by requesting that the Chiefs of Staff secure the services of an economist to serve as a resident adviser to the college. Acknowledging the additional expense entailed, Richmond was at a loss to see how actions proposed in plans could be considered unless their economic import was known. This was particularly so when considering maritime sanctions, but economic understanding was also essential when weighing the impact mobilization had on the commercial and fiscal health of the Empire. That Richmond's request for an economic adviser came as his tenure was ending was recognition that Dickens would have to serve as a true naval member of the directing staff upon the Admiral's departure and time would not allow him to continue researching economic questions for the college. This proposal was confirmed with the proviso that the official hired also be available to support the work of the CID.[109] Fayle assumed the post initially on a part-time basis

---

106 ADM 1/11112, Report of 1936 Course by Commandant of Imperial Defence College.
107 Fuller to Liddell Hart letter of 3 September 1926, Liddell Hart Papers, LH 1/302/99, LHCMA.
108 ADM 116/2273.
109 CAB 53/3, Chiefs of Staff Committee minutes of 14 January 1929.

becoming a resident expert adviser for economic matters in 1935.[110] Indeed, so central was the economic aspect of war planning becoming at the IDC that by 1935 an understudy to Ernest Fayle was tabled.[111] The Chiefs acted with alacrity on this suggestion, endorsing the proposal, and turned to the confidential IIC for assistance. This was a fortuitous move on their part for ill-health was soon to plague Fayle reducing his role in 1937 before forcing his retirement in 1938.

It was during this same session of the COS Sub-Committee when agreement was reached to appoint an economic member to the college's staff that Richmond sought sanction for the qualifiers to travel to Belgium to visit the site of the 1918 St George's Day raid at Zeebrugge along with certain French battlefields. The costs involved, at £5 per officer, were trifling and when approval was at last rendered, a warning that a precedent for future ventures had not been established was felt necessary.[112] Maybe this admonition was only eyewash for the Treasury (or for the First Sea Lord, Admiral Sir Charles Madden), but the previous also highlights that, as with so much of interwar defence affairs, the potential cost of any proposal was an ever-present consideration. This applied, too, to the IDC itself should any war arise, and an early decision reached was that upon mobilization the college would cease operations.[113] The grounds for this decision, of course, were not merely economic, as the officers in residence at the IDC represented a pool of talent singularly qualified for active service. Yet, the fact that the Chiefs of Staff failed to consider a shadow faculty formed from retired or reserve officers to continue its operations indicates that cost was probably a primary reason why this option was not pursued. This decision is all the more surprising as the need for trained staff officers in war would only grow as expansion and casualties working to equal effect consumed the modest cadre available. This had been a sharp lesson of the last war and presumably one that could reasonably be expected to appear in any new conflict.

The Chiefs approved the visit to the continent as a one-off occurrence because it was held to fall outside the scope of the college's curriculum. One can sympathize with Richmond if he felt that he was being asked to make bricks without straw for the approved syllabus was broad in its scope and on the surface provided the faculty plenty of room for interpretation on exactly how to proceed in accomplishing the objectives of the college. The charter of the IDC, which today might be termed an organizational Mission Statement, announced its purpose as:

---

110 Gray, *Imperial Defence College*, p. 8.
111 CAB 53/8, Chiefs of Staff Committee minutes of 22 December 1937.
112 CAB 53/3, Chiefs of Staff Committee minutes of 14 January 1929.
113 CAB 53/2, Chiefs of Staff Committee minutes of 23 February 1928.

The training of a body of officers and civilian officials in the broadest aspects of Imperial Strategy and the occasional examination of concrete problems of Imperial Defence referred to them by the Chiefs of Staff Committee.[114]

The supporting syllabus was equally robust allowing for visits and tours. Though the study of past battles was not expressly mentioned, its drafters presumably understood that recourse to historical example was a proper form of instruction.[115] Brooke-Popham believed such visits acquainted qualifiers with both the problems and the capabilities facing officers of a Service separate from their own.[116] This no doubt was true, but such visits also allowed qualifiers to appreciate the setting for a recurring exercise of the period: war against Germany in alliance with France and Belgium in support of her Locarno commitments.[117]

As it proved, tours to the continent soon became a regular part of the year's educational programme. Beyond the visit made in 1929, ones quickly followed in 1930 and 1931 with General Sir Archibald Montgomery-Massingberd,[118] a future CIGS and late Chief of Staff to General Rawlinson of the Fourth Army, providing expert commentary. The practiced lapsed for a time but in 1935, a return to France and an inspection of the Maginot Line of defensive works, including a talk by General Maurice Gamelin to the qualifiers, followed.[119] Now serving as CIGS, Montgomery-Massingberd was no longer free to support these tours and Haining employed, instead, the services of Lieutenant General Sir John Brind,[120] then overseeing British Saar occupation duties.[121]

The IDC was very much focused on the higher aspects of war and the allocation of resources to commitments was a common topic of investigation. As these subjects appeared regularly in the yearly reports developed by the Chiefs of Staff Sub-Committee, qualifiers were being prepared to consider the very matters that they would have to deal with upon leaving. Towards this end, lectures such as 'The Army in Relation to Military Commitments' and 'Air Strength and Commitments' provided the necessary context for the schemes later considered. Dill, the first Army and senior member of the directing staff, offered as an early contribution a lecture on

114 Gray, *Imperial Defence College*, p. 7.
115 *Ibid*, p. 81.
116 CAB 53/22, Commandant, Imperial Defence College to Secretary, Chiefs of Staff Sub-Committee letter of 11 March 1931.
117 CAB 53/18, 'Exercise No. 7 (1928), British Empire, France and Belgium v Germany'.
118 Later Field Marshal Sir Archibald Armar Montgomery-Massingberd (1871-1947). Staff Course Camberley, 1905-06; Staff College Quetta, 1909-11 and Staff College Camberley, 1911-14.
119 Alanbrooke Papers, 3/17/1, LHCMA and Gray, *Imperial Defence College*, p. 12.
120 Later General Sir John Edward Spencer Brind (1878-1954). Staff Course Camberley, 1913-14.
121 CAB 53/27, Commandant, Imperial Defence College to Secretary, Chiefs of Staff Sub-Committee letter of January 1936.

'Imperial Armies in Relation to the Military Situation' informing the recently assembled qualifiers:

> Our policy, as you have heard, is "to seek peace and ensure it." We plan no conquests, we seek merely to maintain and develop what is ours. We are said to be what Von der Goltz would call a people which, in its historical course of development, has reached a state of rest, and will, therefore, only go to war by compulsion. A people which consequently will confine itself to a strategical defensive. This state of rest has been looked upon as a dangerous state, leading swiftly to decline. However that may be, the fact remains that our Imperial military strategy has for very many generations been almost uniformly defensive and yet during the lifetime of all of us we have, as an Empire, added vastly to our responsibilities – assumed or accepted responsibility for the government of vast territories and huge populations.[122]

Dill's exposition highlighted the risks facing Britain and the Empire and testified that, notwithstanding its defensive orientation, risks were latent.

Yet it was insufficient of the qualifier to appreciate the naval, military, and air strategies and responsibilities of Britain, officers needed to appreciate, too, the machinery governing the control of British war planning, especially, the Committee of Imperial Defence, imperial relations, and their relationship to the broader international system. Here, Haldane and Hankey proved frequent contributors providing background material on the current arrangements and the historical practices which influenced the present organization. Still, Haldane and Hankey being responsible in large measure for the present establishment were not neutral observers and there were limits to what they could say or would offer as recommendations as a way ahead. One must conclude that Hankey, even if believing his own views to be correct, could not escape the interest that he retained in the arrangements of the status quo. Addressing the subject directly whether a Minister of Defence was required, he commented:

> No single Minister of Defence could, in a great war, handle the problems of the sea, the land and the air simultaneously. As one who witnessed these events at very close quarters, as one who was in the intimate confidence of those responsible throughout the War, I say with conviction that if any one Minister or Ministry had been responsible in the Great War for all the Service Departments there would have been a complete breakdown.[123]

122 'The Imperial Armies in Relation to the Military Situation', lecture of 20 January 1927, Dill Papers, 2/3, LHCMA.
123 Maurice Hankey, 'The Higher Direction of War', lecture, 7 November 1927, Alanbrooke Papers, 3/12, LHCMA.

Hankey continued by stressing that only the prime minister of the day was fit to oversee the general direction of war under the British system. A fair sentiment, yes, but he failed to address why a prime minister would not be similarly encumbered while also retaining residual responsibilities beyond the waging of war. Joining Haldane and Hankey in lecturing on 'The Higher Direction of War' during the college's inaugural course were the former premier, Foreign Secretary and First Lord of the Admiralty, Arthur Balfour, Field Marshal Sir William Robertson (the former Chief of the Imperial General Staff for the greater part of the World War) and General Sir Percy Radcliffe.[124]

When the time came to consider the question of the 'Higher Direction of War' during one of the session's several exercises, Brigadier Andrew McNaughton,[125] Canadian Militia, drawing on his experience and knowledge spoke of the changed circumstances now in imperial affairs. Accepting that defence of the Empire was a given, he emphasized that Dominion support was not automatic and that matters in 1927 now approximated for Canada what they had been between Britain and France in 1914—a junior partner in a common endeavour, but still a partner and not a minion. This was all too much for McNaughton's fellow qualifier Brooke, the thoughtful, if rather headstrong gunner. Whether he declaimed, 'I flatly disagree', as was his manner at times is unknown, but he would concede nothing of the sort in the discussions following.[126] In fairness to Brooke and others of his view, McNaughton's argument, if correct, was only so because the strength of the Royal Navy offered cover to Canada and the other Dominions. Ceding strategic direction and control from the center was difficult to imagine when they were all so militarily weak.

The McNaughton-Brooke interchange highlighted some of the pitfalls awaiting those addressing the question of the higher direction in war and why senior military leaders might have been reluctant to tackle the matter. The same caveats ostensibly applied to Fuller when lecturing in November 1933 on the same topic, but discretion was a not suit worn easily by him notwithstanding earlier warnings by the Army Council. Now in his third year without employment, shortly to retire the following month having refused the command of the Bombay District, and not being responsible for present arrangements, Fuller left his audience in little doubt what the risks were and what was required:

> In the last war the supreme struggle was not with the enemy but within ourselves (Great Britain and the Dominions) and between ourselves and our allies. Then, our first problem, was the establishment of imperial unity – a political question, and

---

124 CAB 53/14, Report of the First Session Imperial Defence College, 22 December 1927.
125 Later General Andrew George Latta McNaughton (1887-1966). Staff Course Camberley, 1921 and Imperial Defence Course, 1927.
126 John Sweetenham, *McNaugton, Volume I, 1887-1939* (Toronto: Ryerson Press, 1968), pp. 232-33.

our second, of inter-allied unity – a strategical question. After two years and eight months of chaos the first was solved by the creation of the Imperial War Cabinet; the second does not concern us here; but its lesson was, that unless allied forces are centrally directed unity of action is impossible.[127]

While Fuller was clearly guilty of hyperbole, he nevertheless touched upon a crucial point. Unity of action demanded more than mere cooperation, the principle specified in the *FSR*, the *RAF War Manual* and the *Naval War Manual*, and if he did not formulate a new principle of 'Unity of Command', he was certainly close to appreciating its truth.

The higher direction of war as a topic of study allowed the question of whether a defence minister holding his portfolio separate from the premiership was wise, and, further, if it was desirable for members to belong to a war cabinet void of departmental responsibilities. The arguments for and against both are beyond the remit of this study, however, the subject introduced officers to issues having a direct bearing on strategic affairs indicating that the solution achieved was not simply a matter of what was practical and desirable for operational efficiency's sake. They touched on the very nature of constitutional propriety, the sources of bureaucratic power, and the limits of individuals. An officer preparing a combined appreciation in the loose confederation that was the British Empire needed to be aware of the limitations facing politicians at home and those further afield when set against the freedom of action that a totalitarian state was likely to enjoy.[128] The underlying assumption that a fascist or communist state, subject to the control of one leader, was in a stronger position to direct a war was accepted as a given, and the not so easily solved problem facing the qualifier was emphasized:

> In no existing nation or Empire do the demands of defence present such a vast and complex problem as in the case of the British Empire. No other nation or group of political entities is confronted with the necessity of so co-ordinating its defence requirements as to meet the demands of so many responsible governments.[129]

All the Service Ministries felt the burden of coordinating defence policy against the myriad structure of Empire, though the problem was more acute for the Army and the Air Force, as they were borne and made demands on their hosts on a scale typically beyond the Navy. Whether it was the Colonial Office, the Government of India, the separate Dominions, or the even Foreign Office, a straight line of command did not

---

127 J. F. C. Fuller, 'Our Higher Direction of War', lecture, 15 November 1933, Alanbooke Papers, 3/12, LHCMA.
128 Longmore, *From Sea to Sky*, pp. 186-87.
129 Imperial Defence College, Exercise No. 9 (1933), 'The Higher Direction of War', Staff Solution, p. 1, Alanbrooke Papers, 3/12, LHCMA.

always flow from the Service Ministries in London to its deployed forces. Moreover, the problems arising from this diverse structure were apt to be most severe during the transition from crisis to war when time was critical and control passed from the civil agency responsible in peacetime to the Services.

That a totalitarian state was thought to have a distinct advantage in directing war is somewhat inexplicable as prior historical experience demonstrated Britain's capacity to defeat militarized states as different as Napoleon's France or Wilhelm's Germany. The tensions of the modern empire no doubt influenced the view—witness the Chanak Crisis—and contemporary economic woes and the acceptance that her publics at large desired to escape the responsibilities of Empire and League probably contributed to the assessment. Another possibility is that some influential British officers having pronounced Right Wing views or sympathies naturally assumed the case to be true without ever examining the underlying tensions existing in such states. In any event, without defining how Britain should organize herself in a large-scale conflict, the lesson offered was:

> Success in war has come to depend more and more in recent years on the ability to organise in peace and to develop in war those economic and financial resources which are essential for war expansion and for the maintenance of industry and commerce under war conditions. The higher direction of war is therefore no longer largely confined to the direction of the military machine; it embraces the control of activities of all sections of the community and all branches of industrial effort.[130]

Unlike most exercises at the IDC, the Commandant typically oversaw the 'Higher Direction of War' when it was held towards the end of the final term. This oversight occurred not so much because of its importance but was due mainly to the other members of the directing staff preparing themselves for the next course. Brooke-Popham was not sure the topic's value was commensurate with the effort required to arrange it. After all the discussions held and papers written, the conclusion frequently reached was that the arrangements that had evolved at the end of the World War were apt to be the most suitable.[131] This was not without controversy as it implied the concession of power by Dominions and Ministers alike. Still, Brooke-Popham conceded that having the qualifiers come to this view themselves rather than simply spoon-feed them a pat answer ensured that if the problem ever arose anew they would be cognizant of the pitfalls to be avoided, the aims to be achieved, and the tensions likely existing hindering the establishment of effective control.[132]

---

130 *Ibid*, p. 2.
131 'Higher Direction of War', pp. 1-2, IDC lecture of 16 November 1934, Brooke-Popham Papers, 9/7/2, LHCMA.
132 *Ibid*.

By 1938, Desmond Morton of the IIC had supplanted Fayle as the primary lecturer for economic analysis at the college. Turning to the issue of the war's higher direction, Morton concluded that most of the problems experienced during the World War were fundamentally arguments over resources and, thus, economic in nature stressing to his audience:

> Stripped of its smoke-screen of personal animosities and local conditions it resolved itself into the extent to which a Commander-in-Chief in war may, in pursuit of a plan of campaign, demand the right to drain the nation of its wealth. Contrariwise it concerned the extent to which the elected representatives of the nation may regulate a Military Commander's campaign by denying him supplies of men and material.

Even in 1917 and 1918, therefore, the root of the matter was economic.[133]

Morton's view of the problem may have been necessarily narrow, but his view does indicate what a rich field the higher direction of war was for the course, as it touched upon so many elements of concern to commander and statesman alike. Morton suffused his lecture by drawing on the practices of the United States, France, Italy and Germany in organizing themselves for war leaving it to the qualifiers to consider whether Britain and her consorts could do better.

While it is understandable how Brooke-Popham might conclude studying war's higher direction was an impossible task and consumed too much time, it is hard to see how the IDC could ignore such a vital topic. The experience of the World War highlighted previous perils and the introduction of air power only made the question yet more vital. Difficult as it was to reach consensus merely within imperial circles, operating with allies made a complex issue that much more intractable. Having a field force operate under the command of a foreign entity was probably a necessity, but how an ally felt about having British air forces strike at targets deep inside an enemy state from its territory at the behest of the British Air Staff was an altogether different problem. When the problem arose for real in 1939, the prior and frequent consideration of the question at the IDC provided the Chiefs of Staffs and the CID with ample evidence how to proceed in staff talks with the French as one of the first exercises conducted was war against Germany in concert with France and Belgium.[134]

Amongst the problems needing to be resolved were the collection, analysis, and dissemination of intelligence at the strategic level. Beyond the sources available to the Service Ministries existed the information available to the Foreign Office, the Board of Trade, the Treasury, and the myriad other elements of Government. Arguing against the amalgamation of the disparate offices handling intelligence,

133 Desmond Morton, 'Higher Command in War', lecture of November 1938, Tennant Papers, NMM/TEN/42/4.
134 Exercise No. 5, War Against Germany in Alliance with France and Belgium, Alanbrooke Papers, 3/12, LHCMA.

Brooke-Popham concluded that a clearing house was needed, nevertheless, to oversee the efforts expended and the results obtained. The product developed would then be available to the staff officer for immediate consideration and use.[135] Some problems, it seems, remain eternal.

As for the actual problems considered, in 1927 the IDC qualifier weighed possible wars against Japan, France, Germany, and the Soviet Union while weighing relations with the United States and the importance of Palestine. The conclusions of the four wars considered were forwarded to the Chiefs of Staff Sub-Committee, though not without garnering a mixed reception. Hugh Trenchard, the Chief of the Air Staff, complained that the analysis of war with Russia was fundamentally at odds with a recently completed War Office assessment, the appreciation developed for a Japanese war failed to consider the employment of battle cruisers, and the one reviewing a possible German war totally misused British airpower. Milne, the CIGS, believed the more basic problem was that the IDC was attempting too much in the limited time it had available. Researching fewer schemes and availing themselves of the expert knowledge present in the Service Ministries on those problems then considered, would prove more fruitful in the future.[136] In light of his earlier request to secure access to specific CID papers to inform the development of the college's curriculum and schemes, these criticisms demonstrated a certain lack of tact, to say nothing of sympathy, against the barriers Richmond faced. Moreover, if the IDC failed to appreciate the true role of airpower, then Trenchard shared some of the blame as he had appointed the air member of the directing staff, and he had insisted, too, that the members of the directing staff were to represent their Service's views at the college.[137]

Richmond was every bit aware of these deficiencies, but faced a dilemma. Educating officers to consider problems of the broadest scale and of the myriad possibilities facing the Empire operated against developing papers of the highest quality. His initial approach was to focus primarily on one problem and introduce subsidiary problems to ensure other aspects of British and Imperial defence were recognized and appreciated. For 1927, the detailed problem was an Anglo-French war arising in 1934 over events in Syria.[138] The appreciation and supporting exhibits submitted to the Chiefs of Staff came to 79 pages and consumed over two months of the course. Though the lesser problems weighed that year were not nearly as extensive, the one forecasting a Japanese war was only slightly less daunting at 70 pages and developed over nine weeks.[139] By the end of his tenure, Richmond saw that this approach was defective. The abilities of the qualifiers varied greatly and he came to believe that prior

135 *Ibid*, p. 25.
136 CAB 53/2, Chiefs of Staff Sub-Committee minutes of 23 February 1928.
137 CAB 53/1, Chiefs of Staff Sub-Committee minutes of 2 December 1926.
138 CAB 53/18, Exercise No. 2.
139 CAB 53/15.

attendance at a staff college was essential. Until this happened, the best approach was to examine several problems advising the COS:

> The study of a very restricted number of problems would fail to give a true realisation of the great diversity of the defence problem of the Empire. It would not bring clearly into view, even if it brought out at all, the fundamental principles which should govern the constitution of the forces the Empire needs to maintain, since it would concentrate attention upon a few particular instances only. The principles can only be illustrated, developed, and impressed upon Officers, if the great variety of the functions that fell upon the Services, and the great variety of circumstances in which they may have to be used, are fully explored.[140]

This same observation could also have been leveled against the Staff Colleges repeated survey of a Far Eastern war. For the IDC, the conflict was never truly resolved during the period as the actual number of exercises waxed and waned. Though the evidence is fragmentary, that which exists, suggests that more exercises of a general nature became the norm and specific threat problems were limited to but one a term.

Thus, for the 1936 course, eleven exercises were conducted with the third term having but three with two of those three considering war against Italy and Germany.[141] Real-world scenarios were entertained as they enhanced student interest and were not without some value to the Service Ministries, joint planners, and the Chiefs of Staff, but specific tasking from the COS ceased. By the time of the 1938 course, the last full meeting before the outbreak of war in September 1939, ten exercises were being considered consisting of a balanced mixture of individual effort, group conferences, and syndicate work. Periodic visits to British military, naval, and industrial establishments still featured and the annual summer trip to the battlefields of France continued with nary a word from the COS. Setting the scenes for the problems considered was a strong regime of outside lecturers and, in this instance, amounted to over 120 separate talks. Leading statesmen, experts from the Service Ministries, and area specialists from the Foreign and Colonial Offices predominated. Thus, a reasonable conclusion is that the IDC directing staff remained primarily facilitators and not instructors. Absent were any grounding topics in staff duties, military and naval history, and the writing lessons so prevalent at the Staff Colleges. This is not to say qualifiers did not write; they did. Only now, they were expected to be past masters in slinging ink. Moreover, no longer could a Commandant lament that the naval hierarchy ignored the IDC, for

---

140 CAB 53/17, Richmond to Secretary, Committee of Imperial Defence letter of 27 February 1929.

141 CAB 53/29, Commandant, Imperial Defence College to Secretary, Chiefs of Staff Sub-Committee letter of December 1936.

Admiral Backhouse in common with his fellow Service Chiefs came to the college and addressed the qualifiers.[142]

As for Trenchard's criticisms over the misuse of airpower encapsulated in the appreciations, it was all water off a duck's back to Richmond for the 1928 appreciation covering Germany announced:

> Air attacks on German industrial centres – particularly the Ruhr – by cutting off, or at any rate limiting, the supply of essential munitions of war, would reduce the fighting efficiency of the German armies. It is, however, reasonable to suppose that Germany has sufficient war material to carry her through the first months of war. The material effect of bombing the Ruhr would not, therefore, react immediately upon the armies in the field and would consequently be of little assistance to the Allies in the early stages of their advance into Germany.[143]

Knowing that his tenure was ending and that he would not have to face Trenchard again—at least as Commandant—perhaps permitted him to sail a little closer to shoal water. He had done his best and remained Richmond to the end.

However, it was for his successor, Major General William Bartholomew,[144] a gunner, to prepare the only appreciation especially commissioned by the COS between the wars. This study reviewed the strategic implications of a channel tunnel between Britain and France and the COS solicited the college's views after the prime minister sought their advice on the question. This was not a new topic for Government consideration, as the Cabinet had weighed the question as recently as 1924. Now though, the Chiefs of Staff could avail themselves of the IDC, a course that had not earlier existed when they rendered their views.[145] Bartholomew initially sought to avoid preparing the appreciation.[146] His reservations stemmed in part from the difficulties weighing such a problem posed with new qualifiers, though his own recent appointment, as Commandant, was doubtlessly a further complicating factor. Given the genesis of the project and that a White Paper was to be forthcoming, the Chiefs of Staff would not be denied. They, however, did allow Bartholomew until the end of the year to present his findings. When ultimately presented they were adopted with but minor changes and found highly commendable by the Chiefs of Staff.[147]

142 CAB 53/43, Report by the Commandant of the 1938 Course of the Imperial Defence College, 22 December 1938.
143 CAB 53/19, Exercise No. 7 (1928).
144 Later General Sir William Henry Bartholomew (1877-1962). Staff College Quetta, 1909-10 and Commandant, Imperial Defence College, 1929-31.
145 CAB 23/48, Cabinet minutes 39 (24) of 2 July 1924.
146 CAB 53/3, Chiefs of Staff Sub-Committee minutes of 28 February 1929.
147 *Ibid*, Chiefs of Staff Sub-Committee minutes of 1 May 1930 and CAB 53/21, Deputy Chiefs of Staff Sub-Committee report of 4 April 1930. The actual report is contained in CAB 53/20 and concluded that invasion by France was not a practical proposition and that

The potential scale of attack Britain faced upon completion of any channel tunnel proved the only problem referred to the IDC by the Chiefs during the interwar period. This owed nothing to Bartholomew's abilities as Commandant who was highly thought of by each of the Chiefs of Staff, but rather stemmed from a reluctance of the Service Ministries to share detailed intelligence and their own operational shortcomings with the college. Moreover, employing the IDC in such a manner challenged the role of the Services' Plans Divisions as well as the function of the Joint Planning Sub-Committee (JPC) reporting to the Chiefs of Staff collectively. Significantly, when the channel tunnel appreciation was reviewed, the Naval Staff objected to the study's allocation of naval forces and its minimizing the technical difficulties in conducting an amphibious assault.[148]

In truth, the Commandants did not desire to pursue such tasks and, in 1936, at the behest of Haining, the Chiefs of Staff modified the college's terms of reference removing this aspect of its mission, though reserving the right to forward problems at a subsequent date if circumstances warranted.[149] Such was the background when the Deputy Chiefs of Staff sought the advice of Ernest Fayle of the IDC when reviewing the JPC's formal appreciation of war with Germany prepared that same year. This, though, was more of a back-channel request than a formal tasking to the IDC and demonstrates that the tenor of Haining's request was honored.[150] Even Richmond had come round to this view by 1929 seeing the risks it presented to college. The qualifiers lacked the time required and the information necessary to prepare such appreciations, and, frankly, most were not up to the task. They lacked both the skills and the proper perspective necessary to consider any broad problem, as until trained, they were thought imbued with only their Service's view to any question. Such no doubt confirmed the need for a joint college, but this was a reversal from Richmond's 1923 position to the Wood Committee (the body which ultimately proposed the establishment of a joint defence college) when he believed such appreciations would form an important part of any college's work.[151]

Trenchard's criticism that the college failed to appreciate the role and application of air power was delivered to his fellow Chiefs of Staff in private before Richmond entered the meeting. The charge if not without merit, only highlighted that a vacuum existed in current air doctrine as understood by others. Richmond thought part of the problem was that the Services, though sharing common principles of war, did not share a common understanding as evidenced by their respective doctrinal publications. This required correcting and he proposed:

if a channel tunnel were built forces would need to be stationed nearby to forestall a *coup de main*.
148 CAB 53/21, Deputy Chiefs of Staff Sub-Committee report of 4 April 1930.
149 CAB 53/5, Chiefs of Staff Sub-Committee minutes of 24 January 1936.
150 CAB 54/3, Deputy Chiefs of Staff draft Report of 1 December 1936.
151 CAB 53/17, Richmond to Secretary, Committee of Imperial Defence letter of 27 February 1929.

> I should suggest that, as a first step, towards common thinking, the chapter, in each
> Manual, should be identical throughout, and should in no way be particularistic to
> the individual service, or refer to particular types of war.[152]

Trenchard deprecated any such move believing that time would resolve the matter.[153] As the RAF, at this moment, had the greatest divergence of view on the application of the principles amongst the Services, Trenchard probably believed any resolution of the matter would operate at the expense of the RAF. Richmond agreed that concurrence on the matter should not be forced, but resolution, nevertheless, remained a desirable, even essential, goal.[154]

Fundamentally, Richmond was arguing that before training to a common doctrine could be effective, a common language needed to exist amongst the Services. Their differing views of the matter partially stemmed from the limitations of the day's taxonomy when discussing war. Operations, though discussed in British military literature, were not yet couched in terms of the operational art or a distinct operational level of war. This conceptual void was most noticeable in the *Combined Operations Manual* governing the employment of force by all three Services working in unison. Still, even if this conceptual barrier did not exist, it remains problematic whether concurrence would have followed, as space and time worked to a different scale in air operations when contrasted to the more traditional arms. This difference was recognized, but the implications of the differences from a theoretical viewpoint were not so easily grasped. Richmond, elsewhere, argued that the principles of war only addressed the employment of armed force, but whether this employment worked across the spectrum of war remained unsaid.[155] His lectures on tactics delivered at the Senior Officers' War Course are replete with references to the principles of war. As the IDC did not address tactics, the purpose of studying the principles was to inform strategic decision-making and the aims of war as surveyed in the schemes covering the higher direction of war. Whether the principles of war were universal and applicable to each of the Services in the same way was a debatable point, given the differing environments they operated in and the means at their disposal. If, however, such principles were universal, then it was at the strategic level that this unity was apt to exist as the strategies employed by the different arms merged to meet the needs of policy. This was the crux of the problem for Richmond: to teach the higher direction of war when there was little agreement on the direction to be pursued amongst the Services was a circle that could not be squared. Thus, it is not surprising that when examining the principles underlying imperial defence it had proved impossible to reach agreement during

---

152 CAB 53/14, Report of the First Session Imperial Defence College, 22 December 1927.
153 CAB 53/2, Chiefs of Staff Sub-Committee minutes of 23 February 1928.
154 *Ibid.*
155 Richmond, 'Principles of War: A Criticism', *Journal of the Royal United Services Institution*, p. 714.

the course's first year. Richmond promised to tackle the question anew during the course's second session, but resolution was not certain, as his successors would find.[156]

From as early as 1910, the attempt to establish a series of principles governing imperial defence had been made and the Overseas Sub-Committee of the CID issued an updated memorandum on the matter in March 1928. Their conclusion was, notwithstanding the many events and changes since 1910, the principles then enunciated remained valid.[157] Whether the college's efforts to investigate and resolve the question in 1927 was pursued at the behest of the CID—either in support of or as a contrasting opinion—to the work of the Overseas Sub-Committee is unknown. The likelihood is that it was, else there would have been scant reason to address it specifically in the close of year report. Resolving such differences was never going to be easy, but the problem became manifestly harder when the Chiefs of Staff decided in September 1928 that it would be inappropriate for them to lecture on 'The Higher Direction of War' at the IDC whilst serving in office. This was unfortunate as the question was a key component surrounding the many facets of defence policy in general such as imperial strategy, the relation of the constituent parts of the Empire, the organization of forces, and even the role of the staff officer. The nominal reason specified by the Chiefs for not addressing the topic was that it was for the Government of the day to decide such matters.[158] Until the CID considered a report on the question addressing the options available, it was presumptuous of the Chiefs of Staff to address the question and the better course was for others, including Hankey, to talk to the difficulties. Their decision followed an invitation issued by Richmond seeking their participation in the yearly exercise on the problem.[159]

The Chiefs position while correct masked a larger problem: constitutionally, what was the role of a Service Chief to his Minister and to the Government of the day and did he have a responsibility for voicing an opposing view beyond the confines of the Chiefs of Staff Sub-Committee if action was being contemplated of which he disagreed? The experience of Admiral Lord Fisher and Churchill in the late war demonstrated the potential perils if subservience and silence went too far. For the moment, the position was unclear and only in 1938 did the CID conclude that a Service Chief of Staff had a duty to express his views on a question even if not directly asked.[160] Still, the problem of the higher direction of war was not one limited to the

156 CAB 53/14, Report of the First Session Imperial Defence College, 22 December 1927.
157 *Ibid*, 'Some General Principles of Imperial Defence', Overseas Sub-Committee memorandum of 12 March 1928.
158 Thus, when preparing an appreciation in 1930 for possible war with Russia in the Iran-Iraq area, the section covering command arrangements stated 'The question of command will be for decision by H.M. Government on the advice of the Chiefs of Staff, when the occasion arises.' See CAB 53/22, 'Plan for Subsidiary Operations Against Russia in the Perso – Iraq Area', 11 December 1930.
159 CAB 53/2, Chiefs of Staff Sub-Committee minutes of 28 September 1928.
160 CAB 53/16, 'Supreme Control in War', note by Secretary of 26 May 1938.

point where strategy and policy were resolved. It also concerned the relationship of a Commander-in-Chief to his Service counterparts in a theatre of operations and the proper means of communicating orders to a commander engaged or contemplating operations from the home establishments. These were not necessarily new problems, but modern communications, the lethality of current weaponry, the speed required to deal with a 'bolt from the blue', and the newer agencies of government made the matter more acute than heretofore. It was not for the IDC to resolve these issues though their schemes might indicate the scale of the problem. If a 'staff solution' did not yet exist, the airing of the problems and the close association of officers working the issues remained interest to be collected at a future date.

Only when the directing staffs of the Staff Colleges and IDC met at the time of the 1935 Camberley joint exercise did the implications of the twin problems of operational and strategic direction emerge. That the *Combined Operations Manual* was in need of revision was manifestly clear, given the roles of the CID, the COS, and the JPC and their absence from the publication. Additionally, though the *Manual* recognized that multiple types of combined operations were feasible, it only addressed that of an opposed amphibious landing in detail.[161] However, Haining pointed to another failing—the *Manual* ignored that before an evolution could be considered on its operational merits, it needed to be measured against its place in the wider aim of any war. Writing to the COS, he observed:

> It is, no doubt, extremely important to ensure that the proposed expedition is a feasible operation; the lessons of the Dardanelles bring this home very strongly. But this is only one point of a number of others which it is necessary to decide. What is essential at the outset is that the proposed operation should be reviewed in relation to its place in the war as a whole. There are many examples of operations being proposed during 1914/18, which, attractive in themselves, did not fit in properly with campaigns in other theatres, either actual or projected. It is essential that any proposed operation shall, in the first place, make its proper contribution towards the main object of the war; if it does not, then it is a wasted effort even if the object of the expedition itself can be, and is, attained.[162]

There remains a wonderful symmetry in Haining's observation, as contemporaneously just this problem was bedeviling the Chiefs, the Services, and their commanders in the field for this was the moment of the Italo-Ethiopian conflict. In hindsight, the problem and the palliative are clear, but the conceptual barrier present and the linguistic obstacles to be overcome made resolving difficult. Minor strategy and

161 CAB 54/3, Director of Military Operations and Intelligence to Secretary, Committee of Imperial Defence letter of 9 July 1936.
162 CAB 53/27, Commandant, Imperial Defence College to Secretary, Chiefs of Staff Sub-Committee letter of January 1936, Appendix C to Enclosure.

major tactics might both reference the middle spectrum of war, but absent a common taxonomy, doctrine became like Alice's use of language. Churchill desired that the IDC teach a common doctrine of war. Perhaps a better start would have been writing a common dictionary of war.

If the Chiefs proved reluctant to discuss war's higher direction, then some were not so reticent about discussing matters closer to home. Within a month of rejecting Richmond's request to speak on 'The Higher Direction of War', Trenchard delivered his thoughts on 'The War Aim of the Royal Air Force'. His presence was an outcome of Richmond's earlier meeting when the Admiral drew attention to the fact that officers coming to the IDC required a stronger grounding in International Law. This was especially so when questions covering the employment of air and naval forces in war arose where the rights of the non-combatant or the neutral became acute. This shortcoming in officer education featured again as a criticism in Richmond's second and final report as Commandant and by 1929 the college began to provide lectures on the subject to qualifiers.[163]

Whether Trenchard thought officers required a better grounding in the law of war is unknown, but he certainly believed the college's understanding of air doctrine was faulty. This he sought to put right by circulating his thoughts on the war aims of the Air Force to his fellow Chiefs.[164] In short, the Air Ministry view was that the aerial operations contemplated in any distant war conformed to the law governing air warfare, as it then stood, though the law was 'gravely lacking in precision' on what constituted 'indiscriminate bombing'.[165] Buttressing his argument, Trenchard had previously advised that if only two bombs of a hundred dropped found the correct target this was acceptable as long as the Air Force aimed to hit the target. What it could not do was simply drop a hundred bombs on a city trusting the bombs to find the target by chance. Field Marshal Milne, for one, was not quite so sure this was sufficient.

With this as background, the Chief of the Air Staff appraised the IDC qualifiers on how the RAF would contribute to a future war. Sharing something of the views separating the Service's Chiefs, Trenchard remarked, 'Some while ago it became apparent to me that there was some divergence of view between the three Services as to what should constitute the war aim of the Air Force and what are the best objectives for the force to attack in pursuing that aim.'[166] Trenchard left his audience in little doubt continuing:

163 Gray, *Imperial Defence College*, p. 9.
164 CAB 53/14, Chief of the Air Staff memorandum of 2 May 1928.
165 CAB 53/2, Chiefs of Staff Sub-Committee minutes of 30 May 1928.
166 Air 5/169, Air Staff memorandum No. 43, S. 28279, 'The War Aim of the Royal Air Force', Chief of the Air Staff address to the Imperial Defence College, October 1928, p. 3.

> It is the policy of the Air Staff to exploit to the fullest advantage the high mobility of aircraft over long ranges. The advent of air power now enables us to attack many things which are essential to the continuance of the enemy's resistance, but which we could never attack before. The enemy cannot provide effective protection everywhere, and air power enables us now to select for attack vital centres at which the enemy is at his weakest, but which we have hitherto not been able to reach.[167]

Whether Richmond, after hearing the above, concluded that one person requiring an introduction to International Law was the Chief of the Air Staff is unknown, but he was not pleased when Wing Commander Sholto Douglas attending the prior year's course had suggested an immediate attack on Paris to resolve matters during 'Exercise No. 2, War with France'.[168] This was the premier scheme of the first course lasting over two months and, discounting the deleterious effects if the proposal had been adopted for what was intended, the suggestion was just the sort of evidence Richmond doubtless had in mind when noting members required a firmer grounding in the law of armed conflict.[169] Military and Naval Law and the Law of Armed Conflict were staples of the Greenwich Staff Course and each ship carried a copy of *CB 3012: Notes on Maritime International Law*, but more commonly known as the *Blackbook* specifying what actions a commanding officer could legally take.[170] Protecting non-combatants was a given, but so too was the necessity to eschew wanton destruction. These tenets of warfare seemed lacking in the training of some officers and Richmond deplored their ignorance.

In fairness to Douglas, there was an element of the surreal about the scheme posed. Designed as a limited war between France and the British Empire, the primary theatre was the Mediterranean, even though France was but a short distance across the English Channel. Douglas in his memoirs claimed that Richmond aimed to orchestrate a fleet action in the waters of the Near East and his proposal wrecked that objective.[171] There is a large element of truth in the accusation, as attacking the French fleet at Toulon and establishing an advance base in Corsica to facilitate Royal Navy observation of the port, were two of the several actions weighed.[172] In view of all that had transpired during 1914-1918, the war contemplated owed more to the eighteenth and nineteenth centuries than the times Richmond now occupied. Resort to unrestricted submarine warfare by the French was discounted due to the probable ill-effects of neutral opinion whilst assuming that action would remain limited in a war likely to last at least a year between two first rate powers was optimistic to say the least. Richmond was a very good self-taught historian and the actions contemplated

167 *Ibid*, p. 4.
168 Douglas, *Years of Command*, p. 20.
169 CAB 53/14, Report of the First Session Imperial Defence College, 22 December 1927.
170 ADM 186/79, *CB 3012. Notes on International Maritime Law, 1929*.
171 Douglas, *Years of Command*, p. 20.
172 CAB 53/15, Imperial Defence College, Exercise No. 2.

were entirely possible. Yet, the studying of war must account for not only what is possible, but also what is probable and here Douglas was nearer the mark. British air doctrine did not relish the tasks that Richmond's scenario contemplated and attacks of the enemy's vital centres, of which Paris was the foremost, was a foundational tenet.

Of course, Richmond recognized the problems of posing a war between France and Britain. If the presumed crisis governing the scenario was contrived and known to be so, the problem of French control of Syria had featured as a source of Anglo-French dispute at Versailles—something Richmond would have appreciated, nevertheless, from those earlier Foreign Office contacts who had lectured at the Senior Officers' War Course during his tenure at Greenwich.[173] Fundamentally, though, the exercise served as a vehicle for qualifiers to consider the broader nature of such a war. Dividing the problem of an Anglo-French War into a number of discrete questions, the problem was worked by considering its broadest elements first. Syndicates designated to assess the British and French strategies were first asked to write an appreciation discussing the strategy to be followed using the separate decision trees provided which mapped the courses of action available. Next, the syndicates had to consider the problem from the opposing point of view. Thus, the British syndicates now drafted their appreciation of probable French action whilst those playing France did the same for Britain.

By the exercise's third phase, the syndicates were considering how to achieve command of the sea, if viewing the problem from the British point of view, or how to dispute such command, if evaluating the question from the French perspective. Following on this, the syndicates weighed the benefits and costs of securing an advanced base in Corsica in furtherance of the British objective of securing command of the sea. By the fifth phase of the exercise, British syndicates were evaluating the specific forces and measures needed to defend Egypt and Palestine, in the first instance, and ultimately how to eject the French from Syria. In turn, French syndicates were weighing those measures necessary to deny Britain her objectives and advance the French strategic goal. By the ultimate phase, British syndicates were preparing a detailed general appreciation for Cabinet and CID consideration built upon the earlier problems weighed. Meanwhile, their counterparts were preparing a corresponding appreciation for consideration by the Conseil Supérieur de la Défense Nationale.

A précis outlining the political situation of 1934 was provided to the syndicates, and the forces and defensive preparations that had been adopted from the actual moment of consideration were summarized. Thus, the British garrison of the Rhine was no longer present but had returned to England. In accomplishing all this, the qualifier secured an understanding of the trade-offs, costs, risks, and likely benefits of the plan ultimately prepared. They understood what was required (i.e., the object) and weighed means to ends to accomplish. They also would have an appreciation of what the enemy's goals were, the forces available to him, and his possible actions. For those who had previously studied at the Royal Naval Staff College or taken the SOWC

173 Goldstein, *Winning the Peace*, pp. 269-77.

much was familiar about the above. Yet, where the ultimate aim at Greenwich was to appreciate the problem as a prelude to preparing an operational plan, the general appreciation an IDC qualifier wrote sought to inform the political establishment of what was realistically possible and the risks associated with the courses of action discussed.[174]

The above example depicts in general terms the employment of an exercise to teach the mechanics of preparing a general appreciation. Along the way, the qualifier learned not only the process, but also something of the essential considerations weighing at the strategic level of war. Where the scheme involved Britain in a war with allies, the number of syndicates posing as Britain and the defined adversary were reduced by one. This allowed two sections to address the problem from their perspective and shed light on the complications likely to arise.

Into the 1930s, the defence of British interests in the Near East was receiving attention and, particularly, her place in Palestine and the Trans-Jordan. The problem of war with Russia and Japan continued to be regular item of study, and, in 1938, qualifiers were examining how Britain and the Empire should prepare for a European war arising in 1939. The specific issue used to provide the context for the war's starting point was the Sudetenland, thus, mirroring events of the Munich Crisis.[175] With the qualifiers from the Air Force and even Air Commodore Grahame Donald, the directing staff air member, having departed for operational duties, Longmore, the Commandant, believed the same fate awaited him. Those remaining wrote appreciations for the mock war contemplated and argued over the likely results for the plan developed.[176] In the event, Admiral Hugh Binney, Longmore's designated successor, was already on the scene preparing himself for his forthcoming duties and would have assumed command earlier if Longmore's departure had been accelerated.[177]

This was not the first time when a scheme under consideration at the IDC anticipated life's events, as a similar congruence occurred in 1935 whence at the very moment the college was weighing war with Italy, the invasion of Abyssinia occurred. Again, syndicates considered the problem facing Britain and Lieutenant Colonel Giffard Martel,[178] the gifted sapper, made much of playing the part of Mussolini when relating the story to Liddell Hart.[179] The college's findings were submitted to the Chiefs of Staff and Haining, the Commandant, believed the incident was fortuitous

174 CAB 53/18, Exercise No. 2.
175 Longmore, *From Sea to Sky*, pp. 188-89.
176 *Ibid.*
177 CAB 53/41, Commandant, Imperial Defence College to Secretary, Committee of Imperial Defence letter of 28 September 1938.
178 Later Lieutenant General Sir Giffard Le Quesne Martel (1889-1958). Staff Course Camberley, 1921-23; Staff College Quetta, 1930-33 and Imperial Defence Course, 1935.
179 Martel to Liddell Hart letter of 29 October 1935, Liddell Hart Papers, LH 1/492, LHCMA.

as it was not always possible to correctly anticipate which schemes were in fact the more salient studies for qualifiers and country alike.[180]

Still, the culminating event for the 1938 course was an examination of the overall defence requirements of the British Empire where qualifiers were asked to examine the issue in light of all the previous problems examined during the course. Qualifiers, knowing something more of the means and ends available towards defence, now considered the extent to which Britain could engage in further disarmament agreements. Captain William Tennant directed the exercise and amongst the problems posed for the qualifiers to resolve was the degree the limitations in naval armaments reached with Germany in 1935 should inform British preparations for the negotiations planned.[181]

As was the custom at the Royal Naval Staff College, the assignments of the participants in the game approximated those of their real-life duties with civilian qualifiers playing the roles of key Cabinet members and uniformed personnel acting as one the Chiefs of Staff or their assistants.[182] Thus, the scheme was heavily weighted towards the military problems of defence with four syndicates playing the part of the COS whilst only one acted as the Cabinet.[183] Captain Leonard Murray,[184] RCN, played the part of Chief of the Naval Staff in the first syndicate whilst Captains Richard Symonds-Tayler,[185] Guy D'Oyly-Hughes,[186] and Gerald Warner acted in similar

---

180 CAB 53/27, Commandant, Imperial Defence College to Secretary, Chiefs of Staff Sub-Committee letter of January 1936.

181 Imperial Defence College, Exercise No. 10, 1938, p. 22, Tennant Papers, NMM/TEN/42/3.

182 Ellwood, IWM Sound Recording 3167 of 20 February 1978.

183 'Imperial Defence College, Exercise No. 10 (1938), Syndicate List', Tennant Papers, NMM/TEN/42/2.

184 Later Rear Admiral Leonard Warren Murray (1896-1971). Staff Course, 1928; Naval Staff Officer in HMCS *Stadacona*, 1931-32; Commander (D) Eastern Division in HMCS *Saguenay*, 1932-34; Operations Division, 1936-37; Imperial Defence Course, 1938; Captain, 1938; Rear Admiral, 1941 and retired, 1946.

185 Later Admiral Sir Richard Victor Symonds-Tayler (1897-1971). Staff Course, 1927-28; Naval Mission to Greece, 1930-31; Plans Division, 1931-32; Captain, 1936; Senior Officers' Technical Course, 1937; Imperial Defence Course, 1938; Commanding Officer, *Centurion*, 1939; Director, Training and Staff Duties Division, 1941-42; Rear Admiral, 1946; Vice Admiral, 1949; Commander-in-Chief, America and West Indies, 1949-51; retired, 1952 and Admiral, 1953.

186 Captain Guy D'Oyly-Hughes (1891–1941). Commanding Officer, *C 3*, 1917, *E 35*, 1917-18, *E 49*, 1918-19, and *L 41*, 1919; War Staff Course, 1920-21; War Staff Officer, Atlantic Fleet, 1921-22; War Staff Duties to Rear Admiral, Submarines, 1922-24; Commanding Officer, HMS *Conquest*, 1924-25; Aviation Course, 1930; Directorate of Training, Air Ministry, 1931-33; Captain, 1932; Senior Officers' Technical Course, 1933; Tactical Course, 1934; Commander, First Submarine Flotilla, 1934-36; Chief Staff Officer in *Drake*, 1936-38; Imperial Defence Course, 1938; Tactical Investigation Section in Tactical School, 1939; Higher Commanders' Course Aldershot, 1939; Commanding Officer, *Glorious*, 1939-40 and lost with ship.

capacities for the other three syndicates.[187] Meanwhile, Captains Maurice Mansergh, William Parry[188] and John Howson[189] acted as assistants to the COS within each of the separate syndicates.[190] Murray *et al* were asked to tender advice to the Government being played by Syndicate No. 1 on what Britain could offer and expect in any multi-party disarmament negotiations. Of these officers, Parry was seen by Longmore, as the most exceptional of the naval qualifiers though inclined to be argumentative.[191]

Thus, the syndicate featured as the primary method of parceling out work and considering any problem. For the hypothetical wars under consideration, the syndicates mirrored the Defence Committees of the powers engaged and not the operational staffs of fleets, air forces, and military formations represented at Greenwich. This allowed political and economic factors to receive their due weight in the strategic assessments prepared and took advantage of the civilian qualifiers unique perspectives on the issue at hand.[192] Perhaps the best exercise for which documentary evidence remains of a syndicate's actual appreciation to an IDC scheme is one from 1928 covering the possible lines of war with Japan occurring in 1938. Exercise No. 4, as it was styled, was the primary problem of that year's second term and posited British actions to maintain her presence in North China and the Yangtse River environ against Japanese encroachment. Six syndicates of four members each evenly divided between British and Japanese teams weighed the issue.[193]

187 'Imperial Defence College, Exercise No. 10 (1938), Syndicate List', Tennant Papers, NMM/ TEN/42/2.
188 Later Admiral Sir William Edward Parry (1893–1972). Five firsts in examinations for Lieutenant; Squadron Torpedo Officer in *Calcutta*, 1919-20; War Staff Course, 1922-23; War Staff Officer in *Curacoa*, 1923-24; *Vernon*, 1924-26; Squadron Torpedo Officer in *Curacoa*, 1926-27; Staff Officer to Rear Admiral Submarines, 1927-28; Tactical Division, 1929-32; Tactical Course, 1932; Captain, 1934; Senior Officers' School Sheerness, 1935; Senior Officers'Technical Course, 1936; Short Anti-Submarine Course, 1936; in Command, Anti-Submarine School in *Osprey*, 1936-38; Imperial Defence Course, 1938; Commanding Officer, HMS *Achilles*, 1939-40; Rear Admiral, 1944; Vice Admiral, 1948; Admiral, 1951 and retired, 1952.
189 Captain John Montagu Howson (1893-1959). Flag Lieutenant in *Pembroke*, 1918-20; Squadron Signal Officer in *Barham*, 1923-24; Signal School, 1924-26 and 1927-28; Squadron Wireless Telegraphy Officer in *Warspite* and *Barham*, 1928; Signal Department, 1929-31; Tactical Course, 1931 and 1937; Staff Course, 1932; Commanding Officer, *Witch*, 1933-34; Senior Officers'Technical Course, 1935; Staff Officer Operations and Intelligence in *Victory*, 1935-37; Captain, 1936; Imperial Defence Course, 1938; Tactical Investigation Section of Tactical School, 1939; Flag Captain and Chief Staff Officer in *Effingham*, 1939-40 and retired, 1946.
190 'Imperial Defence College, Exercise No. 10 (1938), Syndicate List', Tennant Papers, NMM/ TEN/42/2.
191 ADM 196/92 (Parry).
192 Longmore, *From Sea to Sky*, p. 177.
193 CAB 53/18, Exercise No. 4, Syndicates.

Syndicate No. 2 composed of the former England rugby player Captain Norman Wodehouse,[194] RN, Lieutenant Colonel John Mitchell-Baker, Union of South Africa Defence Force; Major Henry Loyd, Coldstream Guards, and Frederick Shedden of Australia considered matters. The problem as ever was how to defend Hong Kong and Singapore until the arrival of reinforcements in the theatre of operations in the face of Japanese aggression. The 1928 exercise concluded that Hong Kong was particularly exposed and, here, the restrictions of the Washington Naval Agreement governing the establishment of fixed fortifications were viewed as the primary obstacle to any successful defence. By 1929, the problem was thought more manageable, as Britain began redeploying forces originally sent to Shanghai in 1927 to protect her interests in the International Settlement to the Crown colony.[195]

Shedden's syndicate assumed that the defence arrangements supporting Singapore would be completed as programmed for 1938 and not curtailed between 1928 and the specified date of war. Wodehouse, as the naval member, was responsible for preparing the maritime portion of the appreciation. It advocated attacking Japanese commercial shipping with the two-fold purpose of denying Japan the economic wherewithal to sustain her economy or forcing her to defend such trade by forcing naval battle. Either result was deemed advantageous to Britain though the prospect of naval battle should achieve the desired end-state more rapidly. Mitchell-Baker and Loyd, as the military members, advocated establishing an advance base north of Hong Kong to support the operations contemplated. Wodehouse eschewed this view believing the effort required to acquire such a base was not commensurate with the likely costs and probable benefits following.

The conclusions the syndicate reached was that Singapore was safe as long as the main fleet sailed quickly and arrived promptly. This did not mean that the fortress would not face bombardment, but on the assumption that her fixed defences were as contemplated and air support was available, subjugation was not apt to occur. The plan as developed avoided land operations in China, as Japan could reinforce her units there more easily than could the British. However, enticing Chinese support through an active propaganda operation was considered a promising possibility. Something, in fact, actually considered by the COS when at a future date fiction approached fact.[196]

194 Later Vice Admiral Norman Atherton Wodehouse (1887–1941). Six firsts in examinations for Lieutenant; War Staff Duties in *Hawkins*, 1919-20; Fleet Gunnery Officer in *Hawkins*, 1920-21; Naval Ordnance Department, 1922-24; Captain, 1926; Senior Officers War Course, 1927; Senior Officers' Technical Course, 1927, 1931 and 1935; Tactical Course, 1927 and 1935; Imperial Defence Course, 1928; Commanding Officer, *Ceres*, 1929, Calypso, 1930 and HMS *Champion*, 1931; Royal Naval College Dartmouth, 1931-34; Flag Captain in *Barham*, 1935-37; Rear Admiral, 1937; Vice Admiral and retired, 1940. Recalled for war service and lost at sea whilst serving as Commodore of convoy, 1941.

195 Sir Frederick Geoffrey Shedden Papers, National Archives of Australia, Canberra, Australia, NAA: A5954, 19/5.

196 CAB 53/5, Chiefs of Staff Sub-Committee minutes of 24 July 1934.

Ultimately, the British Empire would prevail as she was not depended on her Far Eastern trade and so Japanese attacks against this commerce would not be critical. As long as her territories remained un-invaded, the conflict would degenerate into a commerce war favouring the British Empire which could upset Japanese trade in a manner not available to her opponent.[197] The syndicate's appreciation did not address how Hong Kong was to be defended and the implied conclusion was that it could not. Most likely, its restoration to British sovereignty was assumed to follow the conclusion of the negotiated peace that would end any Anglo-Japanese War; a view that Arthur Balfour had reached in 1921.[198]

Supporting the conclusions of the appreciation were statistics depicting the nature and volume of trade for Britain and Japan derived from Asia and the Pacific regions. The similarity of Britain and Japan's shared strategic positions and problems (i.e., island empires astride continental powers heavily depended on imported goods to sustain their economies) highlighted that examining a possible Japanese war offered lessons to Britain under other scenarios. Serious gaps remained in the appreciation, such as the exact role air forces were to play, but this probably owed more to the limitations of time and the make-up of the syndicate itself. After all, a syndicate of four cannot be expected to master what takes a properly constituted staff with ample time to resolve. Admiral Richmond thought enough of the syndicate's conclusions to forward them to others and, ultimately, copies were provided to the Plans Divisions of the three Services. Moreover, in the staff solution to the problem forwarded to the Chiefs of Staff Richmond incorporated Shedden's economic analysis so impressed was he by its value.[199]

While fault may be levied with the assumptions made and the conclusions reached, the value of the appreciation to this study is that it demonstrates the qualifiers employing the skills they learned at the IDC and confirms that the subjects covered in lectures were not idle discourses. Yet, while the IDC considered real-world problems, there were limits to how far the Chiefs of Staff Sub-Committee would go in using their talents. Air Commodore Richard Williams,[200] Royal Australian Air Force, attending the course in 1933 saw dated papers that were less recent than those he reviewed previously when serving as his Commonwealth's Chief of the Air Staff. Security was the likely reason, though perhaps not the only one as discussed below, for the IDC's broad membership encompassing as it did members from the Civil Departments posed a dilemma in information sharing.[201]

197 Shedden Papers, NAA: A5954, 19/5.
198 CAB 34/1, Arthur J. Balfour, 'Note on Military Situation in the Far War', 3 May 1921.
199 CAB 53/19, 'Exercise No. 4, Examination of the Economic Position of the British Empire and Japan in a War in 1938', paper by F. G. Shedden.
200 Later Air Marshal Sir Richard Williams (1890-1980). Staff Course Camberley, 1923 and Imperial Defence Course, 1933.
201 Gray, *Imperial Defence College*, p. 33.

Trenchard's criticism that the initial course had attempted too much was accepted by Richmond and in the following years he and his successors sought to emphasize the quality of appreciation over the quantity of appreciation. Bartholomew, Richmond's immediate successor, explained the approach being followed in a letter to Liddell Hart, the journalist and *bête noir* to many in the Army:

> I agree with you that localization of study on one or two campaigns is almost dangerous, that is to say, that war is such a many-sided thing that if study is limited in this way erroneous deductions will be made. I do not, however, in saying this, advocate superficial study of many campaigns. The beginner must learn to study first by concentrating on a few, and gradually expand his knowledge.[202]

During Bartholomew's tenure, the Japanese scheme was the major exercise of the year lasting over four months. When shortly becoming Director of Military Operations and Intelligence at the War Office, Bartholomew drew upon the findings of the IDC scheme when the Chiefs of Staffs weighed the emergence of a sudden Japanese threat in April 1933. When the CIGS noted that an IDC scheme showed how difficult conducting operations beyond Hong Kong against the Japanese would be, Bartholomew concurred citing the shipping required to support any move and the intelligence necessary to develop a detailed operational plan.[203] Of course, such schemes were not an official pronouncement on the problem, but neither were they the ill-considered ramblings of the uninformed and the Chiefs of Staff agreed with Admiral Preston that the release of IDC exercises and appreciations were to be restricted.[204]

Richmond, himself, lectured on the nature of limited and unlimited wars. Sir Henry Tizard spoke to the possible substitutes of war and Austen Chamberlain, the current Foreign Secretary, and Viscount Cecil, the Chancellor of the Duchy of Lancaster, addressed the place of the League of Nations. Though speakers at the 1927 course, Richmond would have secured similar, if not the exact presentations, for Wodehouse's and Shedden's course of the following year. In addition to lecturing on the nature of war, Richmond addressed the object of war in a series of four lectures covering the historical practices of the eighteenth and nineteenth centuries and the principles governing the Empire's defence. Exercise No. 8 covered a possible British Empire war with Russia and, here, Richmond prefaced the qualifiers' work by lecturing on the strategy of the Crimean War.

As to what the qualifiers were reading in support of their studies in 1927, a healthy regime of official and un-official works were prescribed dovetailing with the subjects of

202 Bartholomew to Liddell Hart letter, no date, but *c.* 1929, Liddell Hart Papers, LH 1/45/5, LHCMA.
203 CAB 53/4, Chiefs of Staff Sub-Committee minutes of 11 April 1933.
204 *Ibid*, Chiefs of Staff Sub-Committee minutes of 27 June 1934.

the syllabus. Background theory on naval warfare was provided by recourse to Mahan's *The Influence of Sea Power on History*, Corbett's *Some Principles of Maritime Strategy*, and James Thursfield's *Naval Warfare*. Readings supporting the qualifier's appreciation of the air included an Air Ministry memorandum on 'Air Power and Imperial Defence' issued in 1926, Alan Cobham's *My Flight to the Cape and Back*, and *Aircraft and Commerce in War* by the Air Ministry civil servant James Spaight. Amplifying war's higher direction were readings from Sir Frederick Maurice (*Government and War*), Ellison (*The Perils of Amateur Strategy*), Robertson's multi-volume memoir, *Soldiers and Statesmen* and Sir John Marriott's *The Mechanism of the Modern State*.

Ellison's work is especially noteworthy for the critique it made of British war making during the World War where Ministers ordered military action without receiving professional advice, and the Services, in turn, pursued military strategies without coordinating their plans with the other offices of state. Richmond reviewed the book upon its release and just before becoming Commandant.[205] Typical of him, he did not shy from voicing his own views on the subject whilst capturing Ellison's treatment of the problem. Ellison, only recently retired, was a veteran of the ill-fated Dardanelles venture. That episode, though strongly amplifying his argument, was not the sole argument employed by the author to demonstrate poor defence control during the war and his survey concluded that the faults committed resided as much with the 'brass hats' as with the 'frocks'. Ellison believed that root and branch changes in the machinery of control were required with strategy and administration needing to be treated as separate responsibilities in war. Moreover, an enquiry was required to determine the most effective way of conducting war in a democracy. If not, Britain risked making the same errors again. A view Richmond wholeheartedly endorsed. Given such an indictment from the one-time official historian of the campaign, well could the Chiefs of Staff conclude that their presence at the IDC discussing war's higher direction was a fence that need not be taken.

Works examining the British Empire included A. V. T. Wakely's *Some Aspects of Imperial Communications*, Fuller's *Imperial Defence, 1588-1914*, Alfred Zimmern's *The Third British Empire*, and King-Hall's *Imperial Defence: A Book for Taxpayers*. Other readings offered contemporary views on Russia and Japan whilst a final series published under the general rubric of *To-day and To-morrow* sought to shed light on the future. Amongst these titles were those by the philosophers Bertrand Russell (*Icarus, or the Future of Science*) and Ferdinand Schiller (*Tantalus or the Future of Man*). *Daedalus or Science and the Future* by the early geneticist John Haldane described the influence of science in the World War and the possibilities—good and evil—that it held for the future, whilst Basil Matthews' *The Clash of Colour: A Study in the Problem of Race* addressed the likelihood of conflict arising along racial lines.[206]

---

205 Herbert Richmond, 'The Perils of Amateur Strategy', *The Naval Review*, Vol. XIV, No. 3, August 1926, pp. 548-60.
206 Alanbrooke Papers, 3/7, LHCMA.

Classical allusion in the titles of contemporary works was not a mere literary contrivance at a time when many claimed an education stronger in the knowledge of the Greats than our own. If Major Brooke's reading list does not include Liddell Hart's *Paris or the Future of War*, then the cause may owe something to the fact that he was already familiar with it and the other military works recommended and felt no need to record their reading rather than being an obvious omission. For Brooke's list is striking for the few military works identified with Lawrence's *Revolt in the Desert* proving a rare mention. By all regards, the reading undertaken was a mixture of the practical and the theoretical, stressing the place of the British Empire within the greater political and societal context of its times. It complemented the course's focus on the strategic and political aspects of war, and when contrasted with what was taught contemporaneously at the Services' own schools, was clearly not about operations and tactics.[207] Though borrowing heavily on the Senior Officers' War Course that he previously directed at Greenwich, Richmond took the IDC to a higher plane of learning. Absent were works addressing tactics and maximizing the strategy of a single arm and in their place were those surveying war on the broadest scale. If some of the works used smack of the pseudo-scientific and appear esoteric, then they only reflected the temperament of their age. Science and historical analysis may be neutral disciplines, but their exponents, trapped within the mores of their time and shaped by their own experiences, are rarely so. If a common lament of the General or Admiral is that he all too often fights the next war with the mindset of the last, then the curricula and readings of the course indicate that Richmond, Dill, and the other founding members of the IDC made a conscious effort to avoid such pitfalls focusing their attention on the future.

Richmond, not unlike Fuller, firmly believed that war was capable of scientific analysis and understanding. Given the number of works with 'science' in their title appearing in the IDC reading list (and others that did not, such as, Liddell Hart's *A Science of Infantry Tactics Simplified*), it was a view manifestly shared by many others. Yet, where the engineer and the scientist employed reason and experimentation to test their theories, the strategist and tactician relied upon history.[208] That different laws operated were accepted, but that laws operated was the important thing. This explains why the principles of war were fundamental to early twentieth century military thought, as they provided the foundation upon which war could be understood no less than Newton's Laws of Physics underscored the natural world. Today, military writing is wont to emphasize that waging war is an art and not a science. This is not necessarily an advance upon Richmond and Fuller, and given its heavy reliance on jargon rather than plain language surrounding current military writing, decidedly

<hr>

207 Per Admiral Preston, minor strategy and major tactics were taught at the Staff Colleges. See CAB 53/5, Chiefs of Staff Sub-Committee minutes of 1 November 1934.
208 David French, 'Doctrine and Organization in the British Army, 1919-1932', *The Historical Journal*, Vol. 44, No. 2, p. 503.

this is so. This is not to claim that early twentieth century military writers did not possess their own *lingua franca*, but a key objective of the IDC was to reduce this to a minimum ensuring that leaders understood what was being proposed in any appreciation presented.

That Richmond and his ilk were wrong in believing that war could be made a science is all too true, but the important thing was that they believed it could be better understood. History was a poor foundation upon which to build their science because our understanding of the past is incomplete, at best, and plain wrong, at worse. Yet, for the moment, it was the best foundation possible. If some of the period's leading lights reached their conclusions first and read their history to conform to their hypotheses, others, such as Richmond, truly did aspire to make their art, science. Richmond (and Corbett and John Knox Laughton before him) employed original sources and made their efforts available to a broader audience through the Navy Records Society.[209] Their goal was never to fashion a new navy so much as to better understand the principles governing the present one's proper use. Richmond was not the scientific historian that he claimed. Yet, neither was he a scientific polemicist in the fashion of Fuller and Liddell Hart and not without reason has the latter's official biographer characterized him as the *'Alchemist of War'*.

How effective Richmond was as a Commandant is difficult to measure. There can be no doubt that he brought a degree of academic rigour to its proceedings though he himself was not of the academy. As a layman, his understanding of history, law, economics, statistics, and philosophy was impressive and matched with a brilliant pen, he could convey what he actually meant and thought in a manner more senior officers, at times, found irksome and his juniors found dogmatic and even spiteful. Brooke-Popham weighing the balance on what made a good Commandant once said, 'One does not want a brilliant scholar for the job; Richmond was and did not make a great success of it.'[210] The essence of Brooke-Popham's conclusion being one might be correct in the classroom and wrong on the battlefield. Other attributes were important, too, such as collegiality and a receptiveness to newer ideas, to newer arms.

Speaking many years after the moment, Auchinleck when asked his views on the first course remarked that it was Dill, as lead instructor of the directing staff, who carried the show.[211] This compliment to the very capable Dill slights the contribution of Richmond who took a strong hand in the naval side of the IDC's proceedings— perhaps too strong a hand. Dickens in a position to know the qualities that both officers offered believed that Dill was a 'splendid foil to Richmond and the combination of their differing mentalities and systems of work was a big factor in establishing

209 Andrew Lambert, *The Foundations of Naval History* (London: Chatham Publishing, 1998), pp. 142-43.

210 Brooke-Popham to Sir Edward Ellington letter of 13 February 1936, Brooke-Popham Papers, 4/3/22, LHCMA.

211 Warner, *Auchinleck*, p. 269.

a most successful method of the wide study which the I.D.C. was called upon to undertake.'[212] As with re-establishment of the Staff College and the SOWC and the founding of the Tactical School, sustaining the momentum of the IDC was never going to be easy. Richmond and his staff—Dill, Joubert de la Ferté, and Dickens were a strong cast to follow.

In late 1928 as his period as Commandant was nearing completion, Richmond was informed that he had been awarded the Chesney Medal by the Royal United Services Institution for his work *The War of 1739-48*. Though the parameters of the award required that it be rendered to an author of a particular volume, 'the Council were greatly influenced not only by the merits of the one publication but by your many services in promoting and advancing naval military science and literature.'[213] Though the letter does not specifically mention his formative role at the IDC, the timing was not accidental and the award can be seen as independent recognition in all that he attempted and achieved whilst serving at Greenwich and the IDC.

The value of a school is measured in its students no less than by its staff and the immediate problem was just keeping numbers up. Attendance by the Dominions was one concern, but even the Services—particularly the Air Force—had issues with meeting their quota. Whether the college would have been better served by accepting the 32 most eligible candidates each year regardless of status is difficult to weigh at this point. Against the bias that might have arisen in its proceedings was the problem that having too few present lessened the points of view offered in discussion. One thing was certain; too few Commonwealth officers made consideration of imperial defence a doubtful proposition, as Brooke-Popham warned the Chiefs of Staff in early in 1932.[214]

Following Richmond's departure in late 1928, the breadth of the course grew if not the actual number of participants. British planners were expected to weigh the utility of war against the economic dislocation that might arise and, consequently, the financial costs of any action were stressed. To this end, Fayle lectured on the economic aspects of war. During the 1934 course, the session where the best documentary evidence is available on the scope of studies, Fayle offered no less than nine lectures. He proved to be the year's most frequent contributor offering perspective on the economies of Italy, the Soviet Union, Japan, France, Belgium and more. Indeed, for every problem considered, Fayle provided the economic underpinnings of the analysis. These, then, were complimented by lectures in the political and strategic issues arising for the country in question given by experts from the Foreign Office, academia, or Service attachés.

212 G. C. Dickens, 'Richmond', *The Naval Review*, Vol. XL, No. 3, August 1952, p. 337.
213 Royal United Services Institution to Richmond letter of 3 October 1928, Richmond Papers, NMM/RIC/7/2.
214 CAB 53/22, Commandant, Imperial Defence College to Secretary, Chiefs of Staff Sub-Committee letter of 29 January 1932.

Before all this was attempted, the general situation facing Britain and the Empire featured. Haining, the current Director of Military Operations and Intelligence, spoke to the world situation whilst senior officers from each of the Services addressed the balancing of force commitments to responsibilities. In this case, Vice Admiral Charles Little, the quiet-mannered DCNS and an early member of submarine service, spoke of the naval situation, Sholto Douglas addressed the air, and Haining covered the military problem. Brigadier Brooke discussed how to prepare a combined appreciation, Hankey spoke to the role of the CID, and Major General Arthur Temperley, serving on its Permanent Advisory Committee, surveyed the place of the League of Nations. Mobilization of the industrial base and, the employment of manpower were additional topics necessary for a qualifier to understand before considering specific country problems as were International Law, fuel, finance, science in war, public opinion, and intelligence. Concluding the preparatory portion of the course, Britain and the constituent parts of the Empire were canvassed.[215]

The strategic problems considered in 1934 included the Far East, Egypt and the Middle East, Europe, the United States and South America, India and Afghanistan, and the Soviet Union. Specific lectures followed on the nature of the problems posed for each with its strategic and economical contexts emphasized before considering the matter from the perspective of the British forces likely to be involved. Some problems were thought to be primarily naval (e.g., Far East) whilst others were deemed military (e.g., India and Afghanistan). Of course, some questions were held to be truly joint, such as those concerning Europe where all arms had their part to play. For the naval officer attending, those schemes covering India and the Middle East broke new ground from what was specifically covered at Greenwich, and though the Far East was certainly familiar ground to all attending, the political and strategic breadth of the problems posed was perhaps new for most. Certainly, the final portion of the course where war's higher direction and issues of disarmament now appeared made this true. In 1934, no less than six lecturers surveyed the 'Higher Direction in War', whilst Sir Warren Fisher, the Permanent Secretary to the Treasury, and, thus, the senior British civil servant discussed the machinery of government.[216]

Propaganda as an element of 'soft' power has already been mentioned in passing and though qualifiers and staff did not discuss it in terms of 'information operations' or 'shaping the battle space'—its modern appellations—its role was just as similar. Simply put, Britain did not possess the military and naval forces to be strong everywhere to guard against foreign attack or domestic subversion. Other less overt means of reducing these risks were available to indirectly shape opinions and qualifiers needed to think beyond their previous limited horizons. Favourable foreign newspapers might receive subsidies, academics or public notables could be encouraged to lecture in other

215 CAB 53/5, Chiefs of Staff Sub-Committee minutes of 1 November 1934, Enclosure, Imperial Defence College, Summary of Lectures in 1934.
216 *Ibid.*

countries explaining the British points of view, and reciprocal exchanges promoted. More directly, the broadcasting of news via wireless transmissions with the aim of placing British actions in a positive light and, hopefully, enhancing her perceived prestige might be pursued through the nominally independent, but Government licensed and funded British Broadcasting Corporation. Yet, the propaganda attempted was for naught if not supported by the actions of British officials locally.[217] Educating officers on the role and use of propaganda was thus key. Of course, there were limits to what Britain could attempt in the absence of war or a declared emergency domestically, but externally the options for exercising propaganda were more available with the Foreign Office taking the lead. To this end, the BBC coordinated its transmissions directed at Germany with the Foreign Office.[218] Accordingly and during the latter part of the 1930s, Sir John Reith, the Director-General of the BBC, spoke at the college on multiple occasions to acquaint qualifiers on the role and power of broadcasting.[219] Reith's visit was portentous for another reason. As the Director-General of the BBC, he was likely to lead any Ministry of Information during wartime when the BBC would no longer operate under guidance provided by the Foreign Office but would transfer to full Government control with matters of propaganda coming to the fore.[220]

Though the Staff Colleges had a healthy regime of interchange amongst themselves culminating in the yearly combined exercise, such formal links never developed with the IDC. The most that was attempted was to foster touch amongst the directing staffs of the institutions so that ideas could be exchanged and common problems discussed.[221] To this end, the IDC shared the settings of its major exercises and staff solutions developed with the Staff Colleges, the Service Ministries, and the Royal Naval War College. This included sharing points of special importance highlighted in its exercises with the relevant department concerned.[222] Still, it would not share its work with the Tactical School or with the Army's Senior Officers' School. The omission was deliberate and stemmed from the differing objectives of the Staff Colleges and the IDC to the broader defence educational establishment.[223] Preston was the initiator of this restriction, but Haining took the measure further and expressed

217 CAB 51/11, Committee of Imperial Defence, Standing Official Sub-Committee for Questions Concerning the Middle East, 'Measures to Influence Minor Powers and Arab States' 29 April 1938.
218 CAB 24/281, Secretary of State for Foreign Affairs to Cabinet memorandum C.P. 284 (38) of 8 December 1938.
219 Gray, *Imperial Defence College*, p. 9.
220 Programmes in Wartime Committee minutes, 22 October 1937, BBC Archives. http:// www.bbc.co.uk/archive/ww2outbreak/7950.shtml; accessed 21 December 2011 and CAB 68/1/9, Minister of Information Report, 10 October 1939.
221 CAB 53/5, Chiefs of Staff Sub-Committee minutes of 24 January 1936.
222 CAB 53/43, Report by the Commandant of the 1938 Course of the Imperial Defence College, 22 December 1938.
223 CAB 53/23, Commandant, Imperial Defence College to Secretary, Committee of Imperial Defence letter of 29 May 1934.

reservations about developing even closer ties with the Staff Colleges.[224] Exactly why Haining adopted this attitude is not clear. Probably, it stemmed from the viewpoint that the education afforded by the differing colleges was progressive and not complimentary. As many IDC qualifiers were already graduates of their Service's respective courses, little was to be gained.[225]

This outlook is not without substance, but a further consideration was that he was either trying to protect the independence of the IDC, fearing its greater objective would become circumscribed, or sought to ensure that the Staff Colleges were not compromised, or both. This writer is inclined to think that protecting the independence of the IDC was his primary motivation. Concurrent to his conclusion that its qualifiers should not participate in tandem with the Staff Colleges was his proposal that the IDC should have the review of concrete problems removed from its terms of reference. His fear in this instance was if ever an outside authority audited the workings of the college, the lack of effort in this area would be evidence against the value of the college.[226] This Haining was not willing to risk and a similar fear may have motivated his conclusion that cooperation with the Staff Colleges at the qualifier level was a risk too far.

In any event, his position found favour in some circles, particularly amongst the Air Staff. When the Admiralty proposed that greater emphasis should be stressed on developing the specific means of conducting combined operations, the Air Ministry objected believing too much time already was spent considering the problem at the Staff Colleges when the likelihood of such evolutions against a military power of the first rank belonged to another age.[227] Hindsight would demonstrate this view was premature, but in fairness to the Air Ministry, only so much could be accomplished in their single year course at Andover. Meanwhile, at the IDC the view of naval instructors was that whilst the consideration of the *means* of amphibious operations and minor tactics were not appropriate fields of study, the broader strategic issues of such operations were. Gerald Dickens speaking to the qualifiers expounded the position best:

> Our studies have shown us that in practically all the possible major wars of the future, the necessity for the fleet to remain at sea for long periods stands out with the utmost clearness. Very high endurance in all classes of ships composing a fleet is essential. We see it in a war with Japan where we want to get an advanced base. The army has to be transported to the neighorhood of that base and its lines of communications protected while it is fighting for it. The fleet, then, must be prepared to remain at sea until the base is available – a matter perhaps of some considerable time.

224 CAB 53/5, Chiefs of Staff Sub-Committee minutes of 24 January 1936.
225 CAB 53/27, Commandant, Imperial Defence College to Secretary, Chiefs of Staff Sub-Committee letter of January 1936.
226 *Ibid.*
227 Maund, *Assault from the Sea*, p. 6.

We see it in the same war where we want to cut Japan's vital communications.

We see it in a war with France where the protection of our trade is concerned. We see it in a war with Russia if the Navy is to play its part by operating in the Black Sea with no base nearer than Malta or in the Baltic with the nearest base in the British Isles. We see it in a war with the U.S. The bigger modern units of the fleet have good endurance, but the destroyers have not, and, as the big ships must have destroyers with them, the endurance of the fleet as a whole is inadequate to meet the needs of Imperial Defence….

Another lesson that seems to stand out clearly from our study of war is that, however much we may dislike the idea of opposed landings, we shall have to face them. Operations of this nature are inevitable in the amphibious warfare to which this Empire is always committed.[228]

The quality of those appointed to study at the college was, not surprisingly, high and of the 78 British and Commonwealth officers of the naval service identified as having attended the IDC no less than 61 achieved flag or general rank.[229] As at least six officers died in the Second World War whilst holding an inferior rank, the summary of achievement indicates that the Service and Dominions both took care to appoint up and coming officers to the IDC.[230] Well could lecturers tell Air Marshal Longmore that appearing at the college was such a trying ordeal given the diversity and background of the audience where it seemed every subject had its in-house expert.[231]

The creation of the IDC was an important step in educating and training senior officers and equivalent civilians in defence matters. Where the Staff Colleges addressed operations, tactics, and administration, the IDC, in common with the Senior Officers' War Course, sought to make an officer a better strategist. Yet, the IDC did more and sought to consider strategy at national and imperial levels and not merely the military strategy of a single arm. Beyond making an officer an effective joint staff officer, there were limits to what it could achieve. These stemmed from the vacuum that existed at the highest reaches of the British political and defence hierarchies and the relative lack of seniority of the officers attending. Gifted, talented and promising as many were, it would be many years before the IDC's graduates reached the highest positions of command and training qualifiers of Captain's rank or their equivalent was a 'Ten Year Rule' of another type. 'Tiny' Ironside captured something of this failure when he told Liddell Hart, 'The Defence College doesn't reach the present leaders. We, the Generals are beginning to go down the drain in the opinion of the public.'[232]

<hr>

228 Dickens lecture, no date but *c.* 1928, Dickens Papers, LHCMA.
229 This accounting does not include Rear Admiral Reginald Darke who withdrew early.
230 At least 12 officers holding the *idc* died in the subsequent war. Admittedly, some had been recalled for war service and were then serving in a subsidiary position such as Commodore of Convoys, but as a record of sacrifice, it remains impressive.
231 Longmore, *From Sea to Sky*, p. 179.
232 Ironside to Liddell Hart letter of 4 May 1930, Liddell Hart Papers, LH 1/401/5, LHCMA.

Further, it proved impossible to train officers to a common higher-level doctrine at the IDC because a common higher-level doctrine did not exist amongst the Services or between Britain and her Commonwealth partners. The IDC could suggest the framework of such a common doctrine, but it could not impose it, nor would it have been wise to attempt it, absent a common understanding in political and Service circles. The Chiefs of Staff recognized that the IDC required further guidance on the lines of imperial policy and strategy, but never was able to provide it.[233] This was partly Chatfield's fault. When the Chiefs of Staff interviewed Vice Admiral Lionel Preston, the Commandant of the IDC, on the teaching of doctrine at the college, the First Sea Lord failed to address the question of imperial policy and strategy and merely asked him whether the guidance he had from the Chiefs was clear and adequate. Preston affirmed it was and the balance of the meeting then focused on the degree of overlap between the several schools or the divergence of doctrine amongst the Services. Preston believed a measure of overlap existed amongst the schools as they shared many of the same outside lectures, but thought the college was successful in fostering a common outlook amongst its students.[234]

It is tempting to conclude that an officer of Richmond's experience, reading, and temperament would have rendered a different verdict but that is unknowable. Preston, whilst a capable sea officer possessing joint experience through his command of *Eagle* and his period as Fourth Sea Lord was an able administrator, but was not an naval officer given to theoretical or abstract musings. Given more to art than the art of war—he presented the IDC with an exquisite painting attributed to the Dutch artist Jans Massys titled appropriately, enough, 'The Philosopher'—Preston is not known to have contributed to *The Naval Review* prior to his appointment as Commandant though he did write the Staff History of the World War's minesweeping operations.[235] A previous graduate of the SOWC, Preston's lack of Staff College training or teaching was reflected in the hands-off approach adopted at the IDC. He failed to immerse himself in the details of the IDC's instruction and relied on the college's directing staff, especially Brooke, to shape the course.[236] As part of his evidence provided to the Chiefs of Staff Sub-Committee, Preston submitted a précis of the lecturers and their topics for the recently completed 1934 course. Absent at all is any topic on which he was the responsible speaker. Doubtless, he must have spoken at least at the course's beginning and ending periods, but the record substantiates that Preston was the least prepared academically of the three naval officers to oversee the IDC between the

233 CAB 53/5, Chiefs of Staff Sub-Committee minutes of 9 October 1934.
234 *Ibid*, Chiefs of Staff Sub-Committee minutes of 1 November 1934.
235 http://www.mod.uk/DefenceInternet/AboutDefence/WhatWeDo/TrainingandExercises/
       RCDS/About+Us/SeafordHouse.htm. (Accessed 26 October 2011) and 'Post War Naval
       Histories', no date but between 1939-45, Dewar Papers, NMM/DEW/4. The monograph
       is *C.B. 1553. The History of British Minesweeping* completed in 1920, ADM 189/129.
236 Gray, *Imperial Defence College*, p. 33.

wars.[237] Moreover, Preston was the only naval Commandant not to be a subscriber to *The Naval Review*, a trait he also shared with James Somerville, Arthur Power and Pat Horan.[238]

Preston's selection generated a minor flap when the Navy announced his appointment without forwarding his name to the Chiefs of Staff for approval. Whilst the arrangement practiced was that each Service would name a qualified officer in turn, the position itself fell under the Chiefs of Staff Sub-Committee. Hankey objected to the manner adopted by the Admiralty in announcing Preston's appointment without informing the COS of their intent. On the surface, the Admiralty demonstrated poor administrative practice, but the implication in Hankey's criticism was that the Admiralty was putting forth an unqualified officer. In the event, the representatives from the Army and Air Force did not make an issue of the matter with the CIGS commenting that it would be invidious to criticize a naval appointment. If Field Marshal Milne's point conceded too much as a means to maintain harmonious relations amongst the Chiefs, Hankey, nevertheless, had made his point and all future nominations for Commandant or acts to terminate early an appointment were coordinated through the COS first.[239]

The third and last naval officer to hold the office of Commandant during the period was Vice Admiral Thomas Binney. His nomination to serve was submitted to the Chiefs of Staff Sub-Committee in the summer of 1938 in anticipation of Longmore's pending departure in January 1939. In this case, the Admiralty coordinated the nomination with the War Office and the Air Ministry before submitting Binney's name for consideration.[240] A former Director of the Tactical School, Binney had taken his complement of technical and war courses, though he had not previously attended a Staff Course or passed through the IDC. That he had not previously attended the IDC highlights how the Navy appointment process had to balance the needs of naval training with the essential element of time at sea for an officer to advance in the Service. Attendance at the IDC might indicate an officer was intended for better things, but it was not mandatory and one still needed to prepare oneself for command at sea first. Upon leaving the Tactical School, command of *Hood*, time in *Drake*, and promotion to flag rank quickly followed. Binney does not feature as a contributor to *The Naval Review* nor has he left a memoir. In this, he was not outside the norm as only Richmond wrote at length on professional topics and Longmore was the sole officer to leave a memoir. Still, it was to Binney that Roger Backhouse turned (along with Vice Admiral Sidney Bailey and Rear Admiral Lancelot Holland) to examine

<hr>

237 CAB 53/5, Chiefs of Staff Sub-Committee minutes of 1 November 1934, Enclosure, Imperial Defence College, Summary of Lectures in 1934.
238 'List of Members', *The Naval Review*, May 1934, pp. i–xi.
239 CAB 53/4, Chiefs of Staff Sub-Committee minutes of 29 February 1932.
240 CAB 53/39, First Lord to Secretary of Staff for War letter of 31 May 1938 and CAB 53/9, Chiefs of Staff Sub-Committee minutes of 1 July 1938.

the workings of the Naval Staff in light of the Czech Crisis of 1938 and Dudley Pound touted his breadth of professional knowledge and ease of speaking to effect.[241]

Assessing how the three naval officers who served as Commandant performed their tasks and how they measured-up to their opposite numbers is necessarily a highly subjective assessment. Richmond was certainly the most qualified academically of the naval officers and had no peers in that regard from Bartholomew and Haining his Army successors, nor Brooke-Popham and Longmore his Air Force counter-parts. Preston, as a former Director of the Signal School, was equipped to handle the administrative side of the college, and having commanded the carrier *Eagle* with its embarked complement of RAF personnel was no stranger to ensuring that inter-Service relations and tensions were kept to a minimum. Promoted to Admiral whilst at the college, the appointment was nevertheless his last one retiring, as he did, in 1935. Academically, if not intellectually, he was the least prepared of the three naval Commandants. Binney, a previous Director of the Tactical School had experience of teaching and running an educational establishment. A very competent, seagoing officer the criticism that he was better suited to overseeing the Royal Naval War College than the IDC would not be an unjust verdict. Necessarily, any verdict rendered on Binney is presumptuous as his tenure for all intents and purposes ended in July 1939 at the close of the second term. Students made their way to France for the now yearly battlefield tour but the onset of war saw them posted to operational duties as the college now closed. This included Geoffrey Norman and two other naval officers who were posted to the Admiralty to serve as Duty Captains.[242] As for Binney, he was a good mixer in a social setting, eventually becoming Governor of Tasmania and testimony enough that he was not merely master of the quarterdeck.[243]

Turning to the Army officers appointed, Bartholomew appears an odd choice given his lack of teaching experience at a Staff College and his Quetta provenance when that Service had so many gifted instructors available. As a commander, Ironside rated him poorly while also distrusting him, as he lacked a measure of integrity.[244] Robert Haining similarly had an unconventional background for a Commandant, as he had not served at Camberley or Quetta, nor was he even a *psc*. He, however, had, been a member of the IDC's opening course and was a qualified barrister. Such were his talents that the War Office sought his early release from the IDC, so he might resume

241 Malcolm Murfett, 'Admiral Sir Roger Roland Charles Backhouse (1938-39)' in Murfett, ed., *First Sea Lords*, pp. 174-75 and ADM 196/91 (Binney). Murfett cites Admiral Sir Sidney Barnes in error as the second member of the triumvirate examining the Naval Staff. As no Admiral with that name on either the active or retired list is specified, the most prob-able officer is Bailey then Vice Admiral Commanding, Royal Naval War College.
242 Vice Admiral Sir Horace Geoffrey Norman, Sound Recording 8870 of 1985, Imperial War Museum, London.
243 Gray, *Imperial Defence College*, p. 13.
244 Ironside conversation with Liddell Hart of 6 May 1937, Liddell Hart Papers, LH 1/401/188, LHCMA.

his duties as Director of Military Operations and Intelligence. Ironside who was blessed to have both Fuller and Haining as staff officers at the same moment thought the latter's greatest weakness was his jealously exhibited towards others, especially, Fuller and that he lacked originality of thought.[245] If that was the standard, well could he be envious of Fuller. Notwithstanding his unconventional provenance, Haining was a sound choice as Commandant and Vice Admiral John Durnford,[246] a student of the 1936 course, held him in high regard noting the power of his expositions—a by-product of his legal training.[247]

As a former Commandant of the Royal Air Force Staff College and a prewar graduate of Camberley, Brooke-Popham brought a degree of perspective with him not easily matched by those appointed by naval and military authorities. The same could also be claimed for Arthur Longmore who besides attending Camberley had begun his career in the Royal Navy. Andrew Cunningham saw in Longmore positive traits not present in other officers of Air rank and attributed these to his naval origins. Whether Longmore's views really made him unpopular with his Air Force peers as Cunningham held is problematic, but that he attempted to bridge the deep divide between the opposing schools of thought on the role of air power must rank as his greatest contribution whilst Commandant.[248]

Members of Parliament with an eye ever on finance questioned whether separate Staff Colleges were necessary or even wise once the IDC was established. Duplication seemed to prevail, variances in doctrine abounded, and did they not foster a 'them against us' outlook? This was a fair question to ask and the answer by default was that they remained necessary in the absence of experience proving different to the contrary. Yet, a pristine path to ensuring the proper coordination of defence was not a given. There were limits in how far the Admiralty was willing to extend assistance to the IDC. Not because of what the IDC was doing or pursued, but in what it might learn of internal naval deliberations. Requests received from the IDC for certain publications and files covering past events, whilst highly useful for the consideration of like problems of the present, opened the Service's business up to the eyes of others who might not read the dockets with a neutral eye. This was especially so

245 *Ibid.*
246 Vice Admiral John Walter Durnford (1891-1967). Commanding Officer, *P 39*, 1918, HMS *Porpoise*, 1918-19, HMS *Offa*, 1919, HMS *Volunteer*, 1923 and HMS *Somme*, 1923; to RAN, 1928-30 and 1940-42; Operations Division, 1930-32 and 1935; Tactical Course, 1932 and 1939; Commanding Officer, *Warwick*, 1933-34; Short Anti-Submarine Course, 1934; Commanding Officer, HMS *Echo*, 1934-35; Captain, 1935; Imperial Defence Course, 1936; Flag Captain and Chief Staff Officer to Flag Officer Malta, 1937-39; Commanding Officer, *Suffolk*, 1939-40; Imperial Defence College, 1940; Flag Officer and Chief Staff Officer in *Resolution*, 1942-43; Director, Royal Naval Staff College, 1943-44; Rear Admiral, 1944; Director, Naval Training, 1944-47; retired and Vice Admiral, 1948.
247 'A Shellback Remembers', unpublished manuscript, p. 172, Vice Admiral John Walter Durnford Papers, Imperial War Museum, London, PP/MCR/C10.
248 Andrew Cunningham to Gerald Dickens letter of 29 May 1951, Dickens Papers, LHCMA.

where the record reflected criticisms or judgments not favourable to those outside the Service. The Admiralty previously had shared such files with Greenwich with the understanding that they were to go no further, but it was deeply reluctant to provide non-naval officers and outside civilians this same level of access.[249] This dilemma continues to face organizations whenever realignments occur or new establishments are created. In an age of 'Wikileaks' it easy to conclude that the correct answer is to provide the requested material as its contents will at some point become known in any event. That is doubtlessly the case and the specific papers requested by the IDC in 1932 were those outlining the Service's views on the role and function of the Naval Staff. Containing as they did criticisms of the Weir Committee and its October 1925 report addressing the amalgamation of services common to the Navy, Army, and Air Force an acute measure of embarrassment was possible. Benign seemingly in itself yet, what if the files sought covered the Dardanelles and spoke to the performance of the Army and Dominion forces?

For a time, the IDC maintained a unique position. Constituted as a permanent establishment, its reviews of defence problems as summarized in its appreciations, notwithstanding their limitations in development, were seen as serious works of analysis. The General Staff, for one, accepted a large part of its analysis governing a possible Anglo-Japanese War, citing it with effect as the JPC appreciation supporting the defence of Hong Kong was being prepared in 1929-30.[250] That this position of influence ebbed was only natural as the Deputy Chiefs of Staff Sub-Committee and the JPC matured as staff organizations subordinate to the COS. Its influence may have also waned as a consequence to the many crises arising in the approach to a second global conflict. Conclusions must be more tentative in this regard, but the record indicates that the COS, the body responsible for the IDC, increasingly found their time consumed by deliberating on more weightier matters than the progress of the IDC and references to the college in the yearly defence reviews issued by the COS ceases from about 1935.

Because it was educational and administrative and not political and executive, the IDC could investigate and air many problems facing Britain in a manner which no other Government body could. Its use of economic analysis to inform its instruction and shape its appreciations anticipated the role operational research played in the coming war, whilst its higher-level training to civilians was years ahead of the establishment of a Civil Service College. Its consideration of extending its instruction to partners outside of Government, again, was a move that would escape comment of a modern institution, but in its day, broke fresh ground. The desire that it should accept problems for investigation was an early appreciation that having a ready body of senior officials free to look at issues absent an in-built organizational bias was a strong desire

---

249 ADM 1/8760/217, Mobilization Branch minute of 15 December 1932.
250 CAB 53/21, Joint Planning Sub-Committee, 'Defence of Hong Kong', 31 May 1930, Appendix No. 11, 18 November 1929.

and reflected that the existing means of investigation were less than ideal. That some Commandants eschewed this task merely reflected that qualifiers of the college were not as free of bias as supposed and that such tasks operated against the essential reason of their being at the IDC in the first place.

That the IDC broke new ground and filled an important educational gap is a given. Yet, important omissions remained. Foremost, was the failure to cover the *War Book* and its place in British defence planning and execution. Exactly why this was omitted is not known, but the gap was recognized by William Tennant during his period as naval member of the directing staff.[251] It may be that its contents were thought too sensitive for airing in a class drawn from across the Empire or it could be the case that it was deemed more a matter of detail to be left to the sub-committee responsible for its preparation. This writer tends to the former view, if only because Hankey and the COS were always reticent about sharing information with the greater IDC when such was still held to represent current thinking.

The command arrangements for amphibious operations and the linkages between the political objectives of such movements and their actual execution was another facet missing from the course. This omission owed something to the belief that the Staff Colleges with their heavy emphasis on combined operations had the subject well in hand and a reluctance to be seen treading in similar furrows. Yet, they invariably approached questions from the perspective of Britain acting alone or in consonance with the Empire and did not typically address problems involving allies. Moreover, the Staff Colleges typically focused on the how of combined operations and not the why. This was fertile ground for the IDC and it was ground that remained largely unspoiled by its analysis. This conclusion is not simply being wise after the event. Concerting her actions with allies, including naval actions, had been a problem in the World War and featured in IDC exercises such as the 1927 lines of an Anglo-Japanese war. Then, one of the discrete tasks was to write an appreciation from the perspective of the United States in light of the British war plan shared with her.[252]

Having civil servants study alongside officers was another innovation of the IDC and by 1938 the Admiralty was sending Jack James,[253] a principal secretary, to the course.[254] Though the numbers attending were sometimes below scale and their early removal from the course for new postings was a problem, successive Commandants valued the perspective, knowledge, and skills that they brought to the course.[255] The problem of removal was not unique merely to the civilians attending and the

251 Tennant Papers, NMM/TEN/42/3.
252 CAB 53/15, 'Exercise No. 3, Situation 5, U.S.A. as Ally of British Empire'.
253 Later Sir John 'Jack' Hastings James (? –1980). Imperial Defence College, 1938.
254 CAB 53/43, Report by the Commandant of the 1938 Course of the Imperial Defence College, 22 December 1938.
255 CAB 53/29, Commandant, Imperial Defence College to Secretary, Chiefs of Staff Sub-Committee letter of December 1936. Typically, four civilians attended each year, though in 1929, the number fell to two whilst in 1931 and 1937 it was but three.

War Office and Air Ministry both on occasion withdrew qualifiers prematurely.[256] Moreover, as has been previously recounted, the War Office in one instance reassigned a Commandant in order to fill a more pressing vacancy in its eyes. This problem too faced the Navy on at least two occasions necessitating the separation of an officer from the college. The first instance was during the inaugural course when Henry Moore departed for two months to support the Geneva naval conference. He subsequently returned to complete the balance of his studies when that conference collapsed.[257] Surprisingly, Moore's brief separation did not raise Richmond's hackles unlike the earlier contretemps at the War College—at least this writer has not located any evidence. This suggests, that in this instance, Richmond saw Moore's support of the conference as likely to reflect positively upon the new institution; and so it did when the Foreign Secretary registered his appreciation of Moore's service to the conference to the Admiralty.[258] Meanwhile, the second occasion saw a naval qualifier leave early due to illness. Fortunately, this occurred very early in the course and a suitable alternate was quickly appointed.

However, it was in its employment of economic analysis in furtherance of its exercise schemes leading to the eventual retention of a full-time economist where the college made its greatest contribution to the study of war. This move was anticipated at both the Royal Naval Staff College and Royal Naval War College, but it remained for the IDC to bring it to fruition. This was more Adam Smith than Carl von Clausewitz and owed everything to Richmond. Befitting a nation of shopkeepers, if a 'British Way in Warfare' existed, then its spiritual home was the Imperial Defence College with antecedents at Greenwich. Richmond's efforts to secure a full-time economist was itself a recognition that, notwithstanding his grasp of history, history was not the best guide for investigating the economic implications of contemporary military and naval actions. The rise of finance and credit in the modern era allowed nations to transfer a large portion of a war's cost to future generations and made reliance on historical method in solving current military problems inadequate.[259] Knowing that serving officers were not apt to have the requisite skills to assess these problems, Richmond won his case to secure the services of Fayle and took defence planning to a new plane.

The fallout of this was that soon the college and the Joint Planning Sub-Committee of the CID were demanding statistical information on foreign powers from the Board of Trade that it simply did not possess. During the World War, a War Trade Intelligence Department had met this requirement, but this was a temporary expedient. Its absence

256 CAB 53/43, Report by the Commandant of the 1938 Course of the Imperial Defence College, 22 December 1938.
257 ADM 196/92 (Moore).
258 ADM 196/50 (Moore).
259 Richmond's recognition of the new situation featured in his rebuttal to Liddell Hart, 'Economic Pressure or Continental Victories', *The Journal of the Royal United Services Institution*, Vol. LXXVI, No. 503, August 1931, p. 504.

was soon felt and a joint sub-committee of the CID dating from 1923 was created to fill the void. By 1925, it was extending its work to the consideration of economic pressure on an enemy power during wartime.[260] Its work, though, was not released to the Service Ministries, and, consequently, probably not to the IDC as well. Whether Richmond was aware of the Advisory Committee on Trade Questions in War is unknown, but the answer is probably not. His prompting for economic information, however, moved the CID to form a second committee with representatives from the Treasury, Board of Trade, Foreign Office, and the Services to address the shortfall. This proved insufficient and, ultimately, an Industrial Intelligence Centre created in 1931 arose to gather and collate such data.[261] In truth, the IIC was but a quasi-public face to that portion of the Secret Intelligence Service (Section VI) tracking foreign industrial and commercial intelligence of which Desmond Morton headed. The circle was completed when this new office seconded a member to augment Fayle's work at the college and eventually, the IIC assumed responsibility outright for supporting such analysis upon Fayle's retirement in 1938 with Morton assuming much of the burden.[262] Doubtlessly, the IIC, or something approaching it, would have been formed even in the absence of the IDC, as the Royal Air Force increasingly required similar information to develop its target sets and the Navy sought to understand the effectiveness of any blockade. Yet, as was the case in spotlighting differences in Service doctrine, creation of the college identified gaps and omissions in knowledge sooner than if matters had remained constant.[263]

Though the IDC studied potential problems facing Britain and the Empire, and, at times, the problem under review was mirroring actual events, the IDC made little effort to capture the results of recent crises and incorporate them into its future curriculum. The Shanghai Crisis of 1927 is a case in point. Arising quickly and requiring the deployment of divisional size forces to the International Settlement, the college did not seek to fashion an exercise in 1928 or subsequent years based on the scenario presented. Rather, it continued to pursue the defence of Hong Kong and Singapore in its Far East studies, worked the lines of a Soviet threat to India via Afghanistan, and contemplated German action over the Rhineland. As these were the major problems facing the Empire, their consideration is understandable. Yet, they each posited war with a major power to the exclusion of considering those minor operations which British defence policy claimed was the more immediate problem. To be sure, the war considered may have been limited, but even a limited war with a first-rate power was no minor affair requiring, as it would, large scale forces to be deployed and actions

260   F.H. Hinsley *et al*, *British Intelligence in the Second World War: Its Influence on Strategy and Operations, Volume One* (London: Her Majesty's Stationery Office, 1986), p. 30
261   CAB 53/26, Deputy Chiefs of Staff Sub-Committee report of 1 January 1936.
262   CAB 53/43, Report by the Commandant of the 1938 Course of the Imperial Defence College, 22 December 1938.
263   On Morton and the IIC see Gill Bennett, *Churchill's Man of Mystery: Desmond Morton and the World of Intelligence* (Abingdon: Routledge, 2009), pp. 135-99.

to be weighed in multiple theatres. Of course, the college did weigh minor problems with Palestine and the Trans-Jordan and disarmament questions being notable examples. Still, when measured in its time spent on major war, these schemes were almost an afterthought for the IDC.

Why the IDC fixated on the waging of major war is open to conjecture. The simplest and best answer is probably that major war poses major problems. The principles that the IDC sought to impart were most amenable to understanding by weighing war with a major power. This also ensured that all attending had an interest in the exercises and allowed the broadest of considerations whether—political, military, or economic—to be contemplated. Fundamentally, the IDC was a defence college. While only a statement of the obvious, it also meant that defence problems and not necessarily strategic problems in their broadest sense, were weighed. Matters such as imperial preference or maintaining the gold standard never featured in even its minor exercises. How the college would have differed if it reported to the CID directly or operated under the Foreign Office is impossible to know. Again, the same can be asked if it had had a dedicated civilian member on its directing staff as originally envisioned and did not rely on adjunct lecturers from the Civil Departments. If it accomplished much during the brief period of its operations before the onset of a new war, then it remained that more awaited to be done. If its pace of study was not always strenuous, then it at least benefited from being always serious.

One clear benefit in establishing an IDC is that the Services collectively, and not just those studying at the college, became better aware of the views of each other, the problems facing the British Empire, and how those problems redounded on each of the Services. This is not to say that such an understanding was only possible by an IDC, as other venues existed for airing defence matters such as the committee system in use and independent fora, such as the Royal United Services Institution or Chatham House. Still, the IDC focused these matters in a manner that those venues could not, as once the decision was taken to create a college, the immediate step became deciding what to teach. Richmond's summary of the 1927 course to the Chiefs of Staff highlighted that differences in doctrine existed amongst and, especially, was this so with regard to air doctrine. The upshot was that Chiefs of Staff were soon thrashing out the implications of air doctrine in private and away from the IDC, once Trenchard heard what was being taught. Madden, for one, argued that if Britain adopted the path advocated by the RAF and attacked an adversary's vital centres, she need look no further than to the late war and the German submarine campaign to see the error of doing so.[264]

For the naval and marine officer attending new and different horizons presented themselves. Where previously he had weighed a Service problem through Service channels, now he was expected to appreciate the broader context of any defence

264 CAB 53/16, Chiefs of Staff Sub-Committee, 'The War Object of Air Force', First Sea Lord
    note of 21 May 1928 and Chief of the Imperial General Staff note of 16 May 1928.

question. Officers who had shined at their own Staff College did not always sparkle so brilliantly at the IDC. Outlook and temperament were two reasons why this was the case, but a third was that the heavens now possessed many more stars. Bill Slim, the future Field Marshal and CIGS, was one officer whose luminance paled at the IDC. Following his period there, he concluded he needed a bit of a break and a return to regimental duties was in order having worried so much about fleets and armies.[265] Slim was no mean officer and is considered by many to have been the best British field commander of the Second World War. There are many qualities required of an officer and it may be that those instilled at the IDC from 1927-39 were not necessarily those fit for a commander so much as for his staff officer. That it by and large trained officers below flag and general officer rank—promising as they were—and did not prepare officers actually holding those ranks was a serious omission. Thus, it was leaving to chance that the officers trained would actually rise further in their respective Services. Compromise was inherent in all that the IDC attempted, and, of its graduates, the iconoclast was always the exception. Yet, navies and armies need their share of iconoclasts if the cauldron of military thought is not to grow tepid. Establishing contacts and friendships with civilians and his counterparts from the other Services, he was no longer just a naval man though still only a naval or marine officer. The motto of the college, '*Concordia Artus Roborat*', was loosely translated as strength through unity and adorned a college insignia incorporating the lion of the British Army, the wings of the Royal Air Force, and Neptune's trident. To the students attending, the emblem was simply 'The Beast' and what use the Navy made of its officers who had mastered 'The Beast' and, for that matter, those also trained at Camberley, Andover, Greenwich and Portsmouth, is now ripe for evaluation.[266]

265 Lewin, *Slim*, p. 60.
266 Gray, *Imperial Defence College*, p.64.

7

Of Admirals and Administrators: The Uses of an Educated Officer

An ideal staff should include if possible a man with the best imagination procurable.
Imagination in Great Britain is rarer than a well cooked omelette.[1]

Admiral Sir Gerald Dickens

Preparation for *a* war is primarily the duty of those at the Admiralty, but readiness
for battle is essentially the duty of the Commanders-in-Chief, Flag Officers and
Captains afloat.[2]

Rear Admiral J. E. T. Harper, 1928

Education in the military environment is never pursued as an end in itself, but is
conducted with the broader purpose of making a better sailor, soldier, airman, or
marine. This was true in the period between the wars and it remains the case today.
Accordingly, the necessity for preparing officers in staff duties or in the higher calling
of their profession passes without remark, but the preceding has shown that for
the Royal Navy this was not always the case. Faced with the timeless challenge of
ensuring that the most capable officers rose in the Service, the consensus governing
the exact attributes such officers required was breaking down. Knowledge and skill
were a given, but what skills and what knowledge in the face of the many changes
in technology and the increasing complexity of war were matters not amenable to
definite resolution. Even the human condition was not static, as the norms governing
societal behaviour were evolving and political certainties all around were collapsing.
In the face of this milieu, the Royal Navy adopted a gradualist approach. Finance may
have made this a policy of necessity, but experience made it a policy of choice. Absent
a direct threat and facing the need to severely contract its establishment, or force

1 Admiral Sir Gerald Dickens Papers, Imperial War Museum, London, IWM/90/35/1.
2 J. E. T. Harper, 'Staff or Anti-Staff', *Journal of the Royal United Services Institution*, Vol.
LXIII, No. 492, November 1928, p. 717. Original emphasis.

structure in the language of today, the Navy made gradual and incremental changes to how it trained and educated its mid-level officers in the postwar environment.

A Staff Course was quickly reformed at Greenwich, technical courses for Captains and Commanders soon re-appeared, as did a reconstituted War Course. In time, a certain measure of rationalization occurred as the Intelligence Course was subsumed into the greater Staff Course and a separate Commanders' Technical Course was merged with the course for Captains, becoming now the Senior Officers' Technical Course. Sclerosis of thought was not a problem and creation of a Tactical Course ensured that the Navy trained to a common battle doctrine weighing the best manner of employing its forces in any engagement. Moreover, though creation of the Imperial Defence College was not an Admiralty idea, the Admiralty adapted to its presence by sending strong players to study there. This was not necessarily appreciated at the time, as some officers failed to make the cut to flag rank at a later date, whilst others who did, retired without going perhaps further. The Navy, between the wars, grew its officers with a healthy dose of severe pruning and though some officers displayed a 'Blimpish' attitude, a naval equivalent to Colonel Blimp was noticeably absent.[3] This difference was mainly attributable to the relative rapidity with which the naval officer advanced in rank compared to his military counterpart and his earlier separation from Service if he failed to advance further. Thus, where the Lieutenant Commander who failed to secure his 'brass hat' faced retirement at 45, an Army officer failing to advance beyond Major could stay on until he reached 50 years of age.[4] Nor was the path to their majority an easy one for the military officer. Major General George Richards spent twelve years as subaltern whilst a Sandhurst term-mate recorded seventeen years in grade. Contrasted to the pace of naval advancement, the prospects of the interwar Army officer or, indeed, the Royal Marine officer, were at times a forlorn one.[5]

This contrast is also reflected in the relative ranks held by officers attending the Senior Officers' School of the Army with the naval officer typically sitting the course as a Captain whilst his marine opposite number held his majority. For the marine, time at Sheerness often came immediately after or preceding his appointment to the Chatham depot—cost avoidance and opportunity, in this case, working in unison. Attending the course allowed the marine to approximate the preparation afforded to his Army counterpart at corresponding moments in their respective careers. In contrast, for the naval officer sitting the course, it was frequently the practice of pairing his time at Sheerness with other courses such as the SOWC, the SOTC, the Tactical Course, or even attendance at the Imperial Defence College. Thus, if also a prior graduate of a staff college, then this amounted to nearly two years of professional officer development

3   B. H. S. 'The Musings of Methusaleh', *The Naval Review*, Vol. Vol. XXVII, No. 3, August 1939, p. 420.
4   CAB 24/173, First Lord of the Admiralty to Cabinet memorandum CP 261(25) of 26 May 1925.
5   Richards, Reel 5, Sound Recording 866 of 1976, IWM.

in addition to any specialist training he might have pursued upon leaving Dartmouth and his Lieutenant's courses at Greenwich and Portsmouth.[6] The actual purpose of the course for naval officers is more speculative, as their backgrounds showed more variance given their opportunities for specialization. Members of the gunnery, signals and navigation fields predominated as students, as did the ubiquitous salt horse officer. Notwithstanding any differences in specialization, though, certain common strands appear. Before attending, appointment to a naval barracks, naval college, or to Cambridge University as part of the staff overseeing officer development and training resonates clearly.[7] Additionally, prior service in the Ordnance Department or on the Ordnance Committee was a trait shared by several other officers.[8] Meanwhile, time spent in the Naval Intelligence Division was another variable shared and the number subsequently appointed is noteworthy.[9] Still others could expect to command a flotilla of destroyers or minesweepers when leaving. What the training did not serve was as a preparatory grounding for officers seconded to the War Office as no officer traced saw duty there following completion of the course.

Again, for naval officers, only those of the executive branch attended and, even within this pool of candidates, the net was drawn fairly tight with no aviators appearing. Timing played its part in this as Pilots and Observers of the required rank were only coming to the fore at the close of the period. Whether they would have gone to course if war had not occurred in 1939 cannot be said definitively, but the likelihood is that they would not given the other pressures required of their services. Confirmation, of sorts, is offered by Sidney Bailey's assessment of Captain Maitland Boucher,[10] a flier and former Flag Captain in *Courageous*, during the latter's attend-

6   Officers sitting the course in this fashion included Admiral of the Fleet Viscount Cunningham, Admirals Sir Douglas Fisher, Sir William Parry, and Sir Harold Walker, Vice Admirals Thomas Drew, Lewis Crabbe, Ronald Hallifax, Arthur Lyster, and Patrick Parker, Rear Admirals Alfred Phillips and Captains Sir Hubert Acland, Richard Benson, Guy Coleridge, William Gell, Alexander Hammick, Robert Hamond, Charles Hotham, and Robert Mackay.

7   Representative of this class of officers is Admiral of the Fleet Sir Rhoderick McGrigor, Admirals Sir Cecil Harcourt and Sir Harold Walker, Vice Admirals Sir Francis Austin, Frederick Dalrymple-Hamilton, Arthur Lyster, and Thomas Drew, Rear Admirals the Honourable Claude Hermon-Hodge and Denham Bedford and Captains Guy Coleridge, Geoffrey Cooke. Alexander Hammick, Charles Hotham, and Humphrey Jacomb.

8   For example, Admiral Sir Douglas Fisher, Vice Admiral Arthur Lyster, Rear Admiral Alfred Phillips, and Captain Richard Benson.

9   Included in this group were Admiral Sir William Parry, Vice Admiral Ronald Hallifax, Rear Admirals Sir Richard Bevan, Charles Burke, and the Honourable Claude Hermon-Hodge, Captain Geoffrey Cooke, and Lieutenant Colonels Harold Bird and Francis Bramall.

10 Rear Admiral Maitland Walter Sabine Boucher (1888-1963). Commanding Officer, HMS *Totnes*, 1918-19 and HMS *Leamington*, 1919-21; Local Defence Division, 1922-23; Operations Division, 1923-24; Senior Officers' Technical Course, 1925, 1928 and 1937; Commanders' Short Flying Course, 1925-26; Tactical Course, 1926; *Argus*, 1926-28; Air Ministry, Directorate of Training, 1929; Captain, 1930; Air Ministry, 1930-31;

ance of the SOWC in 1937. Bailey noted that his lack of prior opportunity in studying the higher aspects of his profession limited his overall progress in the course.[11] It also limited his ability to fashion a new department within the Admiralty when Boucher, appointed Director of Air Matérial in early 1938, left his War Course after only two months' attendance. Though probably an ideal officer from both technical and practical considerations for his new appointment, his lack of administrative ability—the strong suit of the trained staff officer—was soon deeply felt.[12] Bailey's testimony also indicates that the SOTC and Tactical Course, if higher level courses were never really about the higher aspects of war.

As for those officers who had passed through Andover, less than 25% saw service with the Air Force either in a carrier or with the Air Ministry following their period of study. The typical first assignment for a graduate was either to return to sea in flying, general service, or staff duties or to serve on a staff ashore. A smaller number (e.g., six) were appointed to command a minor vessel whilst two proceeded from Andover to take the Tactical Course. As trained staff officers, many would subsequently serve at the Admiralty with the Intelligence Division the most prominent destination. Drawing any conclusions is difficult given the lack of documentary evidence specifying why officers were appointed as they were, but a reasonable assumption is that the Navy used these officers to enhance the ability of its staffs to address air matters rather than trained these officers to represent the naval view within the Air Ministry and the greater Air Force.

Boddam-Whetham was, however, one officer who did go to the Air Ministry following his time at Andover. Brooke-Popham valued his assistance during the course recommending him for the *psa*. Following a short course to complete his training as an Observer, Boddam-Whetham spent a brief period in *Argus* before attachment to the Air Ministry where he worked on the development of the FAA from Adastral House. Given his growing knowledge of the air, he was recommended for further employment in air matters and Trenchard himself provided high praise of his talents. Such was not to occur for a return to general naval duties followed. He spent no further time either in liaison work or as a Naval Observer and in due course took command of a flotilla of destroyers where he failed to impress Admiral Joe Kelly.[13] Consequently, his next posting was to Bermuda and its naval dockyard indicating that his career was no longer favoured. This surely was not the best use of his talents in air matters even if the Admiralty decision to remove him from the path of preferment proved essentially correct.

Commanding Officer, *Champion*, 1931-32; Naval Air Division, 1932-35; Flag Captain and Chief Staff Officer in *Courageous*, 1935-37; Senior Officers' War Course, 1937; Director of Air Matérial, 1938; Tactical Investigation Section, Tactical School, 1939; to RAN, 1939-41; Rear Admiral and retired, 1941. Recalled for war service.

11   ADM 196/92 (Boucher).
12   *Ibid.*
13   ADM 196/127 (Boddam-Whetham) and ADM 196/92 (Boddam-Whetham).

Till's *Air Power and the Royal Navy* provides an excellent overview of the many challenges facing the Senior Service between the wars in developing naval aviation. The system in place following the World War was strongest at the practical level but suffered as one left the fleet and approached Whitehall. Many are the reasons why this proved so, but of relevance to this study are three. First, divided control required naval pilots to hold RAF rank also, for Air Ministry regulation required that only an Air Force officer could fly Service aircraft. Perforce, this meant that for a naval officer to succeed in flying duties he needed to secure promotion not only in his own Service, but also within the RAF. This did not always happen and his career prospects, if he chose this path of specialization, faced a needless double hurdle. Secondly, the Navy insisted that its officers return to general duties after five years of flying to round out their professional development. Schools needed to be attended and duty afloat, other than as a pilot was essential, if the officer were to compete against his naval peers for promotion. Of course, this operated against the naval officer maintaining currency in his corresponding air rank and contributed to the problem that the naval officer frequently fell behind his Air Force counterpart, though nominally equal or superior in rank courtesy of their naval standing. Finally, the RAF officer was expected to specialize in another field beyond flying such as engineering or weapons. Consequently, Air Force officers routinely served in the fleet for an initial period, but might never return to do so again. In a very real sense, the RAF was only following a variation of the naval custom where officers returned to general service duties. Yet, this was not necessarily appreciated by the Navy and it did make sustaining progress that much more difficult.[14] The upshot was that no naval officer best qualified by experience (i.e., a pilot or observer) to liaise between the two Services as a staff officer attended both the Greenwich and Andover Staff Courses between the wars though many other specialist officers did.

Fortunately, within the Air Force were several officers of proven ability that had originally entered military service through the Royal Navy or the Royal Marines. In this, they possessed an understanding and sympathy of maritime air requirements. This was all to the good and so was the fact that some naval officers, having brothers in the Royal Air Force, sought attachment to the air service. Of these, Captain Arthur Dowding,[15] brother of Hugh, of Fighter Command and Battle of Britain fame, was

14   Thursfield, 'Fleet Air Arm', 1 August 1936, LH 1/695/6-7, LHCMA.
15   Vice Admiral Sir Arthur Ninian Dowding (1886-1966). Five firsts in examinations for Lieutenant; Training and Staff Duties Division, 1919-21; Air Ministry, Directorate of Training, 1925-27; Captain, 1926; Senior Officers' Technical Course, 1927 and 1937; Commanding Officer, HMS *Sandhurst*, 1927-29 and HMS *Cornwall*, 1929-31; Directorate of Air Operations and Staff Duties Air Ministry, 1933-35; Tactical Course, 1935; Senior Officers' War Course, 1935; Commanding Officer, *Furious*, 1935-36; Rear Admiral, 1936; Rear Admiral Commanding, Tenth Cruiser Squadron, 1937; Army Higher Commanders' Course, Old Sarum, 1937; RAF Army-Air Co-operation Course, 1937; Vice Admiral and retired, 1940. Recalled for war service.

perhaps the most prominent. Yet, those airmen whose roots resided in the Navy, such as Longmore and Samson joined by Edmund Robertson,[16] Raymond Collishaw,[17] and Albert Durston[18] were a diminishing breed. Thus, collateral education and improving the conditions of service were important objects in their own right if dual-control was ever to be a viable method for delivering maritime air support.

Turning to those officers who attended Camberley, the Service at least did not have to face the task of having to balance military proficiency and a dual-rank structure. Accordingly, marine officers who completed the course were able to serve attachments with the Army on a more or less equal footing and at least four did so. This included Captain Roy Smith-Hill,[19] who following two years at Camberley where he was awarded a King George Prize Scholarship in each of his years attending, served four years in Army Southern Command including two years as Brigade-Major to the Devon and Cornwall Light Infantry.[20] For the 38 naval officers traced, a return to sea was the norm with 21 of these Camberley attendees following this path soon after departing. The actual assignment filled was a function of seniority and the relative stage an officer occupied in his career. For the more junior, it was to serve a period as a division officer or to hold a staff appointment afloat, whilst a more senior qualifier went to his ship as the executive or even its commanding officer.

Experience of the Army Staff College proved to be a common trait of several of the officers later to attend the IDC with no less than seven naval and marine officers sharing this background.[21] In comparison, Admiral Charles Daniel proved the lone naval Andover qualifier attending the joint college between the wars. As five of the Camberley officers had also attended the Greenwich Staff Course during this period, significant portions of their interwar career were consumed in professional, rather

16   Air Commodore Edmund Digby Robertson (1887-1956). Commander, School of Naval Co-operation, 1920-21; Fleet Aviation Officer Atlantic Fleet in *Revenge*, 1924-26 and Fleet Aviation Officer China Fleet in *Hawkins*, 1927-28.

17   Air Vice Marshal Raymond Collishaw (1893-1976). Staff Course Andover, 1924-25; *Courageous*, 1929-30; Tactical Course, 1930 and *Courageous*, 1929-32.

18   Air Marshal Sir Albert Durston (1894-1959). *Argus*, 1922; School of Naval Co-operation, 1925; Commander School of Naval Co-operation, 1927-28; *Courageous*, 1931-32; Tactical Course, 1933; Coastal Area, 1934-36; Fleet Aviation Officer Home Fleet in *Rodney* and *Nelson*, 1936-38; Air Staff for Naval Co-operation, 1938-39; Director of Naval Co-operation, 1939-42 and Air Officer Commanding No. 18 Group Coastal Command, 1942-43.

19   Later Brigadier Philip Royal Smith-Hill (1897-1996). Staff Course Camberley, 1932-33; Staff Captain, Army Southern Command, 1935-37; Brigade-Major, 130th (Devon and Cornwall) Infantry Brigade, 1937-39; Acting Lieutenant Colonel, 1940; Brevet Lieutenant Colonel, 1941; Acting Colonel Commandant (Temporary Brigadier), 1944; Colonel, 1946; Brigadier and retired, 1950.

20   'Brigadier Roy Smith-Hill', *The Independent*, 13 August 1996 and ADM 196/65 (Smith-Hill).

21   They were Admiral Sir John Edelsten, Vice Admirals Henry Bovell and John Dundas, Major General Alan Bourne, Captains Augustus Agar and Bernard Warburrton-Lee and Lieutenant Colonel Vincent Brown.

than practical, education.[22] Many at some point would serve at the Admiralty and, especially, on the Naval Staff. Here, the likelihood was that the officer would serve in either the Training and Staff Duties Division or the Plans Division and this says something more about how the Navy employed qualified staff officers within the Admiralty. To wit, officers trained at Camberley migrated to staff and planning duties whilst those officers with an Andover background leaned towards planning and intelligence duties and though the odd trained staff officer appears in the Operations Division, this was typically the preserve of the non-trained staff officer. Durnford confirms that lacking staff qualification was no bar to service in the Operations Division. Following his period in HMAS *Anzac*, he reported to the Naval Staff in early September 1930 assuming responsibility for overseeing the deployment of cruisers in home and foreign waters.[23] Of the seven other officers serving with Durnford, only three were qualified staff officers; this proving a slight improvement on the previous year when only two officers were.[24]

As for those attending the Royal Naval Staff College, the recently designated *ws* and after 1924, *psc*, usually returned to a seagoing assignment which may or not have been on a staff. Often, as the experience of Charles Drage demonstrates, a brief sojourn to the Admiralty was the first assignment for the officer. If on a staff of a flag officer or a Commodore and depending on whether he had previously qualified as a technical specialist, then the officer might serve as the Fleet or Squadron Gunnery, Torpedo, or Navigating Officer.[25] In a sense, these were natural appointments to make, but they tended to divide recent qualifiers into two distinct types: operational and technical staff officers and Astley-Rushton whilst Director of the Staff College reported on

22   Namely, Agar, Bovell, Dundas, Edelsten and Warburton-Lee.
23   '*Shellback Remembers*', p. 154, Durnford Papers, IWM/PP/MCR/C10.
24   *The Navy List* (London: His Majesty's Stationery Office, January 1931), pp. 404 and 314-15 and *The Navy List* (London: His Majesty's Stationery Office, January 1930), pp. 404 and 314-15.
25   In 1919, the styles of staff officer carried in the fleet were Fleet or Squadron Signals Officer, Torpedo Officer, Gunnery Officer, Wireless Telegraphy Officer, Communications Officer, Navigating Officer, Intelligence Officer, Physical and Recreational Training Officer, Flying Officer, Anti-Submarine Officer, Mining Officer, Photographic Officer, Royal Marine Officer, Maintenance Officer, Engineer Officer, Accountant Officer, Medical Officer, Health Officer, Chaplain, Master of the Fleet, and Chief Staff Officer. If, however, the officer occupied a general staff billet, the title War Staff Officer for naval officers and General Staff Officer, if a marine, was employed. If the flag officer held a concurrent appointment from the Foreign Office, such as High Commissioner, a Political Officer holding temporary naval rank might be present. Thus, when Admiral John de Robeck was both Commander-in-Chief, Mediterranean and High Commissioner, Constantinople Commander Harry C. Luke, RNVR, was embarked in *Iron Duke* as liaison to the Foreign Office. In truth, the variety and style of titles adopted was indicative that a systematic basis of staff organization and execution was yet to exist. See Koe, 'Observations Upon the Naval Staff System', *Naval Review*, November 1924, p. 664.

their abilities along such lines.[26] Lieutenant Commander Leslie Brownfield offers a case in point. A gunnery specialist, Brownfield whilst attending the Staff Course, was approached about serving as an exchange officer to the RAN for a two-year period as he neared the end of his studies. Married and with a small child, acceptance came at a price as his family could only accompany him at his own expense. In the end, he accepted becoming Squadron Gunnery Officer in *Canberra*. Completing his time in Australia, Brownfield returned home to serve in the Second Battle Squadron becoming its Squadron Gunnery Officer and Staff Officer Operations (SOO) in the *Royal Oak*.[27]

Concurrent with the change in staff officer designation and qualification, the practice of appointing one as performing Operations Duties or acting as a War Staff Officer ceased.[28] Officers were now styled as SOO, Staff Officer Intelligence (SOI), or noted as occupying one of the technical staff appointments as was the case previously. For the salt horse staff officer, duty as a SOO was a typical assignment, though some believed that the ideal SOO was a suitably trained Navigating Officer.[29] If the appointment was to a major command, such as the Mediterranean Station, the SOO might share the duties with another officer. Chatfield whilst commanding in 1931 had two SOO's with him in *Queen Elizabeth*, as did Michael Hodges in the Atlantic Fleet in *Nelson*.[30] Pound, meanwhile, in command of the Battle Cruiser Squadron did not have a designated SOO though he did have at least three officers performing staff duties and holding the naval *psc*.[31] This was unusual as the staff of ACQ typically included a SOO and would again when Rear Admiral Wilfred Tomkinson[32] assumed command. Others, though, were not so fortunate and during Richmond's period as Commander-in-Chief, East Indies he had but one qualified staff officer holding the *ws* designation, then used to denote a trained staff officer.[33]

26   ADM 196/145 (Stewart).
27   Brownfield, *Sea Cycle*.
28   Officers holding the *ws* distinction were now noted as *psc*. Officers designated *qs* continued to be carried in the *Navy List* as such.
29   John N. Knox, 'Staff Officers (Operations) for Captains', *Naval Review*, Vol. XV, No. 4, November 1927, p. 863.
30   *Navy List*, January 1931, pp. 254 and 261.
31   *Ibid*, pp. 262-63. Officers having qualified at the Staff College and holding squadron appointments included Commanders Arthur W. la T. Bissett, Arthur G. Talbot and William G. Benn. Additionally, Lieutenant Commander George Belben, the *Renown*'s gunnery officer, had completed the Staff Course in 1929.
32   Later Wilfred Tomkinson (1877-1971). Captain, 1916; Flag Captain in *Colossus*, 1917; Captain (D), Sixth Flotilla, 1918-19; Flag Captain and Chief Staff Officer in *Lion*, 1919-20 and *Hood*, 1920-21; Deputy Director, Operations Division, 1921-23 and Director, 1923-25; Tactical Course, 1926; Rear Admiral, 1927; Chief of Staff, Mediterranean Fleet, 1927-29; ACNS, 1929-31; Commander, Battle Cruiser Squadron, 1931-32; Vice Admiral, 1932 and retired, 1935. Recalled for war service.
33   Herbert Richmond, 'Notes Bearing on the Staff Requirements of a Commander-in-Chief', *The Naval Review*, Vol. XXVII, No. 1, February 1939, p. 1.

By 1932, one observer was noting (incorrectly) that all afloat SOO's were graduates of the Staff College, though the corresponding numbers for Gunnery, Torpedo, and Navigating staff appointments stood at only 25%. For afloat Signals and Wireless staff appointments the associated figures were zero.[34] This was progress as many officers holding staff appointments in the period following the war were not graduates of the Staff Course or even designated for staff duties. Commander Edward Thornton is representative. A salt horse with much experience commanding minor vessels, he served as SO&I at Chatham from 1926-27.[35] The paucity of trained staff officers for those positions geared to ensuring the proficiency of the Service's technical means was a reflection that these duties required, after all, technical competence ahead of administrative ability. Where the Admiral commanding the fleet was responsible for the operational efficiency of his squadrons and ships, then a Fleet Gunnery Officer or Fleet Wireless Officer served the supporting purpose of ensuring that the individual departments of the ships serving in that fleet were in proper order.

Still, the anonymous commentator in *The Naval Review* overlooked Commander Reginald Portal,[36] serving as SOO in *Courageous*, but the general thrust of his argument was otherwise sound, as he lamented this state of affairs after twelve sittings of the postwar Staff Course. On cannot draw from the preceding that officers specializing in signals and wireless were absent from the Staff College though. Rather, it was more the case that by the time they took their course other duties awaited them upon leaving than holding the position of a Fleet Signals Officer. Serving as an executive officer or even command of a minor ship, such as a sloop or destroyer, might follow and appointment to the Signal Department of the Admiralty was a frequent follow-on destination. Meanwhile, Commander Claude Graham-Watson[37] became a SOO in *Kent* following his Staff Course.

That Signals Officers did not necessarily proceed to a typical staff appointment was merely an indication of the nature of their prior duties and experience where proximity to command was a constant. These officers routinely worked a step away

34   Anon., 'Royal Naval Staff College', *Naval Review*, February 1932, p. 20.
35   *The Navy List* (London: His Majesty's Stationery Office, February 1926), p. 258.
36   Admiral Sir Reginald Henry Portal (1894–1983). Anti-Submarine Division, 1917-18; to RAF, 1923-25; Pilot in *Eagle* and *Furious*, 1926; Chemical Warfare Course Porton Down, 1927; Squadron Torpedo Officer in *Hood*, 1927; Naval Air Division, 1929-31; Staff Officer Operations in *Courageous*, 1931-32; Captain, 1934; Tactical Course, 1934; Assistant Director, Naval Air Division, 1935-38; Air Matériel Department, 1938; Senior Officers' Technical Course, 1939; Senior Officers' War Course, 1939; Rear Admiral, 1943; Vice Admiral, 1947; Admiral, 1950 and retired, 1951.
37   Later Captain Claud Boothby Graham-Watson (1894–1960). Five firsts in examinations for Lieutenant; Flag Lieutenant in *Colossus*, 1917, in HMS *Caledon*, 1917-20, in *Emperor of India*, 1920-21 and in *Hood*, 1921-23; Signal School, 1923-25; to New Zealand Division, 1925-28; Flag Lieutenant Commander and War Staff Officer in HMS *Constance*, 1928-29; Tactical Course, 1929; Staff Course, 1930-31; Staff Officer Operations in *Suffolk* and *Kent*, 1931-33; retired, 1935 and Captain, 1939. Recalled for war service.

from the commander and his staff and this closeness gave them a strong practical knowledge of the staff and its workings before they ever began their staff education. Commander Ralph Edwards[38] offers an excellent example of this advantage at work. Arriving at Greenwich in January 1936, by this date he had already served on the staff of the Signal School,[39] been Flag Lieutenant to Frederick Field in the Mediterranean Fleet,[40] and acted as Vernon Haggard's Flag Lieutenant Commander and Fleet Wireless Officer on the America and West Indies Station.[41] Upon departing Greenwich, Edwards assumed command of an escort vessel on the China Station before adopting a more staff oriented role when he became a naval liaison officer.

Moreover, it was a custom to employ non-specialist officers as the head of the Communications Departments in private ships, those vessels where a flag officer was not embarked, unlike those departments overseeing gunnery, torpedo, or engineering duties.[42] Consequently, the number of Signals Officers borne was actually quite small. Thus, given their prior experience and pattern of employment, a transition to more command-like responsibilities following their time at Greenwich was the norm if they were to stay competitive with their peers for promotion.

Portal's service as SOO was a consequence of the recent creation of a flag officer billet overseeing carrier operations. It also demonstrates that whilst the duties of a SOO officer were of a common nature, the actual skills necessary to perform an assignment, nevertheless, were highly depended on the type of operation being overseen. Thus, it was insufficient merely to be a *psc* in all cases and Portal's prior service in the RNAS as an Observer and attachment to the RAF fitted him to be SOO in *Courageous* in a manner that another trained as a mere staff officer did not. The brother of a future Chief of the Air Staff, Portal had served as an Observer during the war and sought a transfer to the Royal Naval Air Service. This was denied and he subsequently sought service in submarines before taking the Long Torpedo Course. Demonstrating the pernicious influence of how dual-control worked, Portal resumed torpedo duties

---

38   Later Admiral Sir Ralph Alan Bevan Edwards (1901-1963). Short Course, Cambridge University, 1922; Flag Lieutenant in *Cormorant*, 1925-26; Signals Course, 1926-27; Signal School, 1927-28; Flag Lieutenant and Assistant Fleet Signals Officer in *Queen Elizabeth* and *Warspite*, 1928-30; Flag Lieutenant/Flag Lieutenant Commander and Fleet Signals and Wireless Officer, in *Despatch* and *Delhi*, 1930-32; Signal School, 1934-35; Staff Course, 1936; Commanding Officer, *Sandwich*, 1937-38; Naval Liaison Officer Shanghai, 1938; Captain, 1939; Deputy Director, Operations Division, 1939 and Operations Division (Home), 1940, and Director, Operations Division (Home), 1941; Deputy Chief of Staff and Chief of Staff, Eastern Fleet, 1942-45; Commanding Officer, HMNZS *Gambia*, 1945-46 and HMS *Illustrious*, 1947-48; Rear Admiral, 1948; ACNS, 1948-50; Commander, First Cruiser Squadron, 1951-53; Vice Admiral, 1952; Third Sea Lord, 1953-56; Admiral, 1955; Commander-in-Chief, Allied Forces Mediterranean, 1957 and retired, 1957.
39   *Navy List*, October 1935, p. 287.
40   *Ibid*, January 1930, p. 286.
41   *Ibid*, January 1931, p. 231 and January 1932, p. 230.
42   'Reflections on Invergordon', Drage Papers, Reel 4, IWM//PP/MCR/99.

following his period of flying though he eventually became to all intents and purposes a naval air staff officer if never formally trained as such.[43] Moreover, Portal's background as a Pilot, service with the Air Ministry, and time as SOO were ready assets used by the Admiralty in its negotiations to secure control of the Fleet Air Arm and he was a conspicuous player in these ultimately successful efforts.[44] Though he had sought and was strongly recommended for the Staff Course by Vice Admiral Dreyer, the Navy had other plans for Portal appointing him SOO in *Courageous*.[45]

Replacing Portal as SOO in *Courageous* was Henry Johnstone, an Observer and a trained staff officer via Andover.[46] Two officers, Commander Stanley de Courcy-Ireland[47] and Lieutenant Commander Caspar John,[48] in turn, succeeded Johnstone. Caspar John soon left for attachment with the RAF and Lieutenant Guy Willoughby,[49] a Pilot, assumed the second SOO billet. Though both were aviators, only Willoughby had secured a *psc* or trained to its air equivalent, the *psa*, demonstrating that knowledge in flight operations and procedures was prized above comprehension in formal staff duties but an officer holding both was a prize above rubies.[50]

The SOO's role was to consider strategical and tactical problems and though usually borne in a flagship of a Commander-in-Chief, in time, the SOO began to appear in destroyer and submarine flotillas working similar problems at a lower lever.[51] The title SOO is somewhat misleading to the modern reader as the actual gamut of the position encompassed matters beyond current operations touching as it did upon strategy which typically weighs future events. Yet, a designated Staff Officer Plans at the fleet level did not exist and staff officers of the day had broader responsibilities than any

43 ADM 196/56 (Portal).

44 ADM 196/92 (Portal).

45 ADM 196/127 (Portal).

46 *The Navy List* (London: His Majesty's Stationery Office, January 1934), pp. 225 and 316.

47 Later Captain Stanley Brian de Courcy-Ireland (1900-2001). Short Course, Cambridge University, 1919-20; to RAN, 1922-24; Observer in *Furious*, 1926-27, in *Argus*, 1929 and in *Hermes*, 1929-31; Tactical Course, 1931; Observer in *Furious*, 1931-34 and in *Courageous*, 1934; Commanding Officer, *Calypso*, 1935; Staff Officer Operations in *Courageous*, 1935-36; to Air Ministry, 1936-39; Captain, 1942; Senior Officers' Technical Course, 1948 and retired, 1951.

48 Later Admiral of the Fleet Sir Caspar John (1903-1984). Pilot in *Hermes*, 1926-27, in *Argus*, 1929-30, in *Courageous*, 1930-31, in *Exeter*, 1931-32; Staff Officer Operations in *Courageous*, 1934-35; Pilot in *Glorious*, 1936; Naval Air Division, 1937-38; Air Matériel Department, 1938-39; Captain, 1941; Imperial Defence Course, 1947; Rear Admiral, 1951; Vice Admiral, 1954, Admiral, 1957; Admiral of the Fleet, 1962 and retired, 1963.

49 Later Rear Admiral Guy Willoughby (1902–1987). Pilot in *Egmont*, 1926, in *Hermes*, 1926-27, in *Courageous*, 1927-28; in *Columbine*, 1928; in *Argus*, 1929-30, in *Furious*, 1931-32, in *York*, 1933-34; Staff Course, 1934; Pilot in *Furious*, 1935; Staff Officer Operations in *Courageous*, 1936-37; Senior FAA Officer in *Glorious*, 1937-39; Captain, 1943; Rear Admiral, 1953 and retired, 1956.

50 *Navy List*, October 1935, pp. 225 and 314.

51 Knox, 'Staff Officers (Operations) for Captains', *Naval Review*, November 1927, pp. 863-64.

formal title might convey. During exercises or in the event of actual operations, the SOO was the officer responsible for organizing the strategical plot in a flagship. This plot was maintained from the moment the ship departed port until its eventual return by the SOO and other embarked officers, such as midshipmen or Naval Instructors, employing standardized reference symbols. If contact with an enemy was established, then a subordinate officer, such as the Torpedo or Navigating Officer, initiated the tactical plot tracking the course of his own ship and that of the most likely target.[52] Given the nature of plotting, it is easy to see why Navigating Officers were deemed especially fitted to the role of the SOO.

On aircraft carriers, the Intelligence Officer assisted in the maintenance of the strategical plot, but the requirement to maintain a tactical plot was missing. With both plots in use in a flagship, data from the tactical plot was passed periodically to the strategical plot, and until automated plotters were fitted in ships, this perforce was a manually intensive process prone to error. Additional plots for tracking ship course and speed and controlling torpedo shoots were maintained by Navigating Officers and Torpedo Control Officers and their data was shared with the strategical and tactical plots by voice pipe and phone.[53] Though the officer trained at the Staff College learned much of the theory of maintaining strategical and tactical plots (as did his counterparts at the *Vernon* and the *Dryad*), the skill to truly master these tasks came only from constant practice at sea. By 1931, the SOO ceased to have the explicit duty of maintaining the strategical plot. Now, with six officers and men maintaining it and the associated tactical plot in a flagship the change was understandable: there simply were insufficient SOOs present to perform the duties specified.[54]

Charles Drage's time in the East Indies Squadron offers evidence of the lot of a SOO. Preparing the lines of tactical and strategical exercises was a common duty, as too, the drafting of signals, reports, orders, instructions and appreciations in the first instance for consideration by his flag officer. Tracking the movements of attached units, the fuel expenditure of the fleet, and the status of reserve assets available for consumption or distribution within the station were all details mastered during the Staff Course and forming part of his duties.[55]

If a naval officer had passed successfully as an interpreter, then appointment as a Staff Officer Intelligence (SOI) often followed. Likewise, marines graduating from the Staff Course were ever present as a SOI. Major Sam Bassett,[56] commis-

---

52   ADM 182/85, Admiralty Fleet Order '1611.—Strategical and Tactical Plotting' of 11 June 1926.

53   ADM 182/87, Admiralty Fleet Order '2989.—Strategical, Tactical and Navigational Plotting' of 23 November 1928.

54   ADM 182/90, Admiralty Fleet Order '2462.—Strategical, Tactical and Navigational Plotting—REPORTS', 16 October 1931.

55   Daniel de Pass, 'Hellenic Naval Staff College: Opening Address', *Naval Review*, Vol. XVI, No. 4, November 1928, p. 762.

56   Later Colonel Samuel John Woodruff Bassett (1889-1974). Naval Intelligence Division,

sioned from the ranks in 1915, took his Staff Course following two-years in the Intelligence Division securing his naval *psc* in late 1933.[57] He immediately proceeded to Colombo, Ceylon (present day Sri Lanka) where he served as SOI for three years. Though established at the same time as the SOO billet, the SOI became the common term used to designate both afloat and ashore Intelligence Officers. Meanwhile, Captain and Brevet Major Gordon Seath demonstrated another path to intelligence duties. Following a period at Greenwich where he was responsible for the training of Probationary Second Lieutenants, a status equivalent to the naval midshipman, Seath completed the 1922-23 War Staff Course. Posted for duties as the District Naval Intelligence Officer and then Staff Officer Intelligence at Gibraltar following his course, he departed in 1926 to spend four years in the Naval Intelligence Division. He then returned to duties of a more traditional nature with his Corps.[58]

Until 1922, there had been fourteen separate District Intelligence Offices located in Britain and across the Empire, though the closure of the centre at Ascension Island in October reduced the number by one and its duties were transferred to other stations.[59] This structure remained intact until 1938 when offices were added on the Africa and America and West Indies Stations and the two centres located in Canada became sub-centres reporting through Ottawa to the Admiralty.[60] When serving in home waters, the SOI typically did double duty as the SOO also. In this case, he was not likely to be a graduate of the War Staff Course (now restyled, Staff Course) though this changed over time. Instead, the officer was more apt to be a graduate of the more limited and focused Intelligence Course with Commander Henry Wilson[61] proving indicative of this trend. Completing his Intelligence Course in March 1924, he was immediately appointed to the *Victory* serving as a District Intelligence Officer and then SO&I to the Commander-in-Chief, Portsmouth.[62]

On the Mediterranean Station and in the China Fleet, the SOI served in the fleet flagship and typically held his *psc*. In this regard, King-Hall was one of the first Staff

1926, 1929, 1930-32, 1933 and 1936; Staff Officer Intelligence Capetown, 1926-29; Staff Course, 1933; Senior Officers' Course Browndown, 1933; Staff Officer Intelligence in *Hawkins II* and *Norfolk II*, 1933-36; retired, 1944. Recalled for war service, 1944; Colonel, 1946 and retired, 1960.

57 ADM 196/65 (Bassett).
58 ADM 196/63 (Seath).
59 ADM 182/34, Admiralty Fleet Order 3026, 10 November 1922.
60 ADM 182/98, Confidential Admiralty Fleet Order '127.—America and West Indies Station—Changes in Intelligence Areas and Centres', 27 January 1938 and 'Confidential Admiralty Fleet Order 1735.—Intelligence Areas and Centres—Changes on America and West Indies and Africa Stations', 21 January 1938.
61 Later Captain Henry Percival Wilson (1884-1957). Intelligence Course, 1924; District Intelligence Officer in *Victory*, 1924-26; Senior Officers' Technical Course, 1926; Staff Officer Operations and Intelligence in *Victory*, 1926-27; Commanding Officer, *Winchester*, 1929-30; Captain and retired, 1930. Recalled for war service.
62 ADM 196/126 (Wilson).

Officer's Intelligence serving in the *Queen Elizabeth* until replaced by Lieutenant Commander Malcolm Farquhar[63] upon the shift of the Commander-in-Chief's Mediterranean flag to the *Warspite*. Meanwhile, in the China Fleet, Lieutenant Commander Ivan Franks[64] filled the same capacity in the *Hawkins* until Commander Richard Schwerdt[65] relieved him.[66] Franks posting to China for a second time as an Intelligence Officer had been at the specific request of the Director of Naval Intelligence to fill an important assignment at what was deemed a critical moment. Unfortunately, a recurring medical problem rendered him unfit with retirement soon following.[67] Schwerdt came to the *Hawkins* from Greenwich where he had just finished taking the Intelligence Course before its consolidation with the revamped Staff Course. A member of the 1919-20 War Staff Course, fluent in German and already a proven Intelligence Officer, Schwerdt's taking of the Intelligence Course was probably an effort to appraise him of the unique circumstances and reporting requirements of serving in China.[68] It also demonstrates the limitations of the Staff Course that sought to produce a general type of staff officer and not a specialized one.

The SOI, of necessity, kept on eye on the dispositions of foreign warships and commercial trade operating within his area, but he was responsible also for tracking former naval and marine officers who had settled on the station upon leaving the Service. These officers represented a ready pool of personnel available for recall to serve on a convoy's staff or to round out the compliment of the local element supporting the Naval Control of Shipping.[69] At a level below the SOI or his earlier counterpart,

63   Later Captain Malcom Farquhar (1895-1989). Observer in *Argus*, 1922-25; Staff Course, 1925-26; Naval Intelligence Division, 1926; Staff Officer Intelligence in *Warspite*, 1926-27; Observer in *Argus*, 1929, in *Furious*, 1929-30, in *Courageous*, 1930-31, in Fleet Air Arm, 1931-32; Naval Air Division, 1932-34; Commanding Officer, *Crescent*, 1934-36; Staff Officer Operations in *Courageous* and *Ark Royal*, 1936-39; retired and granted war service rank of Captain, 1946.

64   Later Acting Captain Ivan Bromhead Franks (1893-1960). In Command, HMS *Sandpiper*, 1919-20; Liaison Intelligence Officer Hong Kong in *Tamar*, 1920-22; Naval Intelligence Division, 1922-23; Staff Course, 1924-25; Staff Officer Intelligence in *Hawkins*, 1925-26; Tactical School, 1927-28 and retired, 1928. Recalled for war service and Acting Captain, 1944.

65   Later Captain Charles Maxwell Richard Schwerdt (1889-1969). Qualified War Staff, 1919; War Staff Course, 1919-20; War Staff Officer in *Caledon*, 1920-22; Intelligence Course, 1926; Naval Intelligence Division, 1926; Staff Officer Intelligence in *Hawkins*, 1926-27; Plans Division, 1928-30; Tactical Course, 1932; Naval Intelligence Division, 1932-35; retired and Captain, 1936. Private Secretary to Governor of Newfoundland, 1936-39. Recalled for war service and attached to RCN.

66   *Navy List*, February 1926, pp 244, 261 and 314 and *Navy List*, October 1927, pp. 244, 286 and 314.

67   ADM 196/145 (Franks).

68   ADM 196/127 (Schwerdt).

69   Richmond, 'Staff Requirements of a Commander-in-Chief', *Naval Review*, February 1939, pp. 5-6.

the DIO, non-designated officers served in private ships as Reporting Officers. They supplied information to the SOI as defined in published Admiralty Intelligence Orders specifying the detailed reporting requirements for the station. The information sought covered not only purely naval matters, such as the sailings of foreign vessels, but also items of political and economic interest, notable foreign personalities present on the station, sources of supply, technical intelligence, and corrections in navigational data. The report submitted to the Naval Intelligence Division by the SOI typically was arranged by geographic area and the routine visits that a station's ship made to these locations served the purpose of gathering the necessary data to complete the report. Of course, other sources were used to facilitate fleet reporting requirements and might include data provided by knowledgeable officials, such as local consular officers or a friendly, foreign naval officer encountered during these visits, a review of local news-papers, or the monitoring of *en clair* wireless traffic of nearby vessels or transmitters. The report submitted also noted where information contained in Admiralty publica-tions issued under the Confidential Book or Official Use series was in error or lacking, and though the report was primarily a typed one running easily in excess of twenty pages, drawings often accompanied it providing amplifying information to the main body of the submission.[70]

For all the effort expended by the SOI in researching and drafting his periodic reports, the actual value of the product was thought to be little. Much of the infor-mation gathered was held to be merely encyclopedic in nature, of use perhaps to the Foreign or Colonial Offices, but of no naval value in the event of war. This was the conclusion of not only Vice Admiral Richmond complaining from the East Indies Squadron, but also of Roger Keyes, the Commander-in-Chief, Mediterranean who commented unfavourably on the present nature of the 'Commander-in-Chief's Intelligence Report'.[71] This prompted the Admiralty to modify its reporting require-ments in 1926 to stress the practical requirements for war whilst also noting that the questionnaire used to frame reporting requirements was but a part of the process.[72] The more important step was how the SOI kept his Reporting Officers appraised of the operational requirements unique to them based on the intelligence they would be expected to provide during war. Accordingly, Stephen King-Hall received especial

---

70   The above summary taken from a review of Drage diary for the period May 1925, Drage Papers, IWM/PP/MCR/99, Reel 2.

71   Director of Naval Intelligence memorandum 0123/26 of 7 July 1926, Dewar Papers, NMM/ DEW/4.

72   *C.B. 999. Naval Intelligence* and *C.B. 1518. Intelligence Organisation* (multiple volumes) were originally released in 1919. They described the reporting requirements and duties for Intelligence Officers. *C.B. 1578* was a 1921 update to the *Intelligence Organisation* manual whilst *C.B. 1666* replaced *C.B. 999* in 1924. *C.B. 3000. Naval Intelligence Duties* supplanted the *C.B. 1666 c.* 1926. None of these documents has been traced with their existence and purpose derived from their references in other Admiralty publications and correspondence.

praise for the style, content, and overall value of his report in this regard.[73] Such accolades were not new to King-Hall for a previous report submitted to the Intelligence Division on Kure, Japan in 1922 also received the appreciation of the Their Lordships, as did his appliance to assist in torpedo instruction.[74]

Marine officers attending the Staff College could expect an appointment to a staff position with intelligence duties a strong likelihood, but more difficult was it to find a posting that demanded the skills and knowledge imparted at the War Course or the Imperial Defence Course. One outlet would have been to assign these marines to the Admiralty, but the more typical assignment was a return to normal duties at one of the home shore establishments. Thus, given their seniority, few positions were open to them. Consequently, some soon retired and death through illness claimed others. As for those attending a Staff College, attachment to the Army such as Robin Campbell's experience upon leaving Camberley was another possibility. Yet, as with his naval counterpart who had to secure postings in command billets afloat to be eligible for promotion, the marine needed to return to his duties as well and serving as a Squadron and, then Fleet, Royal Marine Officer was the normal path.

Yet, those naval officers completing the Intelligence Course did not always proceed to an intelligence appointment in home waters. More typically, they went to command a minor ship—often deployed overseas—and this pattern confirms Admiral Henry Oliver's warning that the course provided a necessary and useful seasoning of an officer before assuming command of a minor vessel. Sidney Geary Hill, a qualified Hydrographer, went from his course to commanding the surveying ship HMS *Endeavour*.[75] As he had already commanded *Endeavour* previously and was to retire in 1927 upon relinquishing its command, the course's aim in his case was never about making him a professional Intelligence Officer but rather an officer competent to see the intelligence potential in any survey undertaken. An officer more closely matching Oliver's reminder concerning attendees of the course was Commander Ralph Wilkinson[76] who went to HMS *Chrysanthemum* in the Mediterranean Fleet. Both Wilkinson and Geary Hill were from the same class as Henry Wilson.

Others, though, might hold a dual-assignment as Brownfield had with the Second Battle Squadron. This included Commander Ronald Bayne,[77] a gunnery specialist

73   Director Naval Intelligence memorandum 0123/26 of 7 July 1926, Dewar Papers, NMM/DEW/4.
74   ADM 196/55 (King-Hall) and ADM 196/145 (King-Hall).
75   *Navy List*, July 1925, p. 237 and *Navy List*, January 1930, p. 504.
76   Later Captain Ralph Wilmot Wilkinson (1884–1936). Commanding Officer, *Seawolf*, 1919-21; Senior Officers' Technical Course, 1921; Intelligence Course, 1924; Commanding Officer, *Chrysanthemum*, 1924-26; Naval Intelligence Division, 1926-29; Captain and retired, 1930.
77   Later Acting Captain Ronald Christopher Bayne (1897-1979). Squadron Gunnery Officer in *London*, 1930-32; Senior Officers' Technical Course, 1932; Staff Course, 1933; Fleet Gunnery Officer in *Kent*, 1933-35; Staff Officer Operations and Intelligence in *Drake*, 1938-39; Acting Captain, 1942; Deputy Director, Trade Division, 1942-46 and retired, 1947.

serving in *Drake*, who performed the dual responsibilities of Staff Officer Operations and Intelligence (SOO&I).[78] Meanwhile, Commander Alfred Bingeman,[79] having completed the Long Torpedo Course, acted as SOO and Squadron Torpedo Officer in *Barham* of the First Battle Squadron.[80] Thomas Drew,[81] only recently returned to England and now Flag Captain in *Royal Oak* and designated to be Flag Officer Malaya, asked Brownfield to accompany him to Singapore to serve as SOO and Naval Attaché Siam—an offer the latter quickly accepted.[82] Having an officer perform double-duties may have only reflected that some jobs were less taxing than others—Brownfield's duties as attaché required only quarterly visits to Bangkok, but it was not a practice sustainable in war. Even in peacetime, some officers felt burdened by the practice of having to manage too much and despaired that time was ever sufficient to attend to either requirement adequately. Still, on a minor station there might not be a designated SOO or SOI at all. In this instance, a single officer covered both duties whilst other officers saw to a squadron's or flotilla's specialist duties in gunnery, torpedo, signals, and the like. Bad as the practice was for a staff officer, it was even more pernicious when involving a commanding officer, as events will soon depict.

Other officers performing staff duties were destined to elements where a flag officer was not present, though performing many of the same duties as a SOO or SOI, and were simply carried as Naval Staff Officers or Staff Officers. Such was the case in the Royal Canadian Navy where Commodore Walter Hose,[83] commanding for much of the period, relied on the support of trained officers shorn of the SOO title. Finally, others securing their *psc* served only the briefest of periods as a staff officer. This was most apt to be true of the officer who had previously specialized in gunnery or torpedo. These skills were ever required and aggravating matters was that officers were not electing to specialize in the numbers required.[84] Lieutenant Commander William

78   ADM 196/119 (Bayne).
79   Later Captain Alfred Mervyn Bingemann and, from 1937, Bingeman (1900-1988). Short Course, Cambridge University, 1921-22; Torpedo Officer in *Emerald*, 1927-29; *Vernon*, 1929-31; Torpedo Officer in *Glorious*, 1931-33; *Orion*, 1933-36; Squadron Torpedo Officer in *Orion*, 1936; Naval Equipment Department, 1936-37; Staff Course, 1938; Staff Officer Operations in *Hood*, 1939; Staff Officer Operations and Squadron Torpedo Officer in *Barham*, 1939; Acting Captain, 1943; Captain and retired, 1950.
80   ADM 196/148 (Bingeman) and *Navy List*, March 1939, p. 219.
81   Later Vice Admiral Thomas Bernard Drew (1887–1960). In Command, Coastal Motor Boat Flotilla in HMS *Dolphin*, 1923-25; Captain, 1928; Senior Officers' School Sheerness, 1929; Commanding Officer, HMS *Ambuscade*, 1929-30, *Active*, 1930 and *Vansittart*, 1930; Senior Officers' War Course, 1930-31 and 1936; Senior Officers' Technical Course, 1931 and 1936; Flag Captain in *Kent*, 1932-33; Deputy Director, Personal Services Department, 1933-35; Tactical Course, 1935; Flag Captain in *Royal Oak*, 1936-38; Rear Admiral, 1939; Vice Admiral, 1943 and retired, 1943.
82   Brownfield, *Sea Cycle*.
83   Walter Hose (1875-1965). Captain, 1918; Director, Naval Service, 1921-28; Chief of the Naval Staff, 1928-34; Rear Admiral and retired, 1934.
84   Gamma, 'In Praise of Specialists', *The Naval Review*, Vol. XXV, No. 3, August 1937, p. 478.

Parry upon, completing the War Staff Course in 1923, went to *Curacoa* as War Staff Officer in the Second Light Cruiser Squadron. After a year in *Curacoa*, he was back in the *Vernon* before serving in *Dolphin* for duty with submarines.[85]

In light of the overt information-gathering role that they performed, one assignment where officers previously performing intelligence duties might be expected to feature was as a naval attaché. Surprisingly, this was rarely the case and Captains Gerald Muirhead-Gould, accredited to Berlin from 1933-36, and Desmond Tufnell, serving in Tokyo during 1938-39, were notable exceptions to this pattern of employment. Of the two, only Tufnell possessed the naval *psc* reflecting the preference for assigning capable, generalist officers with a proven background over a formally trained staff specialist. Again, the barrier to non-executive officers attending the Staff College was palpable, as officers of the engineer branch routinely were accredited as an assistant attaché to serve under his executive superior. Still, even an experienced intelligence officer had limitations and Muirhead-Gould, if handy in several, was not a qualified interpreter in any language, let alone German, though Tufnell did at least speak Japanese courtesy of the School of Oriental Studies and time spent in Japan.[86] This handicap was also present in Captain Richard Bevan;[87] he had not mastered Italian before being accredited to Rome, though he did pass as a preliminary interpreter in the language in the final year of his first posting to Italy.[88] This raises the further question whether such limitations appreciably altered the technical intelligence being collected against possible enemy navies before the onset of war. Rank contributed to the problem, as there were only two officers of the executive branch holding Captain's rank qualified in Italian in January 1931 and one of these was Bevan's predecessor in Rome.[89] The case for the Berlin posting at the time of Muirhead-Gould's tenure was even slightly worse; only one Captain of the executive branch was a qualified interpreter in German and he was commanding his cruiser in the Mediterranean.[90] Whilst this attaché position had only recently been upgraded, and, doubtlessly, made the matter more challenging, the conclusion that the Admiralty relied too much on

85   ADM 196/145 (Parry) and ADM 196/127 (Parry).
86   ADM 196/145 (Tufnell).
87   Later Rear Admiral Sir Richard Hugh Loraine Bevan (1885-1976), Flag Lieutenant in *Orion*, 1912-13 and HMS *Charybdis*, 1914-16; Designated War Staff Officer, 1918; War Staff Duties in *Queen Elizabeth*, 1919; Signal Department, 1920-21; Captain, 1923; Senior Officers' Technical Course, 1924; Senior Officers' War Course, 1924; Flag Captain and Chief Staff Officer in *Birmingham*, 1924-27; Senior Officers' School Sheerness, 1927; Naval Attaché Rome, 1928-31 and 1940; Tactical Course, 1931; Commanding Officer, *York*, 1931-34; In Command, Signal School, 1934-35; Rear Admiral and retired, 1935. Recalled for war service.
88   ADM 196/49 (Bevan).
89   *Navy List*, October 1930, pp.315-16A and 421. Bevan succeeded Captain Charles Burke.
90   *The Navy List* (London: His Majesty's Stationery Office, February 1936), pp. 230, 317-19, and 430. Captain George Thomson, the only executive branch Captain qualified in German, commanded the *Devonshire* from 1935-36, being relieved by Muirhead-Gould.

the initiative of its officers to secure proficiency in at least one other language rather than simply ordering an officer to do so is a difficult conclusion to avoid. In time, Muirhead-Gould mastered German though never receiving the pay of a qualified interpreter. That his responsibilities were broader than reporting only on Germany was also a contributing factor and his ease with Russian made him excellent attaché material in any event.[91]

Those that did secure a language qualification often did so whilst waiting to attend their Staff Course or before going to the IDC. With France so near, officers often opted to go there to take a cramming course in the language before returning to Greenwich or London. Thus, Captain John Durnford proceeded to France in the summer of 1935 immersing himself in the language and culture, before sitting his proficiency examination in French. He then entered the 1936 class of the Imperial Defence College.[92] Meanwhile, Muirhead-Gould studied German whilst serving in Malta in 1932 and spent a brief period in Germany to brush up his natural language skills before assuming his attaché duties; a course of action not too dissimilar from Bevan who spent time in Italy immediately before his appointment.[93]

As for the actual work pursued by the staff officer, many felt overwhelmed by the scope of what was required. To this, other officers believed he deserved all that he got as 'the zeal of these officers is usually such that they require about twenty months in every year to fit in all the exercises, practices, sports and other stunts which they think should be carried out. This is all very laudable, but it is bound to lead to the extraneous subjects suffering.'[94] Captain Bent, the critic in this case, was not alone in this view. When at a later moment Cyril Parkinson derived the law for which he is famous upon observing the workings of the British Civil Service, he was merely codifying what victims and practitioners both commonly appreciated was a tradition where work and staffs grew by ever-increasing amounts.

A consequence of such regulation of affairs was that a belief took-hold that the staff system and its practices needlessly stifled the initiative of commanding officers.[95] Common enough in naval circles before the 1939-45 war, this was appreciated by others following it. Field Marshal Wavell, for one, lamented the problem writing to Liddell Hart:

> [O]fficers are constantly being taken away for staff duty, whether they want to or not; and an entirely undue proportion of the Commanding Officer's time is taken up by answering questions from above and getting permission from the staff to do things

91   ADM 196/92 (Muirhead-Gould).
92   *Shellback Remembers*, p. 170, Durnford Papers, IWM/PP/MCR/C10.
93   ADM 196/92 (Muirhead-Gould) and ADM 196/91 (Bevan).
94   Eric R. Bent, 'The Staff and Its Relation to the Work of the Fleet', *The Naval Review*, Vol. XVII, No. 3, August 1929, p. 553.
95   Drax, 'Naval Efficiency and Discipline'. *Naval Review*, November 1944, p. 301.

that he should certainly be able to do on his own initiative, and would have in the old days.[96]

This was a difficult balancing act to secure. Doubtless, the reduced manning of the ships as the period progressed made the problem ever more acute. Returns on fuel expenditure, equipment casualties, reports of proceedings, and analysis of results achieved in tactical exercises had to be prepared at the unit level, reviewed and collated at the squadron and fleet levels, before submission to the Admiralty. The reports submitted to the Admiralty were reviewed at the Secretariat level and shared with the appropriate arms of the staff. A particular point regarding a recent visit to a foreign port might be shared with the Foreign or Colonial Office and comments noting deficiencies were passed to the responsible element concerned with the defect in question. Summaries of tactical exercises were studied by the separate elements of the Naval Staff, particularly, the Gunnery and Torpedo Divisions, until overall responsibility for preparing these annual reports became the responsibility of the newly styled Tactical Division in 1929 and the Training and Staff Duties Division.[97] Here the submitted reports were mined for key findings and used as the primary source material for the annual confidential book, *Progress in Tactics*.[98]

Meanwhile, a regular pattern of evolutions continued such as full-speed trials, gunnery, and torpedo exercises, which further reduced the discretion a commanding officer enjoyed for regulating the internal affairs of his ship with even the seemingly trivial regulated by central direction.[99] In truth, the staff officer was as much a victim to the process as those in command. If oil, coal, and aviation spirit operated the seagoing fleet, then information directed the Navy. Absent the modern computing and communications appliances of today, perforce, the data required a great deal of time and effort to accumulate and assimilate.

The same was true at the fleet level, as the fleet seldom operated as a single entity. Squadrons and ships most times operated independently, showing the flag, yes, but also conducting minor tactical evolutions. Coordinating the constituent parts of the fleet required much advanced planning and the issuing of orders outlining the general training scheme being pursued. Defects in individual ships and the periodic refits in other vessels had to be coordinated with the nearest dockyard. All the while, ships operated to a regular commissioning schedule necessitating either their return to homeports or the transfer of personnel between ships. However, life outside the fleet operated to its own patterns. An emergent crisis or natural disaster put-paid the

96   Wavell to Liddell Hart letter of 29 October 1949, LH 1/733/373, LHCMA.
97   ADM 182/55, Admiralty Fleet Order '3251.—Naval Staff—Revised Peace Organisation', 28 December 1928.
98   'C.B. 3016/30, Progress in Tactics, 1930', p. 3, British Sources, Box 12, Naval Historical Center, Washington, DC.
99   Bent, 'Staff and Its Relation to the Work of the Fleet', *Naval Review*, August 1929, p. 556.

orderly schedule previously arranged, as ships were detailed to meet the problem and staff officers struggled with revising their now outdated plans.

To the modern eye, one is struck not so much by the degree of control and reporting that was required in the interwar Navy, as how this was accomplished absent the modern means of collating, assessing, and transmitting the data once collected. Those contemporary officers complaining of a system running amok were those who had come of age when the telephone was still new, wireless was supplanting the post, and correspondence was drafted in longhand in the first instance. Often, the staff officer and his ways were viewed as needless intrusions into Service life and affairs. Even Rear Admiral John Harper, sympathetic to the need for a properly aligned and functioning staff, conceded that the pendulum might have swung too far observing:

> It may be that the prejudice which still exists against the "staff system" is due in large measure to the unnecessary and often irritating mass of detailed orders and instructions penned by zealous officers and, unfortunately, issued broadcast under the authority of their chiefs. This is a product of a system which has not yet found its proper level; but should not be used as an argument against the existence of a Staff. It may fairly be considered to be caused by the Staff being larger than is required to compete with the *necessary* work. If a staff is too large it is obvious that paper work will assume unnecessarily large proportions.[100]

Such regulation of affairs was thought to skew an officer's proper sense of priorities. This led to an undue emphasis focused on the material side of things at the expense of overseeing the development of good order and discipline amongst the lower deck with morale necessarily suffering as a result. Following the mutiny in the Atlantic Fleet at Invergordon, this view gathered credence, though a certain measure of *post hoc ergo propter hoc* reasoning is manifestly evident by those reaching such conclusions.[101] Paymaster Lieutenant Commander Arthur Duckworth, serving in the *Nelson* and Secretary to Rear Admiral Colvin, the Chief of Staff of the Atlantic Fleet, was one officer believing first principles had been abandoned. Though not quite saying 'Millions for defence, and not a cent for tribute', Duckworth believed the Admiralty had to meet the demands of the men or worse would follow. Behind it all, Duckworth believed the Navy had gotten its priorities terribly wrong with undue emphasis placed on gunnery readiness and fleet action in the false expectation that immediate readiness for war was required when war was nowhere on the horizon. This attitude was nonsense, in his eyes, entirely wasteful and owed everything to

---

100 Harper, 'Staff or Anti-Staff', *Journal of the Royal United Services Institution*, November 1928, p. 718. Original emphasis.
101 Drax, 'Naval Efficiency and Discipline'. *Naval Review*, November 1944, p. 302.

the likes of Admirals Dreyer and Astley-Rushton.[102] This assessment must be seen in the immediate context with which it was rendered. One story in the air was that Astley-Rushton had pressed Tomkinson to take the fleet to sea to continue the series of exercises then planned when the disturbances arose but Tomkinson desisted.[103] If the story cannot be confirmed, it remains in keeping with Rushton's manner and his motives were probably to get the ships out of port, away from any influences operating ashore, and have the men concentrate on their duties and not their pay-packets.

If Duckworth's verdict is somewhat overheated, it, nevertheless, demonstrates that the many strides taken to make the Navy a better instrument of war needed to be balanced by human considerations keeping in mind who actually did the work. Colvin had just gone to London to appraise the Admiralty and, subsequently, the Cabinet, of the problems in discipline now facing the Atlantic Fleet. Grenfell knowing Colvin well and personally liking him, believed his professional success owed much 'to keeping on the right side of authority.'[104] If so, his hour was at hand. With events ongoing and uncertain, Colvin was more than just a conduit between the fleet and the Admiralty though he was certainly that. He was an eyewitness to the disturbances and Ministers relied upon his observations to capture the scale of the problem and the best means of resolving it.[105] Colvin's presence in London suggests that, fundamentally, Tomkinson was a weak leader seeking support that would not have been available to him if faced with similar circumstances on a foreign station. The reliability and security of his communications to the Admiralty may have necessitated Colvin's departure, but the expedient of shifting his flag to a more reliable ship remained a measure Tomkinson failed to adopt first.[106]

How much Duckworth's verdict mirrored that of his superior cannot be said, but his specific references to Astley-Rushton, then serving in the Atlantic Fleet and commanding the Second Cruiser Squadron and, more importantly, to Dreyer, presently on the Board of Admiralty, probably owed something to Colvin. Shortly, the views espoused by Duckworth were mirrored in official circles. Introducing the *Estimates* to parliament in the following March, Sir Bolton Eyres Monsell, the First Lord, pointed to the over-emphasis in material training which plagued Service life remarking:

> I believe that the highest degree of efficiency is attained only when the balance of training in material and *moral* is properly adjusted. Two main results have been produced by this disequilibrium of training. The first is that discipline and *moral*

102 Diary entry of 18 September 1931, Captain Arthur Dyce Duckworth Papers, Imperial War Museum London, IWM/76/207.
103 Diary entry of 14 September 1931, Drage Papers, IWM/PP/MCR/99, Reel 3.
104 Grenfell to Liddell Hart letter of 6 June 1937, Liddell Hart Papers, LH 1/330/11, LHCMA.
105 CAB 23/68, Cabinet minutes of 16 September 1931.
106 On the question of illicit visual and electronic communications during the mutiny, see diary of 15 September 1931, Drage Papers, IWM/PP/MCR/99, Reel 3.

have suffered, because intensified fleet training has left insufficient time to develop in officers and petty officers the arts of leadership and the power of command, and a too centralised system of fleet command, a too highly organized routine, tends to crush initiative, and fails to provide opportunities for the development of individuality.[107]

Monsell's statement taken at face value offers a damning indictment of the senior leadership of the Royal Navy depicting as it does a body more interested in securing material perfection achieved at the expense of personnel satisfaction and contentment. In fairness to Monsell, he was generally capturing the views of those six Commanders-in-Chief who had reported on the state of their fleets, squadrons, or establishments in light of the recent problems experienced. Monsell conceded that the underlying causes of recent affairs were 'too short commissions and too frequent changes of officers and men during those commissions.'[108] Yet, the statement is misleading on a number of accounts.

In the first instance, it fails to consider the proximate cause of the disturbances. This was the pending reduction in pay recommended in the *May Report*. This had been an issue of long-standing exasperated by the differing scales of pay operating within the Royal Navy and dating from 1925. Servicemen engaged before October 1925 received a higher rate of pay dating from 1919 than those subsequently enlisting. Then, the Government was unwilling to concede that a right to a certain scale of pay existed, but fairness alone appeared to dictate that one's pay should not be reduced for those already serving.[109] Astley-Rushton, an officer very much at the centre of events during the Invergordon Mutiny, had forewarned of the problem some years previously during an address to a War Course whilst serving as Director of the Staff College. Outlining why certain non-military subjects were taught at the Staff Course, he explained:

[C]ertain subjects which, though not especially connected with the service, are essential to the intelligent understanding of problems of the day, and others of a professional nature which are not dealt with on similar lines elsewhere. Under the former category comes economics; i.e. pure economics, not applied economics, on which we have two lectures. These help us to understand such questions as reparations, and the relationship between money and real wages. Whether the pay of the lower deck will sooner or later be decreased is a question on which I will not offer an opinion; but should that day come, we are likely to be able to get through a difficult

107 Sir Bolton Eyres Monsell, 'The First Lord's Statement', *The Naval Review*, Vol. XX, No. 2, May 1932, pp. 217-18. *Naval Review* Editor's emphasis.
108 *Hansard*, House of Commons Debates, 7 March 1932, v. 262, cc1493-1573.
109 *Ibid*, House of Commons Debates, 17 September 1931, v. 256, cc1104-1108 and CAB 24/174, 'Second Committee on Marriage Allowance for Naval Officers and Rates of Pay of New Entrants in the Fighting Services. Vested Rights. Report', C.P. 371 (25), 24 July 1925.

process more easily if the naval officer is in a position to explain intelligently the reasons to his men.[110]

If Rushton's cautionary words proved too optimistic at the day of reckoning, then part of the reason must lie in the context in which the cuts occurred, and, most especially, due to the operational tempo of the fleet. A root cause of the imbalance in training was the steady reduction in the *Estimates*, as amplified in Appendix I. Accordingly, the problem was not the intensity of training—after all the Mediterranean Fleet worked to a quicker pace—but rather the compression of training, which aimed to provide the same degree of readiness into a more abbreviated operating cycle. Steaming hours were curtailed in the Atlantic Fleet as a measure of economy and when exercises were scheduled, they frequently followed periods of leave or refit. Often this required many preparatory tasks, such as painting ship, storing ammunition, and calibrating compasses to be completed quickly in order to meet a proposed sailing date. If officers and men serving in ships had remained constant, the ill-effects of this abbreviated cycle would have been less drastic. This was not to be and officers—particularly senior officers—rotated through ships on a yearly basis leading to a severance of ties between those leading and those led.

This may be contrasted with the regimental system where an officer could spend a large measure of his career amongst the same men and fellow officers. The regimental system had other faults, but it had its strengths, too, and Ironside commenting on the recent events within the Atlantic Fleet observed, 'The Navy seem to be in a bad way. I don't think the officers look after their men enough & don't explain things to them. The officers are never in touch with the men in the same way as ours are.'[111] To this there was much truth. Julian Patterson, Flag Captain in *Hood*, dated his appointment only from 27 April 1931 whilst the *Valiant*'s commanding officer, Captain Charles Scott,[112] had assumed his duties on 2 December 1930. Both ships were at the fore in the disturbances and compounding matters for *Valiant* was that her 'Jimmy' had left the ship in July 1931 to undergo an operation and had yet to return. Though Commander Eustace Wace[113] coming from the Royal Naval Barracks Portsmouth as a temporary replacement to fill the gap was held to be a first-rate officer,[114] Drage,

---

110 ADM 203/69, Astley-Rushton, 'Staff College' lecture, no date but *c.* 1924, p. 15.

111 Ironside to Liddell Hart letter of 8 November 1931, Liddell Hart Papers, LH 1/401/72, LHCMA.

112 Captain Charles Arthur Scott (1880-1940). Naval Ordnance Department, 1914, 1918 and 1925-27; Commanding Officer, HMS *Gorgon*, 1918-19; Local Defence Division, 1919; Commanding Officer, HMS *Sterling*, 1920-21; Captain, 1920; Deputy Director, Training and Staff Duties Division, 1922; Commanding Officer, *Calypso*, 1922-24; in Command, *Excellent*, 1928-30; Commanding Officer, *Barham*, 1930 and *Valiant*, 1930-31 and Senior Officers' War Course, 1932.

113 Later Captain Eustace Harold Wace (1888-1951). Air Ministry, Directorate of Instruments, 1924-26; Captain and retired, 1934.

114 ADM 196/127 (Wace).

now serving as head of *Valiant's* Communications Department believed the change, if welcome, would not help really help matters in the ship for the appointment would only last a few weeks.[115]

Nor were circumstances much different in the case of the *Rodney* and HMS *Adventure*, two other ships at the fore in the disturbances. Captain Bellairs' appointment to the former dated from 16 December 1930, whilst the minelayer *Adventure* saw Lieutenant Commander Eric Oatley,[116] an officer entering via the special entry programme and now serving temporarily in command from only 20 August 1931.[117] Oatley was perhaps most unfortunate in these circumstances. *Adventure* normally was commanded by an officer of Captain's rank and Oatley's period of responsibility coincided with a period of maintenance and refit. A gifted Torpedo Officer with a strong academic bent to him—he excelled at mathematics and had previously received recognition from Their Lordships for his proposals for a Torpedo Tactical Plotter. Now, though, he was acting as a caretaker with the responsibilities of command thrust upon him. A non-smoker and teetotal, atypical habits for his times, he was thought a poor leader as he lacked tact and social skills. Indeed, in June 1931, three-months before Invergordon, Roger Keyes had reported against his promotion.[118] Notwithstanding his executive limitations, when *Adventure* re-commissioned in the spring under a new captain, Oatley remained as the ship's Mining Officer; the duties he was imminently fitted to serve.[119]

A pattern of affairs not much better existed in the instance of subordinate officers—those presumably in closest touch with the men—in each of the previously named ships. In the case of *Hood*, she had only come out of a long refit in March 1931 and had just initiated her working up period in June. *Valiant* was of similar circumstances re-commissioning on 2 December 1930 after undergoing a large reconstruction. Compounding the matter was that the ships were also coming off periods of leave granted at the end of the summer cruise. Drage, himself, was absent *Valiant* for such reasons from 19 August to 1 September 1931 and when he returned he noted several changes in the wardroom observing that 'some of the most useless blokes have gone: but there are still far too many duds in the ship.'[120] Clearly, the days when an officer could spend 4½ years in the same ship as happened during the war were no more and creating any reservoirs of mutual sympathy and loyalty in such an environment was never going to be easy.

115   Diary entry of 10 July 1931, Drage Papers, IWM/PP/MCR/99, Reel 3.
116   Later Commander Eric Stanley Oatley (1899–1985). Short Course, Cambridge University, 1920-21; Squadron Torpedo Officer in *Cleopatra* and *Marlborough,* 1926; Flotilla Torpedo Officer in *Broke,* 1927; In Charge, *Adventure,* 1931-32; Naval Ordnance Department, 1936; Torpedo Officer in Reserve Fleet, 1936-39 and retired, 1943.
117   *Navy List,* January 1932, pp. 215, 246, 265A and 276.
118   ADM 196/148 (Oatley).
119   *Navy List,* July 1932, p. 214.
120   Diary entry of 1 September 1931, Drage Papers, IWM/PP/MCR/99, Reel 3.

Meanwhile, the series of evolutions necessary to make a ship and its company proficient continued apace, though many of these now took place in harbour rather than underway as the fuel and ammunition allowances of the fleet were curtailed. This the result of economies previously imposed. Arguing that over-centralized planning and tasking were the root causes of the mutiny avoids the more fundamental problem that such control was required in the first place because of the fragmented and constrained operating cycle, the revolving-door manning policy in officer appointments within the Atlantic Fleet, and the lack of staff officers available at subordinate levels. These issues were directly tied to the level of funding received and this was primarily a political question. Reducing the number of officers and ratings borne and the ships in commission to what was financially affordable would have helped matters. Still, this too was a question of policy which the Government was wont to address committed as it ostensibly was to a single-power standard. Monsell's proposed solutions to the problems raised at Invergordon was to stop the revolving door in manning by keeping ships in commission for longer periods and to increase the promotion opportunities for petty officers by fostering their talents and skills. In short, he was seeking to mitigate the worst effects of the recent pay cuts by providing increased responsibilities and, hence, pay to the lower deck.[121]

Pay though the proximate cause of the mutiny is insufficient to explain why ill-discipline arose in the Atlantic Fleet, but was largely absent in the overseas fleets or in the shore depots of the home establishment. Again, this points more to the operating cycle of the fleets as the underlying problem and the need to complete work rapidly to meet projected sailing dates. Here, the case of the *Lucia* is noteworthy. Fully eight months before the events arising at Invergordon *Lucia* had faced a similar outbreak of ill-discipline. This transpired when her crew worked long hours on an abbreviated schedule to prepare the submarine depot ship for sea to support a series of forthcoming exercises. The commanding officer of the *Lucia* was absent from his ship at the moment of crisis in January 1931, as he also held the dual appointment of Commander, Second Submarine Flotilla and was tending to flotilla duties.[122] These duties included overseeing the training of all crews and ensuring that the equipment of the flotilla's submarines was in working order so that the boats could proceed safely to sea. In January 1931, there were six submarines in the flotilla and when combined with the duties of his own ship, Commander Oswald Hallifax,[123] faced a daunting task.[124] Responsibility for completing *Lucia*'s preparations, thus, fell to the First Lieutenant

---

121 *Hansard*, House of Commons Debates, 7 March 1932, v. 262, cc1493-1573.

122 *Ibid*, House of Commons Debates, 11 March 1931, v. 249, cc1205-75.

123 Later Captain Oswald Ernest Hallifax (1888–1984). Commanding Officer, *B 5*, 1915-16, *D 7*, 1916-17, *E 56*, 1917-18, *J 7*, 1919-22 and *K 6*, 1922-24; Staff Course, 1924-25; Naval Intelligence Division, 1925-27; Plans Division, 1928-30; Commanding Officer *Lucia* and Commander, Second Submarine Flotilla, 1930-31; Captain and retired, 1935. Recalled for war service.

124 The Second Submarine Flotilla consisted of *L 52*, *L 53*, *L 54*, *L 55*, *L 69*, and *L 71*.

and his problems were not lessened by the fact that the ship could not be undocked as originally scheduled due to bad weather. Accordingly, a day was lost and much remained to be done in the brief period remaining before the ship was due to sail. That all this was happening immediately before and after New Year's Day, a period when duties are normally light, only added fuel to an already heated situation. The Court of Enquiry convened to weigh the matter faulted both the commanding officer and the First Lieutenant, but noted the mitigating issues of officers only recently appointed to the ship, the holding of dual-appointments, a lack of tact exercised by officers, and the apparent apathy of petty officers.[125] All factors which were to feature again at Invergordon.

Given the gravity of these events and Hallifax's relief and placing on half-pay as an indication of Admiralty displeasure, questions soon followed in parliament.[126] Seeking assurances that the procedures for dealing with lower deck grievances were understood within the Service, Cecil L'Estrange Malone, a former naval officer and now the first Communist sitting in the Commons, pressed for a judicial inquiry to consider the administration of naval law and to review the machinery for handling grievances.[127] This did not occur and whilst a single event does not necessarily portend a broader problem, hindsight shows that this was an opportunity missed. A measure of complacency and expediency greeted the problem by A.V. Alexander, the First Lord, who assured the House that the Admiralty was on top of the problem. Beyond Hallifax, two other officers serving in the *Lucia* were relieved of their duties and placed on half-pay whilst several seamen where separated without recourse following the practice of 'Services No Longer Required' that allowed the Admiralty to voluntarily terminate a rating's enlistment.[128] As Hallifax was the only qualified submariner in the *Lucia*, his flotilla responsibilities could not be delegated to others, and he had perforce to rely on his own ship's officers to meet their duties with little oversight. Though it is impossible to render a view as to his overall culpability in the matter, the Admiralty must share some of the blame for the events arising in not affording him officers better trained to work in a submarine flotilla. As for Hallifax, his last assignment before retiring in 1935 was to be in charge of the Portsmouth detention cells; an appointment reflecting more poorly on the Admiralty than the man who held it.[129]

Even an officer every bit as committed to the tactical excellence of the Navy recognized that the balance in training was less than ideal. Writing some eight years previous to the events at Invergordon, Russell Grenfell registered the view that:

125 *C.B. 3027(A), Mutiny in the Royal Navy*, pp. 9-11.
126 ADM 196/127 (Hallifax).
127 *Hansard*, House of Commons Debates of 28 January 1931, v. 247, cc949-53.
128 *Ibid.* Though a rating discharged SNLR had no right of appeal, an officer was in a somewhat better position in that he could request a court martial. This was not always granted by the Admiralty and, given the adverse commentary surrounding the *Royal Oak* affair at the time, was increasingly not apt to be granted to forestall a similar situation arising.
129 ADM 196/52 (Hallifax).

> Weapon Technique receives a much larger share than does tactics of the energy and thought devoted to training the Navy for battle. It is unnecessary to enumerate the extent to which weapon technique dominates the yearly timetable, the large staffs engaged in the training of the specialist personnel on shore, and the numerous and costly establishments that exist for that purpose.[130]

Though writing from differing perspectives, both Duckworth and Grenfell saw a Navy consumed with the material side of things. Duckworth castigating the human toll of misplaced priorities whilst Grenfell deprecated its tendency to not see the forest for the trees. Not surprisingly, Grenfell was a salt horse, but even trained specialist officers, such as Richmond and Drax, concurred that the naval officer needed to think about the broader aspects of his profession.

Admiral Sir Barry Domvile, a former Admiral-President of Greenwich, thought things had gone too far within the profession embracing specialization inordinately and firmly held to the salt horse tradition of the Service, though, in fairness, it must be admitted that Domvile was a highly respected gunnery specialist.[131] A natural reactionary, Domvile's politics eventually moved even further to the Right than his naval views as he looked askance at the policy of adopting oil rather than continuing to rely on coal and marking him out as an archconservative in naval developments.[132] Making an officer a better seaman, it was also believed made one a better leader with enhanced man-management skills when contrasted to the specialist officer.[133]

The first claim may well have been correct, but even the specialist officer had to work as part of a team so there was nothing, *per se*, in specialization that impeded an officer's ability to acquire leadership skills. Though Staff and War Course training did touch upon aspects of morale and leadership, the topics were tangential to both and, for the War Course, a qualifying officer usually introduced the subject. For the lower course, this may have reflected nothing more than there were no equivalent positions to SOO and SOI in the staff organization of the fleet to oversee personnel matters. True, chaplains were borne in the larger ships and a Fleet Royal Marine Officer was responsible for arranging the affairs of his men. Yet, the emphasis of the former was not on addressing the implications of Admiralty or Government fiats and the duties of the latter were oriented to ensuring that the military side of things was in order. The Captain of the Fleet and his secretary certainly had a role to play in overseeing the personnel side of the fleet. This focus, however, was more attuned to ensuring the conditions within a ship and the fleet were in order, and though questions of pay came

---

130 Grenfell, 'Training in Tactics', *Naval Review*, p. 681.
131 ADM 196/142 (Domvile).
132 HO 283/31, Advisory Committee to Consider Appeals Against Orders of Internment, 'Notes of a Meeting held at 6 Burlington Gardens, W.1, on Monday, 6 October, 1941.' Some politicians, such as James Maxton, favoured the use of coal as it kept miners employed, but this was not the motivating reason in Domvile's case.
133 Gamma, 'In Praise of Specialists', *Naval Review*, August 1937, p. 478.

under his purview, here, it was about ensuring the correct allocation of established pay and allowances. Beyond these concerns, the Captain of the Fleet ensured the logistical affairs of the fleet were satisfied and that other administrative matters were settled.[134] By regulation, the Captain of the Fleet was responsible for the material condition of the command and this left little time for considering personnel issues not directly related to the fleet and its operational efficiency.[135] Moreover, a ship's Captain was supposed to have a grip on his crew. In short, changes in conditions of service were matters, in the first instance, primarily of interest to the Admiralty affecting as they did the entire Navy. This view was not wrong, but it was narrow and reflected the tenor of the times.

This is not a study of the Invergordon Mutiny, but that episode does raise questions of leadership, culture, education and training, and ultimate responsibility for affairs that are central to this work. It is difficult to imagine that if the Admiralty had warned the Atlantic Fleet that its fuel allowance was to face a reduction similar in scale to that anticipated in pay, a communication would not have passed advising what the projected impact would be on its operational routine. In fact, as Duckworth's record notes exchanges along such lines had occurred. Upon the publication of the *May Report* in July 1931, produced at the request of the Government, a number of savings were specified to bring spending more in line with revenue. In fairness to the conclusions of the May Committee, they did not seek the savings attempted in the fighting services' Votes eventually pursued by the Government, as their initial recommendation was for a reduction of £5,000,000. Yet, a sizeable faction of the Labour Cabinet would not countenance the reductions in unemployment insurance envisioned by the *Report* and so an additional savings of £4,000,000 were sought in the Votes supporting the armed forces.[136]

The men of the fleet began to speak about the implications of the *May Report* even if their officers did not.[137] Assertive leadership, if only of a private nature, between the fleet and the Admiralty might have allowed both elements to get ahead of the problem, but the illness and temporary absence of the Commander-in-Chief, Admiral Sir Michael Hodges,[138] away in hospital created a void. Here the strongest evidence presented itself for why a staff was required and the benefits of a staff system. Though

134 Chatfield, *Navy and Defence*, p. 218.
135 *King's Regulations*, Article 1048, p. 470.
136 CAB 23/67, Cabinet minutes 41(31) of 19 August 1931.
137 *C.B. 3027(A), Mutiny in the Royal Navy*, pp. 16-17.
138 Admiral Sir Michael Henry Hodges (1874-1951). Captain 1908; War Course, 1908; Senior Torpedo and Gunnery Course, 1909; Senior Navigation Course, 1910; Naval Attaché Paris, 1914-15; Rear Admiral, 1920; Chief of Staff to Commander-in-Chief, Atlantic Fleet, 1919-20; Rear Admiral (D), Atlantic Fleet, 1920-22; Naval Assistant to First Lord, 1923-25; Vice Admiral, 1925; Commander, Third Battle Squadron and Second-in-Command, Mediterranean Fleet, 1925-26; Second Sea Lord, 1927-30; Admiral, 1929; Commander-in-Chief, Atlantic Fleet, 1930-31 and retired, 1932. Recalled for war service.

Paymaster Lieutenant Doig might avow that a staff should not serve as a crutch for a weak commander, it certainly could help fill the void of an absent one.[139] Moreover, it could ensure that all issues affecting the fleet were examined for their possible ramifications. To be sure, some contemporary officers viewed the disturbances as confirmation of the problems the staff system had created. William Tennant, though removed from immediate events serving as he was at the Staff College, did not dissent observing:

> It may now be said that the problem has swung too far and that the staffs in the flagships at sea and at the Admiralty are too big.
>
> The only reply to this is that the tail should not wag the dog, and if it does, the dog must be very unsteady on its legs.[140]

Notwithstanding the thoughts of some that the problem was the actual size of staffs, a more modern witness would probably eschew this argument given the proliferation and size of today's military staffs. Certainly, this writer does and the more fundamental point not seemingly raised was whether the personnel filling staff appointments were fit for their duties by experience, temperament, and, yes, training and education. Durnford's observation that prior qualification was not a *sine qua non* for appointment to the Naval Staff has already been registered. So, too, it will be recalled that officers typically served on a staff before receiving appointment to a Staff College. This is not contentious, as a strong case can be made that some people are ill suited to staff work and finding this out beforehand is a sound practice. Yet, having a portion of the staff lacking qualification is not the same as having most of the staff void of preparation. The figures tabulated below depict the relative presence of officers trained at a staff college or having secured staff experience by instructing at Greenwich. The first number presented reflects the total officers borne in an establishment and the second one depicts those holding the *psc* qualification:

139 Doig, 'Command and Staff', *Naval Review*, Vol. XII, No. 4, November 1924, p. 657.
140 'The Low Tide of the Royal Navy', Tennant Papers, NMM/TEN/20.

	1920[141]	1925[142]	1930[143]	1935[144]	1939[145]
CNS/DCNS/ACNS	3/0	3/0	3/0	3/2[c]	3/1[c]
Intelligence	46/2	23/3[a]	16/5	20/14[a]	28/11
Operations	16/1	10/2	8/3	7/4[c]	15/5
Plans	9/2	9/5[b]	12/8	15/11	21/17[c]
Training & Staff Duties	13/2	6/2	11/5	9/3	11/8
Air	–	5/0	5/3	8[d]/3	15/5
Tactical	–	2/1	6/4	5/4	6/2
Trade	9/0	6/1	–	–	–
Local Defence	13/1	–	–	–	–
Gunnery	10/0	6/2	–	–	–
Torpedo	7/0	4/0	–	–	–

(n.b., before 1924, only officers designated *ws* are reflected in the above totals. The numbers include all officers irrespective of branch and those serving as additional to the Division in *President*.)

Notes:

a) including one graduate of the Intelligence Course.

b) including one officer completing the second year of Camberley course.

c) including one officer with the *idc*.

d) including one assigned RAF officer.

From the above, it is apparent why Richmond believed reforming the Staff College was the essential precursor to all other measures under consideration following the World War: so few of the officers serving in the Naval Staff were adequately prepared for their duties. Yet, even from this early date trends of a sort are apparent with those officers formally trained in staff work appearing most often in the Plans and Training and Staff Duties Divisions. To this, another observation becomes manifest: strategic planning and the air were increasingly important to the Admiralty and tactical thinking assumed a more holistic approach—at least bureaucratically. Though the evolving changes in the Naval Staff owed much to the financial stringency of the period, such pressures did force the Admiralty to consider how to meet its requirements in not only a more efficient manner, but also in a more effective one.

The absence of qualified staff officers in the Naval Air Section in 1925 is also noteworthy for by this date opportunity existed to send officers to the Air Force Staff College. In the first instance, though, Commander Boddam-Whetham, an early

141 *Navy List*, December 1920

142 *Ibid*, July 1925.

143 Ibid, October 1930.

144 *Ibid*, October 1935.

145 *Ibid*, March 1939.

graduate of Andover, went to serve in the Air Ministry rather than at the Admiralty, while Lieutenant Commander Alfred Sprott returned to sea in the carrier *Argus*. Meanwhile, Lieutenant Commander Keith Dowding,[146] originally earmarked for Camberley, and Henry Johnstone, both graduates from the second course,[147] saw a similar pattern of employment as both went to aircraft carriers in the fleet. Still, of greater significance is the relative priority that the Admiralty afforded the Air Section and its successor, the Air Division, *vis-à-vis* its bureaucratic counterparts. The clearest evidence that the Fleet Air Arm would have fared better under absolute Admiralty control is the relative emphasis it placed on this Division within the Naval Staff. Granted trained staff officers were lacking, but this was ever a problem given the need to balance flying duties against the imperatives of maintaining a broader naval outlook. As casualties amongst officers of the aviation branch occurred in peacetime at higher rates than for other technical branches, and would likely only be yet greater during war, the problem was not an easy one to remedy. The truth remains that the Service never attempted to take full advantage of the naval officer trained at Andover, but continued to regard him as but a cog in a greater machine.

Yet, one means adopted in remedy to the above was to have senior officers attend an aviation course known as the Commander's Short Flying Course covering theory, groundwork, initial flying instruction and preliminary and final deck-landing qualifications. Rear Admiral Maitland Boucher is representative of this type of officer electing to opt for the FAA. Trained as a Navigating Officer, he reverted to general service in 1919 and following his RAF course spent the remainder of his career in aviation assignments.[148] This included attachment to the Air Ministry, command of *Courageous*, and service at the Admiralty in both staff and supply assignments. Boucher qualified as a Pilot, but due to a pre-existing medical condition (i.e., Scarlet Fever), he was rated for limited flying duties only. Thus, the Admiralty would not designate him with the distinctive (P) next to his name in the *Navy List*—a measure more about allowances than proficiency given that he continued to fly over the course of the next ten years and, most especially, during his command of *Courageous*.[149] Though receiving pilot training, the course was more about providing him and others familiarization

<hr>

146 Later Captain William Keith Dene Dowding (1891-?). Minesweeping Department, 1919; War Staff Course, 1922-23; Assistant War Staff Officer in *Queen Elizabeth*, 1923-24; Staff Course Andover, 1924-25; *Hermes*, 1925-26; Operations Division, 1927-29; Tactical Course, 1934; Senior Officers' War Course, 1934; Staff Officer Operations and Intelligence in *Victory*, 1934; Captain, 1935; Commanding Officer, *Tarantula*, 1936; Naval Intelligence Division, 1936; Commanding Officer, *Londonderry*, 1937; Senior Officers' Technical Course, 1937; Commanding Officer, HMS *Fleetwood*, 1938-39 and HMS *Egret*, 1939 as Senior Naval Officer Red Sea; Imperial Defence College, 1941 and retired, 1944.
147 ADM 196/145 (Dowding).
148 ADM 196/144 (Boucher).
149 ADM 196/127 (Boucher) and ADM 196/52 (Boucher).

of the air and establishing a cadre of senior officers acquainted with these matters.[150] This was a necessity if the Admiralty hoped to frame its operational requirements to the Air Ministry in a form that was technically feasible and was to manage the career development of entry-level aviators. This was a commendable start, but coordination of air requirements needed to be more intimate. It was insufficient to attach officers at ministry level and the Chief of the Air Staff suggested posting naval officers to Headquarters Coastal Area.[151] This proposal was not premature, but it was difficult when unfulfilled requirements existed at the Naval Staff level.

Year on year, it was the Plans Division that showed the most sustained and regular employment of the trained staff officer. Moreover, by the close of the period and if of Captain's rank, the officer was likely a graduate of not only the Staff Course, but of the Imperial Defence Course too. Accordingly, in 1939 each of the six Captains employed within the Division was a graduate of the IDC.[152] The pace of planning corresponding to the expansion of the CID's work and its sub-committees structure necessitated this growth; Greenwich and the IDC made its contributions informed and considered. If trained staff officers from Andover were absent from the Naval Air Division, then their presence was felt in the Plans and Operations Divisions. Commander Colin Bittleston,[153] a salt horse and a member of the Plans Division from 1929-30, attended the Air Force Staff College during 1927-28.[154] Such was the case, too, with Charles Daniel, a Signals Officer who went to Andover's tenth course in 1932, eventually joining the Plans Division in 1936.

As for the Intelligence Division, the largest element of the Naval Staff throughout the period, it never secured the number of trained staff officers it seemingly required. Even when the numbers are adjusted to account for those who had taken the earlier Intelligence Course, the supply of trained officers proved unequal to the demand. The years 1931 and 1932 were particularly acute as the number of officers from the naval service studying at the Staff College was but 22 and 21, respectively.[155] In view of the number of staff officers serving in the Naval Staff, at sea, and in other shore establishments, the numbers within the Division are actually remarkable for what they were. Intelligence work was of a special nature demanding the strongest skills in exposition

---

150 *Navy List*, July 1925, p. 309.
151 CAB 53/5, Chiefs of Staff minutes of 9 October 1934.
152 To wit, Captains Victor Danckwerts, John Edelsten, John Dundas, Guy Russell, Rory O'Conor, and Maurice Mansergh.
153 Later Captain Colin John Lawrence Bittleston (1893-1967). In Command, *P 48*, 1918-19; Staff Course, 1924-25; Staff Officer Operations in *Effingham*, 1925-27; Staff Course Andover, 1927-28; Plans Division, 1929-31; Tactical Course, 1931; Commanding Officer, HMS *Achates*, 1931-32; Royal Naval Staff College, 1933-35; Plans Division, 1937; liaison duties to Merchant Navy, 1937-39; Sea Transport Officers' Training Course, 1938; *Drake* for Naval Control of Shipping, 1939-40; Acting Captain, 1939; Trade Division, 1941-43; Captain and retired, 1943.
154 ADM 196/145 (Bittleston).
155 *Navy List*, January 1930, p. 307 and *Navy List*, January 1931, p. 307.

so that the critical point was made succinctly. Not all qualified staff officers were successful in such work and Robin Jeffreys was one such misplaced officer in Barry Domvile's eyes.[156] Another officer who was less than successful in general intelligence duties was Commander Ralph Wilkinson. Though not a naval *psc*, Wilkinson was a graduate of the Intelligence Course, but he was ill at-ease in preparing minutes and answering the dockets making their way through the Intelligence Division. He was, however, a very successful liaison officer to the Fourth Estate where his bohemian lifestyle was surely an advantage in this regard as many newspapermen where fond of a drink and the social banter seen as part of a journalist's stock-in-trade.[157] Still, the Intelligence Division was not getting its share of trained staff officers and forsaking its dedicated Intelligence Course was, ultimately, a false economy married to an even falser promise. As ever, jam today, jam tomorrow.

This is not entirely correct. Few graduates of the Intelligence Course actually served in the Intelligence Division immediately upon leaving with many going to the fleet as Intelligence Officers. This was largely a function of the course's introductory nature and the inexperience of those attending. Many attending were only recent entrants, especially, was this so in the case of marine officers. If an officer remained in the Service, he might yet land an appointment to the Intelligence Division, but in the interim, a posting to the China, America and West Indies, or East Indies Stations, to say nothing of acting as an Intelligence Officer in a private ship, were his immediate prospects. Still, cancellation of the course precluded the Intelligence Division from even considering the venue as a means to augment its numbers. When it is observed that during its brief postwar existence, more officers were passing through the Intelligence Course than through the Staff College each year, the promise of a better trained Intelligence Officer—all other things being equal—was not really a feasible prospect.

Yet, this was not the only problem facing the Intelligence Division. In 1919, the Naval Staff was a hybrid body organized along both functional, or mission lines, and specialist lines. By 1930, it now had a strong mission orientation, and save for the Naval Air Division, its elements were broad divisions requiring specialists of all styles and not merely of one type. Unfortunately, the numbers engaged did not allow specialists of all hues to be represented in a division. Consequently, this meant that an officer working a particular problem needed to be mindful of his limitations in any matter and seek advice as appropriate. Of the elements of the Naval Staff, the most at risk to this problem was the Intelligence Division. Others, such as the Training and Staff Duties and Tactical Divisions, had access to subordinate elements (i.e., the Staff College and Tactical School) and could make use of their expertise. The Intelligence Division was a production element of a separate kind considering problems of both the broadest and narrowest natures requiring officers suitably knowledgeable. It was weakest in officers possessing aviation and anti-submarine warfare skills and this void

156 ADM 196/127 (Jeffreys).
157 ADM 196/126 (Wilkinson).

weakened any assessments rendered. Creation of the Joint Intelligence Committee (JIC), of which more anon, if engaged properly, might alleviate matters of air intelligence, but the Intelligence Division could find no outside succor resolving its gaps in submarine intelligence.

Turning to 1935, the only Division of the Naval Staff to see a reduction in the number of qualified staff officers from previous years was Training and Staff Duties (TSD). As this date coincides with observations reported previously on the perceived status of the Staff College at the time of the James Committee's sitting, one conclusion possible is that Bertram Watson and others holding such views were unwittingly capturing the declining status of the TSD, the element most clearly identified with the Staff College. For it remained that the Division most responsible for staff development was not led by a qualified staff officer from April 1928 until September 1937 when Louis Hamilton,[158] whilst holding the concurrent position of Captain of the College, temporarily served as Director.[159] Captain William Jackson,[160] another naval *psc*, in turn, shortly replaced Hamilton.[161] Meanwhile, 'Turtle' Hamilton when serving as Captain of the College spent time revising the *Naval War Manual* in preparation for its next release.[162]

That the Naval Staff was populated increasingly with trained officers is self-evident. Yet, an explanation must exist why capable, but not trained officers, were now occupying the DTSD position. The incumbent in 1935 was Captain Frederick Buckley,[163]

158 Later Admiral Sir Louis Henry Keppel Hamilton (1890-1957). Commanding Officer, HMS *Taurus*, 1918-19, HMS *Strenuous*, 1920-22, *Taurus*, 1922, HMS *Wild Swan*, 1927-28 and *Ambuscade*, 1928-29; Tactical Course, 1929; Staff Course, 1930; Second Sea Lord's Office, 1931-33; Captain, 1932; Imperial Defence Course, 1934; Plans Division and Naval Intelligence Division, 1935; Flag Captain and Chief of Staff in *Norfolk*, 1935-37; Director, Training and Staff Duties Division, 1937; Captain of the College, Greenwich, 1937-39; Commanding Officer, *Delhi*, 1939; Rear Admiral, 1941; Vice Admiral, 1943; Admiral, 1947 and retired, 1948.

159 *Navy List*, October 1937, pp. 417 and 453.

160 Later Rear Admiral William Lindsay Jackson (1889-1962). Staff Course, 1923-24; to Chilean Navy, 1926-27 and 1930-32; Naval Intelligence Division, 1928; Captain, 1929; Senior Officers' War Course, 1929-30; New Zealand Division, 1932-34; Tactical Course, 1934; Commanding Officer, *Despatch*, 1934-35, HMS *Galatea*, 1935 and *Despatch*, 1935-37; Director, Training and Staff Duties Division, 1937-39 and retired, 1940. Recalled for war service.

161 *The Navy List* (London: His Majesty's Stationery Office, February 1938), p. 416.

162 ADM 196/53 (Hamilton).

163 Later Rear Admiral Frederick Arthur Buckley (1887-1952). Six firsts in examinations for Lieutenant; Naval Assistant to ACNS, 1921-23; Squadron Gunnery Officer in *Hood*, 1923-25; Gunnery Division, 1925-26; Captain, 1927; Assistant Director, Naval Ordnance Department 1928-30; Senior Officers' War Course, 1930-31; Tactical Course, 1931 and 1936; Commanding Officer, HMS *Shropshire*, 1931-33; Deputy Director and Director, Training and Staff Duties Division, 1933-36; Senior Officers' Technical Course, 1936; Commanding Officer, *Revenge*, 1937-39 and *Malaya*, 1939; Rear Admiral, and retired, 1939. Recalled for war service.

whose rank of Captain dated from December 1927. As seven of the eighteen officers holding their captaincy from that year held the *psc*, the issue becomes whether any were available to fill the position. The most promising candidate was Captain George Thomson,[164] then taking his required senior courses in late 1935, but previously serving as Chief of Staff on the China Station in *Kent*. That posting terminated in October 1934 and the easiest solution, if it were deemed a priority to place a qualified officer in the billet, would have been to extend the appointment of Captain Geoffrey Arbuthnot[165] until Thomson became available.[166] Arbuthnot was, in fact, taking the SOTC and the Tactical Course before reverting to unemployed status, so the extension would not have come at a price in upsetting any other pending appointments. Moreover, Arbuthnot had last completed the Tactical Course in 1931 and would have been available to sit either the 15 October 1934 or 15 January 1935 SOTC class.[167]

As for Thomson, being a graduate of both the Staff Course and the IDC, the DTSD position appeared tailored-made to his talents. However, the Admiralty did not pursue this option and the reason requires answering. The most probable explanation is that, notwithstanding the increasing use the Navy made of its staff officers, the fear was always present, as Tennant reminds us, that the staff tail would wag the naval dog and having a non-qualified officer as DTSD was one means of keeping this at bay. Moreover, Thomson had two drawbacks. First, as a submariner, he had spent much time in this class of vessel at the expense of the surface fleet, and, secondly, it was thought that he had spent too much time in staff positions already and needed more sea experience.[168]

The DTSD assignment was a particularly sensitive one as the incumbent had necessarily to negotiate the pitfalls of arguing his own corner and work well with the rest

164 Later Rear Admiral George Pirie Thomson (1887-1965). Five first in examinations for Lieutenant; Commanding Officer, *E 35*, 1916-17, *J 3*, 1917-18, *K 11*, 1918-19 and *K 16*, 1919; War Staff Course, 1920-21; Naval Intelligence Division, 1921-23; Commanding Officer, *K 26*, 1923-24; Staff Officer Operations in *Revenge*, 1924-26; Commander, Sixth Submarine Flotilla, 1926-27; Imperial Defence Course, 1928; Captain, Submarine Flotilla, 1929-30 and 1931-32; Senior Officers' Technical Course, 1931; Tactical Course, 1931 and 1935; Chief of Staff in *Kent*, 1933-34; Senior Officers' War Course, 1935; Commanding Officer, *Devonshire*, 1935-36; Personal Services Department, 1936-37; to RAN, 1937-39; Rear Admiral and retired, 1939. Recalled for war service.

165 Later Admiral Sir Geoffrey Schomberg Arbuthnot (1885-1957). Naval Ordnance Department, 1921-23; Tactical Course, 1925, 1931 and 1934; Commanding Officer, *Seawolf*, 1925-26; Captain, 1926; Commanding Officer, *Suffolk*, 1929-31; Assistant Director and Director, Training and Staff Duties Division, 1932-34; Senior Officers' Technical Course, 1934; Commanding Officer, *Valiant*, 1935; Flag Captain and Chief Staff Officer in *Valiant*, 1936-37; Rear Admiral, 1936; Senior Officers' War Course, 1937; Fourth Sea Lord, 1937-39; Vice Admiral, 1940; Commander-in-Chief, East Indies, 1941-42; retired, 1943 and Admiral, 1944. Recalled for war service.

166 ADM 196/92 (Thomson).

167 ADM 196/91 (Arbuthnot).

168 ADM 196/92 (Thomson).

of the Naval Staff for his responsibilities crossed so many well-established lines, as Vernon Haggard found during his tenure during the early 1920s.[169] Though William Jackson had done a superb job as Director during his tenure, tact was an essential feature of the position and not all staff officers were blessed with this commodity when it came to questions of staff work and the role of the staff.[170] Frederick Buckley, if not a trained staff officer, brought a very strong gunnery background to the position, served as the Division's Deputy Director before assuming responsibility for the Division, and was esteemed as a tactful officer. Held to be an effective administrator, Buckley proved a successful DTSD.[171] Yet, beginning with Captain Bernard Fairbairn[172] in 1928 and lasting through the stewardship of Captain Arthur Lyster[173] in 1937, in the course of over eight years and six separate Directors, an officer holding the *psc* or *idc* never occupied the DTSD billet.[174] With apologies to Lady Bracknell, to lose one may be regarded as a misfortune; to lose six looks like carelessness—or design. Though Lyster performed adequately in the role of DTSD, he did not shine and he was thought better suited to the command of men than of an 'office' and the suspicion remains that a trained staff officer was appropriately thought a requirement in his successor.[175]

An anomaly of a similar sort is discernible in those officers heading the Tactical Division. By the time of their appointment, most officers serving as its Director had not yet attended the Tactical Course. Though admittedly the class of occupants is much more restricted as the course only began in 1925, Captain Bruce Fraser[176] was

169 'A Narrative of Service of Captain V.H. Haggard, pp. 1-2, R.N', Haggard Papers, IWM/85/21/2.

170 ADM 196/92 (Jackson).

171 ADM 196/92 (Buckley).

172 Later Vice Admiral Bernard William Murray Fairbairn (1880-1960). Naval Ordnance Department, 1915-18; Captain, 1919; Gunnery Division, 1920-21; Commanding Officer, *Dragon*, 1922-24; Senior Officers' War Course, 1924-25; in Command, Gunnery School in *Vivid*, 1925-26; Director, Gunnery Division, 1927-28; Director, Training and Staff Duties Division, 1928-29; Commanding Officer, *Ramillies*, 1929-30; Imperial Defence Course, 1931; Rear Admiral, 1931; Vice President, Ordnance Committee, 1932-34 and President, 1934-36; Vice Admiral and retired, 1936. Recalled for war service.

173 Later Admiral Sir Arthur Lumley St George Lyster (1888-1957). Six firsts in examinations for Lieutenant; Squadron Gunnery Officer in *Barham*, 1922-24; Counter-Battery Course, School of Artillery, Larkhill, 1924; Naval Ordnance Department, 1924-26; Captain, 1928; Ordnance Committee, 1929-31; Tactical Course, 1931 and 1937; Senior Officers' War Course, 1931; Senior Officers' Technical Course, 1931; Senior Officers' School Sheerness, 1931-32; Commanding Officer, *Danae*, 1932 and *Despatch*, 1932-33; Captain (D), Fifth Flotilla, 1933-35; Commander, Gunnery School in *Pembroke*, 1935-36; Director, Training and Staff Duties Division, 1936-37; Commanding Officer, *Glorious*, 1937-39; Rear Admiral, 1939; Rear Admiral, Aircraft Carriers, 1940-41; Fifth Sea Lord, 1941-42; Vice Admiral, 1942; Vice Admiral, Aircraft Carriers, 1942-43; retired, and Admiral 1945.

174 The other officers serving as DTSD during the period were Captains Edward Cochrane, James Ritchie, Geoffrey Arbuthnot, and Frederick Buckley.

175 ADM 196/92 (Lyster).

176 Later Admiral of the Fleet Sir Bruce Austin Fraser, Baron Fraser of North Cape (1888–1981).

a rare exception of an officer actually having passed through the course borne in the *Victory*. Thus, William Parry had not taken his Tactical Course before becoming Director in 1929 doing so only in 1932. Still, having won the Nicholson Prize for Tactics in the Atlantic Fleet in 1923, he was seen as an officer with a good eye for the subject.[177] Captain John Crace[178] was another officer who came to the Naval Staff from command of the Anti-Submarine School in the *Osprey* without previously attending the Tactical Course. Crace, though, is representative of another type of successful officer—those who failed to attend any of the higher courses available. Though a qualified Torpedo Officer, his one concession to post-specialist training was to take a two-week chemical warfare course at Porton Down immediately before becoming Director of the Tactical Division.[179] This practice was not unique to Crace, as Terence Back, also a torpedo specialist, took a similar course at Winterbourne Gunner when joining Crace in the Division in 1932 though neither officer had taken, or would subsequently take, the Tactical Course.[180]

Torpedo specialists, such as Crace, were particularly favoured as the type of officer to lead the Tactical Division (or in its former guise, the Tactical Section) appearing no less than four times as Director. Of the four, two, Crace and Captain Donald Budgen,[181] had also commanded the Anti-Submarine Training School. Another officer, Captain Colin Cantlie, was a highly experienced submariner who secured his brass hat as a Commander after only three and a half years. In fact, only one Director or Head, Fraser, was a gunnery officer which may be thought odd at a moment when the gun was seen as the principle naval weapon. Clearly, Buggin's turn did not operate—unless of a reverse kind—as a means to limit Admiralty interference in tactical thought and development.

Lending credence that this pattern of employment was far from random is to recall the experience of those officers sitting a Staff, War or Tactical Course before

Captain, 1926; Tactical Course, 1926 and 1936; Head, Tactical Section/Division, 1927-29; Flag Captain in *Effingham*, 1929-32; Director, Naval Ordnance Department, 1933-35; Commanding Officer, *Glorious*, 1936-37; Rear Admiral, 1938; Commander, First Battle Squadron, 1938-39; Third Sea Lord, 1939-42; Vice Admiral, 1940; Commander, Second Battle Squadron, 1942-43; Admiral, 1944; Commander-in-Chief, Pacific Fleet, 1944-46 and Admiral of the Fleet, 1948.

177 ADM 196/145 (Parry).

178 Later Vice Admiral Sir John Gregory Crace (1887-1968). Torpedo and Mining Department, 1921-22; Commanding Officer, *Osprey*, 1924-26; Fleet Torpedo Officer in *Warspite* and *Queen Elizabeth*, 1927-28; Commanding Officer, *Valhalla*, 1929-30; in Command, *Osprey*, 1930-32; Director, Tactical Division, 1932-34; Commanding Officer, *Emerald*, 1934-37; Naval Assistant, Second Sea Lord, 1937-39; Rear Admiral, 1939; Vice Admiral and retired, 1942.

179 ADM 196/51 (Crace).

180 ADM 196/128 (Back).

181 Later Rear Admiral Douglas Adams Budgen (1889-1947). Mining School in *Vernon*, 1919-21; Torpedo and Mining Department, 1924-26; Captain, 1930; Senior Officers' Technical Course, 1931; in Command, *Osprey*, 1932-34; Commanding Officer, *Capetown*, 1934-36 and *Defiance*, 1936-37; Director, Tactical Division, 1938-39; Rear Admiral and retired, 1941.

immediately assuming instructional responsibilities in those schools. For the Admiralty, the matter was of a different hue—to minimize the perception within the greater Service that the Naval Staff was dominated by the 'staff type'. This was especially a concern of the Training and Staff Duties and Tactical Divisions where the former had as its *raison d'etre* all things staff and the latter was feared for impinging on tactical issues of direct concern to a fleet's Commander-in-Chief.

Chatfield, Chief of the Naval Staff for much of this period, moreover, confirms that keeping the staff in check and from encroaching on the proper preserve of the flag officer was fundamental lest the admiral become but a mindless cog in the naval machine. Otherwise, his initiative would suffer and suffer most assuredly when required in battle.[182] Chatfield, thus, was alive to the contemporary debate on the role and place of the staff within the naval service. Though not a product of the Staff Course, he did understand the necessity of a staff—and a properly trained and constituted one at that—but the staff was only a means to an end. Ensuring properly considered plans was one thing, but ultimately for the flag officer, it came down to making his decisions in a timely manner and having the moral courage to see them through to a successful conclusion. This was the essence of leadership and command. By the time of the Invergordon Mutiny, Tomkinson, commanding the Battle Cruiser Squadron and the next senior officer after Hodges, needed not so much a staff but powers of command to restore discipline.[183] Certainly, this was the judgment rendered by the Board of Admiralty in early 1932 which held, that if he had acted with promptitude and decisively, events need never have reach the proportions they ultimately did.[184]

As a Torpedo Officer, it remains an imponderable whether Tomkinson's specialist technical training and absent the rounding out effects of a Staff Course were a root cause to his approach and limitations during the mutiny.[185] Significantly, the only advanced class that he took was a Tactical Course in 1926, as he failed to sit the SOWC, the SOTC, the IDC, or a course from another Service. As with so many officers, his appointment to the Battle Cruiser Squadron was a recent one dating from April 1931.[186] Yet, it cannot be said that a salt horse would have done better in the circumstances. For more important than the skills one learned as a junior officer was the attitude one adopted as one progressed in rank. The shortcoming of the salt horse was that his approach to naval affairs owed much to the style and influence of his mentors. If his role models were officers of ability, good teachers, and sympathetic to the progress of those entrusted in their charge, then the absence of formal training was not a serious defect. If, instead, the more senior officer was a poor mentor and failed to instill in the junior a strong desire to learn more of his profession, then the

182 Chatfield, *Navy and Defence*, p. 230.
183 *C.B. 3027(A), Mutiny in the Royal Navy*, p. 21.
184 ADM 196/90 (Tomkinson).
185 ADM 196/141 (Tomkinson).
186 ADM 196/90 (Tomkinson).

damage done might extend beyond the confines of the moment, for as Haldane once so tellingly remarked, 'To think costs nothing.'[187] Tomkinson's career was tied, inordinately, to that of Roger Keyes—a great role model, but not necessarily a great teacher. Consequently, how much of the *Royal Oak* contretemps weighed on Tomkinson at the time of Invergordon is unknown, but Keyes who had acted promptly in that instance and who failed to solicit the advice of the Admiralty was not particularly rewarded for having acted so. Perhaps, this, too, weighed with Tomkinson.

That assertive leadership was a key element missing from many officers at Invergordon is neither surprising nor remarkable. For all the emphasis placed on developing the attributes of initiative in its training and educational regimes, it was the exceptional officer who actually raised his head above the parapet at the moment of crisis, perhaps only proving that not of us all are made of the stuff leading to a Victoria Cross. Captain Wilfred Custance,[188] a gunnery specialist, was one officer at Invergordon rising to the occasion. When elements in the *York* looked like they might opt for trouble, he assembled the crew, read the *Articles of War*, and had marines parade with fixed-bayonets. The crew soon knew where matters stood and returned to duty.[189] This is not to avow that Custance's actions were the only ones appropriate for the moment, but sometimes the tenor of a decision can be as important as its substance and in the judgment of Astley-Rushton, his immediate superior officer, 'He did well.'[190]

That the proposed cuts, painful as they were, failed to ignite similar outbreaks of ill-discipline within the wider Navy and the Army and Air Force speaks primarily to a failure in leadership within the Atlantic Fleet, though this does not absolve the Admiralty for failing to consider the matter more forcefully beforehand.[191] Though Alexander was soon warning that the new scale of cuts envisaged were neither practicable nor sustainable, he did so only after the Cabinet had decided to precede the day before.[192] The Cabinet directed the Services to concert their plans for implementing the cuts, something made the more difficult as the pay and allowances of servicemen were not uniform. That this was so owed much to tradition and the conditions of

187 Haldane cited by Lord Gort. Gort to Liddell Hart letter of 31 October 1937, Liddell Hart Papers, LH 1/322/51, LHCMA.
188 Later Vice Admiral Wilfred Neville Custance (1884-1939). Naval Intelligence Division, 1921-23; Captain, 1925; Senior Officers' War Course, 1925-26 and 1936; Senior Officers' Technical Course, 1926; Tactical Course, 1926 and 1937; Commanding Officer, *Yarmouth*, 1926-27; Senior Officers' School Sheerness, 1928; Commanding Officer, *Castor*, 1928-29 and *York*, 1929-31; Commander Gunnery School in *Vivid*, 1932-34; Commanding Officer, *Rodney*, 1934-36; Rear Admiral, 1936; Commander, Australia Squadron, 1938-39 and Vice Admiral, 1939.
189 Diary entry of 15 September 1931, Drage Papers, IWM/PP/MCR/99, Reel 3.
190 ADM 196/91 (Custance).
191 Only the *Delhi* operating on the America and West Indies Station showed signs of similar disaffection.
192 CAB 23/67, Cabinet minutes 42(31) of 20 August 1931.

service followed by the Royal Navy, the British Army and the Royal Air Force. If these matters were largely arcane to the layman, they went to the heart of what it meant to be a seaman and how he looked on his profession.[193] Overturning a directive is always more difficult than blocking one where consensus is the means of decision-making. To this, swapping horses in mid-stream (Austen Chamberlain replacing Alexander as First Lord in late August) did not abet matters with the new First Lord now merely endorsing an earlier minute to the Cabinet by his predecessor that outlined the Admiralty's objections to the savings in pay, pensions, and dockyard wages proposed.[194]

Further, Cabinet meetings were missed at which the tenor of the changes previously reached were confirmed.[195] Whilst all this amply confirms that there are no votes in defence, a Minister still must have a grasp of his portfolio to argue his case at the vital hour. To be sure, the conditions of service between the forces were not identical and ill luck and proximity played their part in the affair. Yet, the failure to understand and anticipate how the cuts proposed affected a serviceman's pension demonstrates one area where the planned savings had received insufficient study at the Admiralty.[196] Time may have been a factor why the Government was caught unawares—August notoriously being a bad month in Britain to consider any major policy proposal that the *May Report* indeed represented. Still, this only emphasizes the value of a properly aligned staff system properly manned at both the Admiralty and in the fleet.

The observation that the staff system as found in 1931 was at best a half-matured construct does not address whether events would have proceeded to a better end if this had been otherwise. This is unknowable though doubtlessly events would have followed another course. Part of the problem at arriving at a satisfactory assessment, beyond entering the realm of the counterfactual, is that officers still needed to employ any staff provided. Old habits die-hard and many senior officers conducted affairs as if the staff supporting them was something of a fifth wheel. Trust may have been part of the issue. After all, most seniors were *very* capable officers and no doubt believed that they could do what was required with little input from others. The tradition where a Captain was responsible for all affairs—good and bad—arising in his ship made centralization of command the style of leadership most understood. This style

193 More so than his military counterpart, the rating anticipated serving until eligible for a pension. Additionally, tradition maintained that he would receive Prize Money in war based on the commercial value of the shipping and goods liberated. The awarding of Prize Money was modified during the war as it became a common pool across the Service rather than a specified amount to be paid to a ship's company. Even following the war, Lieutenant Augustus Agar had received £5 'blood money' for every Bolshevik killed in the 1919 attack on the *Oleg* earning him £4000. See CAB 23/8, War Cabinet 512 minutes of 17 December 1918.

194 CAB 24/223, Note by First Lord of the Admiralty to Cabinet, C.P. 210 (31) of 27 August 1931.

195 CAB 23/67, Cabinet minutes 54 (31) of 7 September 1931.

196 CAB 23/68, Cabinet minutes 67 (31) of 30 September 1931.

required juniors, even in positions of command to defer to their seniors in matters yet of a routine nature. Decentralization of command and initiative were fine as concepts, but execution was another matter. This tradition, or more appropriately, this common law of naval practice, operated against employing the staff system.

Here it should be noted that the custom of having an officer hold a dual-appointment or in the jargon of the moment, 'dual-hatting', probably accentuated the problem. The mental gymnastics requiring an officer to attend to two distinct duties—one of command and the other of staff—was something that was difficult to prepare one to assume. Those officers employed as a Captain (D) or of a like flotilla, Flag Captain and Chief Staff Officer, or even Vice Admiral Commanding and Second-in-Command needed to weigh differing sorts of problems and the limited exercises gamed at the Staff College and the Imperial Defence College largely ignored this tension of responsibilities. This was partly structural, as the games considered at Greenwich primarily considered operational-tactical problems where executive skills were stressed. The IDC, meanwhile, focused on the strategic-political issues of any situation where the emphasis was on force generation and the objectives of policy. The conflict of executive-administrative functions, at the fore of Ellison's warnings voiced in *The Perils of Amateur Strategy*, do not appear to have resonated during peacetime exercises indicating, perhaps, that experience of war is of a class by itself and not easily replicated. This was reflected also in the education afforded to naval and marine officers during the period, as it focused narrowly on the what and how that needed to be accomplished, but did not address the roles created to accomplish these tasks. Thus, whilst lectures were provided in 'Staff Work in Destroyers', 'Staff Organisation', 'Admiralty Organisation', 'Staff Work in the Mediterranean', 'Staff Work Afloat', 'Staff Work in a Flagship', and 'Organisation of Commands and Staffs', the only lecture that has been discovered addressing the role of a specific type of staff officer was Major Mullins' 'Duties of a District Intelligence Officer'. Indeed, only the Intelligence Course examined the actual duties of a specialist staff officer and this it will be recalled was soon being merged into the greater Staff Course in early 1926. The Intelligence Course offered further instruction on the role of 'Liaison Intelligence Officers in the Dominions', 'District Intelligence Officers Home', 'District Intelligence Officers Abroad', and 'Consular, Colonial', Indian and Dominion Reporting Officers'.[197]

Of course, the Intelligence Officer did not typically enjoy command responsibilities and the senior officer now filling a command and staff role had to balance the responsibilities of both offices. This was more than a question of time-management and some officers were better at meeting multiple responsibilities. In fact, often the cleverest and most experienced officers had the most difficult time adjusting to the situation. Others, recognizing their own limitations turned more naturally to members of their staff. When Lyster became Captain (D) of the Fifth Flotilla in 1933, he leaned on

197 ADM 1/8668/172, Director, Royal Naval Staff College to Secretary of the Admiralty letter of 15 December 1923.

his two Divisional Commanders, as it was his first experience in small ships. He also had an abiding interest in fishing. Frequently, when his ship visited a particularly rich ground, such as Scotland, Denmark or Sweden, offering trout or salmon, Lyster turned his duties over to another officer with one contemporary recording that 'His most spectacular fishing achievement was to land at dawn at Helmsdale and kill a salmon before the flotilla sailed to carry out the most important individual ship exercise of the year, the efficiency test.'[198]

Reginald Henderson,[199] whilst Third Sea Lord and Controller, was notoriously 'impatient of other people's opinions and ideas' even though held as a highly regarded officer on all other accounts.[200] Though this was a personal view from Grenfell, it matches closely what Henderson's superiors thought of him as well at other moments in his career. Henderson proved a very successful Controller to the detriment, on the surface at least, of his broader professional prospects, as Duff Cooper kept him on as Third Sea Lord rather than appointing him to command of one of the main fleets for which he was fitted.[201]

Roger Backhouse was another hard-working senior officer who eschewed delegation and could display a withering temper to the errant officer.[202] Grenfell commenting on that officer and Pound, another arch-centralizer, caustically observed:

> Both remarkably capable administrators. Apart from that, they are both extreme examples of a certain type of mind; the type that do everything itself and can credit no one with any intelligence or even common sense, and cannot imagine that any subordinate can be trusted to do anything without elaborate orders. The effect of this attitude on the training and initiative of those under them can be imagined. B.[ackhouse] is the more extreme case. It is an open secret in the Home Fleet that he attends to every paper himself and knows all the most trivial details of ships (*sic*) defect & refits. His first C.O.S., a Rear Admiral, left him because B. saw every paper himself and left the C.O.S. and the staff with nothing to do. After several vain protests, he had the courage to say that he would prefer to go than to remain as a nonentity.[203]

198 *Shellback Remembers*, p. 161, Durnford Papers, IWM/PP/MCR/C10.
199 Later Admiral Sir Reginald Guy Hannam Henderson (1881-1939). To Greek Navy, 1913-14; Anti-Submarine Division, 1916-17; Naval Assistant to ACNS, 1917-19; Captain, 1917; Flag Captain, *Hawkins*, 1919-22; Senior Officers' Technical Course, 1922 and 1931; Senior Officers' War Course, 1922 and 1931; Royal Naval War College, 1922-25; Tactical Course, 1926; Commanding Officer, *Furious*, 1926-28; Rear Admiral, 1929; Rear Admiral, Aircraft Carriers, 1932-33; Vice Admiral, 1933; Third Sea Lord, 1934-39 and Admiral, 1939.
200 Grenfell to Liddell Hart letter of 6 June 1937, Liddell Hart Papers, LH 1/330/11, LHCMA.
201 ADM 196/90 (Henderson).
202 Murfett, 'Backhouse' in Murfett, ed., *First Sea Lords*, p. 173.
203 Grenfell to Liddell Hart letter of 6 June 1937, Liddell Hart Papers, LH 1/330/11, LHCMA.

Rear Admiral Bertram Ramsay was the unfortunate officer made to suffer from Backhouse's proclivities against delegating routine affairs. In his steps, Henry Moore followed and though Backhouse held that he ran the staff effectively, he also believed Moore lacked a certain amount of initiative. This may have been a reflection that Moore knew the limits of what was possible with Backhouse and had no desire to emulate Ramsay.[204] A qualified, though not a trained staff officer, Backhouse was also a gunnery specialist. How much his technical training shaped his habits of administration and style of command is impossible to judge, but a common contemporary criticism cited was that the rote learning instilled during one's technical training whilst a junior officer cramped one's ability later to weigh matters of a broader nature and, correspondingly, fostered an inability to let go of minor details.

Given the frequency contemporary officers leveled the accusation, it merits consideration as a source of the resistance to adopting proper staff lines and procedures. Gerald Dickens was an early commentator on the conflict of skills required and writing in 1920 observed:

> The training required for the "conceptionist" is principally one that places abstract subjects in the foreground and gives specialisation in things purely material a secondary place. As the study of detail in material things is of paramount importance for the junior ranks, the educational problem lies in so shaping a scheme of training that these officers may become experts in the work that they are every day called upon to perform, and, at the same time, may be so forming their minds as to make them ready to deal with higher things. The trouble is that a long study of detail in mechanical subjects is usually antagonistic to the forming of the conceptionist's mind.[205]

Clearly, some officers were able to rise above the tensions operating between the requirements of specialization and the general outlook desired of the strategist for not all such successful officers were of the salt horse variety. Richmond, Drax, and Bertram Watson were all technical specialists who rose above their initial training and successfully negotiated the transition to the senior ranks of the Navy. In Drax's case, Chatfield thought him a user of weapons rather than a 'material' man possessed of a sharp intellect even if of a retiring personality.[206] He even managed to sidestep a fair portion of any blame over Invergordon, though he had been serving as Director of the Manning Department at the time. How effectively an officer delegated tasks to subordinates and employed his staff was but one skill required of a naval officer, and, this not necessarily even the most important. Pound, the arch-centralizer, was still recognized as a very effective leader and a great man by many naval officers, including Geoffrey Norman, and the burdens that he bore during the nadir of the

---

204 ADM 196/92 (Moore).
205 Dickens, 'Training of Naval Officers, *Naval Review*, February 1920, p. 62.
206 ADM 196/90 (Drax).

next war would have broken many another officer.[207] Moreover, if he worked hard, Pound played harder still and was renowned for his early rising to enjoy a good day's shooting. William Fisher thought him an excellent Chief of Staff when Pound served him in the Mediterranean. This was during the Abyssinian Crisis when his eye for detail and logical mind were traits regarded of no small importance.[208] Still, serving and commanding are two different qualities required of an officer.

Henry Oliver, though, saw the problem and worked to alleviate its worst aspects. Thus, when commanding the Atlantic Fleet between 1924-27 he arranged for a period of independent sailing as it struck him that for the most part everything was a little bit too tidy for ships' commanding officers, as the Captain of the Fleet had a firm grip on proceedings. Thus, he arranged for a period of independent sailing following fleet exercises where captains could exercise command removed from the eyes of the Commander-in-Chief and his Captain of the Fleet.[209] Oliver, too, was an officer who had chosen to specialize—in his case, navigation. However, as the officer responsible for the Admiralty War Staff during the first three years of the World War his senses may have been especially acute in such matters.

Still, any barriers precluding the delegation of tasks or ensuring that subordinates had opportunities for exercising initiative probably owed more to questions of habit than of training. Evidence of this exists in the practice of periodically running fleet exercises without prior scripting and written orders. Such an example was exercise 'M.Z.', a combined Atlantic and Mediterranean Fleet serial conducted in March 1929. Here, the opposing fleets drew their tasks and orders by lot as a means of reducing the advantages that one fleet accrued over another when objectives, forces, and dispositions were generally known beforehand.[210] Meanwhile, Admiral Field, the Commander-in-Chief, Mediterranean, conducted a similar serial the previous January during exercise 'N.Y.' whose purpose was *designed to exercise a surprise operation for which no written orders can be issued, owing to the opposing battlefleets being at sea when war is declared.*[211] Thus, extreme centralized control was recognized as a problem, but it was a problem responsible senior officers were mindful of with attempts made to mitigate its most deleterious effects.

Of course, another reason why centralized control arose was that the information to control all naval forces was not equally held by commanders. This was especially so where intelligence derived from the interception of enemy signals or human sources

207 Norman, Reel 3, IWM Sound Recording 8870 of 1985.

208 Anon., 'Admiral of the Fleet Sir Dudley Pound,' *Naval Review*, November 1943, pp. 284-85.

209 *Recollections, Volume II*, p. 292, Admiral of the Fleet Sir Henry Francis Oliver Papers, National Maritime Museum, Greenwich, NMM/OLV/12.

210 Paul Halpern, ed., *The Mediterranean Fleet 1919-1929* (Farnham: Ashgate Publishing Limited, 2011), pp. 544.

211 ADM 186/146, '*C.B. 1769/29(2), Exercises & Operations, 1929. Volume II,*' Admiralty, Naval Staff, Tactical Division, June 1930.

was involved, but it could also be the case of contact reports made by surface, submarine, or aircraft whose transmissions were not copied by fleet units. Such information might only be held at the Admiralty, and for reasons of operational security, the true origins of some information need not be divulged. This had been a feature of the 1914-18 war and exercises such as 'A.A.' and 'M.A.', further evolutions from 1929, saw separate squadrons working in cooperation with each other, but under Admiralty direction.[212] Conversely, the situational awareness held at the Admiralty might not approximate actual events. This might be for reasons of timeliness—reports not made—or it may be that false conclusions were being painted owing to an enemy's efforts at deception. Decentralization of command had its place, but it also had its limits and the need, as always, was to appreciate the balance required.

Meanwhile, another officer was soon noting that the presence of a staff did not obviate the need for senior commanders to still master the details of naval problems insisting that:

> The Staff exist largely to supply information to the Command; at present, in the insensate orgy of decentralisation which resulted from there being no leaders capable of coping with the sheer size of the Great War, they seem to exist largely to supply information to one another. The extent to which large staffs can exist and to outward appearances flourish by taking in one another's washing needs to be seen to be believed.
>
> The Command, then, has got to be capable of assimilating vast quantities of information and of attending to masses of detail; the idea that the possession of a staff does away with the necessity for these qualities is unfounded.[213]

Here it should be noted that Doig's 'Command' is more than just the Commander-in-Chief and included his personal assistants. The 'Staff' answerable to the Commander-in-Chief was a step removed from the inner sanctum and was dedicated to coordinating the specific functions of the fleet. How intrusively a Commander-in-Chief concerned himself with the running of the greater staff was very much subject to the traits of the individual Admiral, but it was also a function of complexity, operational tempo, and technology. The more technical the work became and the more voluminous the correspondence, the more a greater staff proved a necessity. Moreover, if the Commander-in-Chief had concerns which were more than just purely naval, such as serving in a dual capacity like High Commissioner to Constantinople and reporting to the Foreign Office, the conflict of responsibilities required a measure of delegation to ensure that neither responsibility suffered.

---

212 *Ibid* and '*C.B. 3016/30, Progress in Tactics, 1930*', p. 84, British Sources, Box 12, Naval Historical Center, Washington, DC.

213 Doig, 'Command and Staff', *Naval Review*, November 1924, p. 658.

Circumstances forced Backhouse to alter his work-habits as the external crises of the period and the organizational structure of British defence planning and oversight made it simply impossible for an officer, no matter how capable, from carrying the load. This highlights that external factors were always the greatest impetus in fostering change and the habits that Backhouse practiced in the Home Fleet would not avail him at the Admiralty where the First Sea Lord and Chief of the Naval Staff was also a member of the Chiefs of Staff Sub-Committee. The time available simply did not permit a senior officer at the Admiralty to exercise direct control to the extent permitted within the fleet, and if he tried to force events to conform to his previous work-habits, he was apt to head for a fall.

Here, the closer integration of defence planning which occurred after the World War was a key impetus in breaking down the previous patterns of naval work. The initial changes were not sudden, nor were they revolutionary, but once embarked upon they gathered an ever-increasing momentum. Even after more than four years of war, the system for overseeing the strategic affairs of the country had visible defects. This was most demonstrably shown during the limited, but unacknowledged war against the Soviets arising in late 1918. The Admiralty, for one, desired to see the establishment of a body along the lines of the Eastern Committee. This agency, operating under Curzon and overseeing British interests in the Near East, defined the policy objectives for that region. Such control was missing for British forces operating in the Baltic and, in light of the recent loss of the light cruiser HMS *Cassandra*, the Admiralty pushed for such to be implemented.[214] This desire was in addition to the First Lord's previous request that better coordination machinery was required to process, review, and direct the intelligence efforts of the Secret Service then operating under the direction of multiple offices.[215]

Of course, Russian affairs had been overseen by a succession of sub-Cabinet committees which had arose to deal with coordinating and supporting her involvement in the Great War. These committees demonstrated the British system at its best—flexible, adaptable and ever evolving—but they also demonstrated its greatest weakness: effective political and administrative oversight was essential to ensure that the rationale for their creation still existed and that the objectives being pursued were understood. This was not always provided and as the *ad hoc* committees previously used were wound-up at war's end with responsibilities transferred to more permanent agencies, such as the Foreign Office, this oversight was ever more difficult to effect.[216]

The Chiefs of Staff Sub-Committee dating from 1923 was the first change fundamentally altering the postwar practice of naval administration. In its original guise, the

---

214 CAB 23/15, War Cabinet 545A minutes of 17 March 1919.
215 CAB 24/73, First Lord to War Cabinet memorandum of 16 January 1919.
216 Keith Neilson, 'Managing the War: Britain, Russia and Ad Hoc Government', in Michael Dockrill and David French, eds., *Strategy and Intelligence: British Policy during the First World War* (London: The Hambledon Press, 1996), pp. 97-104.

COS was very much a committee within a committee of the CID with its proceedings chaired by a leading politician, such as the Lord President of the Council or even the prime minister himself. Over time, this changed and by 1926-27, most meetings were being chaired by Beatty or another Chief of Staff in his absence. Why this happened is unknown; it may have been no more than the original driving force leading to the establishment of the COS was no longer present. The effect was that a certain level of independence absent this immediate political oversight now arose within the COS. The Chiefs of Staff did not always see eye-to-eye on fundamental issues with control of naval aviation a notable example.[217] Increasingly, though, they concerted their message in order to forestall a possible divide and conquer approach from their political masters. At the Chiefs of Staff Sub-Committee, the questions considered might come from the Services, the CID or from the Cabinet. If coming from a Service, then the aim was largely to concert a common view before forwarding the matter to the CID. If coming from the CID or Cabinet, the COS typically attempted to reach a consensus view before registering their opinion. This did not obligate the CID or the Cabinet to accept the view provided, but rejection was that much harder as the professional heads of the Services were deemed experts within their lines of authority. If ostensibly neutral in political questions, they nevertheless retained strong influence with others who were not so constrained and politicians were aware of this reserve of influence. Germane to this study was the revisions the Chiefs recommended in the organization and reporting of the IDC to which the Government readily acceded at the time of its establishment.

Whilst the Chiefs of Staff now met on a formal and regular basis, actions—particularly, sensitive ones—could still be handled through private channels. Eric Grove reminds us that Chatfield employed such an approach with both Inskip and Duff Cooper over the question of defence commitments and domestic spending priorities rather than raise a sensitive political issue formally.[218] During the initial workings of the committee, the Chiefs tended to act as representatives of their Service and not as a new controlling body. This was largely because the Chiefs relied on the support of their own Service staffs and did not have a separate staff serving them in their COS role. Though Hankey and his team of Assistant Secretaries supported the work of the COS administratively, the direct tasking of them was not a regular occurrence. Why is a matter of conjecture but a couple of reasons suggest themselves. Firstly, the Chiefs were suspicious of any 'super-staff' and collectively they resisted efforts at developing one. Ceding too many tasks to Hankey and his team risked moves of creating one by default. Rather, the custom was for plans and appreciations to be prepared at a Service level in the first instance by the force having the primary responsibility for the problem at hand with the appreciation then submitted to the other Services for

217 Roskill, *Churchill and the Admirals*, p. 79.
218 Eric Grove, 'Admiral Sir (later Baron) Ernle Chatfield (1933-1938)', Murfett, ed., *First Sea Lords*, p. 166.

review and comment, meanwhile, informing the CID of the action being taken.[219] Secondly, passing an issue to Hankey and his team meant that a certain measure of control was lost by the COS. Consequently, they were more apt to rely on the Joint Planning Committee (JPC) to support them. Where the COS did use Hankey and the Assistant Secretaries was in polishing work prepared by the JPC destined for the COS and to be forwarded to the CID or the Cabinet.

As for the JPC, the pace of work within the COS whilst still executing their own Service responsibilities soon proved too much and by 1927 it was created to tackle many of these problems at an even lower, joint level. The creation of the JPC can also be seen as clear evidence that after four years of meeting at the COS level, the Chiefs were beginning to think beyond their individual Service remits and were now viewing problems collectively. As the JPC was initially composed of *ex officio* planners from the Services, the implications of this change of thought were masked. Indeed, it probably was not even realized. Yet, what it did was tie the principal Service planners to their masters in the joint arena in a manner that previous practice did not. Moreover, that the Chiefs elected to create a JPC under their purview rather than refer matters to another existing body, such as the Overseas Defence Committee of the CID, promised to strengthen their contributions to the CID. It also reflected their now growing independence from ministerial oversight as the decision was taken without reference to a civilian chair or to the greater CID.[220]

This highlights another problem in British defence planning and coordination for the period between the wars. Increasingly, the fulcrum of effort was borne at the sub-committee level with the CID failing to provide consistent and sustained oversight. That this is so can be measured in the number of sittings of the CID. They met only five times for the fiscal year 1934-35, six in 1935-36 though rising to fifteen sessions by 1936-37.[221] Of the subordinate bodies then in existence, the most active during the same period was the COS meeting 20 times in 1934-35, holding 27 sessions in 1935-36 and conducting 33 sessions in 1937-38. Closely aligned with the pace of work being accomplished at the COS level were the efforts of the JPC meeting in five sessions in 1934-35, convening a further nineteen times in 1935-36 and witnessing 54 sessions in 1936-37. Of course, one reason why the work of the JPC grew as it did was the institution of the Deputy Chiefs of Staff Sub-Committee in 1935 now requiring the JPC to serve two masters.[222] Of course, creation of the Deputy Chiefs was symptomatic of the broader effort Britain was now pursuing to put right its long neglected defence requirements.

219 CAB 53/1, Chiefs of Staff minutes of 8 January 1924.
220 CAB 53/2, Chiefs of Staff Sub-Committee minutes of 14 March 1927.
221 Tennant Papers, NMM/TEN/42/2. Though officially dating from 1935-36, in substance the DCOS had already been working on an *ad hoc* basis since 1930 as COS minutes show. See CAB 53/3.
222 *Ibid.*

This venue now saw the Deputy Chief of the Naval Staff meet with Service equivalents from the British Army and the Royal Air Force at a step below the Chiefs of Staff Sub-Committee. In turn, a Joint Intelligence Sub-Committee followed in 1936 where threat information could be weighed in common.[223] Just as the First Sea Lord was an *ex officio* member of the COS and the Director of Plans and his deputy served a similar role within the JPC, the Director of Naval Intelligence acted in a like capacity within the JIC. This ensured that matters considered in a joint arena were afforded the attention of the principal Admiralty agents concerned, but, of necessity, this operated at the expense of existing duties. Consequently, this in time led to the appointment of other ancillary staff officers to support the principals. Criticisms in the quality and timeliness of joint appreciations was one problem evidenced, and, in 1939, three current attendees of the IDC were tipped to augment the JPC with Captain William Andrewes,[224] a torpedo specialist, nominated as the naval officer. He moved from the joint course to the Cabinet War Rooms in June 1939 as that year's session concluded early.[225] This was symptomatic of a broader weakness in the system of British defence coordination during the period. Though the fulcrum of decision-making was shifting increasingly to the center and away from the Services, the center was not yet staffed or effectively run to meet the responsibilities that it was accruing. Nor was the creation of a Ministry of Defence the true solution, for the problems were not simply coordinating the affairs of separate Service Ministries so much as coordinating the greater affairs of government.

Notwithstanding all the committees and sub-committees created to investigate and handle strategic problems, the system was fundamentally advisory in nature with executive power residing at the Cabinet level, and even more so, in the hands of the prime minister. This was no bad thing in a democratic system operating in peacetime but during periods of strained relations or actual war, it demanded the constant attention of the premier. This was not always forthcoming or timely for a host of reasons of which the period's economic woes remained only one. During the Abyssinian Crisis, outside agitation from parliament and the Fourth Estate assailed the lack of grip at the centre with calls again raised for a Ministry of Supply, Minister of Defence, and a true joint staff. To this, the Chiefs of Staff were largely opposed believing the present system, if not perfect, had demonstrated its soundness and shown that a successful

223 Gibbs, *Grand Strategy, Vol. I*, p. 773.
224 Later Admiral Sir William Gerrard Andrewes (1899-1974). Short Course, Cambridge University, 1919-20; Fleet Torpedo Officer in *Kent*, 1931-33; Staff Course, 1934, Fleet Torpedo Officer in *Nelson* and *Rodney*, 1935-36; Tactical Course, 1937; Captain, 1938; Imperial Defence Course, 1939; Commanding Officer, HMS *Albatross*, 1939-40; Imperial Defence College, 1947-48; Senior Officers' Course School of Air-Land Warfare, Old Sarum, 1947; Rear Admiral, 1948; Vice Admiral, 1950; Admiral, 1954 and retired, 1957.
225 Dickson, Sound Recording 3168 of February 1978, Imperial War Museum, London and ADM 196/93 (Andrewes).

organization had evolved since its inception based on the mutual goodwill and experience of its members.

Absent direct evidence that the current system could not be made to work, fundamental change was difficult. The Chiefs conceded the presence of a Minister as long as he held Cabinet rank, but fundamentally desired to retain the present system. Foursquare, they were opposed to an independent, joint staff.[226] Their arguments were entirely logical and entirely wrong; yet, in their defence, it was very easy to see the presumed pitfalls of advice tendered to Ministers without responsibility for execution. Doubtless, they worried that any changes would upset their status, as ultimately, it did. Now, though they were subordinate to the CID, they also were members of the parent body.[227] In truth, they were a committee within a committee. This gave them a degree of independence and power that any Minister appointed to oversee their affairs would fundamentally change. Whilst arguing against the creation of an independent, joint staff, Chatfield, Montgomery-Massingberd and Air Chief Marshal Sir Edward Ellington[228] did agree that the present JPC needed augmentation. They recommended appointing three officers of Commander or equivalent rank to the JPC to handle the increasing workload and the provision of a fulltime Secretary to arrange its work in a more systematic manner.[229] The Government accepted this proposal with alacrity with Viscount Swinton announcing the planned enhancement to the JPC in the Lords on 27 February 1936, whilst the prime minister confirmed the intention of appointing a Deputy Chairman having day-to-day responsibility for the CID and Defence Policy and Requirements Committee.[230] This was movement of a kind, but in many respects it only returned matters to where they stood *c.* 1925.

The appointment of three additional planners and a new Assistant Secretary certainly helped, but even this was insufficient and the quality of work continued to suffer and fall behind. Thus, the JPC was augmented again in 1939. This was the moment when Andrewes *et al* arrived in 1939 and demonstrated that announcing changes remained easier than actually implementing them and perhaps confirmed the fundamental arguments of the critics in the first instance.

226 CAB 24/260, COS to CID memorandum C.P. 36 (36) of 10 February 1936.
227 Technically, only the prime minister was an *ex officio* member of the CID and all other members held their appointment at his pleasure. Since 1924 and as a result of the Salisbury Committee recommendations, it had become the practice to have the Chiefs of Staff as members.
228 Later Marshal of the Royal Air Force Sir Edward Leonard Ellington (1877-1967). War Course, 1909.
229 CAB 24/260, COS to CID memorandum C.P. 36 (36) of 10 February 1936. The officers eventually appointed were Captain Charles Daniel, Colonel Richard Dewing and Wing Commander Hugh Fraser with Major Leslie Hollis acting as the new Assistant Secretary. See James Leasor, *War at the Top: The Experiences of General Sir Leslie Hollis K.C.B, K.B.E* (London: Michael Joseph, 1959), p. 36.
230 *Hansard*, House of Lords debate of 27 February 1936, v. 99, cc807-59 and House of Commons debate of 27 February 1936, v. 309, cc53-57.

In the absence of better bureaucratic machinery, a simple accounting of the work already completed by the Services, the other departments of state and by the CID would have been an improvement and the failure to develop such a register was largely Hankey's insofar as the administrative operations of the CID were his responsibility. This also indicates that the JPC lacked a certain measure of direction, as before one plans, a reasonable measure is to determine the work already accomplished in order to ascertain what more is required. To be fair, Hankey recognized the parameters of the problem as he initiated the first steps to record what appreciations the Services had prepared individually through 1931.[231] Though a beginning, his did not address the true scale of the problem faced and Henry Pownall whilst serving as an Assistant Secretary captured something of the scope of the matter recording:

> It is very strange to me that in the most central office in the kingdom there is no organized record system. C.I.D. minutes of meetings have a full & good index. All the other mass of papers, on all subjects, which we possess and to which we constantly have to refer, are not indexed at all. There is merely a contacts sheet, not in alphabetical order, at the beginning of each volume. It is inevitable that in such circumstances that valuable papers, even the work of whole "ad hoc" committees whose existence has been forgotten, should get overlooked. A notable example is that of the General Staff appreciation on the attack of the Dardanelles (1906) which was never considered in 1915 when the Dardanelles became a project. It is all very well to rely on peoples' memories, as is mostly done here. However good memories, that is not a system….I am told that Hankey is utterly averse to all change in anything, that he will not consider for a moment any system for the whole office.[232]

Moreover, a compendium did not even exist as to what all the committees were and their associated members and would not until one was prepared in September 1939.[233] That one was required is confirmed when Balfour speaking in the Lords in 1926 acknowledged the existence of thirty sub-committees reporting to the CID though of the exact number he did not know.[234] Such then highlights that Ellison's charges of the same year leveled in *The Perils of Amateur Strategy* were far from baseless with much remaining to be corrected.

That a prime minister could not devote all his attention to strategic matters is understandable, as his remit was so broad. Yet, the same applied to Hankey, also, who besides being Secretary to the CID and a host of its sub-committees acted as Cabinet Secretary, and, from 1923, Clerk to the Privy Council. Pownall's frustration

231 CAB 53/3, Chief of Staff minutes 26 January 1931.
232 'Volume I, diary entry of 22-26 May 1933', Lieutenant General Sir Henry Royds Pownall Papers, Liddell Hart Centre for Military Archives, King's College, London.
233 CAB 181/5, 'Precedent Book. Part III. Cabinet Committees', no date, but after February 1950.
234 *Hansard*, House of Lords Debates of 16 June 1926, v. 64, cc415-46.

in bringing matters to the fore was not helped by Hankey's style of work and his near all-encompassing responsibilities. Doubtlessly, many a lesser person would have succumbed to the burden, but not even working seventeen-hour days was sufficient for Hankey to meet all the strands of the responsibilities he carried.[235] (Indeed, Maund upon departing as an Assistant Secretary entered Haslar Naval Hospital to rest indicating perhaps the pace of work.[236]) The result was that, at times, Hankey's CID duties suffered with Pownall observing at one instance:

> Hankey scribbled some general remarks in agreement but it was quite clear that he hadn't considered the subject, and that he wasn't going to give it much thought. His head is now full of the forthcoming Economic Conference – and there's a limit to what even his head can contain. So defence questions are shelved with him as they are with his masters. And that is a pity, for surely one of the most useful functions of the Secy of the C.I.D. is to ensure that defence questions are _never_ shelved – far less to allow them to be so in his own head.[237]

Creation of a Minister for the Co-ordination of Defence, as the Deputy Chairman was soon styled, can be seen as a weigh-stop on the road to the formation of a greater defence establishment, but that is to see in events an inevitability that contemporaries did not enjoy. Now, the Chiefs of Staff became expert advisers and no longer members of the CID.[238] In choosing Sir Thomas Inskip, a King's Counsel but also a man with Admiralty experience, as the first holder of the portfolio a bone was being passed to Chatfield as Chairman of the Chiefs of Staff Committee. If creation of the post met the most strident objections of the Chiefs, then the person most to suffer from the change was Hankey, as having a Minister with the remit to call meetings of the Chiefs of Staff Sub-Committee and force the pace of CID deliberations now meant that he could not perforce cover all bases. Facing this problem, Hankey proposed abolishing the sub-committees on Defence Policy and Requirements and Defence Plans (Policy) and merging their activities into the CID proper from whence they originally derived.[239] This would allow him to maintain personal control of the primary bodies overseeing defence and still allow him to meet his greater responsibilities to the Cabinet and to the prime minister.

On the surface, Hankey demonstrated all the attributes of the centralizing naval officer and, as an officer late of the Royal Marine Artillery and the Naval Intelligence Department, his habits of work must have owed something to his previous duties.

235 Hankey to Liddell Hart letter of 14 July 1933, Liddell Hart Papers, LH 1/352/21, LHCMA. Original emphasis.
236 ADM 196/127 (Maund).
237 Diary entry of 2 June 1933, Pownall Papers, Volume I, LHCMA. Original emphasis.
238 CAB 24/273, Secretary to the CID memorandum of 3 November 1937.
239 _Ibid_, Minister for the Co-ordination Defence covering note of 3 November 1937 to Secretary's memorandum.

Whether attendance at a staff college would have mitigated these traits is impossible to say but a broad purpose of education is to correct the defects of habit and to allow one to think of other modes of action. Certainly, Commander William Friedberger believed so and gave such as his reason for wishing to study at Camberley.[240]

Hankey was by no means the only one supporting the CID and its constituent parts as previous commentary has shown. Augmenting him were a number of civilians along with representatives from the forces. The Treasury typically met the salary and allowances of these officers, if serving as a Commander, and the Admiralty met the difference through a grant-in-aid if the officer received promotion whilst serving,[241] as such occurred with Captain Angus Nicholl.[242] Overall, for the naval officer assigned, the duty was comparable in enumeration to duty in the Admiralty.[243] From 1921, the naval officer serving in Hankey's office was of Commander's rank and a graduate of the Royal Naval Staff College. The only exceptions to this were Henry Moore, who held the *qs* designation and came from the faculty of the Staff College to the CID in late 1921, and Leslie 'Jo' Hollis.[244] Moore's appointment came at short notice and was to have a profound influence on his future career. He replaced Commander Richard Hamer,[245] a Signals Officer, who held neither the *qs* or *ws* designation when appointed to the CID in 1920 and left Hankey and the CID early due to medical problems, which ultimately forced his retirement.[246] Though the record is unclear, it appears Hamer was not an entirely successful destroyer officer and, as a consequence, was recommended for a staff appointment which duly followed when posted to the *President* and duty with the CID.[247] Meanwhile, for Hollis, a Royal Marine Officer

---

240 'Personal Reasons for Joining the Staff College', W.H.D. Friedberger, December 1938, ap Rhys Price Papers, NAM 9204-164-11-5.

241 ADM 196/122 (Nicholl).

242 Later Rear Admiral Angus Dacres Nicholl (1886-1977). Short Course, Cambridge University, 1919; Staff Course, 1927-28; Staff Officer Operations in *Osprey*, 1929-30; Commanding Officer, HMS *Torrid*, 1930 and *Vidette*, 1930-32; to RAN 1932-34; Commanding Officer, *Cairo*, 1936; Assistant Secretary, CID, 1937-41; Captain, 1939; Rear Admiral, 1948 and retired, 1951.

243 ADM 196/51 (Hermon-Hodge).

244 Later General Sir Leslie Chasemore Hollis (1897-1963). Staff Course, 1927-28; Staff Officer Intelligence Capetown, 1929-32; Plans Division, 1932-36; Assistant Secretary, CID, 1936-46; Brevet Lieutenant Colonel, 1937; Temporary Colonel, 1939; Acting Colonel Commandant, 1941; Temporary Brigadier, 1941; Lieutenant Colonel, 1943; Acting Major General, 1943; Colonel, 1946; Major General, 1947; Acting Lieutenant General, 1947; Lieutenant General, 1949; Chief Staff Officer, Ministry of Defence and Secretary to the Cabinet, 1947-49; Lieutenant General, 1949; General, 1951 and retired, 1952.

245 Later Captain Richard Lloyd Hamer (1884-1951). Flag Lieutenant in HMS *Dominion*, 1909-10 and *Hermes*, 1910-12; Commanding Officer, HMS *Rowena*, 1916-19; HMS *Tempest*, 1919 and *Stonehenge*, 1919-20; Assistant Secretary, CID, 1920-21; retired, 1923 and Captain, 1929. Recalled for War Service.

246 ADM 196/49 (Hamer).

247 ADM 196/126 (Hamer).

holding the naval *psc*, coming to 2, Whitehall Gardens, a cul-de-sac off of Whitehall proper, was to prove eventful and rewarding. Entering as a Major in 1936, he departed as an Acting Major General ten years later after working closely with Churchill during the latter's first premiership.[248]

Greenwich and the Staff College were common to the seven officers assigned to the CID and though two could claim prior time with a Dominion, this was not a requirement of the appointment. Following their appointment as an Assistant Secretary, each officer returned to more traditional duties with the most common posting being as an Executive Officer in a ship. Meanwhile, two of the officers who served as Assistant Secretaries (e.g., Moore and Claude Hermon-Hodge[249]) would have experience of working in the sub-committee system itself returning as *ex officio* members to the JPC and JIC. Still, appointment as an Assistant Secretary marked one as a *proto* staff officer and the downside was that the officer had to work that much harder upon returning to the Navy and demonstrate that more remained of one. Henry Moore receiving the thanks of both Hankey and Beatty for his efforts as an Assistant Secretary, in this respect, was also disadvantaged by his short stature and retiring manner. If size did not allow Moore to compete in games with any chance of success, he, nevertheless, was physically fit proving a good rider, a shooter, and a fisherman. He succeeded in making converts of his greater abilities and that his time with the CID benefited the Navy was reflected in the quality of the strategic schemes that he helped develop for fleet exercises.[250]

As with Moore, Hankey, at least officially, rated Hermon-Hodge an excellent Assistant Secretary and could not speak too highly of his abilities when vouchsafing his service following his entry into the zone of promotion for Captain; a view that Beatty endorsed.[251] Still, perhaps the best indication of an officer's success in any appointment are the subsequent assignments received. By this yardstick, Arthur

---

248 Assistant Secretaries serving from the Royal Navy and Royal Marines were Richard Hamer, 1920-21, Henry Moore, 1921-24; the Honourable Claude Hermon-Hodge, 1924-28; Loben Maund, 1928-31; Cecil Allen, 1931-34; Arthur Clarke, 1934-37; Leslie Hollis, 1936-46 and Angus Nicholl, 1937-41. The Cabinet Office moved to Richmond Terrace in August 1938 when the previous buildings (i.e., 2-3-4 Whitehall Gardens) were demolished.

249 Rear Admiral the Honourable Claude Preston Hermon-Hodge (1888-1952). Commanding Officer, HMS *Foxglove*, 1915-17, HMS *Gunnet*, 1917 and HMS *Tonbridge*, 1918-19; Staff Officer in HMS *Impregnable*, 1920-22; Commanding Officer, HMS *Cornflower*, 1922-23; War Staff Course, 1923-24; Assistant Secretary, CID, 1924-28; Captain, 1930; Senior Officers' War Course, 1930-31; Tactical Course, 1931; Commanding Officer, HMS *Versatile*, 1931-32; Senior Officers' Technical Course, 1932; Senior Officers' School Sheerness, 1933; Commanding Officer, HMS *Hastings*, 1934-35; Deputy Director, Naval Intelligence Division, 1935-38; Chief Staff Officer in *Drake*, 1939; Rear Admiral and retired, 1940. Recalled for war service.

250 ADM 196/92 (Moore).

251 ADM 196/127 (Hermon-Hodge).

Clarke[252] was a clear favorite, as Ismay sought his services again in July 1939 to coordinate Anglo-French naval relations within the CID and then the War Cabinet.[253] So, too, was Nicholl, whose appointment was extended upon the outbreak of war. Yet, given that the appointment was made whilst an officer held the rank of Commander, no Assistant Secretary was a prior graduate of the IDC though Moore would later be so. Indeed, his time with the CID made his efforts at the IDC all the more successful prompting Richmond to recommend Moore's later appointment to the college's directing staff.[254] Though this never happened and accepting Richmond's observations of Moore's contribution whilst a qualifier, attending the IDC whilst a Captain operated against appointing officers to one of the positions most relevant to their training. Of course, their Staff College education was helpful, but the vacuum of IDC training led the Service to appoint only holders of the naval *psc*—a move that was against the tradition of appointing the best officer available irrespective of specific qualifications. In hindsight, it is clear that a two-tier level in staff training and command was no less a requirement for the joint environment as it was for the Service itself. That one was not established before the onset of war in 1939, though some clearly recognized the requirement, reflected the problems of conflicting priorities during the period of defence expansion, the thought that the planned expansion of the IDC in 1940 might alleviate the problem, and a certain amount of bureaucratic inertia.[255]

As for those naval and marine officers who did serve as Assistant Secretaries to Hankey, their appointment usually lasted for three years. The three-year limit was the most that was sustainable for a naval officer whilst allowing one to remain competitive with respect to promotion. Army officers employed by Hankey could serve longer and did, but a naval officer typically had to secure his captaincy by the eighth year following his promotion to Commander and any more time away from his parent Service was problematic. As ever, the salt horse having secured his *psc* was the officer most apt to serve at the CID. Discounting Hollis the marine, four officers had not specialized, two had been Navigating Officers, and Hamer was the lone signals specialist. Too much should not be read into this except that it is indicative that the Navy aimed to place its specialist officers afloat or in its technical schools first and, as ever, faced a shortage of qualified specialist officers during the period.

252 Captain Arthur Wellesley Clarke (1898-1985). Short Course, Cambridge University, 1919; Staff Course, 1926-27; Staff Officer Operations in *Queen Elizabeth* and *Warspite*, 1928-30; to New Zealand Division, 1930-33; Assistant Secretary, CID, 1934-37 and 1939-40; Commanding Officer, *Penzance*, 1937-38; Captain, 1939 and retired, 1948.
253 ADM 196/93 (Clarke).
254 ADM 196/127 (Hermon-Hodge).
255 On the planned expansion of the IDC in 1940, see 'Requirements for a Civil Servant on the Directing Staff of the IDC', draft memorandum, no date but *c.* 1939, Tennant Papers, NMM/TEN/20.

Hankey possessed wide latitude in the make-up of his office, though the actual appointment of an Assistant Secretary was subject to prime ministerial approval.[256] For a naval appointee, the normal path followed was of the Admiralty proposing a candidate to the CID and the prime minister concurring in the appointment.[257] Doubtlessly, a word from Hankey would have sunk an officer's proposal or selection and one who sought appointment but deemed unacceptable by Hankey was Richmond who attempted to become a naval assistant in 1917.[258] Having assistants from the separate Services shared the burden of work, but it also meant that note takers and facilitators were on-hand who understood the mission, organization, and language of each arm. Though assistants held equal status, Hankey recognized the abilities and worth of each officer assigning tasks to them based not only on their expertise but also on their availability.[259] Each had responsibility for overseeing the workings of their assigned sub-committees, keeping minutes of their proceedings, and registering decisions made. Ismay, though, makes clear the work proved more than secretarial in nature, as secretaries were expected to initiate actions, see that the committee's work came to a successful conclusion, and de-conflict the committee's efforts with any of the myriad other bodies reporting to the CID.[260] That the naval officers serving came from the executive and not the accountant branch, offers further evidence that they were no mere note-takers. Finally, another category of officers serving within the CID organization were those detailed to support its Historical Section. Archibald Bell and Oswald Tuck, an Instructor Commander, were two such officers, but another was Commander Charles James,[261] an interpreter in Japanese, who eventually found himself in the Admiralty Historical Section following completion of his Intelligence Course in 1920.[262]

As for those officers who did hold the *idc* post-nominal, the challenge remained how best to employ them absent any immediate assignments requiring the level of joint training received. Not surprisingly, many returned at once to sea in command of a warship or a flotilla, while another sizeable contingent continued their period of training by taking one or more of the courses directed to those of their rank. A lesser number were posted to a staff element and two of these officers were employed directly as an Assistant Director of the Plans Division. A further three assumed command responsibilities ashore with a like number becoming attachés—a posting seemingly

---

256 Ismay, *Memoirs*, p. 42.
257 ADM 196/128 (Allen).
258 Stephen Roskill, *Hankey: Man of Secrets, Volume I, 1877–1918* (London: Collins, 1970), p. 42.
259 Ismay, *Memoirs*, p. 52.
260 *Ibid*, pp. 52–53.
261 Later Captain Charles Herbert Neill James (1884–1956). War College, 1908; Intelligence Course, 1920; Historical Section, CID, 1920-22; retired, 1922 and Captain, 1929. Recalled for war service.
262 ADM 196/126 (James).

ideal for their newfound breadth of view. That a fair portion soon left the Service or failed to secure flag rank to the chagrin of some in parliament merely confirms that training was never an end in itself and the Navy still had to match billets to people. Appointing such officers to the Intelligence Division might be considered one measure to marry their skills with an unmet requirement. Yet, given their seniority, this was really not an option whilst they remained on the active list without adversely influencing the promotion prospects of others, though the Service was remiss in not employing them once retired, much as Alfred Dewar supported the Historical Section.

Below the Chiefs of Staff level, concerting actions across the Services within a theatre of operations proved a problem every bit as difficult to resolve. With formal mechanisms lacking and common doctrine fledgling, amiable relations between commanders and staffs became essential. The pitfalls of this void being amply demonstrated during the Abyssinian Crisis of 1935, though fortunately, with no ill-operational effect. Individually, separate appreciations were drafted, but a joint appreciation addressing the problems facing Britain and the objectives to be secured, so much the focus of IDC training, was initially missing.[263] Whether this appreciation should have been drafted by the JPC and provided to the naval, military, and air component commanders, or, instead, have been written within the theatre and endorsed by the COS and CID was ill defined. Ultimately, the COS requested from the theatre commanders that a joint appreciation be prepared, but at this date, November 1935, the latter believed it was a little late in the day for such an assessment.[264]

In retrospect, it can be seen that two types of appreciations were required. The first typed prepared by the JPC for the COS and CID would have established Britain's strategic objectives, the forces available for use, the role of allies and likely attitude of neutrals, and the constraints facing her. A second form of appreciation, developed at the theatre-level, would have specified how the forces available would accomplish the strategic objectives established. This in fact largely happened, but the plans initially developed were drafted in isolation to one another, were intended for separate audiences, and were difficult to coordinate given the distances involved. In short, the planning conducted separately in London and Egypt was intimate, but synchronizing the two was difficult. The crisis also demonstrated that it was easier to develop planners than it was to ordain effective Ministers overseeing the process. As ever, the higher direction of war was the problem most lacking clarification. In truth, this was but a subset of the problems of Cabinet-style government where executive authority is fragmented, decision-making is situational dependent, but responsibility remains collective.

Even the question of how orders emanating from London should be issued to three component commanders (e.g., Commander-in-Chief Mediterranean, General Officer

263 Brooke-Popham, 'Notes on the Emergency in the Near East. September 1935 to June 1936', Brooke-Popham Papers, 4/12/1, LHCMA.
264 *Ibid.*

Commanding Egypt and Air Office Commanding) proved vexing. The CID was only an advisory body operating in peacetime with a War Cabinet supplanting it at the appropriate moment. This had yet to occur and, meanwhile, the COS had no collective authority to command, as this responsibility remained at the Service level for their respective forces. Flexibility, adaptability, and ever resourceful in a country with no single constitutional source, Brooke-Popham recorded that the expedient adopted was for the Admiralty to issue its orders and instructions to Admiral Sir William Fisher with the request that he forward the same to Lieutenant General Sir George Weir[265] and to himself.[266] Actually, the process was even looser than which Brooke-Popham described. The practice became one where the Chief raising an issue was asked by the other Chiefs to forward the item for action to his subordinate commander with instructions that he coordinate, as appropriate, with his theatre counterparts. Thus, when Montgomery-Massingberd, the CIGS, voiced concerns about the defences of the Suez Canal, the COS agreed:

> To ask the War Office to communicate with the General Officer Commanding in Egypt and instruct him to forward a joint appreciation, prepared in conjunction with the Naval Commander-in-Chief and the Air Officer Commanding, on the Italian threat from Libya and proposals for meeting it.[267]

Meanwhile, in the Near East the three component commanders worked to concert their plans. With the Mediterranean Fleet now operating largely from Alexandria due to the proximity of Malta to Italian airfields, the defence of that city was conferred upon a naval officer, Rear Admiral Robert Raikes. This was at variance from previous custom that conceded to military authority such responsibility, as only the local military garrison was apt to be available in all instances to repel attack.[268] Raikes, thus, became responsible for not only the anti-aircraft defences of the port and city, but also controlled No. 29 Squadron Royal Air Force, then operating Bristol Bulldog and Hawker Demon fighters.[269]

Flexibility from formal procedure ever a hallmark of British ways, Admiral Fisher benefited from the presence of Dudley Pound who was designated as Fisher's replacement as Commander-in-Chief, Mediterranean. In the event, it was decided that replacing Fisher in the midst of a crisis was not the most opportune moment and Pound, now on station, acted as Fisher's understudy and adviser. Neither were graduates of the War Staff Course, though Pound had taught at the War College in the year

---

265 Later General Sir George Alexander Weir (1876-1951). Staff Course Camberley, 1912-14.
266 Brooke-Popham, 'Notes on the Emergency in the Near East. September 1935 to June 1936', Brooke-Popham Papers, 4/12/1, LHCMA.
267 CAB 53/5, Chiefs of Staff Sub-Committee minutes of 16 October 1935.
268 CAB 53/1, Chief of Staff Sub-Committee minutes of 4 December 1923.
269 Brooke-Popham, 'Notes on the Emergency in the Near East. September 1935 to June 1936', Brooke-Popham Papers, 4/12/1, LHCMA.

before the World War and Fisher had briefly attended the senior course. Though many officers trained at the Staff College eventually reached positions of high responsibility, the Navy never accepted that such training was a prerequisite to command. Of the nine officers commanding the Mediterranean Fleet between the wars, none held the naval *psc*, though Andrew Cunningham was a holder of the *idc*, and only two had taken a War Course—Chatfield and Pound.[270] Alternatively, of the eleven officers holding command of the Atlantic and Home Fleet between the wars, only two held the naval *psc*: Charles Forbes and William Boyle.[271] Even in these instances, their post-nominal designations arose from service on the directing staffs of Greenwich rather than through attendance as qualifiers. More tellingly, was the place of the SOWC or its prewar predecessor, as eight of the officers commanding had attended the course at least once during their career.[272] Even for those not attending, the impact of war, continued immediate employment and their eventual seniority operated against attendance and accounts for their absence.

The disparity between the numbers attending the SOWC between the two fleets may represent nothing more than chance for even Cunningham never attended a War Course, though he attended more than his share of other courses. What is obvious is that though flag officers increasingly attended the Tactical Course, the SOTC and the SOWC, attendance occurred within the allowances of first meeting other obligations. Though the naval system of staff training and education was heavily influenced by what had previously been developed at Camberley, it was no mere clone of the British Army and observers were wrong to believe that it should be. Most officers eventually commanding squadrons and flotillas, if not fleets, were products of the SOWC, the SOTC, and the Tactical Course, too. The Staff Course would increasingly become an important criterion, but another war and the contractions of the fleet awaited.

Unsurprisingly, many officers gravitated to the Admiralty to serve on the Naval Staff or to one of the technical departments overseeing weapons development or ship-building. If on the Naval Staff, they had to grapple with the steady stream of reports flowing from lower echelons that documented the many aspects of naval routine. This frequently resulted in the publication of the many yearly and semi-annual publications produced during the period capturing the essential points of progress (or lack thereof) in weapons use. Others had responsibility for overseeing their section's technical publications revising them and issuing temporary addenda until a fully revised edition was forthcoming. Consequently, a great deal of their time was consumed with researching, writing, and editing and those skills honed at the Staff College in exposition and clarity of thought at last came to fore.

270 Namely, Admirals Gough-Calthorpe, de Robeck, Brock, Keyes, Field, Chatfield, Fisher, Pound and Andrew Cunningham.
271 The eleven officers were Admirals Beatty, Madden, de Robeck, Oliver, Brand, Chatfield, Hodges, John Kelly, Boyle, Backhouse and Forbes.
272 Those absent were Admirals Madden, de Robeck and Henry Oliver. .

Sitting on committees was another call on the staff officer's time whilst at the Admiralty and though this may have been a temporary *ad hoc* group to work through a specific issue, such as defining the manning bills within ships for a given Department, others were joint committees. Chartered under the auspices of the CID to work through a problem common to many, frequently the skills required leaned more to those of his technical specialization than to what he had acquired in staff training and education. Commander Edward Dangerfield[273] offers an example. A trained Signals Officer and earning his naval *psc* in 1933, Dangerfield whilst serving in the Signal Department also served as the Admiralty's representative to the Sub-Committee on the Control of Radio Transmission.[274] Here, his technical knowledge was the more salient requirement and that he had served in the Experimental Branch of the Signal School previously especially fitted him to his committee work.[275]

Whether a specialist or a salt horse, the trained staff officer following his service on a staff would in the course of time return to sea in a non-staff position. This was essential if he desired to advance further. It has already been seen where officers in flying duties returned to a period of general service and much the same pattern was followed for submariners, anti-submarine specialists, and even the salt horse who spent much time at sea, but frequently in destroyers. Each could find appointment in a billet outside their normal provenance and this leavening added to an officer's experience and professional development. It also acted as a brake on keeping staff sizes down as the trained staff officer would know something of affairs outside his own immediate technical field if he returned to staff duties.[276] Of course, it was easier for the specialist to play the part of the generalist and for the salt horse, his cross-training might be no more than moving to a large ship where the difference in his duties proved more one of scale, than of substance. Consequently, the practice operated against the salt horse officer and the place of a Captain (D) of a flotilla is illustrative. Frequently filled by a specialist officer, it was the salt horse officer whose prior experience in minor vessels best fitted him to the role of commanding a destroyer flotilla.

Thus, the preparatory courses required of an officer before assuming command of a ship, a flotilla, or a squadron were a means of providing a broadening of outlook before embarking on a new phase in his career. The question arises whether the belief that a good officer, notwithstanding background, could perform most any task was

---

273 Later Captain Edward Dangerfield (1899-1941). Short Course, Cambridge University, 1919; Signals Course, 1924; Flag Lieutenant and Fleet Wireless Telegraphy Officer in *Hawkins*, 1924-26; Signal School, 1927-29; Fleet Wireless Telegraphy Officer in *Nelson*, 1930-32; Staff Course, 1933; Signal Department, 1934-36; Tactical School, 1936; Commanding Officer, HMS *Basilisk*, 1936-38; Captain, 1938 and Commanding Officer, HMS *Dido*, 1940.

274 CAB 24/278, CID, Sub-Committee of the Control of Radio Transmission, *Report* of 18 July 1938.

275 *Navy List*, October 1927, p. 282.

276 Comma, 'Training of Officers', *Naval Review*, February 1935, p. 41.

wise. Certainly, it was deemed necessary, else, attracting officers to some specialist duties given the narrow and limited confines of their work would prove impossible. This was especially true for aviators, submariners, anti-submarine specialists, and, more especially, those qualified as Physical and Recreational Training officers. Nor was it possible to abolish specialization—bane though it was. It arose when the appliances of war provided by a modern, industrial and scientific age, to which the Navy was but a part, went to sea. Turning the clock back to a simpler time was not a feasible option. Thus, cross-training filled an important organizational objective by allowing officers opportunities for greater advancement, but the risk was that one might be found wanting through insufficient preparation at a critical moment. A case in point is Brian Egerton, the esteemed torpedo specialist and capable staff officer. His command of *Courageous* from 1933-35 was not a marked success and his lack of experience or training in carriers or aviation, in general, was part of the root cause. A retiring personality reinforced the image that in this appointment Egerton was out of his depth. If the opinion of Reginald Henderson was harsh, then it remained a fair reading of Egerton's limitations shortly after his appointment and the limitations, too, of insisting upon a good all-round officer exercising responsibility along lines not appropriate to one's talents.[277]

Command of a carrier was the greatest challenge to face a prospective commanding officer of the interwar Navy and Egerton's limitations must be seen in that light. Weather was always a factor, ship-handling was a handmaiden to air and fleet operations, and fatal accidents were common. With a mixed crew of naval and RAF personnel, the administrative burdens coupled with the operational challenges previously mentioned called for skill, judgment and tactful temperament of the highest order. Arthur Dowding, more fitted than most officers to the challenges of carrier command, had also found the experience trying in 1935 when in the *Furious*.[278] The case of Maitland Boucher in *Courageous* offers a similar story. A highly regarded technical officer who still flew whilst serving as Flag Captain in the carrier, his relations with embarked aircrew were highly strained and his judgment was questioned by superiors.[279] Accordingly, Egerton's shortcomings may reflect more upon the limitations of the system than the limitations of the man. Still, two non-flying officers who did do well commanding carriers were Arthur Lyster in the *Glorious* from 1937-39 and Noel Laurence as Vice Admiral Aircraft Carriers during 1935-37. As a trained submariner, Laurence had nerves of steel and was not rattled by the perils of flying operations and Backhouse recommended him for any command position in aviation, accordingly.[280] His experience as Rear Admiral Submarines, if of an entirely different class of vessel was strong grounding for assessing the limitations of subordi-

277 ADM 196/91 (Egerton).
278 ADM 196/91 (Dowding).
279 ADM 196/92 (Boucher).
280 ADM 196/91 (Laurence).

nate commanding officers. Meanwhile, Lyster got on well with the crew, was able to sustain their morale during a period of tragic accidents, and ultimately achieved a high state of efficiency in his command. Lyster and Laurence both proved that a capable all-round officer could succeed in a carrier command, but the challenge remained the greater for such officers.[281]

Though the Navies of the Empire concerted their training and practices to conform with the greater Royal Navy, this was no simple matter and became increasingly more difficult as the period progressed as the aims of Britain and her partners were never entirely congruent. Increasingly, entry-level training was provided by the Dominion, but higher-level training and experience was secured through the Royal Navy. This operated in two ways. First, Commonwealth officers traveled to Britain to receive training and follow-on assignments not available at home and, secondly, senior British officers served in the Dominion force filling key command and technical postings. In the latter case, these were billets that the newer navy was not yet ready to fill whilst conforming to the professional experience expected of an imperial navy operating along common lines. Here, Australia had made the greatest effort and progress, and, as such, paradoxically, made the greatest call for British training and support amongst the separate establishments. Others, such as the India, Newfoundland, and the Union of South Africa relying more on the direct support of the Royal Navy. Thus, whilst accepting the presence of British officers within the RAN, the Commonwealth emphasized that its ultimate policy was to develop a truly stand-alone force able to fulfill the imperial obligations acceded to in 1923 and confirmed again in 1932.[282]

Commander Ralph Binney was one such officer who elected to serve with the RAN upon completing his War Staff Course, but his options for such service did not end there, as it was not merely to the broader outposts of Empire that British naval officers found employment. Between the World Wars, British naval missions operated in Greece, Chile, Poland, Persia, Romania, and China whilst other officers upon retiring, including Binney, were seconded to Columbia with the active assistance of the Admiralty.[283] Such advisory work was beneficial to the host navy, furthered British interests, and also allowed officers to find employment rather than spend a period on half-pay or even retire. Those officers spending time with a foreign navy reads as a veritable naval *Who's Who* indicating that the Royal Navy, ever mindful of its status and prestige, did not stint in its appointments to these missions. Richard Webb, Henry Thursfield, Harold Baillie-Grohman, Clarence Howard-Johnston, and Charles Turle each featured at one time or another in such appointments and William Jackson was especially esteemed in Chile serving two assignments with that country's navy.[284] The work was frequently demanding, as an officer had very much to rely on his

281 ADM 196/92 (Lyster).
282 'Australian Naval Defence' memorandum of 16 August 1935, NAA A5954, 1005/13.
283 Binney Papers, IWM/75/98/1.
284 ADM 196/127 (Jackson).

own resources and initiative and not without a measure of personal risk, as political instability was not uncommon. Moreover, as such service operated outside the direct oversight of the Royal Navy coming as it did under the purview of the Foreign Office, an officer risked being marginalized from his own Service.

Some officers owed their assignment to the influence of a senior who requested them by name to the Admiralty. If the officer was available, the request was likely to be filled, but the Admiralty showed little difficulty in refusing such a petition if the officer was noted for other for duties. The extent and prevalence of such treatment is difficult to judge, but that it existed is confirmed by many officers themselves in their memoirs. Nor was such a practice viewed as necessarily a bad thing. Senior officers needed to know that a subordinate officer was not just capable but also loyal and a good mixer in the close-knit community that the staff became. One observer suggested:

> It should therefore be the clear duty of those who are destined for high command to be on the lookout for useful staff officers for themselves: loyal, active, fearlessly critical, and individually able men who will be prepared to throw in their lot with the command when the time comes even at the cost of some self-sacrifice. And these they should attach to themselves at the earliest possible period in their careers....
>
> A personally selected group of men, accustomed to working together and for the same man, is the best basis on which to start the formation and organisation of a staff. And, since you are then working less in the dark than if you were organizing for a set of "average" men, better results can be achieved by giving the command a freer hand as to their employment—the subdivision of work between them. The great principle all men are unequal always makes staff organisation on paper for the "average" staff officer an ineffective proceeding.[285]

As flag officers were also subject to being passed-over, the system described might not always work to an officer's advantage. In cases such as these, the only course was to find another rising star. This was Paymaster Lieutenant Douglas Doig's fate, as the two flag officers appointed to *Coventry* during his time in the ship soon retired. Ever resourceful, Doig secured appointment to the *Nelson* wearing the flag of Vice Admiral Chatfield, Commander-in-Chief, Atlantic Fleet.[286]

Chatfield was one officer who sought to retain a close personal staff through multiple appointments. Leaving aside an officer of the accountant branch such as Paymaster Captain Rowland Jerram who followed Chatfield numerous times from ship to ship, when Chatfield required a new Chief of Staff in 1931 he opted for Rear Admiral Sidney Bailey. Bailey had served with Chatfield during the war in *Queen Elizabeth* as her gunnery commander. Now he replaced William James, another Chatfield

---

285 Doig, 'Command and Staff', *Naval Review*, November 1924, p. 659.
286 *Navy List*, January 1931, p. 255.

protégé who had also been his Chief of Staff in the Atlantic Fleet.[287] More tellingly, coming with him from his time in *Nelson* and command of the Atlantic Fleet were Lieutenant Cecil Brown, acting as his Flag Lieutenant, Captain Robert Davenport[288] having served previously as Captain of the Fleet and presently Flag Captain in *Queen Elizabeth*, and, now, Doig.[289] That said, the second tier of staff officers such as SOO, SOI, Fleet Gunnery Officer, and their counterparts did not follow Chatfield from the Atlantic to the Mediterranean Fleet. This is not surprising for a Commander-in-Chief's contact with these officers was more remote and formal even though embarked in the same ship. Here, the presence of the Chief of Staff and the Captain of the Fleet were pivotal and it was through these two officers that the Commander-in-Chief principally exercised control over the staff and, consequently, the fleet.[290]

Thus, the Admiralty allowed a certain amount of discretion in a senior officer's ability to organize the affairs of his personal staff for these officers were an extension of his position with both command and social responsibilities. Not all officers relished such an appointment as it forced one to work against type in many respects. For those who did enjoy the close proximity to command with its social demands a friendly lunch with a flag officer was an excellent way to test the mettle of the aspiring young officer. Lieutenant Ralph Fisher whilst serving as a Term Lieutenant at Dartmouth was afforded the opportunity of naming his next posting, a privilege normally extended to holders of his present position. When Fisher stated a preference for the China Station, Rear Admiral 'Biff' Rose[291] and his wife invited him to a luncheon to determine whether he met with the Admiral's requirements for a Flag Lieutenant.[292] He did not, but he still found himself on the China Station serving as

287 Chatfield, *Navy and Defence*, p. 242.
288 Vice Admiral Robert Clutterbuck Davenport (1882–1965). Chief Staff Officer in *Julius*, 1919-20; Captain, 1922; Senior Officers' Technical Course, 1923; Senior Officers' War Course, 1923-24; Commanding Officer *Despatch* and Senior Officer Yangtse, 1924-26; Captain of the College Greenwich, 1926-28; Captain of the Fleet in *Nelson*, 1928-30; Flag Captain in *Queen Elizabeth*, 1930-33; in Command, Naval Barracks in *Pembroke*, 1933-35; Rear Admiral, 1934; Commanding Coast of Scotland in *Greenwich*, 1935-37; Vice Admiral and retired, 1938. Recalled for war service.
289 *Navy List*, p. 261 and *Navy List*, January 1932, p. 261. A number of other accountant branch officers also followed Chatfield from the *Nelson* to the *Queen Elizabeth*, but their transfer lies outside the scope of this study.
290 Chatfield, *Navy and Defence*, p. 218.
291 Later Vice Admiral Sir Frank Forester Rose (1878-1955). Commanding Officer, HMS *Nubian*, 1911-14; Mobilisation Division, 1914-15; Captain, 1918; Senior Officers' War Course, 1920 and 1930-31; Captain (D), Third Destroyer Flotilla, 1921-23; Flag Captain in *Iron Duke*, 1923-25; Deputy Director, Operations Division, 1925-26 and Director, Operations Division, 1926-28; Rear Admiral, 1929; Tactical Course, 1929-30; Senior Officers' Technical Course, 1930; Rear Admiral Destroyers, 1931-33; Vice Admiral, 1934; Commander-in-Chief, East Indies, 1934-36 and retired, 1937.
292 Fisher, *Salt Horse*, unpublished manuscript, p. 36, Fisher Papers, I, LHCMA.

First Lieutenant in a destroyer, though Rose ultimately went to the Mediterranean instead.[293]

An officer working in his immediate office and from the accountant branch could expect to follow an admiral from ship to ship, command to command, if he was deemed a valued member of the team. Likewise, a Commander-in-Chief in a seagoing billet had a measure of latitude in naming his own Chief of Staff and Captain of the Fleet at the appropriate time. If a position was filled already, it would be exceptional to prematurely terminate an officer's appointment, as this could reflect badly on his service record and hinder future advancement. When the moment was ripe for replacement, a choice officer suitable in rank and experience could appear. Of course, much of this would have been coordinated behind the scenes, through personal correspondence or meetings, to ensure that the potential officer desired the appointment before the senior officer ever considered approaching the Admiralty. Doig believed it was not only right, but also essential that an Admiral employ an element of 'inspired favouritism' in selecting his personal staff for trust, confidence, and compatibility were traits of no little importance and necessary to ensure the effective working of any staff.[294] This is surely so, but the risk was always that compatibility might trump competency.[295]

Lest one believe that Chatfield was perhaps unique in his ability to organize his immediate staff, the record is abundantly clear he was not. Admiral Boyle whilst in command at Greenwich and directing the War Course admitted the appointment allowed him to gather amongst him congenial friends and former shipmates whilst making new ones to his present benefit and that of the future.[296] Amongst those friends of previous acquaintance soon appearing were Captains Algernon Willis, later described by Lord Tedder as 'a dry little man without much humour, but very competent', as a member of the SOWC directing staff and Thomas Calvert, the Director of the Royal Naval Staff College.[297] Calvert had served previously with Boyle in *Vivid*, the Devonport naval training establishment, in the early 1920s when Boyle commanded the naval barracks and Calvert was overseeing the attached Petty Officers' Course.[298] With his appointment predating Boyle's to the *Vivid* by a year and having previously served in submarines, Calvert's association with Boyle was of a normal pattern to this date. By 1926, Boyle was Rear Admiral, First Cruiser Squadron flying his flag in *Frobisher* with Calvert now serving as his Flag Captain.[299]

Calvert's appointment to Greenwich again predated Boyle's and was initially to its War College following completion of his IDC course. Shortly, he became Director

293 ADM 196/45 (Rose).
294 Doig, 'Command and Staff', *Naval Review*, November 1924, p. 658.
295 Not in Doig's case, though, as he continued to advance and prosper in the Service and eventually became an Acting Captain.
296 Cork and Orrery, *Naval Life*, p. 155.
297 Tedder, *With Prejudice*, p. 47.
298 *The Navy List* (London: His Majesty's Stationery Office, July 1921), pp. 891-93.
299 *Navy List*, October 1927), p. 240.

of the Staff College succeeding Charles Little upon the latter's departure to the *Warspite* and command of the Second Battle Squadron. In September 1933, Boyle now Commander-in-Chief, Home Fleet, hoisted his flag in the *Nelson*, but not before Calvert preceded him to his designated flagship to serve as Chief of Staff.[300] When Calvert left Boyle and the *Nelson* in May 1935, he turned matters over to—Captain Algernon Willis. Willis held the briefest of appointments, as Roger Backhouse assumed command of the fleet bringing his own team with him in August.[301] Willis and Boyle also first appeared together by happenstance. This was at the Tactical School in 1925 when Willis was the assigned torpedo specialist at the school and Boyle, a junior flag officer, was taking the normal routine of courses, including the Tactical Course, before proceeding to sea in command of his cruiser squadron.[302] In 1929, the roles reversed with Willis now attending the War Course under Boyle; where he remained following completion of the course to become a member of the War College directing staff.[303] From Greenwich, Willis went to the China Fleet to serve as Flag Captain in the *Kent*, but his tour ended early there returning as he did to the *Nelson* and his patron, Boyle, where he continued to display a flair for handling his ship and who led by force of example.[304]

Another favourite of by Boyle's was a more junior officer, Royer Dick,[305] though their paths did not cross at Greenwich during Boyle's tenure there. A young Lieutenant and soon to specialize in signals, Dick served briefly with Boyle when the latter commanded the battle cruiser *Tiger* during 1919-20.[306] Upon completing his Visual and Wireless Signaling courses, Dick was with Boyle again first as Flag Lieutenant and then Flag Lieutenant Commander and Squadron Signals and Wireless Telegraphy Officer in the *Frobisher*.[307] Dick and Willis—absent Boyle now ennobled as the Earl of Cork and Orrery—had not previously served together though both were beneficiaries of Boyle's interest.[308] During the coming war, they were to serve

300 *Navy List* January 1934, p. 256.
301 *Navy List*, October 1935, p. 255.
302 *Navy List*, February 1926, p. 282a and ADM 196/52 (Willis).
303 *Navy List*, January 1930, p. 307 and *Navy List*, October 1930, p. 444.
304 ADM 196/92 (Willis).
305 Later Acting Vice Admiral Royer Mylius Dick (1897-1991). Short Course, Cambridge University, 1919; Flag Lieutenant and Squadron Signal Officer in *Revenge* and *Resolution*, 1924-25; Flag Lieutenant and Assistant Fleet Signals Officer in *Queen Elizabeth*, 1925; Flag Lieutenant and Squadron Signals and Wireless Telegraphy Officer in *Frobisher*, 1926-28; Signal School, 1928-30; Fleet Signals Officer in *Queen Elizabeth*, 1931-32; Signal Department, 1933-34; Staff Course, 1935; Tactical Course, 1936; Short Anti-Submarine Course, 1936; Commanding Officer, *Basilisk*, 1938-39; Plans Division, 1939; Captain, 1940; Chief of Staff to Commander in Chief, Mediterranean, 1942-44; Director, Training and Staff Duties Division, 1948; Rear Admiral, 1949 and retired, 1952. Recalled for service; Acting Vice Admiral, 1953 and retired, 1955.
306 *Navy List*, September 1920, p. 874.
307 ADM 196/146 (Dick).
308 ADM 196/120 (Dick) and ADM 196/52 (Willis).

together under Admiral Andrew Cunningham in the Mediterranean during some of the warmest actions in that theatre. In time, Dick followed Willis' example becoming Chief of Staff to the Commander-in-Chief indicating perhaps that patronage has its own order of effects.[309]

To avow that a system of patronage existed is not to avow that all those benefiting were somehow lacking or that the officer receiving was measured to a lesser standard. Davenport, a Chatfield associate, though seen as an excellent Captain of the Fleet was thought to lack interest in the social affairs of the fleet and did not take exercise. Consequently, he was not likely to advance to the very top with shortcomings such as these.[310] Meanwhile, Willis, a graduate of the reformed War Staff Course, was an early member of the Tactical School's faculty and attended the SOWC before becoming an instructor at the War College. Seen as a very capable and clever officer, the chronicler of the *Vernon* avows that he was '[r]egarded with apprehension by his subordinates and envy by his peers'.[311] Pound from the Mediterranean Fleet and Somerville from the Eastern Fleet both regarded Willis as a first-rate officer who handled the fleet well when acting in command. Even Dreyer who could be extremely critical of an otherwise sound officer was given to singing Willis' praises when he served him as Captain of the Fleet on the China Station.[312]

Meanwhile, Calvert was an experienced submariner before serving in the Training and Staff Duties Division and becoming Head of the Air Section of the Admiralty; in time, he would attend the IDC. Though it is possible to see how Calvert benefited from Boyle's favour, it is also true that he was seen as a highly capable officer before that favour was ever displayed. Still, Calvert had his detractors as previous discussions have demonstrated.

Royer Dick's time in *Tiger* probably marked him as a promising young officer to Boyle but only completion of his signals training allowed him to be considered for his duties in *Frobisher*. Duty in *Frobisher* brought contact with Calvert, and, in time, Willis. The best test for the capability of an officer is to examine their career once their beneficiary departs. Calvert, though reaching flag rank, succumbed in 1938 to illness. Willis, however, became an Admiral of the Fleet and might have gone perhaps further except to his ties to Clement Atlee through marriage indicating that closeness can also come with a risk. Dick's horizons were not quite so broad, but eventually he became a Director of the Training and Staff Duties Division and a flag officer on the active list. Thus, the interest of a senior officer was of use, but more useful yet was to demonstrate that one possessed the requisite skills required of any assignment proffered.

Boyle, it appears, had an excellent eye for spotting talent or maybe he simply benefited from the luck naturally given to his compatriots. One officer not so blessed

309 ADM 196/93 (Dick).
310 ADM 196/91 (Davenport).
311 Poland, *Torpedomen*, p. 116.
312 ADM 196/92 (Willis).

in measuring a friend's talent was Roger Keyes who favoured Wilfred Tomkinson. Tomkinson was Keyes' friend of longest standing with their association dating to their time on the China Station in destroyers with Tomkinson serving as First Lieutenant to Keyes at the beginning of the century.[313] In 1914, Keyes was set to command *Tiger* and had secured approval for Tomkinson to follow, but the appointment was cancelled and Keyes remained with submarines. Shortly, his command received a pair of destroyers and Keyes was soon asking for Tomkinson to take command of one. Not surprisingly, when Keyes secured the Dover command, Tomkinson was by his side and was there, too, during the Zeebrugge raid of 1918.[314]

Following the war, Keyes became ACQ, or Commander of the Battle Cruiser Squadron, with the ubiquitous Tomkinson now acting as Flag Captain and Chief Staff Officer in the *Hood*.[315] After command of his squadron, Keyes shifted to the Admiralty becoming DCNS and though Tomkinson was not a qualified or designated staff officer, shortly he, too, appeared serving in the Operations Division before eventually becoming its Director.[316] In May 1925, Keyes became Commander-in-Chief, Mediterranean wearing his flag in *Queen Elizabeth*. Tomkinson, for the moment, found himself commanding the naval barracks at Devonport. In January 1927, though, he returned to service with Keyes, becoming his Chief of Staff with promotion to Rear Admiral occurring in August of the same year. After the *Royal Oak* affair and Keyes early departure from the Mediterranean, Tomkinson no longer was able to rely on Keyes's patronage. His career continued to prosper, and though a cautionary note might have sounded on what the Chief of Staff perforce was advising his Admiral at the time of the *Royal Oak* denouement, no blame for the matter fell his way. Tomkinson was not the only officer Keyes helped, but his case is illustrative as his sole concession to the higher educational courses of the Service was his attendance at a single sitting of the Tactical Course.[317] Clearly, Tomkinson benefited from all this and Keyes valued having a congenial and reliable officer at close-hand, but the same cannot be claimed for the Navy as a whole. To Boyle's credit, at least the officers he favoured were highly proficient and *trained*.

With some senior officers publicly acknowledging that they practiced a form of largesse, it was natural that officers not benefiting felt aggrieved. Grenfell was one such officer and being extremely talented probably resented the penalty that he endured

313 Aspinall-Oglander, *Roger Keyes*, p. 31.
314 *Ibid*, pp. 90-91.
315 *Navy List*, September 1920, p. 788.
316 *Navy List*, April 1924, pp. 402 and 404.
317 Keyes' two-volume memoir of his service during the World War makes frequent references to Dudley Pound and Victor Crutchley as two other officers benefiting from his interest. See Roger Keyes, *The Naval Memoirs of Admiral of the Fleet Sir Roger Keyes: The Narrow Seas to the Dardanelles, 1910-1915* (London: Thornton Butterworth, Ltd., 1934) and *The Naval Memoirs of Admiral of the Fleet Sir Roger Keyes: Scapa Flow to the Dover Straits, 1916-1918* (London: Thornton Butterworth, Ltd., 1935).

whilst lesser lights progressed. In this regard, he was particularly severe towards Dudley Pound believing that he carried out the practice of blessing his friends to an unhealthy degree—and this when Pound's judgment in such matters was thought poor.[318] Writing of Pound's time as a Lord Commissioner Grenfell complained:

> [W]here he could have done so much to reorganize the most out of date department in the Admiralty, his passion for detail led him personally to sign all sorts of trivial matters that should have been left to subordinates, and I know that he spent hours personally investigating the case of every able seaman and stoker whose Commanding Officer had recommended him for discharge (S.N.L.R.). Meanwhile, although the Sea Lords are always complaining that they are not acquainted with many of the officers whom they have to judge for promotion, Pound, as head of naval personnel, never interviewed a single officer who came into the Admiralty in connection with an appointment.[319]

The degree to which rancor influenced Grenfell's view is unknown, but it must weigh in the equation for following service at the Staff College on its directing staff, his career entered the doldrums. Unemployment, command of a cruiser in caretaker status, and then retirement were hardly resounding flourishes to a career previously full of promise. Pound was Second Sea Lord and responsible for officer appointments when Grenfell would have been seeking further employment after Greenwich, though the immediate problem for Grenfell was that while his superiors conceded his excellent qualities as a staff officer, they noted also a consistent lack of tact married to a temperamental nature.[320]

Within Grenfell's critique of Pound lies the primary reason why patronage existed and was accepted. The size of the officer corps and the far-flung expanse of naval service made it extremely difficult to weigh the relative merits of officers competing for assignments and promotion. Thus, the views of senior officers willing to speak on an officer's behalf carried much weight. The annual report of an officer was a guide, but censure of a subordinate was frequently lacking and how accurately it reflected an officer's merit was difficult to judge. Compounding the problem, the narrative portion of an officer' report represented a subjective and not an objective judgment. This problem was recognized and an endorsing officer might note that the ranking superior was prone to rendering views outside the norm practiced by his peers. Pound, for one during his period commanding the Mediterranean Station, was not afraid of mentioning this issue when endorsing evaluations rendered by Rear Admiral Wells.[321]

318 Grenfell to Liddell Hart letter of 6 June 1937, Liddell Hart Papers, 1/330/11, LHCMA.
319 Grenfell to Liddell Hart letter of 8 June 1937, Liddell Hart Papers, 1/330/12, LHCMA. S.N.L.R., i.e., services no longer required.
320 ADM 196/127 (Grenfell) and ADM 196/145 (Grenfell).
321 ADM 196/92 (Hutton).

Thus, an officer requesting a specific subordinate did so in the knowledge that the officer's failure to perform to the required standard would also reflect poorly on the sponsor. Yet, Grenfell was unfair to Pound regarding his handling of 'Services No Longer Required' cases. Separating a man if a transfer to another ship might alleviate the problem protected the Service's investment in the rating. Moreover, the rating once outside his ship remained a subject with rights. Having no right of appeal to the Admiralty, if he elected to make his circumstances known to his member of parliament, then questions arising in the Commons over the case could cause another sort of problem for the Admiralty. The matter was particularly acute for the senior rating dismissed before completing his time for a pension when all he would have to show for his service was a token payment allowing him to purchase a civilian suit.[322]

If Pound played favorites, then Roskill also records that the blade was honed on both sides for he also punished those who went afoul of him. Rear Admiral Arthur Pridham[323] felt such wrath when he turned a Nelsonian eye to an Admiralty order originating from the Second Sea Lord's Office. Pound, the putative author of the memorandum that Pridham failed to follow, took his revenge when he assumed command of the Mediterranean Station where Pridham was serving and his career now suffered, accordingly.[324] If Roskill is correct in the preceding, then a measure of vindictiveness governed Pound in this matter. Yet, Pridham's service record and simple chronology do not confirm the allegations, as the officer was promoted to flag rank in 1938 on the active list. Clearly, if Pound had wanted to derail Pridham's career he had ample opportunity following *Hood*'s straddling of HMS *Protector* in June 1937 during a practice shoot. As it was, Pound only cautioned him to be more careful in the future; a measure endorsed by the Admiralty.[325] Pound, though, did reduce his numerical rating from the 84 rendered in March to one of 81 in July, which in all events was a modest reduction when life had been imperiled and remained within the margins to allow promotion to flag rank.[326]

Naval appointments much like naval education were matters where all officers felt qualified to voice an opinion, as all had direct experiences of both. Contemporary commentary must be treated with caution because the reasons why an officer holds an opinion are not always expressed. Drage testified to the unthinking, if not

322 In the immediate aftermath of the Invergordon Mutiny, 24 ratings and marines were discharged 'Services No Longer Required' with their circumstances soon taken up in parliament. See *Hansard*, House of Commons Debates, 12 November 1931, v. 259, cc269-70.

323 Later Vice Admiral Sir Arthur Francis Pridham (1886-1975). Torpedo and Mining Department, 1920-22; Plans Division, 1924-27; Captain, 1926; Imperial Defence Course, 1927; Commanding Officer, *Curlew*, 1928-30; Naval Ordnance Department, 1930-33; Tactical Course, 1933; in Command, *Excellent*, 1933-35; Senior Officers' Technical Course, 1935; Flag Captain and Chief Staff Officer in *Hood*, 1936-38; Senior Officers' War Course, 1938-39; Rear Admiral; 1938; retired and Vice Admiral, 1941. Recalled for war service.

324 Roskill, *Churchill and the Admirals*, pp. 111-12.

325 ADM 196/92 (Pridham).

326 *Ibid.*

incompetent, ways of the Admiralty when conducting officer assignments relating the case of Commander Daniel de Pass.[327] De Pass, serving with Drage in *Renown* and one of the few Jewish officers in the Service, explained over an evening meal how following the war he had served in HMS *Glowworm*, a gunboat operating on the Danube. To Drage, it was probably the only appointment in the Navy where the fact one was Jewish would prove a serious handicap. Could it be that Second Sea Lord's Department was not merely foolish but actively malignant?[328] Drage's view, if ostensibly from a neutral observer, was registered when the prospects of further employment remained uncertain, so his opinion of the Second Sea Lord's office at that moment were not of a charitable kind. In fairness to that office, De Pass though not rated as an interpreter, spoke German and French, languages of some use in his posting and, if Jewish, had nevertheless married outside his religion and his synagogue.[329] Yet, for that, there is something essentially correct in Drage's view that the case of Eric Oatley does nothing to dissuade.

Chatfield in his memoirs expounded at some length on the trouble that the Admiralty took in the immediate aftermath of the World War when facing the problem of reducing officer numbers insisting that merit was the only consideration weighed. Chairing a committee with Rear Admirals John Kelly[330] and Arthur Waistell[331] sitting as additional members, Chatfield recounted the heavy scrutinizing performed of every executive officer's record. Having received special reports on the officers affected, redundancies were confirmed only after the senior naval members of the Board of Admiralty agreed to the recommendations submitted.[332] On the surface,

327 Later Captain Daniel de Pass (1891-1963). War Staff Course, 1921-22; Staff Officer Operations in *Iron Duke*, 1926; Naval Mission to Greece, 1927-29; Tactical Course, 1929 and 1937; Staff Officer Operations in *Revenge* and *Resolution*, 1929-31; Captain, 1934; Senior Officers' Technical Course, 1934; Senior Officers' War Course, 1935; Deputy Director, Personal Services Department, 1935-37; Commanding Officer, *Cossack*, 1937-39 and retired, 1944. Recalled for war service.
328 Drage diary entry of 29 March 1932, IWM/PP/MCR/99, Reel 4.
329 ADM 196/53 (De Pass).
330 Later Admiral of the Fleet Sir John (Joe) Donald Kelly (1871-1936). Captain, 1911; War Course, 1912; Director, Operations Division (Home), 1919-20 and Operations Division, 1920-22; Rear Admiral, 1921; Commander, Fourth Battle Squadron, 1922-23; Fourth Sea Lord, 1924-27; Vice Admiral, 1926; Commander, First Battle Squadron, 1927-29; Admiral Commanding Reserves, 1929-31; Admiral, 1930; Commander-in-Chief, Atlantic and Home Fleets, 1931-34; Commander-in-Chief, Portsmouth, 1934-36 and Admiral of the Fleet, 1936.
331 Later Admiral Sir Arthur Kipling Waistell (1873–1953). Royal Naval War College, 1907-10; Captain, 1910; Commanding Officer, *Benbow*, 1917-19; Director, Torpedo Division, 1920-22; Rear Admiral, 1921; Rear Admiral (D), Atlantic Fleet, 1922-23; ACNS, 1923-24; Commander, First Cruiser Squadron; 1924-26; Vice Admiral, 1926; Tactical Course, 1928; Commander-in-Chief, China, 1928-31; Admiral, 1930; Commander-in-Chief, Portsmouth, 1931-34 and retired, 1934.
332 Chatfield, *Navy and Defence*, pp. 198-202.

the process appears fair with sufficient allowance for checks and balances to ensure that a particularly grievous error would not occur. It is impossible to render a valid judgment on the degree to which the committee succeeded in removing patronage as a factor and this observer is willing to take Chatfield at his word that the effort was made. That said, the dynamics of the committee must also owe something to the backgrounds and positions of those rendering judgments. Chatfield was then serving as ACNS, Kelly was the Director Operations Division and Waistell oversaw the Torpedo Division. Let it be said that these were three highly capable and esteemed officers respected by superiors and subordinates alike; qualities that allowed the work of the committee to be received with grace, if not with favour. Still, whether intentional or not, the officers analyzing the problem in the first instance were those whose day-to-day duties focused very much on present operations and on material factors. Others who considered longer-term questions, such as the Directors of Plans or Naval Intelligence were noticeably absent and it was they who were apt to value most an officer having attended the higher courses of naval training. This conclusion is offered notwithstanding that Waistell had served as an instructor of the War Course from 1907-10.[333] This emphasis was not wrong, as it kept the interests of the fleet foremost, but it was incomplete.

Many factors influenced officer advancement and retention. Specialization played a role and though the Royal Navy had to account for this in the decisions rendered, it too had its limits. The Service favoured officers of proven ability more than it favoured officers of a stereotyped-pattern and this accounts for the practice of having officers command vessels that on the surface appeared outside their lines of specialization. Accordingly, Captain Roderick Miles,[334] a torpedo specialist, commanded the *Revenge* in 1937 though the torpedo had ceased to be central to battlefleet tactics and Captain Reginald Henderson commanded the carrier *Furious* in 1926 and later became Rear Admiral, Aircraft Carriers though not an aviator himself. Admittedly, these two officers were exceptions and, in Henderson's case, exceptional. To him belonged a major portion of the credit for bringing about the system of convoy during the last war when others more senior castigated its use. Yet, the practice of serving outside one's normal career path was common and was a means of allowing judgments to be rendered on the merits of officers coming from different backgrounds. Indeed, Miles having spent a disproportionate amount of time considering torpedoes both ashore

333 ADM 196/89 (Waistell).

334 Later Rear Admiral Roderick Bruce Tremayne Miles (1887–1963). Six firsts in examinations for Lieutenant; Torpedo and Mining Department, 1920-23; Operations Division, 1924-26; Commanding Officer, *Hollyhock*, 1926-29; Captain, 1927; Senior Officers' Technical Course, 1929; Senior Officers' War Course, 1929; Commanding Officer, *Comus*, 1929-30; Tactical Course, 1930 and 1935; Captain (D), Third Flotilla, 1930-32; Commanding Officer, *Defiance*, 1932-34; in Command, *Vernon*, 1934-35; Flag Captain and Chief Staff Officer in *Renown*, 1935-36, in *Royal Sovereign*, 1936, in *Revenge*, 1937; Commanding Officer, *Defiance*, 1938; Rear Admiral and retired, 1939. Recalled for war service.

and in small ships was especially recommended for a large ship to broaden his professional experience.[335] This he received, but Miles displayed no marked personality and, consequently, was thought lacking power of command.

That an informal system of interest and patronage existed is clear and for many officers it was no more subversive than asking a former commander perhaps now occupying a position of influence to lend a hand. Henry Fancourt was one such officer who sought recourse to this method of favour. In 1920, he sought the help of Captain Joe Kelly, the Director of Operations, in finding new employment. Kelly, his former commanding officer in *Princess Royal,* might have influence in getting his appointment to the pre-dreadnought battleship *Commonwealth* terminated and placing Fancourt back to a small ship. *Commonwealth* then had the reputation of being a ship where all the bad eggs of the Service served. Fancourt having received 'mentioned in dispatches' during the war, felt the posting beneath him.[336] Whether this reputation existed during Richmond's time in command in 1917 is something to ponder, but the prickly attitude the Admiral displayed at times did not always serve him well. Roskill informs us that his departure from the Admiralty in 1915 was due to his growing unpopularity and his time with the Italian Navy following demonstrated that his penchant for violent mood swings had not abated.[337]

This attitude was on offer again when Richmond sought unsuccessfully to secure Brian Egeton's posting to the IDC. Clearly, some officers were better at playing the game than others. Richmond, though, upon appointment to command the East Indies Squadron in 1923 did secure Maund, then studying at the Staff College, to serve as his War Staff Officer. Upon leaving Greenwich, Maund spent two months with Richmond in London gathering background information about the station paying acute attention to its trade.[338] Here, the purpose was not so much to obtain the details of that trade but rather to acquire a picture of its pattern and its relative importance to the Empire. Though their period together at Greenwich briefly overlapped, Richmond probably sought the best officer available from that year's class rather than specifically seeking out Maund. After all, there is a difference between asking for an officer of noted abilities and repeatedly asking for a notable officer. General Ironside sought to secure Fuller for the Staff College following the war whilst Commandant at Camberley, and whatever shortcomings Fuller possessed as a soldier, a critical faculty was never one of them. That Fuller complimented the talents of Ironside—and a qualified interpreter in seven languages demonstrates they were of no mean order—was a point both officers readily conceded.[339] In securing Maund, Richmond found an

335 ADM 196/92 (Miles).
336 Fancourt, IWM Sound Recording 12274, Reel 3, 30 September 1991.
337 S. W. R. [Stephen Roskill]. 'Portrait of an Admiral', *The Naval Review*, Vol. XL, No. 3, August 1952, p. 341.
338 Richmond, 'Staff Requirements of a Commander-in-Chief', *Naval Review*, February 1939, p. 1.
339 Ironside to Liddell Hart letter of 12 September 1933, Liddell Hart Papers, LH 1/401/104, LHCMA and Fuller, *Memoirs*, p. 416.

officer with a prodigious appetite for work and believed he was the best staff officer he had ever seen.[340]

Typical of many officers completing the Staff Course, Maund spent a short period at the Admiralty before taking up his next posting. In his case, the time passed in the Training and Staff Duties Division preparing himself for his pending appointment with Richmond and the East Indies Station. This it doubtlessly did, but it also allowed the recent qualifier to gain experience of the workings of the Admiralty and the Naval Staff with the theory of instruction now meeting the reality of practice. Departing the Admiralty and Richmond, Maund arrived on station to work alongside the current War Staff Officer before assuming his new responsibilities.[341] This period was essential for Richmond only arrived on station the day before Maund's predecessor departed and the Commander-in-Chief relied on his newly minted staff officer to acquaint him with present affairs.

A key member of Richmond's staff was his personal secretary, Paymaster Commander William Skinner, who worked closely with Maund and had responsibility per naval custom for providing any guidance on international law to the Commander-in-Chief. Though trained accordingly in his Secretary's Course, Richmond deprecated that this responsibility resided with a non-executive officer and not with the squadron's War Staff Officer who had been trained along like-lines during his Greenwich course. For Richmond, given the ongoing debate about the purpose and use of the staff it was imperative that the staff officer stay abreast of international law else his standing be eroded and power and responsibility pass by default to non-executive officers with the implication that the staff system would become still-borne if officers did not uphold their end of the bargain.[342]

Though working the many problems of the moment, the staff officer had of necessity to keep his eye on the calendar and anticipate the schedule of his Admiral. Drage faced this very problem whilst working out the details of the revised naval serials that the East Indies Squadron would operate to when the time came to support the combined exercise with the Indian Army Staff College. Likewise, Maund had to visit a number of ports and identify deficiencies in their defensive measures ahead of Richmond's calls on the local military commanders having responsibility for their protection. Of course, the actual employment of a staff officer or any member of a staff, such as a Flag Captain or Chief of Staff, was always dependent on the attitude of the ranking officer and Ramsay's experience with Backhouse in the Home Fleet is evidence that a wide measure of latitude existed in practices. Beyond the duties enumerated for an officer in command or holding a staff assignment, other calls were made on his services. Frequently, lectures were provided to members of the lower

340 ADM 196/145 (Maund).
341 Richmond, 'Staff Requirements of a Commander-in-Chief', *Naval Review*, February 1939, p. 2.
342 *Ibid*, p. 3.

deck, gunroom, and wardroom, and in a period when departure from port meant immediate links with home and home authorities were separated or constrained, such talks beyond fostering instruction served the purpose of sustaining morale.

One does not have to beg the issue to realize that patronage can only exist if there is a patron. If something untoward were to snare a sponsor, the knock-on effect to favoured subordinates might stall an otherwise promising career. Boyle destined to reach the highest of ranks and no stranger to a Board of Enquiry in matters of ship handling recalled another spell of misfortune in his memoirs. In the course of a single month in the autumn of 1920 whilst commanding the *Tiger* he struck and severely damaged the *Royal Sovereign* when entering Portland Harbour. The customary Court of Enquiry convened to investigate such incidents reported unfavourably on Boyle leading the Commander-in-Chief to order a court martial so that the charge of negligence in hazarding the *Tiger* could be refuted.[343] Boyle escaped serious censure, his charmed life continued; and so, consequently, did it continue for others.

Of course, not all officers upon leaving Greenwich were destined for staff duties. There were only so many billets and other venues needed filling. The need to serve at sea was ever a consideration and for those of Commander's rank, posting to a large ship as its executive officer, or 'Jimmy', was a typical assignment and command of minor ship, as Ralph Edwards' case demonstrates, was a further possibility. Moreover, the Admiralty, too, required the services of trained staff officers though not all were necessarily destined for duty with the Naval Staff. Those trained staff officers who were also naval aviators were especially prized by those portions of the Admiralty dealing with air matters, of which, only the Naval Air Division was part of the Naval Staff proper. Some officers served in the Air Personnel or Air Matériel Departments, dealing with questions of manning policy and equipment, and even those officers serving within the Naval Air Division, such as Robert Ellis,[344] had but limited scope to influence the air's contribution to current operations or future plans, as these tasks were handled by non-aviators in other staff elements.[345]

Often, the Navy is depicted as the odd Service out in matters of staff education and training.[346] This is unfair for the record is clear it developed a strong regime of higher

343 Cork and Orrery, *Naval Life*, pp. 127-28
344 Later Captain Robert Meyrick Ellis (1901-1981). Five firsts in examinations for Lieutenant; Sidney Sussex College, Cambridge, 1921-22; Observer's Course, 1924; Observer in *Argus*, 1924-25 and in *Eagle*, 1925-26; School of Naval Co-operation, 1926-28; Observer in *Courageous*, 1928-31; Staff Course, 1932; Tactical Course, 1933; Staff Officer Operations in *London*, 1933-34; Naval Air Division, 1934; Commanding Officer, HMS *Delight*, 1935-36; Assistant Director and Deputy Director, Naval Air Division, 1937-41; Captain, 1938, Commanding Officer, *Suffolk*, 1941-42; Combined Operations, 1942-43; Combined Operations Division, 1943-45; Commanding Officer, *Queen Elizabeth*, 1945-46 and HMS *Howe*, 1947 and retired, 1947.
345 *When the rain's before the wind*, unpublished manuscript, Captain Robert Meyrick Ellis Papers, Churchill Archives Centre, Cambridge, ELLS/3, p. 6.1.
346 Slessor, *Central Blue*, p. 87.

education for its officers and was desirous of going further. Yet, its requirements were not entirely congruent to those of the Army and the Air Force though there were similarities, to be sure. One of the most striking differences in approach was that the Navy did not concede that formal staff training was necessary to command at the highest levels. This did not mean that preparation was not required before an officer was fitted to command, but the nature of that preparation was of a kind different from staff training emphasizing, as it did, technical proficiency, tactical acumen, and strategic thought. Even here, though, some officers advanced without completing their foundational courses. The numbers were not great and why this occurred is examined in Appendix V. Now, though, it is appropriate to summarize how effectively the Royal Navy trained its executive officers in the period between the wars.

8

Final Thoughts

He would be invaluable in any hazardous operation of war, but lacks the patience and judgment for the best service in peace.[1]

Admiral Sir William Fisher, 1933

Not brilliant but is the type of man a Captain can sleep happy with.[2]

Captain the Honourable Edward Bingham, VC, 1924

The many changes that the Royal Navy instituted in the higher training and education of its officers following the World War is testament that the warnings of the faction of mid-career officers—of which Richmond was at the fore—did not go unheeded. Yet, the conclusion remains that for all their sagacity of vision, theirs, too, was a perspective of limited scope. War might be the final rationale of the Royal Navy, but other tasks, perforce, remained and the Service was constrained in how much of the country's strategic policy it could fashion by itself. Constrained, too, by finance—an aspect rarely appearing in the forefront of their thinking, change came incrementally and through consensus. This is only natural for a Board of Admiralty controlled the Navy and not an all-governing Admiralissimo. The one exception to this consensus was the establishment of the Tactical School, but in this case the underlying reasons for its establishment were not advertised.

The Royal Naval Staff College operated in the shadow of its military counterpart and for the period under review never achieved the standing and prestige of the latter. Those graduating with the naval *psc* were some of the most talented officers of the Service, but their attendance at the Staff College was not confirmation of that talent, but rather only an indication of it. Others who failed to secure the naval *psc* had careers every bit as rewarding as its best graduates. That said, attendance at the course was increasingly a norm desired by executive officers and a quality sought by others when

1   ADM 196/91 (Turle).
2   ADM 196/148 (Bingeman).

considering nominations. This was true of Cunningham in the Mediterranean when acknowledging the work of a young Commander, but still desired to have a graduate of the Imperial Defence College on his staff to support joint planning.[3] Vice Admiral Randle Ford[4] was expressing much the same view at the same approximate moment in Malta when he lamented the failings of his Chief Staff Officer, Captain Edward Denison.[5] Not a trained staff officer, Denison required the support of one who was.[6]

Between the Staff Course and the Senior Officers' War Course, successive Directors of Training and Staff Duties often deemed the former the more important. This importance, though, was primarily a reflection of its role in preparing staff officers. The War Course retained a value that the Staff Course could not aspire given the seniority of those officers attending and directing it. Comparisons of the separate Staff Colleges to each other are difficult because culturally and professionally the schools were as different as chalk to cheese, though nominally having similar educational missions. The War Course for the Royal Navy served a purpose not unlike Camberley and the Senior Officers' School: a clear perquisite for achieving command. Though many graduates of the Royal Naval Staff College achieved command, too, the course was primarily about providing specialized instruction in staff duties and, in that respect, may be seen as a counterpart of the *Excellent* or the *Vernon* save that it aimed to produce specialists in staff work.

The question arises whether the regime of education and training as finally established was too abbreviated and duplicative. When viewed *in toto* it was not too abbreviated though regrettably not all officers attended all courses. As for being duplicative—it was, but this was no bad thing as this was the means for conveying a common doctrine even if the Service frowned on the actual use of the term. Still, a rose by any other name remains a rose. Whether the Navy would have been better served by requiring its officers to attend the Staff Course much as they required senior officers to attend their quota of War, Technical, and Tactical Courses is a more difficult judgment. The answer is probably yes, but it remains that some are not really suited to staff work and the strength of the interwar Service was that it did not expect

3   Michael Simpson, ed., *The Cunningham Papers: Selections from the private and official Correspondence of Admiral of the Fleet Viscount Cunningham of Hyndhope, O.M., K.T., G.C.B., D.S.O.and two bars, Volume I, The Mediterranean Fleet, 1939-1942* (Aldershot: Ashgate, 1999), p.134..

4   Vice Admiral Wilbraham Tennyson Randle Ford (1880-1964). Captain, 1920; Senior Officers' War Course, 1922; Commanding Officer, *Calliope*, 1924-25; Director, Physical Training, 1926; Commanding Officer, HMS *Ganges*, 1927-28, *Royal Oak*, 1929-30; in Command, *Dryad*, 1930-32; Rear Admiral, 1932; Senior Officers' Technical Course, 1933; Commander, Australian Squadron, 1934-36; Admiral-Superintendent, Dockyard Malta, 1937-41; Admiral, 1941 and retired, 1944.

5   Captain Edward Conyngham Denison (1888-1960). Commanding Officer, HMS *Serapis*, 1925; Naval Equipment Department, 1925-27; Captain, 1931; Tactical Course, 1932; Deputy Director, Personal Services Department, 1934-36 and retired, 1941.

6   ADM 196/144 (Denison).

all officers to try. Even an officer who had been trained and then trained others in staff work, such as Tennant, was operating against type, if Pound's evaluation of him is any guide. For Pound believed Tennant an 'able officer but not a good staff officer as he is much too sketchy in his ideas and prefers writing brochures of a general nature, leaving the thorough investigation which is necessary to someone else, rather than delving himself.'[7] Of course, Pound's judgment may also reflect that the First Sea Lord fully expected an officer to master detail where Tennant by training was now inclined to work with others to produce any appreciation or product.

When the above verdict was rendered Tennant was a Captain and as he rose to become an Admiral, Pound's mild stricture proved no bar to his advancement and further success. Though it is always dangerous to draw from a specific case a broader conclusion, if an officer was viewed as capable, then a single adverse report need not be a career killer. Certainly, one officer who suffered a more grievous report than Tennant was Captain John Godfrey. In this instance, the reporting officer was Admiral Joe Kelly and what he saw of Godfrey was not to his liking observing:

> I cannot call to mind, since I reached Admiral's rank, any Captain who impressed me less favourably. A real "Heavy Weather Jack". He is reputed to be "Brainy": the only evidence I have had of this is his propensity for seeing difficulties and obstacles which less "Brainy" people like myself are unable to see....
>
> I was extremely glad to see the last of him and, consequently, of his Ship.[8]

Any report rendered speaks to its author as much as to its subject and with the above views so outside the norm of what Godfrey had secured previously including his own endorsement of his service in the Plans Division, Chatfield, the First Sea Lord, took the wise step of seeking a special report on Godfrey in a further six-month's time.[9] Those reporting subsequently on Godfrey assessed him in different tones than Kelly and he weathered the earlier blast with his career prospering. A lesser officer than Godfrey might not have overcome such a verdict and another First Sea Lord might have been less charitable, but, as a rule, a single report was insufficient evidence to wreck an officer's prospects.

The case of Pat Horan also offers supporting evidence of this, too, if of a negative kind. Viewed as a thoroughly capable seagoing officer—remember Chatfield's praise—he came to be viewed as only that and was seen as lacking in the broader qualities necessary for a flag officer. Most especially, he was seen by Rear Admiral Arthur Murray[10] as someone who was unable to think coherently or hold an argument

7   ADM 196/92 (Tennant).
8   ADM 196/92 (Godfrey).
9   *Ibid.*
10   Later Admiral Arthur John Layard Murray (1886–1959). Captain, 1927; in Command, *Osprey*, 1928-30; Director, Signal Department, 1932-34; Tactical Course, 1935; Commanding Officer, *Dorsetshire*, 1935-37; Rear Admiral, 1939; Senior Officers' War Course, 1939; Vice Admiral, 1942; retired, 1943 and Admiral, 1945.

and could not see how Horan had been selected for the IDC course (*sic*). Meanwhile, Rear Admiral Ralph Leatham believed him conceited playing a little too much to the gallery recommending he be passed over for flag rank. Consequently, Horan soon retired.[11]

Ultimately, training officers to a certain standard becomes wasteful if there are not positions for them to occupy, as the skills acquired will atrophy. Surely, it would have been better to allow the best naval officers to attend the Staff Course and not limit its entrants to merely those of the executive branch. To their credit, the British Army and the Royal Air Force were ahead of the Royal Navy in this regard and the decision owed something to the efforts of many to maintain the status of the executive branch. Yet, atrophy implies that a quality exists already and this risk may be the lesser evil than not having the quality at all, as the Navy realized during the Spanish Civil War when it found it had too few staff officers.[12]

By examining one essential variable in the development of naval and marine officers—education—it is easy to forget that important as this element was in preparing an officer for future duties, it was never more than ancillary to this end. A recurring theme in this study has been the conclusion that a single path to promotion failed to operate between the wars. A subsidiary one is that the reasons why certain officers 'got on' and others did not had as much do about whether one was thought to possess 'officer-like qualities'—so difficult to define but understood nevertheless within the ethos of the Service. How else to explain one Captain's report on a subordinate during the period when he remarked, "An excellent scrum-half. Should be promoted' and a year later, 'Still an excellent scrum-half. Can't understand why he has not been promoted.'"[13] The Navy was not alone in sharing this view—witness the Army's emphasis on polo, mountaineering, and pig-sticking which may say more about early twentieth century concepts of British manliness and leisure than warfare. Still, even Hobart the prototype unconventional officer was moved to comment adversely on an officer's potential by noting a predilection for more genteel competition, 'This officer plays cricket. Need I say more!'[14] Perhaps Hobart should have been a bit more tolerant in his judgment given that a previous superior had concluded derogatively in an annual report on him whilst a junior sapper that he was given to reading poetry.[15] Thus ever it was.

So pronounced was this emphasis on athleticism that an officer who did not play games risked sacrificing an otherwise promising career and the quality was the more prized the further one advanced and the older one became as it demonstrated general physical health. Vivian Voelcker, as a young officer, was especially suspect in this

11   ADM 196/92 (Horan).
12   ADM 196/92 (Fellowes).
13   John Winton, 'H.M.S. *Dryland*', *The Naval Review*, Vol. 63, No. 4, October 1975, p. 301.
14   Cited in Macksey, *Armoured Crusader*, p. 103.
15   Lewin, *Slim*, p. 52.

regard with one commanding officer recording in 1922, 'Does not play games.' The following year his S.206 report now curtly noted 'Plays games' which was progress of a kind. By 1925, Voelcker was said to be 'Keen on games' and promotion to Lieutenant Commander came in due course in 1926, something not all assured if he had maintained his passivity.[16] This emphasis was always more just than demonstrating athletic prowess and though not all might play to an international standard, as Tovey did in football as a youth, it did indicate whether an officer was a team player, physically fit and able to demonstrate powers of endurance.[17] After all, if one was unable to stand the strain of a day's shooting, a round of golf, or even 80 minutes of rugger, then the likelihood existed that an officer could not face the rigors of battle. Sport also showed that an officer had interests outside the narrow confines of one's profession, demonstrated one's overall fitness to subordinates and indicated whether one was of youthful, rather than of aged, disposition.[18]

Fuller possessing a discerning eye and one of the most critical minds of the interwar British Army did not play games and faulted his Service's obsession with sport and its 'cricket complex'.[19] His observations on the faults of British generalship in the late war, if excessively harsh, were not without merit and the balance that the Royal Navy sought was to stress physical fitness, relative youth and technical competence. These were important qualities as influence and tact, rather than recourse to a bald command, were better forms of leadership when handling subordinate officers and ratings and any indication of poor health, beyond a temporary setback, often foreshadowed an early retirement for an officer. This was always more than just 'muscular Christianity' for the number of officers who died whilst near the pinnacle of their profession is striking. Though it was not predestined that Rear Admiral Wilfrid Egerton would have gone further, death by septic poisoning at age fifty in 1931 robbed the Service of an officer seemingly with much promise.[20] To the modern observer such an emphasis on games may appear skewed, yet this is to forget just how tenuous sound health could be in an era when malaria, diphtheria, enteric fever, dysentery, influenza, tuberculosis, smallpox and venereal diseases to name but the most common were not only prevalent, but frequently lethal. To these, of course, must be added a particular malady of the Service, which if non-lethal, was nonetheless debilitating: the duodenal ulcer. As such, many excellent officers failed to progress further for the simple reason that they became medically unfit. This is not to say that an officer who eschewed games would never advance. Charles Daniel certainly did, but he did play a very good piano and his brilliance as a Signals Officer and pronounced officer-like qualities marked him out as

16   ADM 196/146 (Voelcker).
17   R.W.P. 'Tovey', *Naval Review*, Vol. 68, No. 3, p. 205.
18   ADM 196/91 (Horton).
19   Brian Holden Reid, 'Major-General J. F. C. Fuller and the Decline of Generalship: The Lessons of 1914-18' in Dockrill and French, eds., *British Policy during the First World War*, p. 198.
20   ADM 196/46 (Egerton).

a rising officer.[21] Still, if only of average abilities it was better to sparkle elsewhere and the playing field was that one possible place.

Moreover, it was necessary to exhibit competence in the practical execution of one's work offering more than an understanding of only a theoretical nature. Any deficiency in mental acumen was more easily forgiven than a perceived weakness as a seagoing officer. Accordingly, while Ragnar Colvin assessed that Captain Gresham Nicholson,[22] a salt horse, was quite out of depth in his War Course being below average in ability and expression, he suffered no ill-effects in his professional prospects, as his abilities as a seagoing officer were abundantly clear.[23] He had been earmarked to attend the IDC in 1940, though the onset of war forestalled an opportunity to improve upon his prior war studies record.[24] Meanwhile, Charles Drage was recognized as a naval officer outside the norm given his demonstrated academic prowess and strong social skills. Yet, he was prone to severe bouts of seasickness when serving in minor ships necessitating a reassignment to a cruiser during the World War and placing a cloud over his future.[25]

Upon leaving the Service, Drage became an officer of the Secret Intelligence Service not unlike several of his counterparts. Norman Dixon avowed that this sinecure was presumably manned by below average officers failing as they did to advance further in their first career.[26] That is one interpretation, but perhaps a more honest one is that fewer billets existed for senior officers than for junior ones and the skills desired were not all of the academic variety. For instance, how one bore the strain of command was ever an important consideration in an officer's ability to succeed. Here, the officer commanding a shore establishment was at a disadvantage to his peer commanding a ship or flotilla, for though the former assignment was important, commanding the latter was the essence of naval command.[27] Though slightly deaf—ever the curse of the gunnery specialist—Captain Patrick Macnamara was eyed as an officer destined for the highest ranks. Yet, a poor report by Boyle whilst serving as his Flag Captain in *Nelson* ending with a recommendation to serve in any non-seagoing appointment and not in a large ship doomed his prospects. Following his time with the James

21   ADM 196/145 (Daniel).
22   Later Admiral Sir Randolph Stewart Gresham Nicholson (1892-1975). Royal Naval College Dartmouth, 1920-22 and 1929-31; Flag Lieutenant Commander in *Iron Duke* and *Barham*, 1925-27; Tactical Course, 1929 and 1931; Senior Officers' Technical Course, 1929 and 1932; Captain, 1934; Senior Officers' War Course, 1934-35; Commanding Officer, *Pegasus*, 1935 and *Curacoa*, 1936-37; Assistant Director, Naval Equipment Department, 1937-38; Captain (D), Second Tribal Flotilla, 1938-39; Rear Admiral, 1943; Vice Admiral, 1948; retired, 1950 and Admiral, 1951.
23   ADM 196/92 (Nicholson).
24   ADM 196/55 (Nicolson).
25   ADM 196/146 (Drage).
26   Norman F. Dixon, *On the Psychology of Military Incompetence* (London: Futura, 1988), p. 293.
27   ADM 196/93 (Cunninghame-Graham).

Committee reviewing the outlines of higher officer education, he sought appointment to Trinity House and retirement.[28] If the system was unforgiving, it was not wrong when so much weighed on one's perceived ability to meet the unknown. To this, zeal, knowledge, special skills, of temperate habit, powers of command, tact, and social bearing were every bit as important in measuring the potential capacity of an officer for future advancement. Though a superior officer might appreciate an officer of strong views, if backed by a clear and logical mind, he was not likely to appreciate a rude subordinate and tact as a quality was especially prized. Why this was so is not hard to imagine as the senior naval officer could expect to have frequent contact with officials where skills of a less bellicose sort were needed in the first instance. Yet, the Service did not appreciate shrinking violets, either, no matter how talented. Captain Kenneth Mackenzie,[29] a very capable officer and an excellent Navigating Officer, was viewed less positively when attending his War Course by William Boyle. Possessing a retiring manner, Boyle did not change this view of Mackenzie when later reporting on him when commanding the Home Fleet.[30] Another flag officer shared this assessment and though giving Mackenzie every credit for his undoubted ability, Vice Admiral Sir George Chetwode believed he lacked the personality to advance further.[31]

Thus, for officers of a certain calibre and manner a lot could be forgiven—recall Boyle's run of bad luck in *Tiger*. Still, even Boyle might have thought Captain Piers Kekewich[32] was pushing matters too much as he needed to negotiate the shoals of a civil arrest in a foreign country for discreditable conduct, two courts-martial, and a Court of Enquiry all the while receiving glowing reports from his superiors.[33] Vice Admiral Sir Howard Kelly in 1931 claimed he was 'one of the coming men' whilst Captain Percy Noble,[34] commanding the training establishment in *Ganges*, was recommending Kekewich for early promotion in 1927 and marking him out as 'a reader, a

28   ADM 196/91 (Macnamara).
29   Later Rear Admiral Kenneth Harry Litton Mackenzie (1889-1947). Senior Officers' Technical Course, 1925; Squadron Navigating Officer in *Barham*, 1926-27; Tactical Course, 1927; Captain, 1930; Senior Officers' War Course, 1931; Commanding Officer, *Guardian*, 1933-35; Rear Admiral and retired, 1941. Recalled for war service.
30   ADM 196/92 (Mackenzie).
31   *Ibid.*
32   Later Rear Admiral Piers Keane Kekewich (1889-1967). Captain, 1928; Flag Captain and Chief Staff Officer in HMS *Bee*, 1929-31; Tactical Course, 1932; Senior Officers' Technical Course, 1934; Commanding Officer, *Frobisher*, 1934-36; Rear Admiral and retired, 1940. Recalled for war service.
33   ADM 196/144 (Kekewich).
34   Later Admiral Sir Percy Lockhart Harnam Noble (1880-1955). Captain, 1918; Commanding Officer, *Calliope*, 1918-19 and *Calcutta*, 1919-20; Flag Captain and Chief Staff Officer in *Barham*, 1922-24; Commanding Officer, *Ganges*, 1925-27 and *St Vincent*, 1927-28; Director, Operations Division, 1928-30; Rear Admiral, 1929; Senior Officers' War Course, 1930; Senior Officers' Technical Course, 1930; Director, Naval Equipment Department, 1931-32; Vice Admiral, 1935; Fourth Sea Lord, 1935-37; Commander-in-Chief, China, 1938-39; Admiral, 1939 and retired, 1945.

thinker, & an athlete.'[35] Powers of command, a sure manner with others, boundless energy, and a strong athletic abilities were Kekewich's strong suits; however, he was not to rise to the top and poor handling of the cadet training ship *Frobisher* on two occasions during his time commanding proved his undoing.[36] This demonstrates the major distinction between naval and military practice and its effect on officer education and training: the naval officer exercised hands-on tactical command even late in his career and was required to exhibit competency in that discipline before all others including staff work.

Kekewich's appointment to the training cruiser *Frobisher* was an indication that he was highly regarded, as the Royal Navy made it a practice of selecting outstanding officers to those appointments where junior officers were developed. Modern personnel practices might record that this was just a case of mentoring, but to the Navy it was very much leadership by example. Whether the venue remained Dartmouth, *St Vincent*, Cambridge University, whilst that short course was in operation, Greenwich, or the *Frobisher*, an officer excelling in this appointment usually had established himself on a sure path to preferment. This talisman failed to hold the same power for one appointed as an instructor to a technical school or to a school of advanced learning—excepting the IDC—perhaps owing something to the view that by this stage in an officers' career, the habits and knowledge acquired of the mid-level or senior officer made suasion less an option.

One weakness of the system practiced was that it attempted to treat all executive officers as equals, though their work was inherently unequal. This was no bad thing in a peacetime professional Service if officers were to have careers beyond the immediate moment. However, flying duties and commanding submarines were talents best practiced young. What was good for those serving in the broader Service did not necessarily apply to those employed in such duties. Only war would demonstrate this conclusively, though a closer examination of recent experience might have foretold of the problem. This issue is one that remains with us today and is a peril whenever jointness for jointness' own sake is stressed. Jointery is but an end to a means—ensuring unity of purpose, concentration of effort, and economy of force. Doubtless, our age will only appreciate the peril at some later, inopportune, date.

Another shortcoming of interwar staff training was that it minimized matters of logistics or, more appropriately, it assumed that the logistical practices of the past were sufficient for the future. Criticism here must be tempered, as contemporaries lacked the gift of foresight and practical limitations were present. Foremost, financial stringency made it difficult to contemplate otherwise. The much-delayed effort in developing and fielding the Mobile Naval Base Defence Organization is symptomatic of this limitation and attempting to do more would have faced a tiresome battle from both the Exchequer and from the Navy's own financial authorities. Further, the need,

---

35   ADM 196/92 (Kekewich) and ADM 196/127 (Kekewich).
36   ADM 196/51 (Kekewich).

though pressing, was now less than when the fleet relied on coal, and the programme for pre-positioning oil supplies along its axis of approach to the Far East mitigated its worst defects. To its credit, the Navy recognized that it had short-changed logistics and administration as keystones in its planning efforts and modified its *Naval War Manual* to fill the void. Unfortunately, it took another global war to make good this deficiency.[37]

Organizationally, logistics were divorced from the Naval Staff proper residing as it did with the Controller and the Fourth Sea Lord. As so often happens, the pressure of outside events forced changes, as the need to concert her operational plans with the other Services and allies brought logistics matters to the fore in a manner that prewar planning had not.[38] There was no Staff Officer Logistics operating on the staffs of fleets, squadrons and flotillas, indicative that the administrative planning conducted in peacetime was of a narrow and short-term nature within the bounds of the Captain of the Fleet. Of course, a large measure of the logistics work between the wars was the province of the accountant branch and came under the purview of the Fleet Accountant Officer. This worked for routine operations, but operational logistics for any war plan was a nexus where the responsibilities of the executive and accountant branches coincided. That the relationship between these two branches was not as close and as intimate as required is suggested by Paymaster Captain Hugh Miller. Miller, Fleet Accountant Officer in the *Warspite* whilst Admiral Field commanded the Mediterranean Fleet, surveyed the supplies held in Malta for war provisioning. He concluded that insufficient stocks were on hand to outfit the ships of the fleet to their required level of emergency loading informing Paymaster Director-General Eldon Manisty of the problem.[39] As inventory control is even now a problem for navies, it may be that Miller was only remarking on its interwar variant. Yet, it is also true that if he had been intimate with actual war plans, he would have had a strong sense of what specific deficiencies existed and not merely surmise. This was why the accountant officer needed to attend the Staff Course while the executive officer needed to think more about logistics. Finally, the Navy was probably guilty of a degree of mirror-imaging believing that its potential adversaries were similarly constrained in its logistical affairs and not nearly as blessed as the Royal Navy. Again, to its credit, tentative steps were made before the onset of another war to correct matters, as Manisty was seconded to the Plans Division in 1936 upon his retirement.[40]

37 *Naval War Manual*, 1946, (Draft), Robertshaw Papers, IWM/72/75/1. This edition was released subsequently in 1947.
38 C.C.H.H. [Cecil Hughes-Hallett], 'Naval Administration Planning—A Short History', *The Naval Review*, Vol. XXXIII, No. 2, May 1945, p. 118.
39 Paymaster Captain Hugh Miller to Paymaster Rear Admiral William Manisty letter of November 1929, Paymaster Rear Admiral Hugh Miller Papers, Imperial War Museum, London, 73/11/2.
40 *Navy List*, March 1939, p. 273.

Ideally, an officer once selected for Captain would have attended the three foundational courses, i.e., the SOTC, the Tactical Course, and the SOWC, not to say the Staff Course, as well. Though many did many, others did not complete the required regime. Conflict in scheduling set against the demands of sea-shore rotation was certainly one reason why this was so but illness and branch of specialization were two others. As any course offered represents the confluence of subject material, instructors and qualifiers, the absence of an officer rebounded on those attending no less than on the one missing. Here, the failure of officers specializing in the submarine service was especially acute. For though the dearth of naval pilots or observers was partially rectified by the presence of Air Force officers at the Staff College, War Course, and even to a very limited degree at the Tactical Course, there were no compensations for the missing submariner. Of course, the true impact of this problem cannot be measured, but a hint exists in the correspondence of Rear Admiral Henry Thursfield. When his friend and military counterpart at *The Times* wrote that the problem of the submarine was likely to be only the more acute in any future conflict Thursfield took Liddell Hart to task:

> I am, nevertheless, much surprized (*sic*) to learn that you have found naval officers today who attribute to the submarine power greater than it possessed in 1914-18. I have heard much discussion on the subject amongst naval officers – I was nearly for 5 years on the staff of the Staff and War Colleges during my service on the active list – and I do not recollect that position being maintained by any of those who passed through either of those establishments. The influence of air action upon naval operations is, of course, a much less known factor, and it would be unwise to assume that it can today be eliminated too accurately. Nevertheless, there are certain data to go upon, and it may be, I think, just as dangerous to overestimate that influence as minimise it.[41]

Thursfield might have added that he was also a former Director of the Tactical Division and a graduate of the Tactical Course.[42] His comments to Liddell Hart demonstrate that even the most thoughtful of officers could yet learn something new and that what was learnt at Greenwich and Portsmouth depended in no small measure on the quality of officers and the backgrounds of those attending and instructing. In truth, it was a blessing if a single submariner was present in any course offered. What was true of submariners and pilots was also true in a lesser degree of other specialist officers, as the Service had to ensure that key billets were filled first. Alfred Bingeman, a torpedo specialist, first sought the Staff Course in 1929 as a senior Lieutenant and only arrived at Greenwich in 1938 as a Commander. Sound in all regards and regarded as an

<hr>

41   Thursfield to Liddell Hart letter of 18 November 1936, Liddell Hart Papers, LH 1/695/20-21, LHCMA.
42   ADM 196/91 (Thursfield).

exceptional Torpedo Officer, the delay owed everything to serving in his technical field with the Service unable to release him for training.[43]

Measuring the effectiveness of the Navy's interwar training regime is an impossible task and in a sense is unfair. History tells us what enemies arose and when. Thus, the pitfalls facing the Service are more obvious to the later observer than to the contemporary participant. Some observations, though, are not out of place on the interwar experience as a signpost of possible lessons for the contemporary officer. First, the danger is ever present that repeatedly gaming or exercising a problem at some point probably does more to reinforce an existing prejudice than elucidate a path ahead. This is especially a concern when finance is constrained and the means of solving a problem appear intractable within present resources. Here, the repetitive analysis of the Far East problem at the yearly tri-Service combined exercise offers a warning. The isolation of Hong Kong, the presumed safety of Singapore given its distance from Japan, and the sufficiency of time to allow the fleet to arrive were key elements requiring examination. Yet, they were elements considered largely in an operational context without the strategic assumptions underlying the exercises ever seeming to alter. This was a mistake and the most serious of ones as that context evolved over the period largely to the detriment of Britain.

A problem of another sort was the British Army's focus on the defence of India and the maintenance of its internal security—a staple of its staff training. Though offering an officer a deep understanding of the practical problems involved, as a narrow problem it did not force one to look beyond the immediate issue. Montgomery in his lectures of 1922-23 warned the qualifiers of this telling them that Indian warfare provided little 'scope for elaborate strategy; lines of operation usually dictated by nature of country. Objectives of a punitive nature; destruction of villages, crops, etc.'[44] At least the Far East problem for the Navy represented a difficulty of the broadest magnitude forcing it to consider many of the subsidiary issues involved in any lesser action. To that extent, its focus was never so myopic as the Army's commitment to the Northwest Frontier, but it did pose problems of another sort. Given the distances involved in any Far Eastern scenario, the air threat did not feature in a marked degree as when operating in the narrow waters closer to home. Where such waters did influence Far East planning (i.e., the Sunda and Malacca Straits), the aim was to minimize the risk by making the passage at night.

The COS increasingly failed to provide effective oversight of the IDC as the period progressed. This was due to the many crises of the period, the scale of their other obligations, and the practice of spending their time reviewing papers that properly should have been dealt with at a lower level. To this end, was it really essential that officers of their stature proofread appreciations and plans for errors and did this not

---

43   ADM 196/148 (Bingeman).
44   'Montgomery notes for Staff College lecture', *c.* 1922-23, Liddell Hart Papers, LH 1/519/1, LHCMA.

point to a failure in the administrative side of the Committee allowing documents to be considered that were not ready for approval? For this state of affairs, Hankey shares a large measure of the blame.

Still another problem that plagued the Navy was that sending senior officers to courses alleviated periods of half pay, but it also meant that whatever benefit the Service ought to have derived from the course was prone to being soon lost as subsequent employment was often lacking. This was most particularly egregious for the IDC and SOS courses. What was missing was a clear exposition on who should take any course, the point they should aim to attend, and the subsequent use the Service would make of the qualifier so trained. To be sure, *King's Regulations* did address the first two factors in passing, but a thoughtful plan tying education and employment was lacking.

Those historians who claim the Service ignored a significant portion of their recent operational history need to cast their eyes beyond the confines of Greenwich, for the Staff College was but a portion of the education and training regime of the interwar Royal Navy. In every respect, the SOTC and the Tactical Course were just as vital in preparing a naval officer for higher command, and, indeed, the more so as these courses along with the War Course were mandated for officers reaching their captaincy. For though many might achieve flag rank following attendance at the Staff Course, the naval *psc* was never a talisman guaranteeing promotion. More still would fly their flag afloat without following that path, yet having attended their share of courses borne in the *Victory*. It is dangerous to generalize why certain officers advanced beyond their peers, but more than any other consideration those of professional competence, perceived suitability, and a distinct style, even pizzazz if you will, were strong factors weighing in the balance. The foibles an officer displayed such as Beatty's monkey jacket and cover worn at a severe angle, Mountbatten's signalman adorning the bonnet of his automobile, and Teddy Evans[45] tossing his 'Sub' over the side in full view of the visiting commanding officers of the Battle Cruiser Squadron are remembered because they were outside the norm, but not outside the bounds.[46] A good competent officer was seen as such and he advanced and received appointments reflecting his abilities relative to his peers. To be a commander of one of the main fleets, an officer needed to be professionally competent, socially adept, keen on the Service and display a measure of élan.

45   Admiral Sir Edward Ratcliffe Garth Russell Evans, later Baron Mountevans of Chelsea (1880-1957). Captain, 1917; Flag Captain to Vice Admiral Dover, 1917-18; Commanding Officer, *Active,* 1918-19; Captain, Auxiliary Patrol, 1923-25; Senior Officers' Technical Course, 1926 and 1932; Commanding Officer, *Repulse,* 1926-28; Rear Admiral, 1928; Senior Officers' War Course, 1928 and 1931-32; Commander, Australia Squadron, 1929-31; Tactical Course, 1932; Vice Admiral, 1932; Commander-in-Chief, Africa, 1933-35; Commander-in-Chief, The Nore, 1935-39; Admiral, 1936 and retired, 1941.

46   Brewer, 'Melody Lingers On – II', *Naval Review,* January 1974, pp. 51-52.

Of course, demonstrating prior success allowed one to display a certain degree of brashness. Before the fame of 'Evans and the *Broke*', when he torpedoed one destroyer and rammed a second in a desperate action off Dover on 20-21 April 1917 against six German ships, Teddy Evans was already a part of contemporary lore. His miraculous survival of the Antarctic environment when consumed by scurvy was due entirely to the selfless devotion of two naval ratings. A public eager to hear from one actually surviving such conditions in the wake of Captain Robert Scott and his party's demise stoked the legend and Evans did not shy from its telling.[47] Such aggrandizement may have won him few plaudits amongst his peers, but it did little harm with his superiors.[48] As for Beatty, an early DSO and promotion to Captain marked him as a coming officer, but financial security allowed him a measure of independence which a more modest officer dependent on his career could not afford to exhibit. This, too, was true of Mountbatten who like Beatty, if not marrying well, married wealth. Thus, is it the case that initiative, that quality so highly prized in a naval officer—particularly in peacetime when prospects and reward are typically poor—is at the end of the day borne of character anchored in financial security no less than in training and experience? The academy prizes tenure for the independence that it affords the intellectual to break new ground. The Naval Prize Court allowed earlier officers to secure a footing that normal pay and allowances could not. Indeed, the prize system provides the best example of performance-based pay. Today, are moral courage and training enough to provide the independence of thought the public requires in their officers? If Lord Fisher had been wealthier, might he have been bolder and actually voiced the reservations he held regarding Churchill's Dardanelles plan? Certainly, though in a period removed from this study, Mountbatten counseled against invading Egypt in 1956. The answers are unknowable and highly contingent on the personality of the Minister briefed, but for that, their implications require consideration. In short, at the highest levels of war, merit without means may prove insufficient with dependency fostering conformity.

In examining the regime of training and education as followed by the interwar Navy there is much to lament and but even more to praise. The true measure of its effectiveness is indicated by the actions of the Service in the Second World War. Not all officers when challenged proved equal to the task. Yet, thankfully, such occasions remained the exception and not the rule. Andrew Cunningham's, 'It takes the Navy three years to build a ship. It would take three hundred years to build a reputation. The evacuation will continue.' made when prospects were their darkest off Crete in 1941 echoes with the spirit of the salt horse.[49] Still, ABC was a salt horse highly seasoned by instruction. It is tempting to conclude that Cunningham would have been better placed if he had relocated to Cairo to act alongside his military and air counterparts

---

47   '*Shellback Remembers*', p. 152, Durnford Papers, IWM/PP/MCR/C10.
48   Grenfell to Liddell Hart letter of 2 June 1937, Liddell Hart Papers, LH 1/330/9, LHCMA.
49   Cunningham cited in Simpson, ed., *Cunningham Papers*, *II*, p. 428.

rather than remain in Alexandria.[50] Offering further evidence of this viewpoint, he controlled operations off Crete from Alexandria choosing not to wear his flag at sea. As Tedder's relationship was better, certainly smoother, with his fleet liaison officer than with the naval Commander-in-Chief, this is a doubtful conclusion.

Cunningham had worked well with Longmore when the latter was the Air Officer Commanding and General Haining, following his visit to Egypt and assessment of command relationships, believed cooperation amongst the Services was effective.[51] Resources were ever the problem hindering operations, but conflicting styles and temperament cannot be dismissed. Still, modern practice argues otherwise. In any event, naval liaison officers in Cairo and the presence of Group Captain Claude Pelly[52] and Major General John Evetts with Cunningham provided opportunity for ensuring requirements and views were aired at both headquarters.[53] Cunningham, if not a perfect leader in the joint environment, remained a perfect sailor and his leadership inspired many serving beneath him.[54] This was no small feat. Indeed, it was the *measure*. For, as ever, knowledge, skill and opportunity count for little without internal fortitude to see through any problem. This was something ashore training, whether conducted at Greenwich or Portsmouth, did not attempt to replicate being best measured by observing the officer on the bridge at sea. Here, any shortcomings in powers of command or weaknesses in ship handling were certain to feature in one's confidential report and did.[55]

Ashore training remained, though, an important grounding. For if one's personality, including an ability to lead are largely fashioned in youth, the naval officer must yet learn his calling before he can apply what lies innate. Those problems investigated at the Staff and War Courses and in the Tactical Course were not idle exercises. So, too, the principles instilled and the historical problems examined were the foundations upon which future success would be built. If an officer leaving his course had learnt much, then a good one also understood that much remained unknown including the actual higher direction of war to be followed and argued so extensively at the War Course and the IDC. The next war would contain its share of surprises and mistakes *were* made—most, especially, at the confluence of policy and strategy. If bad tactics might cost a battle, the price of bad strategy remained a lost war. It is avowed that Waterloo was won on the playing fields of Eton. Perhaps, but the Battles

---

50   *BR 1736(2). Naval Staff History Second World War. Naval Operations in the Battle of Crete, 20th May—1st June 1941 (Battle Summary No. 4)*, Admiralty, Historical Section, 1960, p. 6.
51   Norman interview, 1985, IWM Sound 8870, Reel 3.
52   Later Air Chief Marshal Sir Claude Bernard Raymond Pelly (1902-1972). Staff Course Andover, 1936 and Imperial Defence College, 1951-53
53   *Naval Operations in the Battle of Crete*, p. 24.
54   Norman interview, 1985, IWM Sound 8870, Reel 3.
55   For example, Rear Admiral Arthur Jackson was noted as being a very clever officer but lacked powers of command being more suited to a smaller ship. He spent a career in the surveying service as a hydrographer. See ADM 196/51 (Jackson).

of the Atlantic and the Mediterranean were won on a table in Portsmouth—and in arguments and schemes played at Greenwich, Camberley, Andover, Quetta, and 9 Buckingham Gate.

Commodore Henry Harwood[56] knew the America and West Indies Station intimately from the preceding three years he had spent in its waters in the cruiser *Exeter*. Of his tactics against the *Graf Spee*, a debt owed to *Victory* and the Tactical School attended in 1930 and to the War College where he refined his thoughts on how a cruiser force should fight a pocket battleship was amply repaid with interest. His success was not ordained, but it was deserved owing everything to initiative— that principle missing from the *Naval War Manual* but ever so present in the *Battle Instructions*, its amplifying authority. Writing to his wife soon afterwards Harwood explained, 'It's the old story—the offensive. He [Langsdorf] expected us to run away and when we attacked and chased him instead, he lost the initiative.'[57] Dudley Pound, praising Harwood's fortitude, believed his actions were entirely correct even if it had meant the destruction of his entire force for a standard of conduct had been set for others to follow.[58]

Expunging any shadows of the *Goeben* and *Breslau*, that perennial search exercise of the interwar Staff Course, would take more than a single engagement. Thus, on Empire Day 1941 *Hood* and HMS *Prince of Wales* met *Bismarck* and *Prinz Eugen* in the Denmark Straits. Ralph Kerr, no stranger to command at sea and now Flag Captain in *Hood* serving Vice Admiral Lancelot Holland where he wore his flag. Holland, though a highly proficient gunnery officer, was new to *Hood* and his only service previously in battle cruisers was a short period in *Renown* in 1935 when serving as the Chief Staff Officer to Commander, First Battle Squadron when that officer briefly shifted his flag to her.[59] Meanwhile, Kerr, the erstwhile salt horse, had become an anti-submarine specialist late in his career attending his course in the *Osprey* in 1931.[60] Kerr had at least served in battle cruisers though, in his case, it had been 27 years previously as a very new Lieutenant in HMS *Inflexible*.[61] The talents of both officers,

56   Later Admiral Sir Henry Harwood Harwood (1888-1950). Six firsts in examinations for Lieutenant; Fleet Torpedo Officer in *Southampton*, 1919-20; War Staff Course, 1921-22; Plans Division, 1922-24; Squadron Torpedo Officer in *Iron Duke*, 1925; Fleet Torpedo Officer in *Queen Elizabeth*, 1925-26; and *Warspite*, 1926-27; Captain, 1928; Commanding Officer, *Warwick*, 1929-30; Tactical Course, 1930; Imperial Defence Course, 1931; Senior Officers' Technical Course, 1932; Flag Captain in *London*, 1932-34; Senior Officers' War Course, 1934; Royal Naval War College, 1934-36; Commanding Officer, *Exeter*, 1937-39; Rear Admiral, 1939; Acting Admiral, 1942, Vice Admiral, 1943; Admiral and retired, 1945.
57   Henry and Stephen Harwood Sound Recording 28811, 2005, Imperial War Museum, London.
58   *Ibid*.
59   ADM 196/92 (Holland).
60   ADM 196/53 (Kerr).
61   *Ibid*.

however, did complement each other. Holland the capable staff officer at home with strategy and Kerr the tactician with an eye for the sea, the turn of the helm, and speed of handling. Yet, was *Hood* the right ship for them and was the pairing of *Hood* to *Prince of Wales* the wisest tactical combination? At least Harwood was fortunate that *Achilles* and *Ajax*, both of the same class, formed a natural gunnery combination. Still, that combination on the day of reckoning was no more egregious than *Bismarck* to *Prinz Eugen*, but so much depended on the tactics adopted.

It is easy to be wise after the event and this observer will eschew commentary on the approach adopted, the time of day battle was offered and risking a battle cruiser against a battleship—something interwar exercises demonstrated repeatedly was a dangerous move.[62] Of more salience to this study and the present, did the Admiralty consider the strengths *and* limitations of officers when making appointments? Was Kerr the best Captain available, or was it merely a case of Buggin's turn with his time arriving to serve in a heavy ship? If ninety percent of what a ship captain does at sea is common irrespective of platform, then how much consideration should apply to the remaining ten percent? To the interwar Royal Navy, the answer was frequently little as a sound, capable officer was always preferred over the narrow, specialist officer. This view had much to commend itself in peacetime as developing officers with broader views was an important consideration. The issue really becomes should practices of employment sound in peacetime continue during war? The conclusion of this observer is yes, but the screening of a ship's officer complement must be pursued with a holistic view. Selecting the best officer fitted for the position in question is safe for the moment, but appointing officers should always think of the future. Considering an officer's merits in relation to the other officers serving alongside can ensure the whole remains more than the sum of the parts. This is done, but rarely acknowledged publicly as charges of preference soon follow.

Selecting the best officer poses another sort of risk as an officer's career may become narrowly focused. Thus, the temptation to keep a good officer in place filling a difficult assignment rather than risk a change becomes the result. John Durnford having served primarily in destroyers, was recommended for a position requiring greater responsibility when serving as a Lieutenant Commander in the cruiser *Durban*. The verdict of the Commander-in-Chief, China held otherwise arguing that simply because an officer had proven himself in one direction was not a sound argument for pushing him in another direction.[63] The traditions of the Service did not agree, for the greatest of commands required the broadest of views and this was not secured by following the well-worn path. As navies desire to have a strong reserve of officers against which to make their appointments, a certain measure of risk is accepted in allowing officers to serve outside their specialties. This cuts against the grain of pure meritocracy, but, as

62   See ADM 186/154, *C.B. 1769/33(2), Exercises & Operations, 1933, Volume II*, Admiralty, Naval Staff, Tactical Division, April 1934, p. 5.
63   ADM 196/145 (Durnford).

ever, quantity has a quality all of its own. Grenfell writing in 1942 believed Admiral Sir Max Horton had 'been too longed concerned with submarines only' and this view was expressed before he had even arrived at the command where his second reputation was won.[64] Given the price of failure in the Battle of the Atlantic, though, appointing Horton and keeping him in place made perfect sense.

Lance Holland paints a slightly different picture. A graduate of the Long Gunnery Course and conversant with tactics through his time as Deputy Director of the Royal Naval Staff College, his first sitting of the Tactical Course ended after only four weeks study when he secured appointment to the *Vindictive* as Chief Staff Officer in 1929; Holland returned and completed the course in 1934. This mirrored his attendance of the SOWC in 1926 when he only attended for its first eight weeks before assuming his instructional duties at the Staff College. Returning to take his Tactical Course in May 1934, he missed the first month of this session.[65] Notwithstanding such checkered studies, Holland had graduated from the IDC where Vice Admiral Preston rated him the best naval student of his class. Nor was Preston's view an isolated evaluation. Admiral of the Fleet Sir Charles Forbes believed him to be one of the best Vice Admirals of the Service. Holland's successful contrast to his naval peers at the IDC owed much to the fact that working in the joint arena was not a new endeavour for him. He had supported the Army's exercises of 1927, and rare for his time, as a naval officer, he held a private pilot's license. Consequently, Holland was thought to be a naval officer with a sound appreciation of the air.[66] In both the Turkish and Abyssinian Crises, he registered strong impressions with his commanders for the 'good' and then 'exceptional' services rendered and command of a major fleet was recommended at the appropriate time.[67]

Indeed, Holland shares a career pattern reminiscent of another officer coming to grief in 1941: Admiral Tom Phillips. Both of these officers spent time away from the Service advising others—Holland to the Hellenic Navy and Phillips to the League of Nations. Time as Flag Captain in *Hawkins* was another trait common to them and both were trained staff officers: Holland courtesy of the IDC and Phillips via his War Staff Course at Greenwich. Holland had only recently departed as ACNS when Phillips was appointed to the Naval Staff as DCNS. Phillips was recognized as a superb staff officer, so much so, that Vice Admiral Dreyer when serving as DCNS had warned against the temptation of curtailing his sea time and returning him to staff duties of which he excelled recording:

<hr>

64   Grenfell to Liddell Hart letter of 1 September 1942, Liddell Hart Papers, LH 1/330/66, LHCMA.
65   ADM 195/51 (Holland).
66   Holland's naval counterparts attending the IDC were Captains Edward Dicken, George Fraser, William Gell, Charles Kennedy-Purvis, Percy Nelles, RCN and Ion Tower.
67   ADM 196/51 (Holland) and ADM 196/92 (Holland).

An <u>exceptionally</u> able staff offr. Very clever & a tremendous worker. His judgment is excellent. I think he is at times a little intractable in regard to his opinions, but that is due to firmness of purpose. A v.[ery] quick offr. with a sense of humour. I strongly recom<u>nd</u> that this v.[ery] valuable offr. should be given plenty of employment in command at sea; there will be a great temptation to bring him back to staff work – too soon for his good & that of the service; this shd. be resisted[68]

Not blessed with foresight any more than the next flag officer, Dreyer, nevertheless, appreciated where the strengths *and* weaknesses of Phillips were when compared to his peers even in 1932.

Thus, one assessment might conclude that just as Holland was ill-suited to command a battle cruiser squadron in 1941, Phillips was not the ideal officer for a seagoing command direct from a shore appointment and absent current war experience, if the prospect of battle was thought near. For the moment, though, his mission remained one of deterrence and his ability to speak with effect and authority in local waters certainly made his choice appropriate from that perspective. Of course, hindsight is a wonderful perspective not shared by contemporaries and even Vice Admiral Sir Martin Dunbar-Nasmith,[69] perhaps the most critical officer of Phillips' abilities, noted his ship-handling and staff skills when serving him as Flag Captain in *Hawkins*.[70]

John Slessor said of Phillips that notwithstanding his many outstanding qualities, his one shortcoming was his attitude to airpower and his belief that navies could meet the threat on equal terms.[71] A more damning verdict could hardly be rendered and given their close association whilst considering joint problems late in the interwar period, Slessor's views cannot be discounted. Perhaps if he had handled a squadron off Norway, in the Channel, or in the Eastern Mediterranean during the war Phillips would have acted differently off Malaya. Yet, arguably, Slessor's verdict remains unfair and unjust all the same. Sailing into harm's way is what navies do and Phillips' fault was not in attempting to reach the Japanese invasion force threatening Malaya, but in failing to concert his actions with the local air commander—the failure of a staff officer to appreciate the situation. Other faults were present in Phillips' action including a tendency to alter his courses of action unnecessarily and failing to carry through a decision to its appropriate end. Better intelligence might have led to better decisions at the tactical level, but strategically, the position was well-neigh hopeless

68 ADM 196/92 (Phillips). Original emphasis.
69 Later Admiral Sir Martin Eric Dunbar-Nasmith (1883-1965). Captain, 1916; Commanding Submarine Flotilla, 1918-20; Senior Officers' Technical Course, 1921; Director, Trade Division, 1924-25; Rear Admiral, 1928; Senior Officers' War Course, 1931-32; Vice Admiral, 1932; Commander-in-Chief, East Indies, 1932-34; Second Sea Lord, 1935-38; Admiral, 1936; Commander-in-Chief, Plymouth, 1938-41 and retired, 1941. Recalled for war service.
70 ADM 196/92 (Phillips).
71 Slessor, *Central Blue*, p. 277.

given the other commitments Britain and the Commonwealth faced and the extremis conditions presented.

Before Holland and Phillips, though, was the unfortunate circumstances surrounding the loss of the carrier *Glorious* in 1940 during the Norwegian debacle. Guy D'Oyly-Hughes was thought a highly effective, if unconventional officer. If not brilliant intellectually, he was viewed as a real 'thruster' who had had an outstanding record in the World War having earned a DSO and Bar, the Distinguished Service Cross, and the Silver Medal of the Royal Humane Society for attempting to save a life at sea. From command of multiple submarines, D'Oyly-Hughes attended the second War Staff Course.[72] Admiral Sir William Fisher praised his command of the First Submarine Flotilla in the Mediterranean Fleet during the crisis with Italy in 1935-36 and Admiral Drax thought he would surely rise to the very summit of his profession. Still, always lurking in the background were D'Oyly-Hughes' shortcomings in broader naval work and judgment in matters of discipline.[73] As with Kerr, he was now operating outside his platform of choice. Prior duty in the carrier *Courageous*, attendance of the Senior Commanders' Aviation Course and attachment to the Air Ministry made him familiar with air operations and procedure, so unlike Holland and Kerr, he did not face learning the capabilities and limitations of a new class of ship. Captain Henry Fancourt who knew D'Oyly-Hughes well conceded that he marched to a different tune, but possessed more than his share of 'guts and go'.[74] This commodity, known in conventional terms as initiative, was prized more than any other quality in an officer. It might not always lead to success, but it remained the only path to victory. This trait, more than any other, accounts why some officers could progress outside the beaten path.

To be sure, the *Hood* was avenged and it was a team effort. Writing to his friend Gerald Dickens, Wilfrid Patterson,[75] Flag Captain in *King George V*, recorded his impressions of the denouement calling attention to the years of training which played the essential part:

> *Rodney* did awfully well, like a rather slow & elderly beagle they kept an inner circle & were in at the kill, (and very glad I was to see her). Everyone, cruisers, destroyers F.A.A. all did their stuff in the grandest way & without any extra instructions. All the war games & exercises of the last twenty years have born fruit.[76]

72   ADM 196/145 (D'Oyly-Hughes).
73   ADM 196/92 (D'Oyly-Hughes).
74   Fancourt, Sound Recording 12274 of 30 September 1991, IWM.
75   Later Admiral Sir Wilfrid Rupert Patterson (1893-1954). Squadron Gunnery Officer in *Hood*, 1928-29, in *Repulse*, 1929 and in *Revenge*, 1929-30; Tactical Course, 1931 and 1936; Captain, 1933; Senior Officers' Technical Course, 1934; Senior Officers' War Course, 1934; Senior Officers' School Sheerness, 1934; Flag Captain and Chief Staff Officer, Reserve Fleet, 1936-38; to RAN, 1938-40; Commanding Officer, *King George V*, 1941-42; Rear Admiral, 1942; ACNS, 1943-45; Vice Admiral, 1946; Admiral, 1949 and retired, 1950.
76   Patterson to Dickens letter of 20 June 1941, Dickens Papers, LHCMA.

Patterson's comments if effusive, nevertheless, are not wrong for he knew of what he spoke having been praised by the Their Lordships in 1921 for introducing new, abbreviated fire control signaling arrangements.[77] If the hero of Trafalgar was correct that only numbers can annihilate, it remains that one still must know how to employ those numbers. In the modern era, training and education were the foundations of later success. Holland, Phillips, Kerr and D'Oyly-Hughes were all highly esteemed, superbly trained, and most unfortunate. The margin of difference separating them from Harwood, who on the day was right and lucky, is much finer than commonly appreciated.

The Navy asked much when it expected an officer to be both an effective commander and an effective staff officer—perhaps too much. Certainly, the Army did not expect this level of genius in their officers and was willing to prize an effective staff officer on his own terms. General Ironside, commenting on the merits of Stephen King-Hall following his time at Camberley, said he was more suited as a staff officer than serving as a commander.[78] To Ironside and the Army this was no criticism, but as a report on a naval officer's fitness for further employment, it struck a false note where the all-rounder was valued. King-Hall may be contrasted with Lieutenant Commander Maximilian Despard[79] who failed his Staff Course but a little over a year later was being strongly recommended for promotion by Teddy Evans when serving in the *Harebell* on the strength of his leadership skills, tact, and ability to get along with foreigners and Britishers alike. It should be noted that the pairing of Evans and Despard was not a new one, as Despard won his DSC in the fierce action of the *Broke* in 1917. He was especially recommended for command of a destroyer and would doubtlessly have gone further if tragedy had not struck suffering a compound fracture of the femur in 1925. Despard, a nephew of Field Marshal Sir John French, was invalided from the Service, but, in time, found employment in Finland as a naval adviser. Evans' appreciation of Despard, though, was incorrect in one aspect—he did not mix well with Astley-Rushton, the Director of the Staff College. Rushton noted the marked personality of Despard, but also concluded that he was poor on paper and intolerant of the views of others. Such a conclusion probably owed something to two strong-willed officers simply rubbing each other the wrong way, but the verdict would have been the more damning if it had found that while excellent on paper Despard lacked any powers of command.[80]

Victor Danckwerts was another officer of marked intellectual ability who shined in staff appointments. He also proved a fine captain of destroyers though seen as 'not "cut-out" by nature for such an appointment.' Somerville assessing Danckwerts' performance

77   ADM 196/56 (Patterson).
78   ADM 196/145 (King-Hall).
79   Later Acting Captain Maximilian Carden Despard (1892-1964). Staff Course, 1922-23; Senior Officers' Technical Course, 1927; Retired, 1928 and Acting Captain, 1933. Recalled for war service.
80   ADM 196/145 (Despard).

and abilities praised the handling of his flotilla yet concluded he approached matters more 'scientifically rather than with a natural eye for the situation.'[81] This was hardly a ringing endorsement though it was sufficient to ensure promotion.

William Chalmers was less fortunate. A highly trained staff officer who seemed destined to advance yet further, his time as Flag Captain in *Delhi* did not go well. Initially, he was seen as easily rattled on the bridge during fleet exercises or when passing close to other ships. Though the problem eased over time, other superior officers had shared this assessment. This placed a real cloud over his future, and so pronounced was his condition that for a time it appeared Chalmers was in danger of being superceded in command of *Delhi*. Fortunately, by the end of his appointment, Chalmers was seen as a very effective Flag Captain and had grown into the job, but Rear Admiral D'Oyly Lyon[82] felt appointment as Director of the Royal Naval Staff College suited his talents and this was duly arranged following a slight detour to Haslar Hospital first.[83]

An officer ever more marked by his staff abilities was Roger Bellairs, Beatty's erstwhile assistant and, latterly, Director of Plans. Returning to the fleet and command of a battleship, he had the misfortune of serving during the Atlantic Fleet mutiny with the *Rodney*, a ship at the fore of the disturbances. Lack of experience in commanding a heavy ship, absent powers of command, and relying too much on subordinate officers were not qualities apt to foster advancement. Never mind his obvious administrative abilities, appointing Bellairs a permanent Director of Plans, as suggested by one flag officer was hardly a serious consideration and retirement quickly followed.[84]

Meanwhile, Robert Raikes is an example of an officer highly esteemed and effective in all his appointments who, nevertheless, suffered from the narrowness of his career. As a member of 'The Trade', or a qualified submariner and an early graduate of the IDC, he transitioned to heavy ships to broaden his professional development. Unfortunately, the Abyssinian Crisis found him co-opted to act as fortress commander of Alexandria where he proved himself a fine administrator in organizing the city's defences with the award of a C.B. soon following. This time of trial and unconventional duty, if duly rewarded in the immediate period, proved detrimental to his broader career as he failed to accrue sufficient time in a heavy ship. William Fisher, consequently, believed command of the submarine force, a War College, or of

81   ADM 196/92 (Danckwerts).
82   Later Admiral Sir George Hamilton D'Oyly Lyon (1883-1947). Captain, 1922; Local Defence Division, 1923; Plans Division, 1923-25; Anti-Submarine Course, 1925; Captain (D), Second Flotilla, 1925-27; Director, Physical Training, 1927-29; British Naval Mission to Greece, 1929-31; Senior Officers' Technical Course, 1931 and 1937; Senior Officers' War Course, 1931 and 1937; Tactical Course, 1932; Commander, Third Cruiser Squadron, 1932-35; Rear Admiral, 1934; Vice Admiral, 1938; Commander-in-Chief, Africa, 1938-39; Commander-in-Chief, South Atlantic, 1939-40; Admiral, 1942 and retired, 1943.
83   ADM 196/92 (Chalmers) and ADM 196/52 (Chalmers).
84   ADM 196/91 (Bellairs).

a lesser foreign station was more appropriate employment for Raikes in the future than commanding a squadron in the main fleet.[85] In fact, his next appointment was Rear Admiral Submarines showing that recommendations from a fleet commander carried influence when weighing the future employment of officers. Though Fisher did not call attention to an earlier role in the collision of the cruisers *Sussex* and *Devonshire* occurring in 1929, it is difficult to avoid the conclusion that Fisher may have had this in mind when remarking on Raikes in 1936. His lack of heavy ship experience was not a factor easily dismissed as a contributing factor then and command of a squadron of battleships might tempt fate once too often at another moment.

Ion Tower paints a more difficult picture. An officer of considered intellect who made a very favourable impression in this regard on his Air Force counterparts when posted to the Air Ministry from 1924-26, Astley-Rushton believed he could have gotten more out of his Staff Course, if he had applied himself to the fullness of his abilities. This opinion mirrored Lionel Preston's when commenting on Tower's performance at the IDC. Why this impression should be so pronounced when taking these courses is impossible to know. It may be that Tower thought a breather was in order following sea time or it could be he thought the coursework beneath him. Another likelihood is Tower did not desire to be thought of as a very clever staff officer at the expense of his seamanship qualities though even Vice Admiral Sir Charles Little, when reporting on Tower as Flag Captain in *Kent* felt he lacked a certain drive, though his ship and his performance had been excellent in all regards.[86] If a firm conclusion is difficult to render, it remains that officers understood those elements of their evaluation most critical of meeting satisfactorily.

Thus, the officer trained in staff and war studies was prized, but the seaman remained the essential element. This was why it was possible for an officer to advance without taking a higher-level course, but it was not possible for a staff officer to avoid sea-time and still prosper. As ever, those who can, not only do, they must. Yet, most truants are destined to have an unfulfilling life because they lack the necessary discipline to recognize when it is better to conform than not. Others, the elect, succeed because married to their unconventional ways is the inner strength to recognize which battles are worth fighting and which are not. This is good, but it is not sufficient. Competence remains an essential ingredient whether acquired through formal education and training or demonstrated innately. Few officers will be a gifted amateur or possess the genius of Nelson. For the rest of us, training and education remain.

Thomas Drew was a *very* good practical seaman who shined as Flag Captain in the *Royal Oak* during 1936-38. Though believed a poor student by Boyle when taking his 1930-31 War Course the Admiral-President conceded he had character and would do better in more active employment. Drew's second sitting of the War Course in 1936 was a slight improvement on his earlier efforts with Colvin recalling that he was 'keen

85   ADM 196/91 (Raikes).
86   ADM 196/92 (Tower).

and practical'. Howard Kelly concurred in these assessments and marked Drew out as a future flag officer. He needed only a leavening of experience at the Admiralty to enhance his knowledge of administrative affairs. This duly followed with appointment to the Personal Services Department.[87] Accordingly, the Royal Navy esteemed competence of practical work over brilliance of mental acumen and an officer still needed to demonstrate three other vital characteristics to progress further: good health, a certain measure of moral rectitude and loyalty to one's superiors. General health has already been remarked upon, but mental health and one's overall outlook were important too. Sometimes the most brilliant of officers suffered in this respect, as they tended to overwork, thus, running themselves down in the process and suffering the less capable badly. This was Robert Ellis' failing and probably denied him his chance for reaching flag rank.[88]

As for moral rectitude, given the scandal of the Abdication Crisis of 1936, re-establishing the moral climate of the age was a core value to the contemporary Establishment, including the Navy, and became an unspoken, but central tenet in assessing an officer's future potential. Though a certain measure of moral laxness could be accepted in wartime, an officer's conduct in peacetime was non-negotiable and enjoying the society of women too much—particularly, if one was already married, earned an officer no credit.[89] As for divorce, much depended on the actual circumstances surrounding the case and even an innocent absent any wrongdoing was still expected to meet his naval responsibilities which such marital pressures did nothing to abet.

Finally, loyalty was a trait firmly prized the higher one advanced in the Service and correctly so. An officer needed to command, yes, but he also at times had to implement decisions that were personally painful, seen as misguided, or simply wrong. Such was ever a possibility, but if he spoke to freely against his superiors in this regard, then the greater discipline of the Service was possibly at risk. Captain Charles Larcom[90] was deemed a very capable gunnery officer, who had shined in staff appointments and even succeeded in commanding the *Courageous*. Both Admirals Chatfield and Dreyer recognised his superior ability, but so too others his habit of criticizing the methods of his flag officers. If this was done to curry favour with his subordinates, it had the exact

87   ADM 196/92 (Drew).
88   ADM 196/93 (Ellis).
89   ADM 196/92.
90   Captain Charles Arthur Aiskew Larcom (1891-1964). Gunnery Division, 1920-21; War Staff Course, 1923-24; Operations Division, 1926-28; Captain, 1932; Senior Officers' War Course, 1933; Intelligence Division, 1933; Chief Staff Officer to Flag Officer Yangtse, 1933-35; Flag Captain and Chief Staff Officer in *Courageous*, 1937-39; Director Naval Air Division, 1939; Commanding Officer, HMS *Sheffield*, 1939-41 and retired, 1942. Recalled for war service.

opposite effect on those reporting on him and Larcom was forced to retire in 1942 even though gaining a well-deserved DSO in the *Bismarck* encounter.[91]

As ever, the randomness of life, or luck, in other words played its part in answering why certain officers advanced and others did not. Rushton's fatal automobile accident in 1935 did nothing to advance his career but William James' change of ship from the *Queen Mary* to the *Benbow* in March 1916 with the former's loss at Jutland saved him for another day and further employment.[92] Good timing, too, no doubt played a hand in Charles Little's success. Commanding the Second Battle Squadron in the Atlantic Fleet from April 1930 to April 1931, he hauled down his flag to take his Tactical Course before becoming Rear Admiral, Submarines. Thus, he was untainted from the Atlantic Fleet mutiny and his career continued to prosper.[93]

Meanwhile, from the experience of the Invergordon Mutiny, another legacy emerged—a stronger appreciation for the naval rating and marine. For now gracing the cover of the 1939 edition of the *Naval War Manual* was a Petty Officer in square rig with the admonition, 'The biggest single factor in war.'[94] This may have been hyperbole, but, if so, it was not misplaced. Officers of the seagoing fleet long-appreciated this truth even if those closer to Whitehall seemingly did not. The Service also appreciated that reflection needed to be the handmaiden to action. Having the perfect solution was nugatory if it was not acted upon at the appropriate moment. That was why leadership, initiative and power of command weighed more than formal training and at the critical moment the need was ever to 'think wisely, plan boldly and act swiftly'.[95] Practical training in the fleet reinforced the last named quality; the Greenwich and Portsmouth courses reinforced the rest.

91   ADM 196/92 (Larcom).
92   ADM 196/90 (James).
93   ADM 196/90 (Little).
94   G. A. F. 'In Which We Have the Honour to Serve', *The Naval Review*, Vol. 67, No. 2, April 1979, p. 92. A slightly different rendition of this quote is given by the same writer in 'Men in Uniform', *The Naval Review*, Vol. XLIX, No. 4, October 1961, p. 419.
95   This tag, adopted by *The Naval Review*, adorned the cover of its journal from 1942 until 1973.

Appendix I

The *Navy Estimates* for 1918-19 – 1939

The sources for the below funding levels of the Royal Navy are taken from the *Navy Estimates* of the period. The possibility that what was actually spent in any one year may vary from the below figures whilst accepted does not detract from the trend represented. The financial year in the United Kingdom operated from 1 April through 31 March of the following calendar year. Beginning with the 1926 *Estimates,* the funds requested were specified for a single year.

1918-19	£334,091,227
1919-20	154,084,044
1920-21	92,505,290
1921-22	75,986,141
1922-23	57,492,389
1923-24	54,064,350
1924-25	55,693,787
1925	60,004,548
1926	57,142,862
1927	58,123,257
1928	57,139,146
1929	55,987,770
1930	52,274,186
1931	51,014,752
1932	50,164,453
1933	53,443,545
1934	56,616,010
1935	60,050,000
1936	69,930,000
1937	96,117,500
1938	125,307,500
1939	147,779,000

Appendix II

The War College Estimates for 1919-20 – 1937

1919-20	£500
1920-21	550
1921-22	550
1922-23	—
1923-24	6,157
1924-25	6,017
1925	6,061
1926	4,594
1927	4,431
1928	4,286
1929	4,649
1930	4,406
1931	4,383
1932	4,320
1933	4,279
1934	4,220
1935	4,350
1936	4,030
1937	4,170

Appendix III

The Staff College Estimates for 1919-20 – 1937

1919-20	£10,459
1920-21	10,175
1921-22	13,572
1922-23	—
1923-24	13,812
1924-25	14,011
1925	14,258
1926	12,532
1927	12,833
1928	12,585
1929	13,043
1930	13,055
1931	12,672
1932	12,977
1933	12,104
1934	12,360
1935	12,420
1936	12,400
1937	12,550

Appendix IV

The Imperial Defence College Estimates for 1927 – 1937

1927	12,000
1928	12,000
1929	—
1930	11,900
1931	12,672
1932	11,985
1933	11,773
1934	11,747
1935	11,703
1936	12,700
1937	7,070

Appendix V

The Truants: Naval Officers Who Avoided the Higher Education of the Service

---

The preceding discussion has shown that the Navy had established a stronger regime of higher officer education in both practice and in regulation than was commonly appreciated at the time or, subsequently, by its critics. If far from perfect, it still proved a good and effective arrangement when measured against the conflicting requirements faced and the strength of its general soundness is how much of its style, if not its content, the Army seemingly adopted by late in the period. Weaknesses were ever present and seemingly how to account for the unique requirements of the air and submarine communities were always two of the most prominent. The submarine arm was perhaps the most amenable to resolution as it operated within its the bounds, but even communities such as gunnery and signals posed their own management difficulties, if a common system was to prevail. Meanwhile, officers assigned to intelligence duties—particularly, if possessing unique language skills—faced a similar hurdle for the tendency became to employ them based on their current skills at the expense of developing any future ones. Here, Desmond Tufnell comes to mind. His period of service with the GC&CS as an instructor in Japanese in 1929, though doubtlessly of benefit to the Foreign Office and the greater defence establishment, operated against his naval career with Domvile, the Director of Naval Intelligence, recording of his merit that 'it is not possible to report on him usefully'.[1] Though Tufnell had commanded several destroyers and would eventually command even heavy ships, his lack of higher technical training, other than the Greenwich Staff Course, would eventually operate against his career. The lesson a modern officer perhaps should draw from Tufnell is that by all means educate yourself, but do not educate yourself outside the norm, as others may view you as an indispensable commodity not to be spared for greater, broader duties.

By and large, though, the Navy did a commendable job at balancing its officer education and training requirements and the measure of its success was that most

1   ADM 196/127 (Tufnell).

executive officers of Captain's rank had attended at least one of the required courses before assuming command of a major ship, though not so for more junior officers commanding a minor one. Indeed, many attended their Tactical and Technical Courses more than once though fewer tended to sit the War Course multiple times. This merely reflected that tactics and technology are the aspects of the profession more subject to change and that the higher examination of war now benefited, after 1926, from the introduction of a joint course to compliment the SOWC.

Still, not all Captains attended even a single course before the onset of war and an explanation is required. Discounting those officers promoted to the rank between the years 1935-39 inclusive, in 1939 there remained 133 Captains on the active list.[2] Of this total, three failed to attend at least one of the required courses specified in *King's Regulations and Admiralty Instructions*. Nor had these three attended a Staff College, the Army Senior Officers' School, or the Imperial Defence College. By any yardstick, the failure of only three officers to attend one of the required courses reflects creditably on the mandatory nature the courses actually achieved following publication of the Confidential Admiralty Fleet Order necessitating attendance in 1930.[3] Admittedly, a sample of three is a very small pool to survey and may reflect nothing more than chance and missed opportunity. This observer accepts this, however, any review of officer education would be incomplete without considering the cases of Vice Admiral Sir John Crace, Rear Admiral Arthur Jackson[4] and Captain Edward Law.[5] This is especially so when it is recalled that though officers needed to take their mandatory courses when selected for Captain, many had actually taken at least one of the classes whilst holding a lesser rank.

The first observation to record is that all three officers successfully negotiated the hurdle of further advancement at least twice. Law and Jackson were both members of the small, close-knit Surveying Service and, in many respects, Law's career had echoes of Jackson's. Jackson had taken one of the last sessions of the Intelligence Course offered and the suspicion is strong that Law would in due course have taken

2   Officers promoted during 1935-39 are discounted merely to control for any impact the operational commitments the Service faced, beginning with the Abyssinian Crisis, may have had on officer training.

3   ADM 182/89, Confidential Admiralty Fleet Order, '131.–Captains, R.N.–Technical and War Courses', 17 January 1930.

4   Rear Admiral Arthur Lambert Jackson (1887-1956). Intelligence Course, 1924; Hydrographic Department, 1924-25; Commanding Officer, HMS *Iroquois*, 1925-27 and 1931-32; Naval Assistant to Hydrographer, 1928-29; Superintendent of Charts, 1929-31; Captain, 1930; Commanding Officer, HMS *Challenger*, 1932; Meteorological Course, 1936; Rear Admiral and retired, 1940. Recalled for war service.

5   Captain Edward Francis Bold Law (1891-1977). Commanding Officer, HMS *Garry*, 1917-18, *Scourge*, 1918, HMS *Pylades*, 1919, HMS *Flinders*, 1926-27, *Endeavour*, 1927-28 and *Iroquois*, 1930; Superintendent of Charts, 1929-31; Captain, 1933; Commanding Officer, HMS *Herald*, 1933-35; Meteorological Course, 1936; Assistant Hydrographer, 1936-39; Surveying Officer, China Fleet, 1939 and retired, 1943.

the same, too, except for its termination. As surveyors and members of such a small professional community, Jackson and Law found themselves serving very much a Navy within a Navy. With few ships comprising the Surveying Service, officers faced repeated postings to the vessels of the same type, if not actually of the same class. Surveyors were closely related to Navigating Officers by training without the same opportunity present, though, for more diversified service. Not all officers found surveying work rewarding and enjoyable and Richmond early in his career left this branch of specialization to make his mark elsewhere.[6] This, too, was a trait shared by many Navigating Officers who elected to leave their specialist trade behind to return to general naval duties. Why they elected to take this action is understandable as any untoward event involving a ship usually revolved around piloting and charts with the Navigating Officer featuring in any Court of Enquiry. Though a gunnery or torpedo specialist might receive a bad report following a shoot, they were not apt to face the same or even a court martial for their misfortune. If the surveyor did not quite face this prospect, then his life remained a typically isolated one where service on detached operations away from the main fleet with its regularized routine including its strong social side of sports, regattas, and 'At Homes' was absent.

If peacetime duties saw Surveying Officers exercising independent duty, wartime could see such specialists assigned to a fleet staff to oversee shore bombardments, minelaying operations, or arranging the net defences for anchorages. These were important tasks, but they were also tasks by and large neglected in most fleet exercises with the policy of not training for shore bombardments until circumstances required already cited. By 1921, one officer lamented the lack of ongoing specialist training for the community's officers and recommended that officers take the SOTC every five years and that at least one Hydrographer attend the Staff Course every other year.[7] As this recommendation appeared in *The Naval Review*, a safe conclusion is that the Admiralty was aware of the general problem, but the need to meet operational commitments was the primary obstacle to ensuring as much in all instances.

As for Crace, he was a torpedo specialist qualifying in 1912. Though born in New South Wales, Australia he entered the Royal Navy in 1902 and was promoted to Lieutenant in 1908. Crace spent most of the World War in the battle cruiser *Australia* and then was appointed to the *Vivid*, as part of the pre-commissioning crew destined to serve in the newly completed *Hood*. A short period at the Admiralty in the Torpedo and Mining Department and in the cruiser *Danae* followed before posted to *Osprey* in command of the Anti-Submarine training establishment. By 1927, Crace was back at sea as Fleet Torpedo Officer in the Mediterranean Fleet and then spent a year in command of a destroyer before returning to the *Osprey* in command once more.[8]

6   Marder, ed., *Portrait*, p. 16.
7   Rolf Viney, 'The Surveying Service', *The Naval Review*, Vol. IX. No. 4, November 1921, pp. 681-82.
8   ADM 196/51 (Crace).

Being born in 1887, Crace's ideal window for attending the Staff Course would have been during 1920-23. Unfortunately, a series of minor medical problems—initially an infected finger, followed by haemorrhoids and latterly an abscess on the neck plagued Crace during the latter part of 1921 and early 1922.[9] Accordingly, he was not pronounced fit for duty until 13 April 1922 when he took up his appointment to the *Danae* as her executive officer on the ship's recommissioning. As he had not sought an appointment to the Staff College, this was no hindrance on that score, but it would have made attending the SOTC problematic during the period. He completed the one-week Anti-Gas Course in April 1923 whilst in *Danae* and went direct from her to the *Osprey* and command of the Anti-Submarine Establishment. From *Osprey*, Crace served as Fleet Torpedo Officer in the Mediterranean Station before taking command of his own ship, the *Valhalla*. By 1930, Crace had returned to *Osprey* to command the establishment a second time. Departing in 1932, he first went to Porton Down to take a special course before becoming Director of the Tactical Division from 1932-34. Flag Captain to 'Biff' Rose of the East Indies Squadron in *Emerald* came next for Crace including a period in temporary command of the station when Rose was invalided home.[10] Returning to England for duty as Naval Assistant to the Second Lord, after two years at the Admiralty, Crace secured flag rank and an appointment to the Royal Australian Navy where he assumed command of the squadron shortly after the outbreak of war in Europe.

Notwithstanding the painful if minor ailments Crace suffered earlier, he was held to be in good health and did not lack for zeal when engaged in his duties. He was an extremely clever officer and had earlier received the thanks of the Their Lordships for his invention of a safe range indicator for use in torpedo control.[11] Consistently receiving glowing performance reports, Rear Admiral Tomkinson elected not to leave matters to chance in June 1928 whilst Crace was in the zone of consideration for Captain going so far as to recommend his immediate promotion; this duly happening at the end of the month.[12]

Though a torpedo specialist, Crace was now tied to the *Osprey* and the anti-submarine warfare community and with expansion now ongoing in the fleet, his appointment to the necessary training classes was ever problematic. In the event, upon leaving *Osprey* for the second time he came to the Admiralty to head the Tactical Division. Sidney Bailey, the ACNS, was critical of his performance believing that, if clever, he relied too much upon the support of his staff officers while deferring to the wishes of his superiors.[13] If Bailey's view is correct, then one reason why Crace may have

9   The number of officers admitted to hospital with abscesses is striking. The causes of such are never cited in the officers' personnel record, but the suspicion remains that exposure to chemical agents was the probable reason.
10 ADM 196/51 (Crace).
11 *Ibid*.
12 ADM 196/127 (Crace).
13 ADM 196/92 (Crace).

relied too much on his subordinates is that in 1933-34 the Tactical Division was over-manned with torpedo specialists such as himself. Of the five officers then within the Division, four were such specialists.[14] This did not abet working any issues outside his field of expertise (and many were the tactical questions outside of torpedo, mining, chemical warfare and anti-submarine work) and failure to attend the SOTC did not afford him the corrective of securing more update technical knowledge in those other disciplines.

Crace, himself, offers indirect confirmation that the criticism of deferring to supe-riors was not unjust. Reporting on his assistant, Captain Richard Scott, during that officer's period with the Tactical Division, the former Navigating Officer was praised for subordinating his interests before the other elements of the Naval Staff.[15] Where Crace believed this made for a good team player, Bailey expected a Director to show a bit more spine. This view is easily understood, for Bailey, as the previous Director of the Tactical School, entertained strong views whilst ACNS on tactical subjects and of the role the Division should be playing in fostering doctrine. The possibility also remains that Bailey viewed Scott as the more capable officer with Crace leaning too heavily on his subordinate's bat noting that Scott 'was an officer of exceptional mental calibre' when endorsing Crace's assessment of him.[16] Scott had already attended the Tactical Course twice before coming to the Division and continued his strong interest in tactics when posted to the Mediterranean Station's War Game Room in Malta in 1929, though nominally borne in *Warspite*, the fleet flagship.[17]

As a Lieutenant, Law had damaged his destroyer, HMS *Scourge*, twice. The first time was when entering port at low water and the second was in a collision with another vessel. A Court of Enquiry convened for the second incident, issued a caution to be careful in the future.[18] At this point, he returned to surveying duties where he spent the balance of his peacetime career earning laudatory reports along the way.

If the submariner and the aviator could expect a return to general service duties after five years in their arm, the Hydrographer seemed destined to spend a life in minor ships. This doubtlessly made him less competitive for promotion to flag rank where his absence as 'jimmy' in a large ship or even command of one marked his outlook as narrow. This narrowness of opportunity was a major reason why officers frequently applied to return to general service duties and leave navigating and hydrog-raphy behind.

Henry Moore, also a prior Navigating Officer, though a graduate of the IDC and holding the naval *psc*, never took the courses required of Captains before reaching flag rank in January 1938. Moreover, he owed his *psc* to his time as an instructor on

14   Namely, Crace, Captain Denis Boyd and Commanders Francis Vaughan and Basil Willett. Commander Terence Back, the exception, was a signals officer.
15   ADM 196/92 (Scott).
16   *Ibid.*
17   *Ibid.*
18   ADM 196/145 (Law).

the Staff College and, it will be recalled, he suffered a two-month hiatus in his IDC studies when he went to the Geneva naval talks of 1927. The explanation in Moore's case is that repeated staff appointments and requisite ship time left little opportunity for attending though it must also be recorded that he took a period of leave lasting over eight months immediately after his Imperial Defence Course.[19]

Finally, to the above mentioned officers, the case of Howard Kelly would not be out place either. Reaching Admiral in 1931, his career was marked by a succession of senior staff appointments that in their style and substance mirrored Henry Wilson and Edwards Spears of the British Army, though as was the custom of the Navy, he returned to sea and command at regular intervals. Kelly did not attend any of the higher courses required and, though a respected staff officer, he was not designated as such. He was, however, fluent in Spanish and French and he, too, was a Navigating Officer. His success owed everything to his language skills which the outbreak of war in 1914 fully engaged.[20] Perhaps if Jackson and Law had passed as interpreter, the absence of higher training would have been less a barrier to a fuller career.[21]

19   ADM 196/92 (Moore).
20   ADM 196/43 (Kelly) and ADM 196/89 (Kelly).
21   ADM 196/51 (Jackson) and ADM/53 (Law).

Bibliography

Official Papers

Archives New Zealand, Wellington, New Zealand.
Francis Bernard Dwyer Papers.

The National Archives, Kew, Richmond, Surrey, United Kingdom.
Admiralty 1. Correspondence and Papers. 1/8549/20, 1/8558/133, 1/8564/210, 1/8567/250, 1/8585/64, 1/8591/119, 1/8597/9, 1/8598/14, 1/8628/120, 1/8640/112, 1/8644/169, 1/8650/244, 1/8657/44, 1/8658/69, 1/8659/74, 1/8668/172, 1/8668/177, 1/8685/151, 1/8687/178, 1/8705/184, 1/8710/138, 1/8714/169, 1/8735/71, 1/8737/97, 1/8740/69, 1/8744/125, 1/8744/140, 1/8760/217, 1/8762/265, 1/8776/146, 1/9007/79, 1/9041, 1/9181, 1/9387, 1/9466, 1/9999, 1/11112, 1/19839.
Admiralty 116. Record Office Cases. 116/2273, 116/2284, 116/2394, 116/2471, 116/6364.
Admiralty 182. Admiralty Fleet Orders. 182/31, 182/34, 182/40, 182/42, 182/55, 182/58, 182/66, 182/72, 182/74, 182/76, 182/83, 182/84, 182/85, 182/86, 182/87, 182/89, 182/90, 182/98.
Admiralty 186, Publications. 186/66, 186/72, 186/75, 186/76, 186/78, 186/79, 186/86, 186/106, 186/145, 186/146, 186/152, 186/153, 186/154, 186/155, 186/244, 186/323, 186/636, 186/637.
Admiralty 196, Officers' Service Records (Series III). 196/43, 196/44, 196/45, 196/46, 196/47, 196/48, 196/49, 196/50, 196/51, 196/52, 196/53, 196/54, 196/55, 196/56, 196/62, 196/63, 196/64, 196/65, 196/88, 196/89, 196/90, 196/91, 196/92, 196/93, 196/117, 196/119, 196/120, 196/122, 196/123, 196/125, 196/126, 196/127, 196/128, 196/141, 196/142, 196/143, 196/144, 196/145, 196/146, 196/147, 196/148, 196/171.
Admiralty 203, Correspondence between Admiralty and Royal Naval Colleges Dartmouth and Greenwich. 203/47, 203/69, 203/73, 203/74, 203/84, 203/85, 203/86, 203/87, 203/90, 203/100.
Admiralty 239, Confidential Book Series. 239/142, 239/261.
Air Ministry 1, Air Historical Branch Papers, Series I. 1/516/16/4/17.

Air Ministry 2, Registered Files. 2/526, 2/2376.

Air Ministry 5, Air Historical Branch Papers, Series II. 5/169, 5/299, 5/339, 5/486.

Air Ministry 10, Air Publications and Reports. 10/123, 10/973.

Cabinet 16, Committee of Imperial Defence, Ad Hoc Sub-Committees: Minutes, Memoranda and Reports. 16/61.

Cabinet 23, War Cabinet and Cabinet: Minutes. 23/8, 23/15, 23/21, 23/24, 23/25, 23/37, 23/48, 23/53, 23/54, 23/67, 23/68.

Cabinet 24, War Cabinet and Cabinet Memoranda. 24/1, 24/4, 24/42, 24/60, 24/73, 24/87, 24/92, 24/94, 24/114, 24/121, 24/158, 24/161, 24/167, 24/173, 24/174, 24/185, 24/193, 24/223, 24/260, 24/273, 24/278, 24/281.

Cabinet 34, Committee of Imperial Defence, Standing Defence Sub-Committee Memoranda. 34/1.

Cabinet 51, Committee of Imperial Defence, Standing Ministerial and Official Sub-Committee for Questions concerning the Middle East: Minutes and Memoranda. 51/11.

Cabinet 53, Committee of Imperial Defence: Chiefs of Staff Committee: Minutes and Memoranda. 53/1, 53/2, 53/3, 53/4, 53/5, 53/6, 53/8, 53/9, 53/10, 53/12, 53/14, 53/15, 53/16, 53/17, 53/18, 53/19, 53/20, 53/21, 53/22, 53/24, 53/25, 53/26, 53/27, 53/28, 53/29, 53/33, 53/34, 53/39, 53/41, 53/43, 53/48.

Cabinet 54, Committee of Imperial Defence: Deputy Chiefs of Staff Committee: Minutes and Memoranda. 54/3, 54/11, 54/13.

Cabinet 68, War Cabinet Memoranda (WP(R) Series). 68/1/9.

Cabinet 181, Cabinet Office: Precedent Books: 181/5.

Defence 2, Combined Operations Headquarters, and Ministry of Defence, Combined Operations Headquarters later Amphibious Warfare Headquarters: Records: 2/709.

Home Office 283, Defence Regulation 18B, Advisory Committee Papers: 283/31.

National Archives of Australia, Canberra.

Frederick Geoffrey Shedden Papers. NAA: A5954

Naval Personnel Records. NAA: A6769, Burnett, Farncomb, H.B., J., Moore, G. D., Oldham, G. C.

Naval Historical Center, Washington, DC, United States.

British Sources, Box 12.

OFFICIAL PUBLICATIONS

The Army List.

B.R. 1736(1), Naval Operations Off Dakar, July–September, 1940, Admiralty, Historical Section, 1959.

B.R. 1736(2). Naval Staff History Second World War. Naval Operations in the Battle of Crete, 20th May—1st June 1941 (Battle Summary No. 4), Admiralty, Historical Section, 1960.

B.R. 1736(5), Chase and Sinking of German Battleship Bismarck, 23–27 May 1941, Admiralty, Tactical and Staff Duties Division, Historical Section, 1950.

B.R. 1736(8), Loss of HM Ships Prince of Wales and Repulse, 10th December, 1941, Admiralty, Historical Section, 1955.

Carter, G.B. *Porton Down: 75 years of Chemical and Biological Research*. London: Her Majesty's Stationery Office, 1992.

C.B. 3027(A), Mutiny in the Royal Navy, Volume II, 1921-1937, Admiralty, Training and Staff Duties Division, 1955. Privately held.

Command Paper 467, An outline of the Scheme for the Permanent Organisation of the Royal Air Force, 25 November 1919.

Field Service Regulations, Volume II, Operations, 1929. London: His Majesty's Stationery Officer, 1929.

Gray, T. I. G. *The Imperial Defence College and the Royal College of Defence Studies, 1927-1977*. Edinburgh: Her Majesty's Stationery Office, 1977.

The Hansard.

Hinsley, F.H. *et al, British Intelligence in the Second World War: Its Influence on Strategy and Operations, Volume One*. London: Her Majesty's Stationery Office, 1986.

The King's Regulations and Admiralty Instructions. Volume I. London. His Majesty's Stationary Office, 1938.

The King's Regulations and Admiralty Instructions. Volume II. Appendices. London: His Majesty's Stationery Office, 1937.

The London Gazette.

Navy Estimates.

The Navy List.

Narrative of the Battle of Jutland. London: His Majesty's Stationery Officer, 1924.

Semi-Official Publications

The Hawk: The Journal of the Royal Air Force Staff College.

PERSONAL PAPERS

British Library, London.
Admiral of the Fleet Earl Jellicoe Papers.

Churchill Archives Centre, Churchill College, Cambridge.
Admiral Sir Frederic Charles Dreyer Papers.
Captain Robert Meyrick Ellis Papers.
Lord Hankey of the Chart Papers (Maurice Hankey).
Admiral Sir Reginald Aylmer Ranfurly Plunkett-Ernle-Erle-Drax Papers.
Captain Stephen Wentworth Roskill Papers.

Imperial War Museum, London.
Captain Ralph Douglas Binney Papers.
General Sir Eric de Burgh Papers.
Admiral Sir Harold Martin Burrough Papers.
Admiral Sir Gerald Dickens Papers.
Commander Charles Hardinge Drage Papers.
Captain Arthur Dyce Duckworth Papers.
Vice Admiral John Walter Durnford Papers.
Admiral Sir Vernon Haggard Papers.
Paymaster Rear Admiral Hugh Miller Papers.
Vice Admiral Sir Ballin Robertshaw Papers.
Engineer Rear Admiral George Campbell Ross Papers.
Rear Admiral Ottokar Harold Mojmir St John Steiner Papers.
Air Commodore John Francis Titmas Papers.
Captain Ronald Adair Trevor Papers.
Midshipman John Waldron Papers.
Captain Gilbert Ridley Waymouth Papers.
Major General Douglas Wimberley Papers.

Liddell Hart Centre for Military Archives, King's College, London.
Field Marshal 1st Viscount Alanbrooke Papers.
Air Chief Marshal Sir Henry Robert Moore Brooke-Popham Papers.
Air Commodore Colin Simpson Cadell Papers.
Sir Julian Corbett Papers.
Admiral Sir Gerald Dickens Papers.
Field Marshal Sir John Dill Papers.
Rear Admiral Ralph Lindsay Fisher Papers.
Air Marshal Sir Gerald Ernest Gibbs Papers.
General Lord Hastings Lionel Ismay Papers.
Sir Basil Liddell Hart Papers.
Colonel Rory Macleod Papers.
Lieutenant General Sir Henry Royds Pownall Papers.

National Army Museum, London.
Brigadier Leslie Skipp Lloyd Papers.
Brigadier Frederick James Moberly Papers.
Brigadier Meyric Home ap Rhys Pryce Papers.

National Maritime Museum, Greenwich.
Vice Admiral Sir Geoffrey Blake Papers.
Admiral of the Fleet Lord Chatfield Papers.
Vice Admiral Kenneth Gilbert Balmain Dewar Papers.
Vice Admiral Sir Robert Francis Elkins Papers.

Admiral of the Fleet Sir Henry Francis Oliver Papers.
Captain Allan Thomas George Cumberland Peachey Papers.
Admiral Sir Herbert William Richmond Papers.
Admiral Sir William Eric Campbell Tait Papers.
Admiral Sir William George Tennant.
Rear Admiral Henry George Thursfield Papers.

Privately Held.
Lieutenant Commander Gordon Colin Campbell Campbell-Johnston Papers.

ORAL HISTORIES

Imperial War Museum, London.
Major General Horace Leslie Birks.
Air Chief Marshal Sir William Alec Coryton.
Marshal of the Royal Air Force Sir William Forster Dickson.
Commander Charles Hardinge Drage.
Captain Charles Hawken Drake.
Major General Nigel William Duncan.
Air Marshal Sir Thomas Walker Elmhirst.
Air Chief Marshal Sir Aubrey Beauclerk Ellwood.
Captain Henry Lockhart St John Fancourt.
Air Marshal Sir Robert Victor Goddard.
Field Marshal Lord Harding of Petherton.
Henry and Stephen Harwood.
Air Vice Marshal Francis William Long.
Major General James Louis Moulton.
Vice Admiral Sir Horace Geoffrey Norman.
Major General George Warren Richards.
Captain Robert John Shaw.
Marshal of the Royal Air Force Sir John Cotesworth Slessor.
Group Captain Frederick William Winterbotham.

Memoirs
Chamberlain, Austen. *Down the Years*. London: Cassell and Company Limited, 1935.
Chatfield, Lord. *The Navy and Defence*: *The Autobiography of Admiral of the Fleet Lord Chatfield*. London: William Heinemann Ltd., 1942.
Cork and Orrery, *My Naval Life*: *1886-1941*. London: Hutchinson & Co. Ltd., 1942.
Douglas, Sholto. *Years of Command*: *The Second Volume of the Autobiography of Sholto Douglas, Marshal of the Royal Air Force Lord Douglas of Kirtleside, G.C.B., M.C., D.F.C.* London: Collins, 1966.
Fuller, J. F. C. *Memoirs of an Unconventional Soldier*. London: Ivor Nicholson and Watson Limited, 1936.

Gibbs, Gerald. *Survivor's Story*. London: Hutchinson & Co. Ltd., 1956.

Haldane, Richard Burdon. *An Autobiography*. London: Hodder and Stoughton Limited, 1929.

Harris, Arthur. *Bomber Offensive*. London: Greenhill Books, 1990.

Ismay, Hastings. *The Memoirs of General the Lord Ismay, K.G., P.C., G.C.B., D.S.O.* London: Heinemann, 1960.

Keyes, Roger. *The Naval Memoirs of Admiral of the Fleet Sir Roger Keyes: The Narrow Seas to the Dardanelles, 1910-1915.* London: Thornton Butterworth, Ltd., 1934.

——. *The Naval Memoirs of Admiral of the Fleet Sir Roger Keyes: Scapa Flow to the Dover Straits, 1916-1918.* London: Thornton Butterworth, Ltd., 1935.

King-Hall, Stephen. *My Naval Life, 1906-1929.* London: Faber and Faber, 1951.

Longmore, Arthur. *From Sea to Sky: 1910-1945.* London: Geoffrey Bles, 1946.

Maund, L. E. H. *Assault from the Sea*. London: Methuen & Co. Ltd., 1949.

Slessor, John. *The Central Blue: The Autobiography of Sir John Slessor, Marshal of the Royal Air Force*. New York: Frederick A. Praeger, 1957.

Smith, Humphrey Hugh. *An Admiral Never Forgets: Reminiscences of Thirty-Seven Years on the Active List of the Royal Navy*. London: Seeley, Service & Co. Limited, 1936.

Tedder, Lord. *With Prejudice: The War Memoirs of Marshal of the Royal Air Force Lord Tedder G.C.B.* London: Cassell, 1966.

SECONDARY SOURCES

General Reference

Who's Who. London: Adam and Charles Black, 1945.

Other

Adams, R. J. Q. *Bonar Law*. Standford: Standford University Press, 1999.

Altham, Edward. 'Some Naval Work in North Russia, 1918-1919'. *The* Naval *Review*. London: The Naval Society, 1921, v. 9, pp. 85-155.

——. 'Imaginary Tactics and the Battle of Jutland'. *The Naval Review*. London: The Naval Society, 1924, pp. 608-618.

Anon. 'Admiral of the Fleet Sir Dudley Pound, G.C.B., O.M., G.C.V.O.', *The Naval Review*, Vol. XXXI, No. 4, November 1943, pp. 283-286.

——. 'List of Members', *The Naval Review*, Vol. XXII, No. 2, May 1934, pp. i-xi.

——. 'Naval Staff College Training', *The Naval Review*, Vol. XX, No. 3, August 1932, pp. 489-91.

——. 'Notes Bearing on the Staff Requirements of a Commander-in-Chief'. *The Naval Review*, Vol. XXVII, No. 1, February 1929, pp. 1-10.

——. 'Royal Naval Staff College, The', *The Naval Review*. Vol. XX, No. 1, February 1932, pp. 6-34.

——. 'Some Notes on the Early Days of the Royal Naval War College. *The Naval Review*, Vol. XIX, No. 2, May 1931, pp. 237-47.

——. 'Specialization'. *The Naval Review*, Vol.. XXII, No. 3, August 1934, pp. 445-50.

——. 'The Study of Strategy by Junior Officers'. *The Naval Review*, Vol. XIX, No. 4, November 1931, pp. 605-06.

——. 'Urgent Need for Economy, The'. *The Naval Review*, Vol. XIX, No. 4, November 1931, pp. 663-70.

A.R.W. (Anthony Wells). 'Staff Training and the Royal Navy—I'. *The Naval Review*, Vol. 64, No. 1, January 1976, pp. 9-17.

Aspinall-Oglander, Cecil. *Roger Keyes*: *Being the Biography of Admiral of the Fleet Lord Keyes of Zeebrugge and Dover*. London: The Hogarth Press, 1951.

Barker, David Wilson. 'The Royal Naval Reserve and its Future'. *Journal of the United Services Institution*, Vol. LXXII, No. 488, November 1927, pp. 699-718.

Beach, Jim. ed., *The Military Papers of Lieutenant-Colonel Cuthbert Headlam, 1910-1942*. Stroud: The History Press, 2010.

Beagle, 'Specialization and Promotion'. *The Naval Review*, Vol. XXXVI, No. 2, May 1948, p. 168-71.

Beaverbrook, Lord. *The Decline and Fall of Lloyd George*. London: Collins, 1963

Beesly, Patrick. *Room 40*: *British Naval Intelligence 1914-18*. New York: Harcourt Brace Janovich Publishers, 1982.

Bell, Christopher M. 'Winston Churchill and the Ten Year Rule'. *The Journal of Military History*, v. 74, no. 4, October 2010, pp. 1097-1128.

——. 'The Royal Navy and the Lessons of the Invergordon Mutiny'. *War in History*, v. 12, No. 1, 2005, pp. 75-92.

——. *The Royal Navy and Strategy Between the Wars*. Basingstoke: Macmillan Press Ltd., 2000.

Bennett, Geoffrey. *The Battle of Jutland*. London: B. T. Batsford, Ltd., 1964.

Bennett, Gill. *Churchill's Man of Mystery*: *Desmond Morton and the World of Intelligence*. Abingdon: Routledge, 2009.

Bent, Eric R. 'The Staff and Its Relation to the Work of the Fleet'. *The Naval Review*, Vol. XVII, No. 3, August 1929, pp. 546-56.

B. H. S. 'The Musings of Methusaleh'. *The Naval Review*, Vol. Vol. XXVII, No. 3, August 1939, pp. 417-25.

——. '"Tim Harington Looks Back."' *The Naval Review*, Vol. XXIX, No. 2, May 1941, pp. 364-69.

Bialer, Uri. *The Shadow of the Bomber*: *The Fear of Air Attack and British Politics 1932-1939*. London: Royal Historical Society, 1980.

Bidwell, Shelford and Graham, Dominick. *Fire-Power*: *British Army Weapons and Theories of War 1904-1945*. Boston: George Allen & Unwin, 1985.

Black, Nicholas. *The British Naval Staff in the First World War*. Woodbridge: The Boydell Press, 2009.

Bond, Brian. *British Military Policy between the Two World Wars*. Oxford: Clarendon Press, 1980.

——. *The Unquiet Western Front*: *Britain's Role in Literature and History*. Cambridge: Cambridge University Press, 2002.

———. and Roy, Ian., eds. *War and Society, Vol. II*. London: Croom Helm, 1977.

Bond, Brian and Robbins, Simon. Eds. *Staff Officer: The Diaries of Walter Guinness (First Lord Moyne) 1914-1918*. London: Leo Cooper, 1987.

Boyle, Andrew. *Trenchard*. London: Collins, 1962.

Brassey, Earl. ed. *The Naval Annual 1919*. London: William Clowes and Sons, Ltd., 1919.

Brendon, Piers. *The Decline and Fall of the British Empire, 1781-1997*. New York: Alfred A. Knopf, 2008.

Brewer, Godfrey. 'The Melody Lingers On – II', *The Naval Review*, Vol. LXII, No. 1. January 1974, pp. 50-59.

Brighten, E.W. 'The Senior Officers' School: The Case for a College of Tactics'. *The Royal United Services Institution*, Vol. LXXIII, No. 490, February 1928, pp. 24-27.

Brown, Judith and Louis, Wm. Roger. eds. *The Oxford History of the British Empire, Volume IV, The Twentieth Century*. Oxford: Oxford University Press, 1999.

Burney, Charles. *The World, the Air and the Future*. London: Alfred Knopf, 1929.

Cathcart, Brian. *The Fly in the Cathedral: How a Small Group of Cambridge Scientists Won the Race to Split the Atom*. London: Viking, 2004.

C.C.H.H. [Cecil Hughes-Hallett]. 'Naval Administration Planning—A Short History', *The Naval Review*, Vol. XXXIII, No. 2, May 1945, pp. 118-21.

Central. 'Some Reflections on the Direction of War'. *The Naval Review*, Vol. XXVI, No. 3, August 1938, pp. 417-28.

Chalmers, W.S. *The Life and Letters of David, Earl Beatty*. London: Hodder and Stoughton, 1951.

Charlton, L . E. O. *The Menace of the Clouds*. London: William Hodge and Company Limited, 1937.

Churchill, Randolph S. *Winston S. Churchill: Young Statesman, 1901-1914*. Boston: Houghton Mifflin Company, 1967.

Churchill, Winston S. *The Great War, Volume II*. London: George Newnes, 1934.

C.K.S.A. [Charles Aylwin]. 'A Naval Visit to Quetta, November, 1934'. *The Naval Review*, Vol. XXIII, No. 1, February 1935, pp. 71-74.

Cohen, Kenneth H. S. 'Functions of Destroyers'. *The Naval Review*, Vol. XV, No. 4, November 1927, pp. 810-12.

Colvin, Ragnar. 'Naval Education' *The Naval Review*, Vol. XV, No. 2, pp. 428-29.

———. 'Jutland'. *The Naval Review*, Vol. XV, No. 4, pp. 861-63.

———. 'Propaganda on Board Ship in Action, *The Naval Review*, Vol. VII, No. 2, pp. 225-27.

Comma. 'The Training of Officers'. *The Naval Review*, Vol. XXIII, No. 1, February 1935, pp. 37-51.

Corbett, Julian. *Naval Operations, Vol. III*. London: Longmans & Co., 1923.

'Courses at Army Schools', *Journal of the Royal United Services Institution*, Vol. LXXI, No. 482, May 1926, p. 387.

Danchev, Alex. *Alchemist of War: The Life of Basil Liddell Hart*. London: Weidenfeld & Nicolson, 1998.

Davison, Robert L. 'Striking a Balance between Dissent and Discipline: Admiral Sir Reginald Drax'. *The Northern Mariner*, Vol. XIII, No. 2, April 2003, pp. 43-57.

——. *The Challenges of Command: The Royal Navy's Executive Branch Officers, 1880-1919*. Farnham: Ashgate, 2011.

Dawson, C. M. 'Royal Navy College Greenwich Centenary 1873-1973'. *The Naval Review*, Vol. LXI, No. 4, October 1973, pp. 330-39.

de Pass, Daniel. 'Hellenic Naval Staff College: Opening Address'. *The Naval Review*, Vol. XVI, No. 4, November 1928, p. 762-68.

Dewar, Kenneth. 'The Influence of Commerce in War,' *The Naval Review*. London: The Naval Society, 1914, Vol. 2, pp. 245-53.

Dickens, Gerald C. 'Naval Education—Admirals or Lieutenants?' *The Naval Review*, Vol. XIV, No. 1, February 1926, pp. 138- 43.

——. 'Richmond'. *The Naval Review*, Vol. XL, No. 3, August 1952, pp. 335-38.

——. 'The Training of Naval Officers'. *The Naval Review*, Vol. VIII, No. 1, February 1920, pp. 60-68.

Dixon, Norman F. *On the Psychology of Military Incompetence*. London: Futura, 1988.

Dockrill, Michael and French, David. Eds. *Strategy and Intelligence: British Policy during the First World War*. London: The Hambledon Press, 1996.

Doig, D.H. 'Command and Staff'. *The Naval Review*, Vol. XII, No. 4, November 1924, pp. 657-62.

Dreyer, Frederic. *The Sea Heritage: A Study of Maritime Warfare*. London: Museum Press, 1955.

Eeyore Smith, '"On the Bridge"', *The Naval Review*, Vol. XXXIX, No. 3, August 1951, pp. 328-29.

Egerton, Wilfrid A. *Contraband of War*. London: Gieve, Matthews and Seagrove, 1915.

Ellison, Gerald. *The Perils of Amateur Strategy: As Exemplified by the Attack on the Dardanelles Fortress in 1915*. London: Longmans, Green and Co. Ltd., 1926.

English, Allan D. 'The RAF Staff College and the Evolution of British Strategic Bombing Policy, 1922-1929'. *Journal of Strategic Studies*, Vol. 16, No. 3, pp. 408-31.

Evans, David C. and Peattie, Mark R. *Kaigun: Strategy, Tactics, and Technology in the Imperial Japanese Navy, 1877-1941*. Annapolis: Naval Institute Press, 1997.

Ferris, John Robert. *Men, Money and Diplomacy*. Ithaca: Cornell University Press, 1989.

First Principles. 'Specialists'. *The Naval Review*, Vol. XXVII, No. 1, February 1939, pp. 14-18.

Fort, Adrian. *Archibald Wavell: The Life and Times of an Imperial Servant*. London: Jonathan Cape, 2009.

Fortescue, The Honourable J.W. 'Field-Marshal French's "1914', *The Quarterly Review*, No. 461, October 1919, pp. 352-63.

Franklin, George. *Britain's Anti-Submarine Capability 1919-1939*. London: Frank Cass, 2003.

Fraser, David. *Alanbrooke*. London: Harper Collins Publishers, 1997.

French, David. 'Doctrine and Organization in the British Army, 1919-1932', *The Historical Journal*, Vol. 44, No. 2, pp. 497-515.

———. *Military Identities: The Regimental System, the British Army, and the British People c. 1870-2000*. Oxford: Oxford University Press, 2005.

———. *Raising Churchill's Armies: The British Army and the War against Germany 1919-1945*. Oxford: Oxford University Press, 2000.

Fuller, J. F. C. *Empire Unity and Defence*. Bristol: Arrowsmith, 1934.

———. 'The Fiends of the Air: The Influence of Aircraft on Imperial Defence', *The Naval Review*, Vol. XI, No. 1, February 1923, pp. 81-115.

Furse, Anthony. *Wilfrid Freeman: The Genius Behind Allied Survival and Air Supremacy 1939 to 1945*. Staplehurst: Spellmount Limited, 2000.

G. A. F. 'In Which We Have the Honour to Serve', *The Naval Review*, Vol. 67, No. 2, April 1979, pp. 91-93.

———. 'Men in Uniform', *The Naval Review*, Vol. XLIX, No. 4, October 1961, pp. 418-19.

Gamma. 'In Praise of Specialists'. *The Naval Review*, Vol. XXV, No. 3, August 1937, pp. 478-82.

Gardiner, Leslie. *The Royal Oak Courts Martial*. Edinburgh: William Blackwood & Sons, Ltd., 1965.

German, Tony. *The Sea is at Our Gates: The History of the Canadian Navy*. Toronto: McClelland and Stewart, 1990.

Gibbs, N. H. *Grand Strategy, Volume I, Rearmament Policy*. London: Her Majesty's Stationery Office, 1976.

Glenton, Robert. *The Royal Oak Affair: The Saga of Admiral Collard Bandmaster Barnacle*. London: Leo Cooper, 1991.

Godwin-Austen, A. R. *The Staff and the Staff College*. London: Constable and Company, Ltd., 1927.

Goldrick, James and Hattendorf, John. eds. *Mahan is Not Enough: The Proceedings of a Conference on the Works of Sir Julian Corbett and Admiral Sir Herbert Richmond*. Newport: Naval War College Press, 1993.

Goldstein, Erik. *Winning the Peace: British Diplomatic Strategy, Peace Planning, and the Paris Peace Conference 1916-1920*. Oxford: Clarendon Press, 1991.

Gordon, Andrew. *The Rule's of the Game: Jutland and British Naval Command*. London: John Murray, 1996.

Gough, Barry. *Historical Dreadnoughts: Arthur Marder, Stephen Roskill, and Battles for Naval History*. Barnsley: Seaforth Publishing, 2010.

Graham, Dominick. *The Price of Command: A Biography of General Guy Simonds*. Toronto: Stoddart Publishing Limited, 1993.

Graham, Dominick and Bidwell, Shelford. *Coalition, Politicians and Generals: Some Aspects of Command in Two World Wars*. (London: Brassey's, 1993.

Gray, Peter. *The Leadership, Direction and Legitimacy of the RAF Bomber Offensive from Inception to 1945*. London: Continuum, 2012.

Green, Andrew. *Writing the Great War: Sir James Edmonds and the Official Histories, 1915-1948*. London: Frank Cass, 2003.

Grenfell, Russell. *The Art of the Admiral*. London: Faber and Faber, Limited, 1937.

——. 'Training in Tactics'. *The Naval Review*. London: The Naval Society, 1923, Vol.. XI, pp. 681-86.

Gretton, P. W. 'Why Don't We Learn from History?' *The Naval Review*, Vol. XLVI, No. 1, January 1958, pp. 13-24.

Grey, Jeffrey. *A Military History of Australia*. Cambridge: Cambridge University Press, 1999.

Grove, Eric. ed. 'Introduction,' in Corbett, Julian. *Some Principles of Maritime History*. Annapolis: Naval Institute Press, 1988.

——. ed. *The Battle and the Breeze: The Naval Reminiscences of Admiral of the Fleet Sir Edward Ashmore*. Stroud: Sutton Publishing Limited, 1997.

Hallam, Richard and Beynon, Mark., eds. *Scrimgeour's Small Scribbling Diary, 1914-1916, The Truly Astonishing Wartime Diaries and Letters of an Edwardian Gentleman, Naval Officer, Boy and Son*. London: Conway, 2008.

Halpern, Paul. ed. *The Keyes Papers: Selections from the Private and Official Correspondence of Admiral of the Fleet Baron Keyes of Zeebrugge, Volume II, 1919-1938*. London: George Allen & Unwin, 1980.

——. ed. *The Mediterranean Fleet 1919-1929*. Farnham: Ashgate Publishing Limited, 2011.

Hamilton, Nigel. *Monty: The Making of a General, 1887-1943*. New York: McGraw-Hill Book Company, 1981.

Harding, Edward or 'C.Q.I.' 'Studies in the Theory of Naval Tactics. III.' *The Naval Review*. London: The Naval Society, 1913. Vol. 1, pp. 208-23.

Harper, J. E. T. 'Staff or Anti-Staff'. *Journal of the Royal United Services Institution*, Vol. LXIII, No. 492, November 1928, pp. 716-19.

——. *The Truth About Jutland*. London: John Murray, 1927.

Hill, J. R. ed. *The Oxford Illustrated History of the Royal Navy*. Oxford: Oxford University Press, 1995.

Hinsley, F.H. *et al, British Intelligence in the Second World War: Its Influence on Strategy and Operations, Volume One*. London: Her Majesty's Stationery Office, 1986.

Holmes, Richard. *The Little Field Marshal: A Life of Sir John French*. London: Weidenfeld & Nicolson, 2004.

Horner, David. *Blamey: The Commander-in-Chief*. St Leonards: Allen & Unwin, 1998.

Hoy, Hugh Cleland. *40 O. B.: How the War Was Won*. London: Hutchinson & Co., 1932.

Hughes, Matthew. *Allenby in Palestine: The Middle East Correspondence of Field Marshal Viscount Allenby, June 1917-October 1919*. Stroud: Sutton Publishing, Ltd., 2004.

Hunt. 'The Specialist Problem'. *The Naval Review*, Vol. XXX, No. 4, November 1942, p. 305.

Hunt, Barry D. *Sailor-Scholar*: *Admiral Sir Herbert Richmond, 1871-1946*. Waterloo: Wilfrid Laurier University Press, 1982.

Hythe, Viscount. ed. *The Naval Annual 1913*. Portsmouth: J. Griffin and Co., 1913.

Jackson, Robert. *Strike from the Sea*: *A Survey of British Naval Air Operations, 1909-1969*. London: Arthur Barker, Limited, 1970.

Jackson, William and Bramall, Lord. *The Chiefs*: *The Story of the United Kingdom Chiefs of Staff*. London: Brassey's, 1992.

Jellicoe, Admiral Viscount. *The Grand Fleet, 1914-16*: *Its Creation, Development and Work*. London: Cassell and Company, Ltd., 1919.

J.M.G.G. 'The University in Naval Education'. *The Naval Review*. Vol. XXXII, No. 2, May 1944, pp. 149-50.

Kay, 'Tactics and the Two-Stripper', *The Naval Review*, Vol. XXVII, No. 2, May 1939, pp. 239-42.

Kennedy, Gregory C. and Neilson, Keith. *Military Education*: *Past, Present, and Future*. Santa Barbara: Praeger, 2002.

Kennedy, Paul M. *The Rise and Fall of British Naval Mastery*. London: The Ashfield Press, 1990.

——. *The Rise and Fall of the Great Powers*: *Economic Change and Military Conflict from 1500-2000*. New York: Random House, 1987.

K. G. F. 'New Schemes', *The Naval Review*, Vol. XXXIII, No. 2, May 1945, pp. 132-33.

Kinvig, Clifford. *Churchill's Crusade*: *The British Intervention of Russia 1918-1920* London: Hambledon Continuum, 2006.

Knox, John N. 'Some Comments on the Relations between the Services'. *The Naval Review*, Vol. XVI, No. 3, August 1928, p. 536-46.

——. 'Staff Officers (Operations) for Captains'. *The Naval Review*, Vol. XV, No. 4, November 1927, pp. 863-65.

Koe, Walter P. 'Some Observations Upon the Naval Staff System and Organisation'. *The Naval Review*, Vol. XII, No. 4, November 1924, pp. 663-72.

Lambert, Andrew. *The Foundations of Naval History*. London: Chatham Publishing, 1998.

——. *Nelson*: *Britannia's God of War*. London: Faber and Faber, 2004.

Lambert, Nicholas A. *Sir John Fisher's Naval Revolution*. Columbia: University of South Carolina Press, 1999.

Lavery, Brian. *In Which They Served*: *The Royal Navy Officer Experience in the Second World War*. London: Conway, 2008.

Leasor, John. *War at the Top*: *The Experiences of General Sir Leslie Hollis K.C.B, K.B.E.* London: Michael Joseph, 1959.

Lewin, Ronald. *Slim*: *The Standardbearer, A Biography of Field-Marshal The Viscount Slim, KG, GCB, GCMG, GCVO, GBE, DSO, MC*. London: Leo Cooper, 1990.

Lewis, Geoffrey. *Carson*: *The Man Who Divided Ireland*. London: Hambledon and London, 2005.

Lewis, Peter. *Squadron Histories: R.F.C., R.N.A.S. and R.A.F. 1912-59*. London: Putnam, 1959.

Liberal Unionist. 'Conservatives and Reformers'. *The Naval Review*, Vol. XXVII, No. 2, May 1939, pp. 219-33.

Liddell Hart. *The Defence of Britain*. New York: Random House, 1939.

——. 'Economic Pressure or Continental Victories', *The Journal of the Royal United Services Institution*, Vol. LXXVI, No. 503, August 1931.

Longsdon, Edwatrd. 'On the Course at Cambridge University for Naval Officers'. *The Naval Review*. London: The Naval Society, 1919, Vol. VII, pp. 228-33.

McDonald, Andrew. 'The Geddes Committee and the Formulation of Public Expenditure Policy, 1921-1922'. *The Historical Journal*, Vol. 32, No. 2, pp. 643-74.

McKibbin, Ross. *Parties and People: England 1914-1951*. Oxford: Oxford University Press, 2010.

Mackenzie-Grieve, John S. M. 'The Axe'. *The Naval Review*, Vol. XVII, No. 3, August 1929, pp. 581-82.

Macksey, Kenneth. *Armoured Crusader: The Biography of Major-General Sir Percy 'Hobo' Hobart*. London: Grub Street, 2004.

Marder, Arthur J. ed. *Fear God and Dread Nought: The Correspondence of Admiral of the Fleet Lord Fisher of Kilverstone, Volume II, Years of Power, 1904-1914*. London: Jonathan Cape, 1956.

——. *From the Dardanelles to Oran: Studies of the Royal Navy in War and Peace 1915-1940*. London: Oxford University Press, 1974.

——. *From the Dreadnought to Scapa Flow, Volume III, Jutland and After: May 1916-December 1916*. London: Oxford University Press, 1966.

——. ed., *Portrait of an Admiral: The Life and Papers of Sir Herbert Richmond*. London: Jonathan Cape, 1952.

Medlicott, W.N. *et al. Documents on British Foreign Policy 1919-1939, Naval Policy and Defence Requirements July 1934-March 1936*. London: Her Majesty's Stationery Office, 1973.

——. *Documents on British Foreign Policy 1919-1939: European and Security Questions 1927-8, Series Ia, Volume IV*. London: Her Majesty's Stationery Office, 1971.

Millett, Allan R and Murray, Williamson. eds. *Military Effectiveness, Volume II, The Interwar Period* London: Unwin Hyman, 1990.

Milner, Marc. *Canada's Navy: The First Century*. Toronto: University of Toronto Press, 2000.

Monsell, Sir Bolton Eyres. 'The First Lord's Statement'. *The Naval Review*, Vol. XX, No. 2, May 1932, pp. 215-22.

Montrose *et al*, Duke of, 'Naval Education'. *The Naval Review*, Vol. XV, No. 1, February 1927, pp. 146-54.

Moretz, Joseph. *The Royal Navy and the Capital Ship in the Interwar Period: An Operational Perspective*. London: Frank Cass, 2002.

Murfett, Malcolm H., ed., *The First Sea Lords: From Fisher to Mountbatten* Westport: Praeger, 1995.

Murray, Oswyn A. R. 'The Administration of a Fighting Service'. *The Naval Review*, Vol. XXV, No. 2, May 1937, pp. 241-56.

Murray, Williamson and Millett, Allan R. eds. *Military Innovation in the Interwar Period*. New York: Cambridge University Press, 1996.

'Naval Notes, Cambridge Course Ending'. *Journal of the Royal United Services Institution*, Vol. LXVIII, No. 469, February 1923.

'Naval Notes, Candidates for Staff Training'. *Journal of the Royal United Services Institution*, Vol. LXXII, No. 487, August 1927.

'Naval Notes, Commanders Technical Courses'. *Journal of the Royal United Services Institution*, Vol. LXVII, No. 468, November 1922.

'Naval Notes, Engineers from Universities'. *Journal of the Royal United Services Institution*, Vol. LXXXII, No. 527, August 1937.

'Naval Notes, Naval Staff College'. *Journal of the Royal United Services Institution*, Vol. LXXI, No. 481, February 1926.

'Naval Notes, Navy War Staff Course'. *Journal of the Royal United Services Institution*, Vol. LXVIII, No. 470, May 1923.

'Naval Notes, New War Course Opens'. *Journal of the Royal United Services Institution*, Vol. LXV, No. 458, May 1920.

'Naval Notes, Officers' Technical Course'. *Journal of the Royal United Services Institution*, Vol. LXV, No. 459, August 1920.

'Naval Notes, Paymasters as Naval Attaché'. *Journal of the Royal United Services Institution*, Vol. LXVI, No. 464, November 1921.

'Naval Notes, Senior Officers' Technical and War Courses'. *Journal of the Royal United Services Institution*, Vol. LXXI, No. 481, February 1926.

'Naval Notes, Study of Naval History' *Journal of the Royal United Services Institution*, Vol. LXXVII, No. 506, May 1932.

'Naval Notes, University Entrants'. *Journal of the Royal United Services Institution*, Vol. LXXX, No. 520, November 1935.

Neilson, Keith and Kennedy, Greg. eds., *The British Way in Warfare: Power and the International System, Essays in Honour of David French*. Farnham: Ashgate, 2010.

Nicholson, Douglas R. 'The Staff System in the Navy: Prussian or British'. *Journal of the Royal United Services Institution*, Vol. LXXIII, No. 490, May 1928, pp. 243-46.

Olsen, John Andreas and Creveld, Martin van. eds. *The Evolution of Operational Art: From Napoleon to the Present*. Oxford: Oxford University Press, 2011.

Onlooker. 'The Navy at Cambridge: 1919'. *The Naval Review*. Vol. LV, No. 3, July 1967, pp. 227-31.

Orange, Vincent. *Dowding of Fighter Command: Victor of the Battle of Britain*. London: Grubb Street, 2008.

———. *Park: The Biography of Air Chief Marshal Sir Keith Park, GCB, KBE, MC, DFC, DCL*. London: Grubb Street, 2001.

'Oscar Parkes, O.B.E., M.B., Ch. B.', *British Medical Journal*. Vol. 2, 5 July 1958, p. 52.

Paret, Peter, ed. *Makers of Modern Strategy: From Machiavelli to the Nuclear Age* Princeton: Princeton University Press, 1986.

Parkhill, Archdale. 'Australian Co-operation in Imperial Defence'. *The Naval Review*, Vol. XXV, No. 2, May 1937, pp. 234-40.

Partridge, Michael. *The Royal Naval College Osborne: A History, 1903-21*. Stroud: Sutton Publishing, 1999.

Peden, G. C. *The Treasury and British Public Policy 1906-1959*. New York: Oxford University Press, 2000.

Perry, Nicholas. ed., *Major General Oliver Nugent and the Ulster Division 1915-1918* Stroud: Sutton Publishing, 2007.

Philpott, Ian M. *The Royal Air Force: An Encyclopedia of the Inter-War Years, Volume One, The Trenchard Years, 1918-1929*. Barnsley: Pen and Sword, 2005.

Plunkett-Ernle-Earl-Drax, Reginald (R.X.). "Home Defence. A Reply by R.X..' *The Naval Review*. London: The Naval Society, 1914, Vol. 2, pp. 254-63.

——. 'The Influence of an Efficient Home Defence Army On Naval Strategy'. *The Naval Review*. London: The Naval Society, 1914, Vol. 2, pp. 23-38.

——. 'Naval Efficiency and Discipline'. *The Naval Review*, Vol. XXXII, No. 4, November 1944, p. 301-04.

——. 'The Staff College'. *The Naval Review*. London: The Naval Society, 1915, Vol. 3, pp. 97-8.

——. 'With the Grand Fleet (4th October, 1915.)'. *The Naval Review*. London: The Naval Society, 1915, Vol. 3, pp. 539-46.

Poland, E. N. *The Torpedomen: HMS Vernon's Story 1872-1986*. Privately published, no date, *c.* 1993.

Porthole. 'Staff Courses'. *The Naval Review*, Vol. XLI, No. 3, August 1953, pp. 303-05.

Powers, Barry D. *Strategy Without Slide Rule: British Air Strategy 1914-1939*. London: Croom Helm, 1976.

Probert, Henry. *Bomber Harris: His Life and Times, The Biography of Marshal of the Royal Air Force Sir Arthur Harris, the Wartime Chief of Bomber Command*. London: Greenhill Books, 2001.

Rainbow. 'Red. White and Blue'. *The Naval Review*, Vol. XXII, No. 2, May 1934, pp. 234-49.

Ramsden, John. ed. *The Oxford Companion to Twentieth-Century British Politics*. Oxford: Oxford University Press, 2002.

Ranft, Bryan. ed. *The Beatty Papers: Selections from the Private and Official Correspondence and Papers of Admiral of the Fleet Earl Beatty, Volume II, 1916-1927*. Aldershot: Scolar Press, 1993.

Reid, Brian Holden. *J. F. C. Fuller: Military Thinker* (Houndmills: Macmillan Press, Limited, 1987.

R.G., 'The Principles of War'. *The Naval Review*. Vol. XXII, No. 2, May 1934, pp. 227-33.

Richmond, Herbert. 'Home Defence, Some Historical Aspects of. A Reply to R.X.'. *The Naval Review*. London: The Naval Society, 1914, Vol. 2, pp. 141-59.

———. 'Introductory'. *The Naval Review*. London: The Naval Society, 1913. Vol. 1, pp. 1-4.

———. 'Notes Bearing on the Staff Requirements of a Commander-in-Chief'. *The Naval Review*, Vol. XXVII, No. 1, February 1939, pp. 1-10.

———. 'Order and Instructions'. *The Naval Review*. London: The Naval Society, 1913, Vol. 1, pp. 163-73.

———. 'The Perils of Amateur Strategy', *The Naval Review*, Vol. XIV, No. 3, August 1926, pp. 548-60.

———. 'Principles of War: A Criticism'. *Journal of the Royal United Services Institution*, Vol. LXXIV, No. 496, November 1929, pp. 714-20.

Roskill, Stephen. *Churchill and the Admirals*. New York: William Morrow and Company, 1978.

———. *Hankey: Man of Secrets, Volume I, 1877-1918*. London: London: Collins, 1970.

———. *Hankey: Man of Secrets, Volume II*. London: Collins, 1972.

———. *Naval Policy between the Wars: Volume I, The Period of Anglo-American Antagonism, 1919-29*. New York: Walker and Company, 1968.

———. (S.W.R.). 'Portrait of an Admiral', *The Naval Review*, Vol. XL, No. 3, August 1952, pp. 339-43.

Royle, Trevor. *Orde Wingate: Irregular Soldier*. London: Phoenix, 1995.

R.P.T. 'Nine Days with the Army'. *The Naval Review*, Vol. XXI, No. 4, November 1933, pp. 737-40.

R.W.P. 'Admiral of the Fleet Lord Tovey of Langton Matravers, G.C.B., K.B.E., D.S.O., D.C.L.'. *The Naval Review*, Vol. 68, No. 3, Jul 1980, pp. 205-11.

Sandilands, H.R. 'The Case for the Senior Officers' School'. *The Journal of the Royal United Services Institution*, Vol. LXXIII, No. 490, May 1928, pp. 235-38.

Schofield, Brian. *British Sea Power: Naval Policy in the Twentieth Century*. London: B. T. Batsford, 1967.

Scott, John M. 'Retirement'. *The Naval Review*. Vol. XVIII, No. 1, February 1930, pp. 135-36.

Schurman, Donald. *The Education of a Navy: The Development of British Naval Strategic Thought, 1867-1914*. London: Cassell, 1965

S.D.S., 'The Canadian View of Naval Policy', *The Naval Review*. Vol. XXII, No. 4, November 1934, pp. 628-34.

Self, Robert. ed. *The Chamberlain Dairy Letters: The Correspondence of Sir Austen Chamberlain with His Sisters Hilda and Ida, 1916-1937*. Cambridge: Cambridge University Press, 1995.

Simpson, Michael. ed. *Anglo-American Naval Relations 1917-1919*. Aldershot: Scolar Press, 1991.

———. *The Cunningham Papers: Selections from the private and official Correspondence of Admiral of the Fleet Viscount Cunningham of Hyndhope, O.M., K.T., G.C.B.,*

D.S.O. and two bars, Volume I, The Mediterranean Fleet, 1939-1942. Aldershot: Ashgate, 1999.

——. *The Cunningham Papers: Selections from the private and official Correspondence of Admiral of the Fleet Viscount Cunningham of Hyndhope, O.M., K.T., G.C.B., D.S.O. and two bars, Volume II, The Triumph of Allied Seapower, 1942-1946*. Bodmin: Ashgate, 2006.

Sinbad. 'Record of a Britannia Term'. *The Naval Review*, Vol. 15, No. 4, November 1927, p. 865.

Siney, Marion C. 'British Official History of the Blockade of the Central Powers during the First World War.' *The American Historical Review*, Vol. 68, No. 2, January 1963, pp. 392-401.

Sirius. 'Ten Days with the Army' *The Naval Review*, Vol. XXI, No. 2, May 1933, pp. 291-97.

Slessor, John. *Air Power and Armies*. Tuscaloosa: University of Alabama Press, 2009.

Smith, Adrian. *Mountbatten: Apprentice War Lord*. London: I. B. Tauris & Co Ltd, 2010.

Smith, H. H. 'The Education of the Naval Officer to Fit Him for Eventual Higher Command'. *The Naval Review*, Vol. XVI, No. 4, November 1928, pp. 710-27.

Spaight, J.M. *The Beginnings of Organised Air Power*. London: Longmans, Green and Co. Ltd, 1927.

Stevens, ed., *H.M.S. "Enterprise": Story of the First Commission, April, 7th 1926 to December 19th 1928*. Aldershot: Gale & Polden, Ltd., 1929.

Stoehr, R. E. 'Combined Staff or Combined Staff College, *Journal of the Royal United Services Institution*, Vol. LXVI, No. 463, May 1921, pp. 477-80.

Stuart, Campbell. *Secrets of Crewe House: The Story of a Famous Campaign* London: Hodder and Stoughton, 1920.

Sumida, Jon Tetsuro. *In Defence of Naval Supremacy: Finance, Technology and British Naval Policy, 1889-1914*. Boston: Unwin Hyman, 1989.

Sweetenham, John. *McNaugton, Volume I, 1887-1939*. Toronto: Ryerson Press, 1968.

Tarrant, V. E. *Battleship Warspite*. Annapolis: Naval Institute Press, 1990.

Taylor, A.J.P. *English History 1914-1945*. New York: Oxford University Press, 1965.

Thursfield, Henry George as H. G. T. 'Admiral Sir Richard Webb', *The Naval Review*, Vol. XXXVIII, No. 1, February 1950.

Till, Geoffrey. *Air Power and the Royal Navy 1914-1945: A Historical Survey*. London: Jane's Publishing Company, 1979.

——. ed. *The Development of British Naval Thinking: Essays in Memory of Bryan Ranft*. London: Routledge, 2006.

——. 'The Impact of Airpower on the Royal Navy in the 1920s'. Unpublished Ph. D. Dissertation, University of London, March 1976.

Tracy, Nicholas. ed. *The Collective Naval Defence of the Empire, 1900-1940*. Aldershot: Ashgate Publishing Limited, 1997.

Trythall, Anthony John. *'Boney' Fuller: Soldier Strategist and Writer 1878-1966*. Baltimore: Naval & Aviation Publishing Company, 1977.

Usborne, C. V. 'The Principles of War: Another Dialogue'. *Journal of the Royal United Services Institution*, Vol. LXXV, No. 49, February 1930, pp. 60-69.

Villiers, Gerald B. 'The New Scheme Lieutenant (S)'. *The Naval Review*, London: The Naval Society, 1914, Vol. 2, pp. 357-60.

Viney, Rolf. 'The Surveying Service', *The Naval Review*, Vol. IX. No. 4, November 1921, pp. 681-82.

Vlahos, Michael. *The Blue Sword: The Naval War College and the American Mission, 1919-1941*. Newport: Naval War College Press, 1980.

Walder, David. *The Chanak Affair*. London: Hutchinson & Co., 1969.

Warner, Philip. *Auchinleck: The Lonely Soldier*. London: Cassell & Co, 1981.

Wester Wemyss, Lady. *The Life and Letters of Lord Wester Wemyss, G.C.B, C.M.G., M.V.O. Admiral of the Fleet*. London: Eyre and Spottiswoode, 1935.

Wickham Legg, L. G. and Williams, E.T. eds. *Dictionary of National Biography 1941-1950*. Oxford: Oxford University Press, 1967.

Williams, E. T. and Palmer, Helen M. eds. *Dictionary of National Biography 1951-1960*. Oxford: Oxford University Press, 1971.

Williams, Mark. *Captain Gilbert Roberts R. N. and the Anti-Uboat School*. London: Cassell, 1979.

Winton, John. *Carrier Glorious: The Life and Death of an Aircraft Carrier*. London: Cassell, 1999,

——. 'H.M.S. *Dryland*'. *The Naval Review*, Vol.. 63, No. 4, October 1975, pp. 297-301.

Young, F. W. *The Story of the Staff College, 1858-1958*. Aldershot: Wellington Press, 1958.

Young, Robert Travers. *The House that Jack Built The Story of H. M. S. Excellent*. Aldershot: Gale and Polden Limited, 1955.

Zuckerman, Lord. *Six Men Out of the Ordinary*. London: Peter Owen Publishers, 1992.

Newspapers

'Brigadier Roy Smith-Hill.' *The Independent*, 13 August 1996.

Internet References

British Broadcasting Corporation Archives: http://www.bbc.co.uk/archive

Ministry of Defence, United Kingdom: http://www.mod.uk/DefenceInternet/About Defence/WhatWeDo/TrainingandExercises/RCDS/About+Us/SeafordHouse. htm

The National Archives, Kew, Surrey, United Kingdom: www.nationalarchives.gov.uk

The Naval Review: http://www.naval-review.org/about.asp.

Vice Admiral Leslie Newton Brownfield, unpublished memoir, *Sea Cycle*. http://www.sandylane.plus.com/brownfield/

Sir Frederick Shedden biography. http://adb.anu.edu.au/biography/shedden-sir-frederick-geoffrey-11670/text20853

Index